FUNDAMENTALS OF ELECTRONICS

Second Edition

DOUGLAS R. MALCOLM, JR.
GMFanuc, Inc.
formerly of Macomb Community College

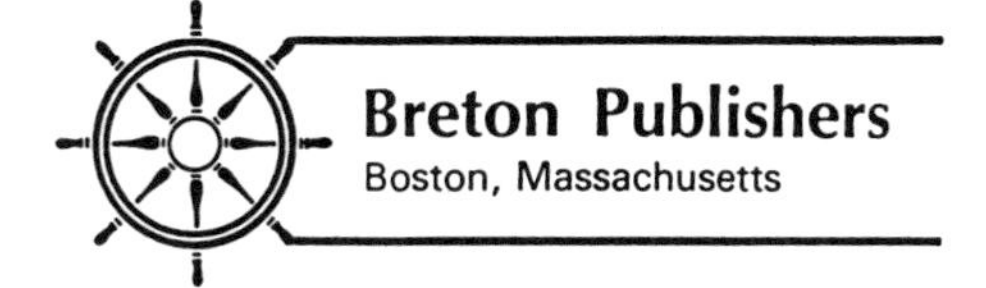

PWS PUBLISHERS

Prindle, Weber & Schmidt • Duxbury Press • PWS Engineering • Breton Publishers •
20 Park Plaza • Boston, Massachusetts 02116

PWS Publishers is a division of Wadsworth, Inc.

Library of Congress Cataloging-in-Publication Data

Malcolm, Douglas R.
Fundamentals of electronics.

Includes index.
1. Electronics. I. Title.
TK7816.M2 1987 621.381 86–26809
ISBN 0–534–06408–6

Printed in the United States of America
1 2 3 4 5 6 7 8 9—91 90 89 88 87

Sponsoring editor: *George J. Horesta*
Editorial assistant: *Susan M. C. Caffey*
Production: *Technical Texts, Inc.*
Production editor: *Jean T. Peck*
Interior and cover design: *Sylvia Dovner*
Cover photo: *Steve Grohe/The Picture Cube*
Composition: *Modern Graphics, Inc.*
Cover printing: *New England Book Components, Inc.*
Text printing and binding: *The Maple-Vail Book Manufacturing Group*

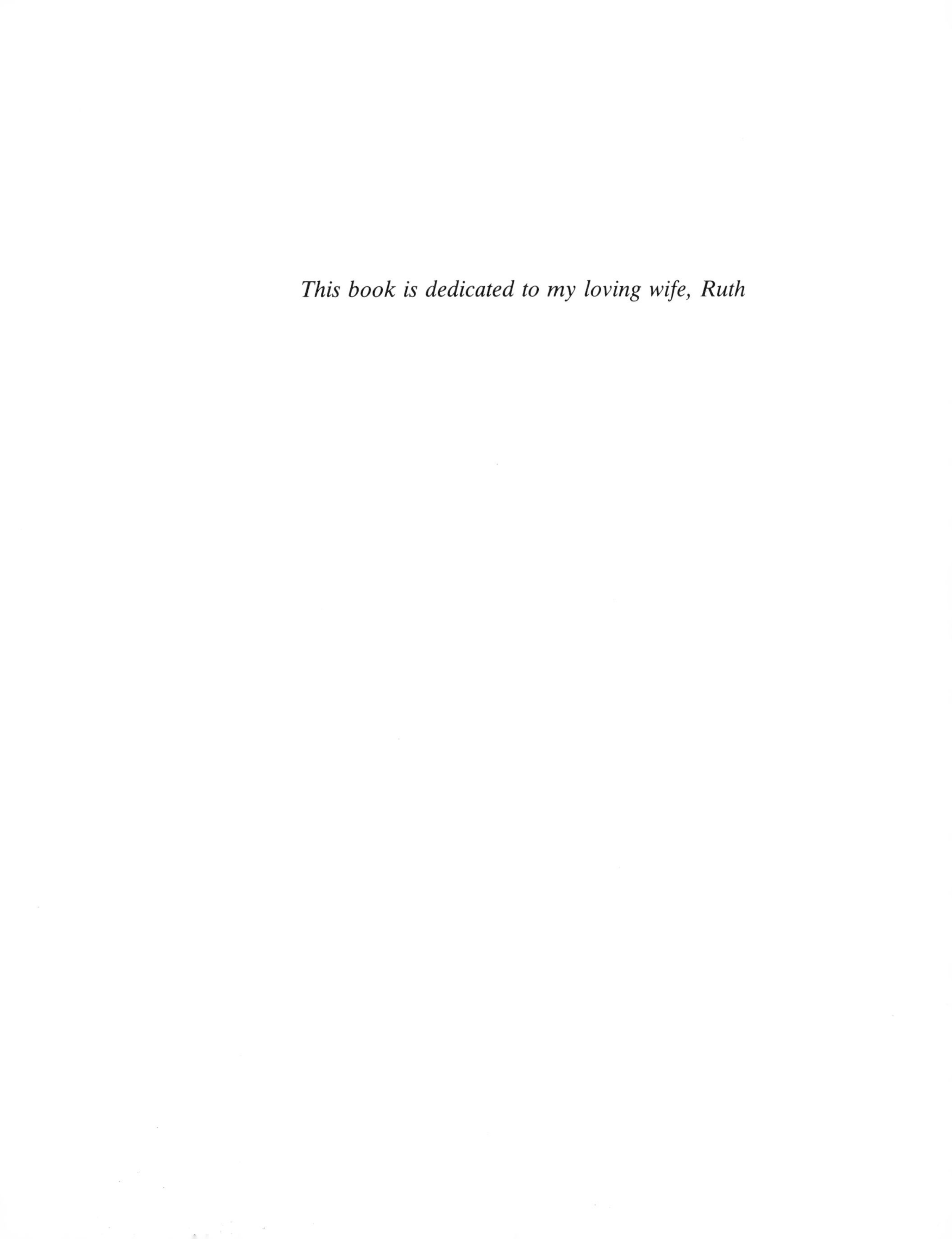

This book is dedicated to my loving wife, Ruth

CONTENTS

PREFACE

Fundamentals of Electronics is an introduction to active electronic components for students enrolled in electronics training. The presentation of material is suited for associate-degree programs on the post-secondary level or for technical-level instruction in vocational/technical high schools. Preparation for this material should include the study of dc/ac fundamentals, as well as an understanding of basic electronics mathematics. Since the text contains an extensive amount of circuitry found in all fields of electronics, it should prove especially useful to individuals undergoing training in service-technology electronics.

The second edition of this text provides a great deal of "real-world" electronics technology. Balanced coverage of discrete and integrated circuits is provided. Also, coverage of materials studied by the student in previous electronics courses is provided as a review before discussions of active electronic components are undertaken. Vacuum tube technology is covered briefly to give the student a background in the development of active electronic components. After completion of this text, it is assumed that the student will continue into the study of advanced electronics technology.

A review of basic electronic circuits and terminology is covered in Chapter 1. Chapter 2 introduces the basics of vacuum tube technology (material that would be vital to students who plan to continue into the area of communication electronics) and semiconductors. The heart of any electronic circuitry—the dc power supply—is the subject of Chapter 3. Basic circuitry found in dc power supplies is discussed in relation to the ac input, the rectifier, the filter, the regulator, and, finally, the load. Also, troubleshooting of the various components of the power supply is detailed in this chapter.

Chapters 4 and 5 discuss the fundamentals of active devices known as bipolar transistors. Chapter 4 details the current-controlling capabilities of these transistors, as well as simple resistance checks used to detect faulty transistors. Covered in Chapter 5 is the dc biasing of the transistor and the selection of its correct dc operating point. Also covered are the methods used to select an amplifier's operating point, bias stabilization, and, finally, troubleshooting dc biasing problems.

Amplifier principles are presented in Chapter 6. The function of amplifier gain, impedance characteristics, and the different classes of amplifier operation are explained.

Small-signal and large-signal amplifiers are studied in Chapters 7 and 8. A description of feedback, frequency response, and distortion in small-signal amplifiers is given in Chapter 7—a discussion that will also apply to large-signal amplifiers. In Chapter 8, where the operation of large-signal amplifiers is presented, the student is introduced to the different classes of power amplifiers. Troubleshooting of these amplifiers is also discussed.

An introduction to special types of electronic components found in electronic circuitry is given in Chapters 9 and 10. The operation of field-effect transistors and MOSFETs is explored in Chapter 9, and the biasing of these components is explained. Chapter 10 details thyristors, including the SCR, the triac, and the diac. The use of these components as they relate to the electronic circuitry is the main thrust of this chapter.

Chapter 11 details the operation and troubleshooting of oscillators. The different oscillators used in electronic circuitry are described along with the waveforms and components necessary to generate oscillation.

Integrated circuits are introduced in Chapters 12 and 13. Chapter 12 discusses the overall operation of the operational amplifier and of 555 timers. Chapter 13 focuses on digital circuitry.

Application of all the electronic components is detailed in the final chapters of the text. The circuits that are studied are those found in the AM-FM receiver: Chapter 14 covers basic receiver operation, while Chapter 15 focuses on the troubleshooting methods that can be applied to AM and FM receivers.

Throughout the presentation of the material in this second edition of the text, an emphasis on practical application is maintained. Thus, the presentation is targeted toward individuals who will be maintaining, troubleshooting, and repairing electronic equipment. The result is a thorough and extremely useful introduction to the theory of electronic components and their circuits.

The material presented in this second edition of *Fundamentals of Electronics* has been developed and class tested by me, and I accept full responsibility for the content. Nevertheless, a large number of people participated in the development and production of this revision, and their assistance is gratefully acknowledged. I would like to thank the following people for their review of the various versions of the second edition manuscript, in whole or in part: Michael R. Light, Northwestern Electronics Institute, Minneapolis, Minnesota; Duane H. Henninger, Lincoln Technical Institute, Allentown, Pennsylvania; Melvin R. Houck, Greenville Technical College, Greenville, South Carolina; Ed Waller, Stautzenberger College, Toledo, Ohio; and Frank Mann, Southern Technical College, Little Rock, Arkansas. Thanks go also to George J. Horesta, Editor, PWS–Kent Publishing Company and to Sylvia Dovner and Jean Peck at Technical Texts, Inc., for their efforts in behalf of this project. Finally, I extend my sincere appreciation to my family—my children, Pamela, Douglas, and Robert, who understood that their father had to spend a great deal of time preparing this second edition, and to my wife, Ruth, for her constant encouragement and love. Without their support, the second edition of *Fundamentals of Electronics* would not have been completed.

Douglas R. Malcolm, Jr.

1 BASICS OF ELECTRONICS

OBJECTIVES

Upon completing this chapter, you should be familiar with:

—Voltage
—Current
—Resistance
—Circuits
—Ohm's law
—Voltage dividers
—Power
—Capacitance
—Inductance
—Transformer
—Impedance
—Resonance

INTRODUCTION

Before starting an in-depth study of electronic circuits, it is always a good idea to review the fundamentals. Electronics is a building-block process in which the technician must constantly study the area. Before starting an in-depth study of electron theory, the basics of voltage, current, and resistance should be understood thoroughly.

VOLTAGE

Voltage (V) is the electrical pressure found in a circuit. This electrical pressure pushes electrons through the circuit. Such movement of electrons performs work in the circuit. The voltage in a circuit, also called the *electromotive force,* is measured in volts (V).

Voltages come in two basic forms. One form is called **direct current (dc),** and the other is called **alternating current (ac).** The dc voltage can be generated by many different sources, such as batteries, solar cells, and dc power supplies. The ac voltage is usually generated by a generator.

As the term *dc* implies, there is always a direct, constant pressure being applied to the circuit. This constant pressure is ideal for operating oscillators, amplifiers, and digital circuits. However, *ac* voltage is a little different. As its name implies, the electrical pressure in the circuit is constantly changing. Figure 1–1 illustrates this continually changing nature of ac voltage.

The two types of voltage are measured differently. Measurement of dc voltage requires a simple dc meter movement, whereas ac voltage is measured with equipment such as an oscilloscope, or an ac voltmeter.

When applied to the circuit, the dc voltage gives constant pressure. An example of how dc voltage remains constant while the switch is closed is shown in Figure 1–2. When the switch is opened, the voltage falls to zero. On the other hand, when ac voltage is applied to the circuit, the voltage increases to a positive peak value, then falls toward zero, and finally reverses the movement of electrons in the circuit when it reaches a negative peak. These voltage readings from zero line to peak (V_p), and from peak to peak ($V_{p\text{-}p}$), are important measurements for the ac voltage. Figure 1–3 illustrates these measurements diagrammatically.

Other measurements for the ac voltage are just as important.

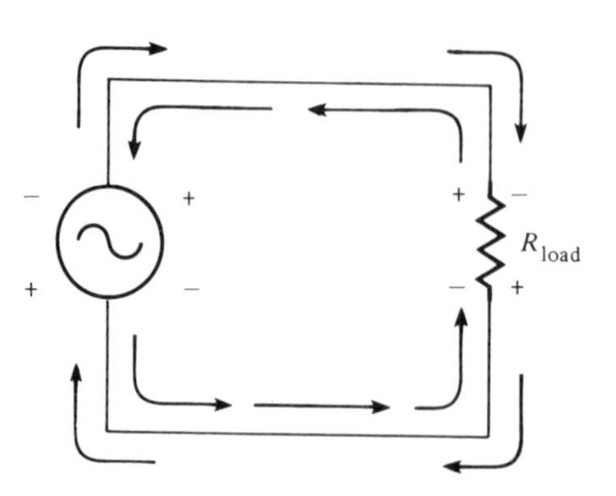

FIGURE 1–1
Alternating Current Changing Direction through the Load

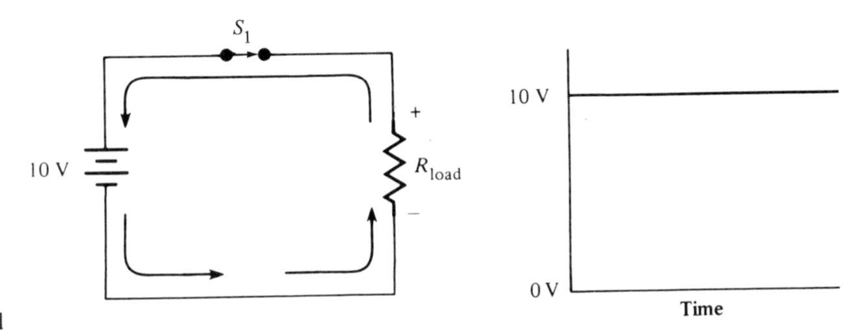

FIGURE 1–2
Direct Current Remaining Constant through the Load

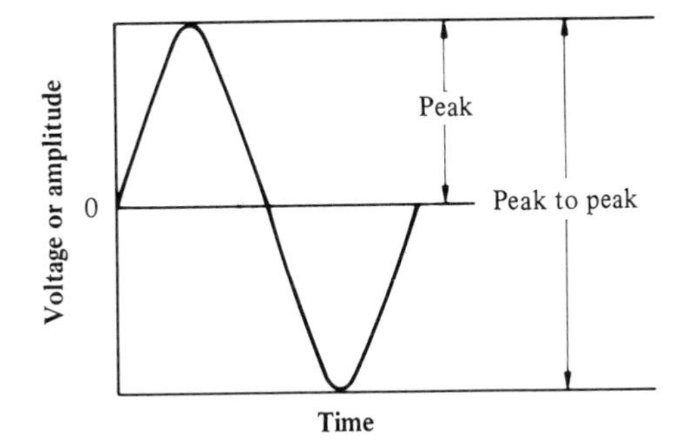

FIGURE 1–3
Sine Wave Measurements for ac

These measurements are the **root mean squared (RMS) value (V_{RMS})**, and the **average value (V_{av})**. The RMS value is used to determine at which point in the circuit the ac and dc power are equal, that is, where the two voltages are delivering the same amount of power to the load. Figure 1–4 shows two examples of sine waves and the formulas used to calculate the RMS and average value of a sine wave. The sine wave in Figure 1–4a has a 10 V peak value or a 20 $V_{p\text{-}p}$ value. Figure 1–4b identifies the formulas to find the average value and the RMS value. Notice, to find the RMS value, the peak value of the sine wave is multiplied by a 0.707 value. Therefore, the RMS value is 7.07 V. The second equation shows the formula and the method used to calculate the average value of the sine wave. Note that the peak value is multiplied by a 0.637 factor. Thus, the average value of the waveform in Figure 1–4a is 6.37 V. It is important to note that if the waveform is measured with a dc voltmeter, the average value of the waveform is equal to zero. Figure 1–4b illustrates the overall relationship of the RMS and the average value of the waveform. These formulas will be used later to find output dc voltage of power supplies.

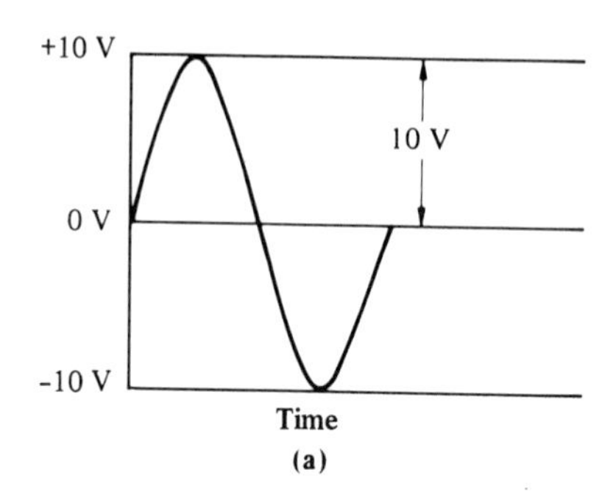

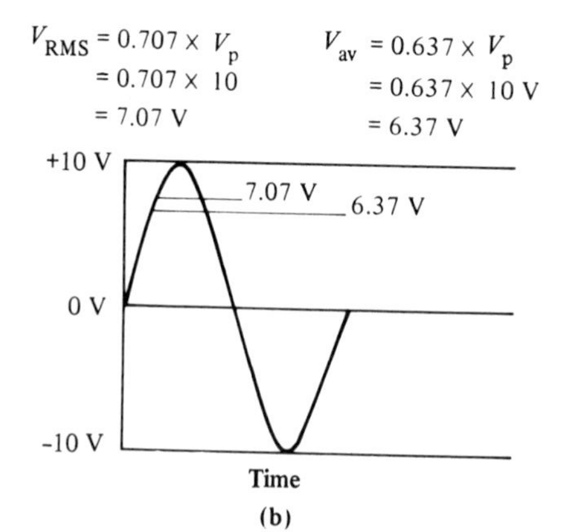

FIGURE 1–4
Calculating RMS and Average Values of an ac Sine Wave

CURRENT

Current (*I*) is the movement of electrons in the circuit. For the electrons to flow in the circuit, the switch must be closed. Current will continue to flow as long as the switch remains closed. Once the switch is open, the electrons cease to flow. Current flow is measured in amperes (A). An example of current action is given in Figure 1–5. Figure 1–5a shows the closed switch and current path through the load; part b shows the open switch and no current through the load.

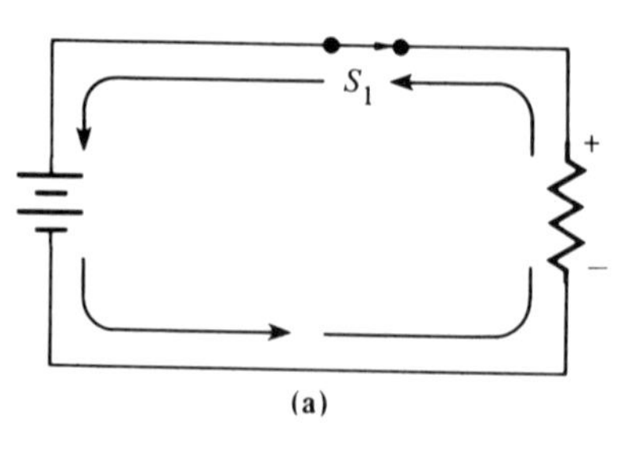

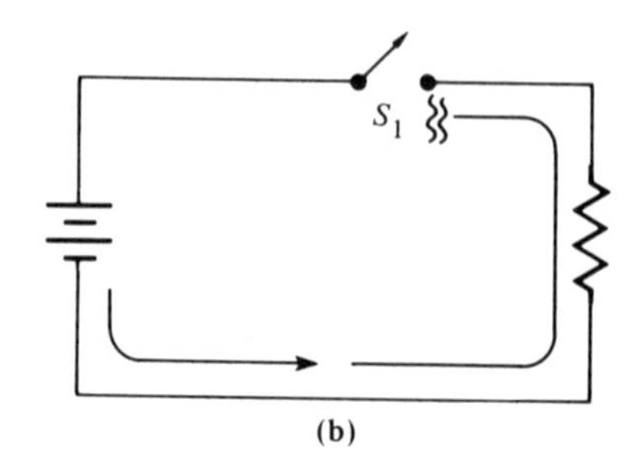

FIGURE 1–5
Current Path Developed in a Series Circuit: (a) Current Path When S_1 Is Closed, (b) No Current Path

RESISTANCE

Resistance (*R*) is the third component found in circuits. Resistance opposes the current flow in the circuit, and is measured in ohms (Ω). An ohm is equal to the resistance of a circuit in which 1 V of electrical pressure causes 1 A of current to flow.

Fixed Resistors

In most instances, the ohmic value of the resistance is color coded on the resistor. Figure 1–6a shows an example of such a color code. The resistor is read from left to right. In this case, the first color is red. As noted in the key on the right side of the figure, this color associates with the number 2. The second color is orange and associates with the number 3. The third color (red) means add two zeros to the first two numbers. Thus, the ohmic value of resistance here is 2300 Ω. The fourth color, silver, indicates the *tolerance band.* Silver tells the technician that the 2300 Ω value has a ± (plus or minus) 10% tolerance.

Figure 1–6b illustrates the fifth band found on resistors. This band is identified as the *failure rate band.* This band is found next to the tolerance band, on the right side of it. Fixed resistors are manufactured in batches that have been tested for reliability. The color of the failure rate band gives a description of how reliable the resistor will be. The rating is a percentage of failure based on 1000 hours of use. Note that brown indicates a 1% failure rate per 1000 hours, red

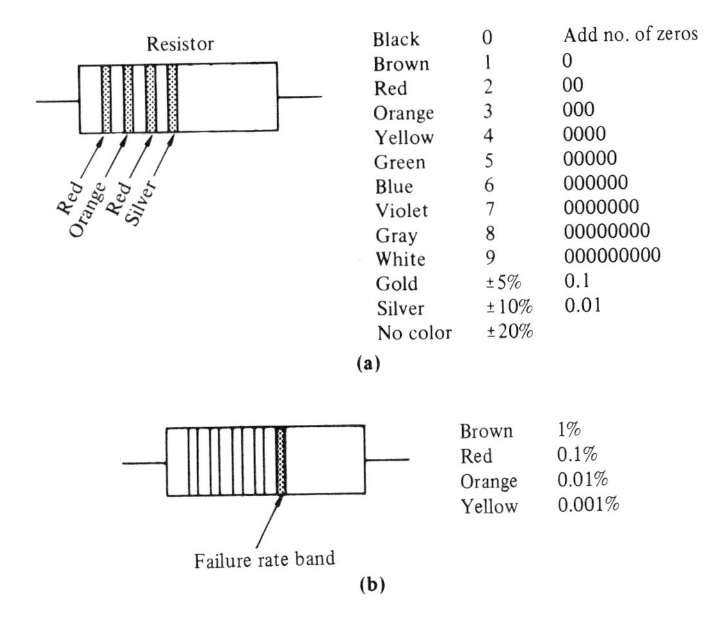

FIGURE 1–6
Resistor Color Code: (a) Ohmic Value, (b) Failure Rate per 1000 Hours

indicates a 0.1% failure rate, orange indicates a 0.01% failure rate, and yellow indicates a 0.001% failure rate per 1000 hours.

The ohmic value of resistance may be very large. Resistances might be as high as thousands or even millions of ohms. To eliminate the zeros necessary to represent thousands or millions, an abbreviation system has been established. In this system, the letter k (for *kilo*) is used to represent thousands, and the letter M (for *mega*) is used to represent millions. Therefore, the value 10,000 Ω can be rewritten as 10 kΩ. The value 1,000,000 Ω can be rewritten as 1 MΩ.

Any time a current passes through a resistance, friction occurs. This friction causes heat to develop in the circuit. Resistors must be able to dissipate this heat. Heat in a circuit is measured in *wattage;* therefore, each resistor has a wattage rating. This rating ranges from 1/8 watt (W) to several hundred watts. A general rule is that the larger the resistor, the more heat the resistor can dissipate. Where large amounts of heat are seen in the circuit, special wire-wound resistors can be used. The wire-wound resistor, or **fixed resistor,** is wrapped with wire and covered with a ceramic material. The common symbol for the fixed resistor is given in Figure 1–7a. Fixed power resistors usually have a 2 W or greater rating. The technician will find the ohmic value of resistance and the wattage rating stamped on the side of the resistor.

Variable Resistors

In addition to the fixed resistor, there is the **variable resistor.** The schematic symbol for the variable resistor is shown in Figure 1–7b. The variable resistor comes in two forms, depending on whether the resistor controls voltage or current. If the variable resistor is used to control voltage, it is called a **potentiometer.** If it is used to control current, it is a **rheostat.**

Also of importance is the way the resistance in a variable resistor changes as the shaft of the resistor turns. If the shaft is turned to midpoint and the resistance there is equal to one-half the value of the total resistance, the resistor is called a **linear tapered resistor.** For example, when the shaft on a linear resistor with a total resistance of 6 kΩ is moved to the halfway point, the resistance will read 3 kΩ.

The other type of variable resistor is the **nonlinear tapered resistor.** In the case of this variable resistor, shaft location has no bearing on the amount of resistance delivered. An example of this

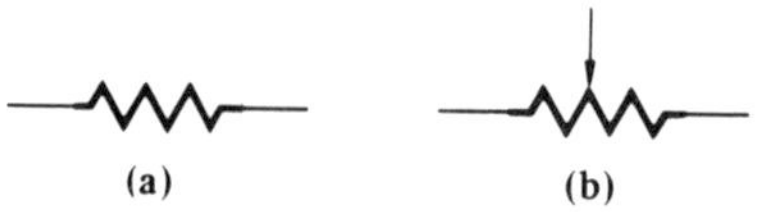

FIGURE 1–7
Schematic Symbol for Resistors: (a) Fixed Resistor, (b) Variable Resistor

(a) (b)

would be a 5 kΩ nonlinear variable resistor. With the shaft at midlocation, the resistance is 1.5 kΩ. Therefore, the placing of the shaft has no bearing on what the output resistance will be.

A good practice to follow when replacing any resistor—fixed or variable—is to use a new resistor with the same value as the old one. The wattage rating is also important. A resistor should never be replaced by a lower-wattage resistor.

Thermistors and VDRs

Two other types of resistors found in consumer products are the **thermistor** and the **voltage-dependent resistor (VDR).** Thermistor resistance is controlled by temperature. As the temperature increases and decreases around the device, the resistance offered by that device will change. When the temperature is low, the resistance is high. As the temperature increases, the value of resistance decreases. The thermistor's counterpart, the VDR, changes the value of resistance by great amounts as the voltage changes across the component. When voltages across the VDR are low, the resistance is low. As the voltage increases, so does the resistance. The VDR is also called a **varistor.**

Both the thermistor and the VDR can be found in consumer electronic devices. These devices are used in amplifier circuits to stabilize the bias; they are also found in color TV receiver degaussing circuits.

CIRCUITS

Voltage, current, and resistance need something within which to work. This something is called a *circuit.* All circuits must have the following characteristics:

—A power source (voltage)
—A path for electrons (current)
—Opposition to current (resistance)
—A controlling device (switch)

There are several types of circuits with which the technician must become familiar: the series circuit, the parallel circuit, and the series-parallel circuit.

Series Circuits

The **series circuit** has only one path for current flow through the load—negative to positive. Figure 1–8 illustrates a simple series circuit and the path for current flow.

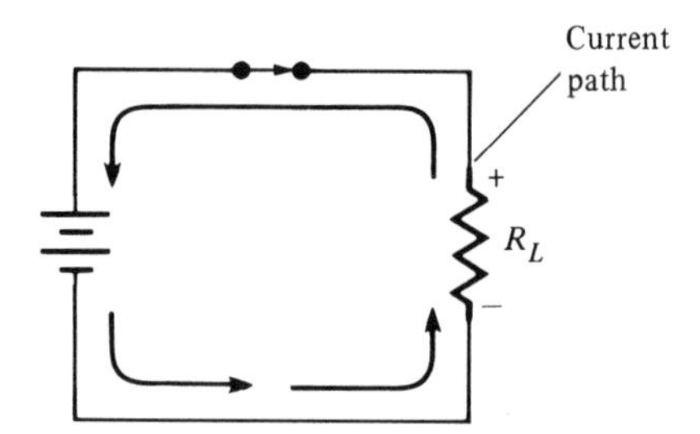

FIGURE 1–8
Simple Series Circuit

An important characteristic of the series circuit is that the resistances found within the series line will add up to the total resistance. Equation 1–1 shows this simple method of finding total resistance in series.

$$R_{\text{total}} = R_1 + R_2 + R_3 + \cdots + R_{n+1} \quad \textbf{(1–1)}$$

An example will be helpful.

EXAMPLE 1

Find the total resistance of the series circuit in Figure 1–9.

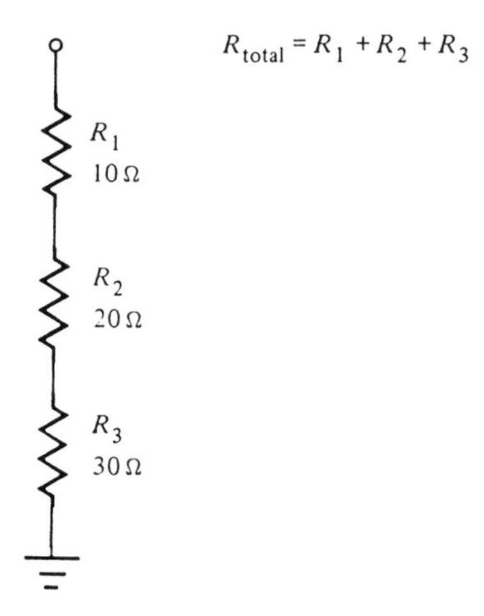

FIGURE 1–9
Calculating Total Resistance in a Series Circuit

Solution:

$$R_{\text{total}} = R_1 + R_2 + R_3 = 10 + 20 + 30$$
$$= 60\ \Omega$$

Therefore, if the circuit were measured with an ohmmeter, a total of 60 Ω would be read.

This procedure holds true for any series circuit.

The current in a series circuit remains constant throughout the circuit. This means that no matter where current is measured in the series circuit, the value will be the same.

Parallel Circuits

The technician also must be familiar with the **parallel circuit.** A sample parallel circuit is shown in Figure 1–10. Notice that there is more than one path for current flow from the negative to the positive pole of the battery. This is the distinguishing feature of the parallel circuit.

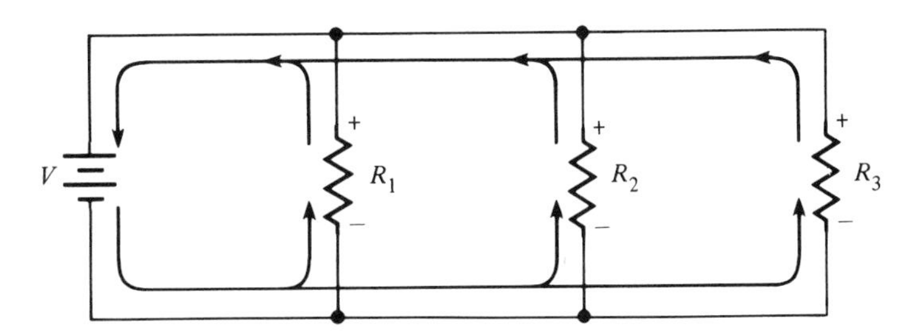

FIGURE 1–10
Simple Parallel Circuit

In a parallel circuit, the total resistance is always less than the smallest resistor in parallel. This means that if three resistors with values of 5 Ω, 10 Ω, and 20 Ω are connected, the total resistance in the circuit will be less than 5 Ω. This provides an approximation of the total resistance.

To find the exact value of total resistance when two or more resistors are connected in parallel, Equation 1–2 can be used.

$$R_{\text{total}} = \frac{1}{\dfrac{1}{R_1} + \dfrac{1}{R_2} + \dfrac{1}{R_3}} \tag{1–2}$$

Again, an example will make things clearer.

EXAMPLE 2

A parallel circuit has three resistors connected in parallel. Their resistance values are 5 Ω, 10 Ω, and 20 Ω. What is the total resistance of this circuit?

Solution:

$$R_{total} = \frac{1}{\frac{1}{5} + \frac{1}{10} + \frac{1}{20}} = \frac{1}{0.2 + 0.1 + 0.05} = \frac{1}{0.35}$$
$$= 2.85\ \Omega$$

This calculation fits the earlier generalization that the circuit's total resistance will be less than 5 Ω.

Another equation can be used to find total resistance in parallel when two resistors are connected. This formula is shown in Equation 1–3.

$$R_{total} = \frac{R_1 \times R_2}{R_1 + R_2} \qquad \textbf{(1–3)}$$

Example 3 is a sample calculation using Equation 1–3.

EXAMPLE 3

Find the total resistance of the circuit in Figure 1–11.

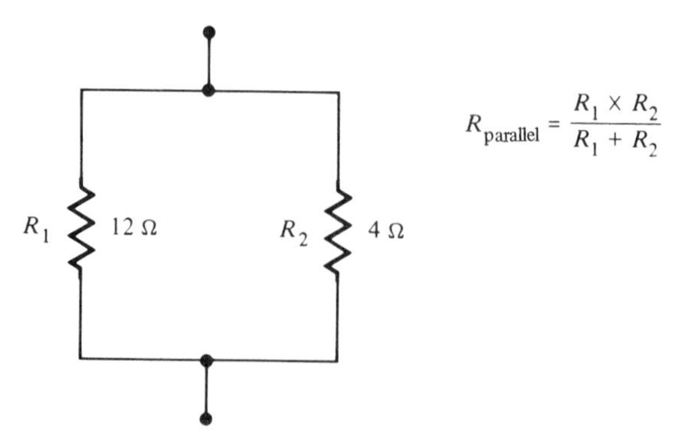

FIGURE 1–11
Calculating Total Resistance of Two Resistors in Parallel

Solution:

$$R_{total} = \frac{R_1 \times R_2}{R_1 + R_2} = \frac{12\ \Omega \times 4\ \Omega}{12\ \Omega + 4\ \Omega} = \frac{48}{16}\ \Omega$$
$$= 3\ \Omega$$

Voltage and current behave differently in parallel circuits than in series circuits. The current in parallel will divide the total current between the loads. The load with the smallest resistance will allow the largest current flow. On the other hand, the voltage in a parallel circuit will remain constant. This means that voltage across each load will remain the same.

Series-Parallel Circuits

The **series-parallel circuit** combines the characteristics of voltage, current, and resistance found in the series circuit and the parallel circuit. To begin with, the circuit's total resistance can be found by calculating the resistance for each part individually. An example circuit is given in Figure 1–12. The parallel branch should be solved first. This can be done by applying the earlier formulas for parallel circuits. The 12 Ω and 4 Ω resistors combine to equal a total resistance of 3 Ω. This 3 Ω resistance can be placed in series with the 3 Ω resistor. Figure 1–13 shows the new circuit resistance. Because these two resistances are in series, they can be added together. This gives a total resistance of 6 Ω.

The current in this circuit also follows the characteristics of the series and parallel circuits. In Figure 1–12, the total current flows

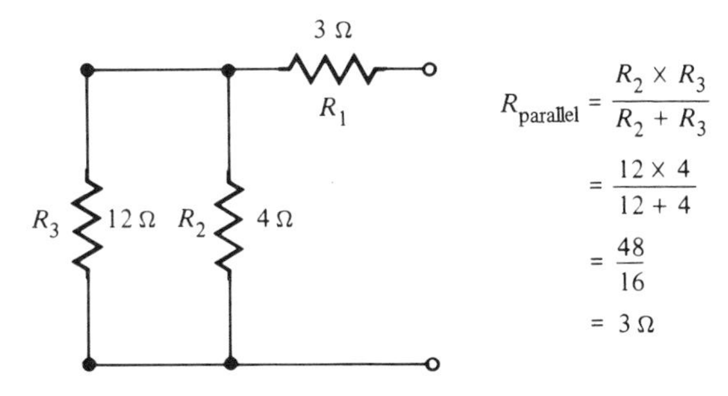

FIGURE 1–12
Series-Parallel Circuit

3 Ω

R_1

3 Ω R'
Combination of R_2 and R_3

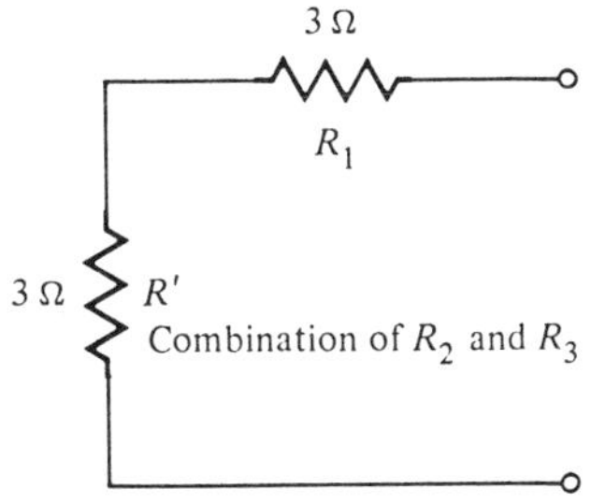

$$R_{total} = R_1 + R' = 3 + 3 = 6\,\Omega$$

FIGURE 1–13
Combination of Parallel Resistors into One Equivalent Resistance

through R_1 because R_1 is in series. The current divides between R_2 and R_3 because these two resistors are in parallel.

OHM'S LAW

It may often become necessary to predict what will happen in circuits. Such predictions can be made using the mathematical expression called **Ohm's law,** which gives the relationship between voltage, current, and resistance in an electrical circuit. With the knowledge of any two factors of this equation, the third can be found. The basic equation for Ohm's law is given in Equation 1–4.

$$E = I \times R \tag{1-4}$$

where:

E = voltage
I = current
R = resistance

By rearranging the elements in Equation 1–4, two other formulas can be derived:

$$I = \frac{E}{R} \quad \text{and} \quad R = \frac{E}{I}$$

To remember these mathematical relationships, the *Ohm's law triangle* can be used, as shown in Figure 1–14. When a finger is placed over the quantity to be found, the mathematical relationship appears. For example, if you want to find resistance, cover up the R, and the formula E/I appears. This process holds for any of the three Ohm's law equations.

The following simple examples will aid in your understanding of Ohm's law.

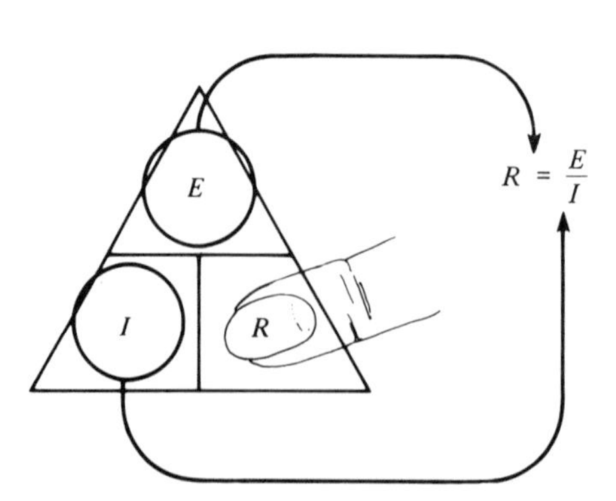

FIGURE 1–14
Ohm's Law Triangle

EXAMPLE 4

A circuit contains a 20 V supply and a 5 Ω load. What is the total current in this circuit?

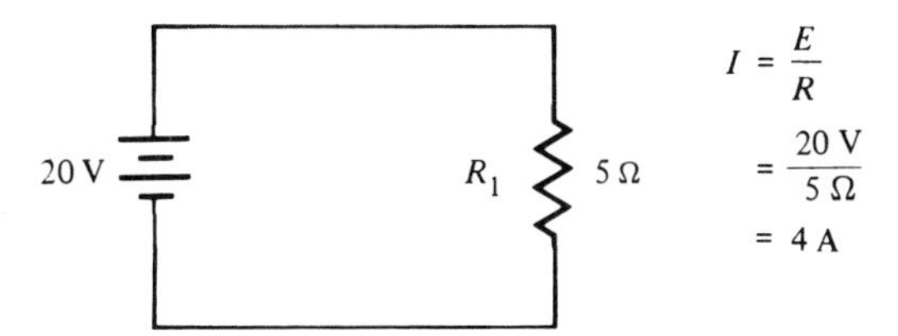

Solution:
Because there is only one load, the current equation of Ohm's law can be used. As the solution in the figure shows, the total current for this circuit is 4 A.

EXAMPLE 5

A series circuit has two loads connected as shown in the accompanying figure. What is the voltage drop developed across R_2?

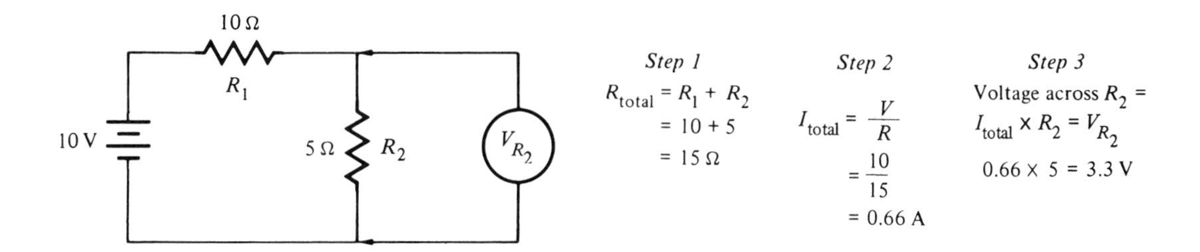

Solution:
Solving this problem involves several steps. The first step is to find the total resistance. Using the series resistance equation, the total resistance is calculated as 15 Ω. Next, the total current can be found. Because both total resistance and total supply voltage are known, these values can be placed into Ohm's law. The total current for this circuit is therefore 0.666 A. Now the voltage drop developed across R_2 can be found. Using Ohm's law again, the voltage drop across the 5 Ω resistor is calculated as 3.33 V.

EXAMPLE 6

Given the series-parallel circuit in the accompanying figure, find the following:

1. Current through R_1,
2. Current through R_2,
3. Current through R_3,
4. Voltage across R_1 and R_3.

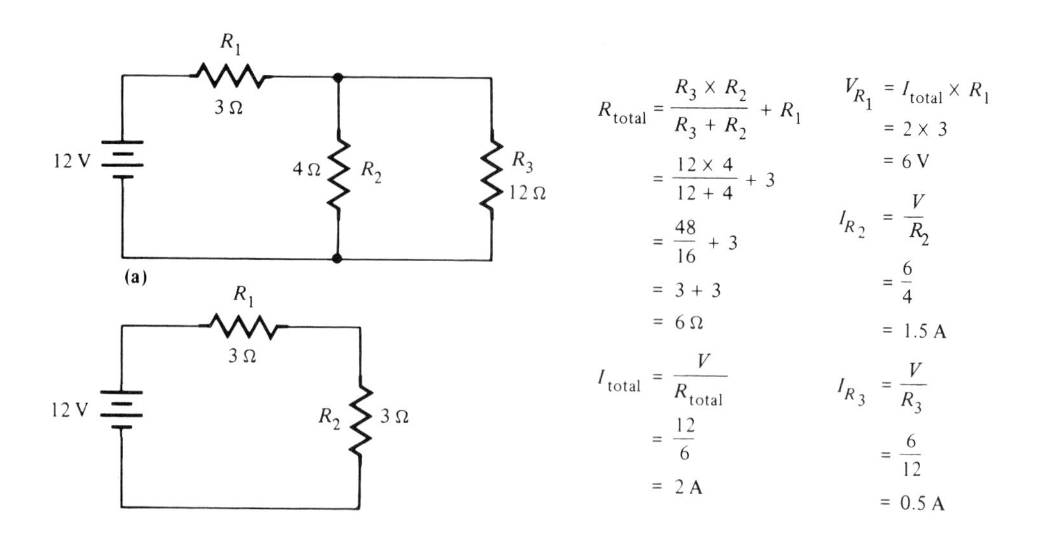

Solution:

Previous calculations will help in solving this problem. In Figure 1–11 a parallel branch method was used that can be employed in this solution. This calculation gives a parallel resistance of 3 Ω. Since the parallel resistance and R_1 are in series, they can be added together. This gives the series circuit a total of 6 Ω resistance.

In further solution of this problem, the voltage drops across each resistor can be calculated. Applying Ohm's law, the voltage across R_1 would be equal to $E_{R1} = I_T \times R_1$. Completing the calculation, the voltage drop across the R_1 would be equal to 6 V. Since these two resistors are in series, the voltage across the parallel branch will be equal to 6 V also.

The current through the R_2 resistor can now be identified. Again, the Ohm's law calculation can be used, but this time the current through the resistor must be found. Use the formula $I_{R2} = E/R_2$. Since the voltage remains constant in parallel, the voltage across R_2 will be 6 V.

The value of R_2 is 4 Ω. This gives the current flow through R_2 as 1.5 A. R_3 current flow can be found in this same manner. The solution shows that the calculated current flow is 0.5 A. These two current flows added together will give the current flow developed through the series resistor R_1, which will be equal to the total current of 2 A.

Therefore the four problem questions have been answered:

1. Current through R_1 = 2 A,
2. Current through R_2 = 1.5 A,
3. Current through R_3 = 0.5 A,
4. Voltage across R_1 and R_3 = 6 V.

Ohm's law is a useful tool for the technician. It provides a good approximation of the voltage, current, and resistance in a circuit. When using Ohm's law, always work totals, and then break them down to find individual values.

VOLTAGE DIVIDERS

More than one resistor connected in series is called a **voltage divider.** Voltage dividers are often used in amplifier, oscillator, and digital circuits to drop the supply voltage to a usable level.

The voltage from a voltage divider can be found by a mathematical process on the order of Ohm's law. One of the simplest methods is called the *voltage proportionate method.* An example of it is given in Figure 1–15. Using this circuit with loads of 10 Ω for R_2 and 5 Ω for R_1, the voltage developed across R_2 can be found. As can be seen, 8 V are being developed here. The remaining voltage is developed across R_1.

An important rule to remember when calculating voltage drop is that the voltage drops across the loads in any closed loop must add up to the total supply voltage. This rule is known as **Kirchhoff's voltage**

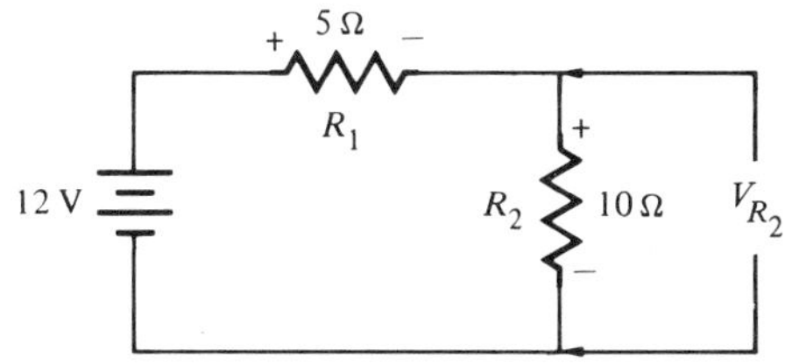

$$V_{R_2} = \frac{R_2 \times V_{total}}{R_1 + R_2} = \frac{10 \times 12}{10 + 5} = \frac{120}{15} = 8\text{ V}$$

FIGURE 1–15
Voltage Divider Network

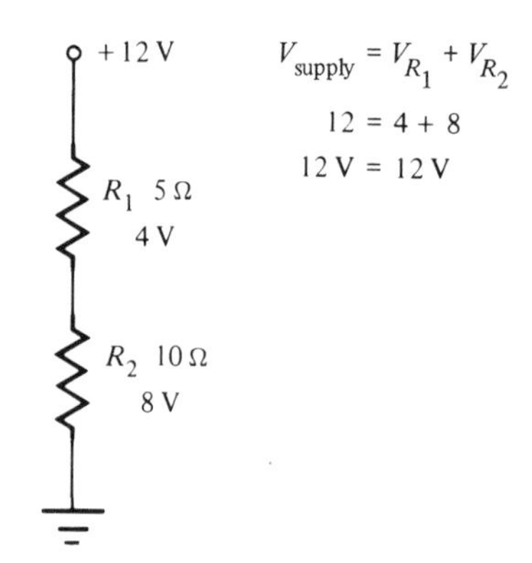

FIGURE 1–16
Kirchhoff's Voltage Law

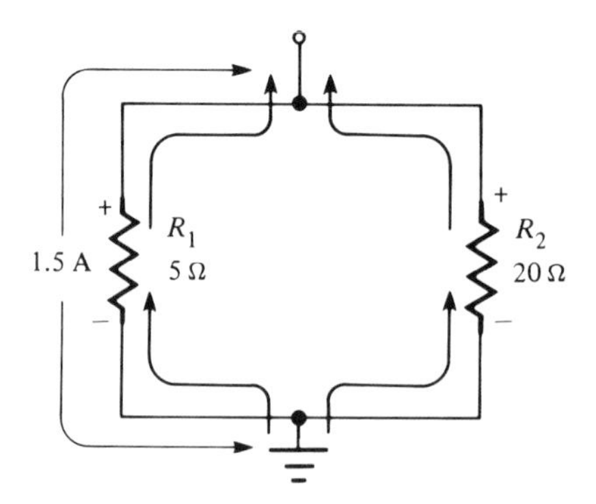

FIGURE 1–17
Kirchhoff's Current Law

law. Therefore, if R_2 has an 8 V drop, the remaining 4 V are seen at the other load. An example of Kirchhoff's voltage law is shown in Figure 1–16. No matter how many loads are connected in series, the voltage drops will always add up to the supply voltage.

Kirchhoff's current law is used in parallel circuits. This law states that in any parallel branch, the current entering the junction is equal to the current leaving the junction. Figure 1–17 shows an example of this action. Notice that the total current is 1.5 A. This 1.5 A will be divided between R_1 and R_2. The R_1 resistor would allow the larger current flow because it is the smaller resistor.

POWER

Power (*P*) is the ability to do work, or the rate at which the work is done, in an electrical circuit. It is measured in watts (W). There are

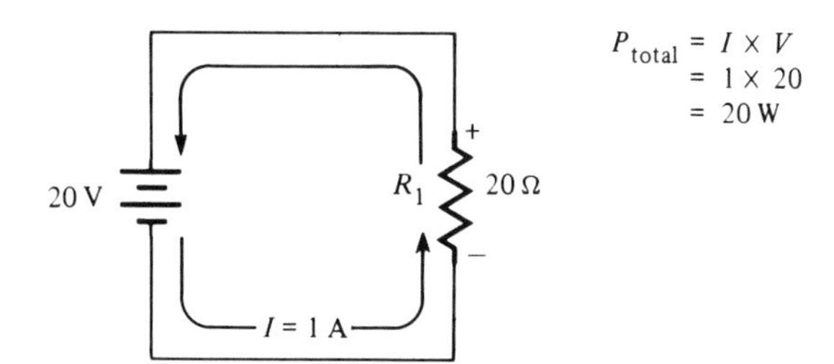

FIGURE 1–18
Power Calculated in a Simple Circuit

several formulas used to find the total power consumed by a circuit. One such formula is the following:

$$P = I \times V$$

where:

P = power
I = current
V = voltage

This power formula can be placed in the same triangular arrangement as Ohm's law. If values for two of the three factors are known, the value of the third factor can be found.

A sample power problem is given in Figure 1–18. Other mathematical equations have been developed from this formula. Examples are $P = I^2 \times R$, $P = V^2/R$, and $I = \sqrt{P/R}$, to name a few. Depending upon the parameters of a given circuit, additional power formulas can be derived.

The total power dissipated in a series or parallel circuit is the sum of the powers being dissipated across each load, as shown by Equation 1–5.

$$P_{total} = P_1 + P_2 + P_3 + \cdots + P_{n+1} \quad \textbf{(1–5)}$$

CAPACITANCE

Two conductive plates separated by some type of insulation describes the passive device known as a **capacitor.** The insulation used to separate the two plates is called the **dielectric material,** or simply **dielectric.** Several types of dielectrics are used in capacitors. These dielectrics can be air, wax, paper, oil, ceramic, or mylar. An example of capacitor plates and dielectric separation is given in Figure 1–19.

Capacitance (*C*) is measured in farads (F). Because the farad is

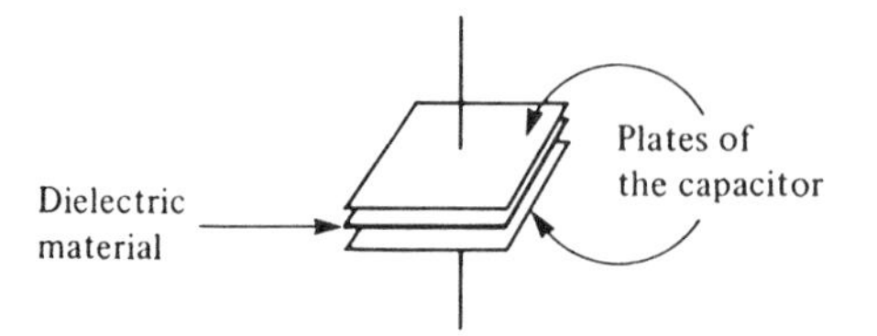

FIGURE 1–19
Structure of a Capacitor

an extremely large unit for consumer products, the sizes of capacitors have been reduced to microfarads (μF) and picofarads (pF).

Not only is the unit of capacitance important, but the amount of voltage that a capacitor can safely handle before it destroys itself is also important. This voltage rating is found on the capacitor and is called the *working volts direct current* WVDC. When the WVDC is exceeded, the dielectric no longer acts as an insulator but becomes a conductor.

As is the case with all components of electronic devices, the capacitor is affected by temperature change. As temperature increases and decreases around the capacitor, the capacitance can change. This shift in capacitance could cause circuit performance to be altered drastically. On most capacitors, there is a temperature coefficient rating. This rating will read either "+" (positive), "−" (negative), or "NPO" (negative-positive-zero). The positive rating means that as the temperature increases, the capacitance increases. The negative rating signifies that as the temperature increases, the capacitance decreases. The NPO temperature rating indicates that if the temperature rises or lowers, there will be no change in the value of capacitance in the circuit. The NPO rating is usually found on ceramic capacitors.

When replacing capacitors in the circuit, the capacitor should be checked for its temperature coefficient rating. A capacitor with a certain rating should be replaced by a capacitor of the same rating.

Series and Parallel Capacitors

Capacitors, like resistors, can be connected in series or parallel. When capacitors are connected in series, the amount of total capacitance decreases. Two capacitors connected in series and the method used to find total capacitance are illustrated in Figure 1–20.

To find the capacitance in parallel, the values of the capacitors are added together. Therefore, finding total capacitance in parallel is like finding total resistance in series. See Figure 1–21 for an example.

Figure 1–22 shows what happens when dc voltage is applied to a capacitor. When the switch is closed, electrons start to flow from the negative side of the dc supply. These electrons build a negative charge on the bottom plate of the capacitor. Electrons that were present on

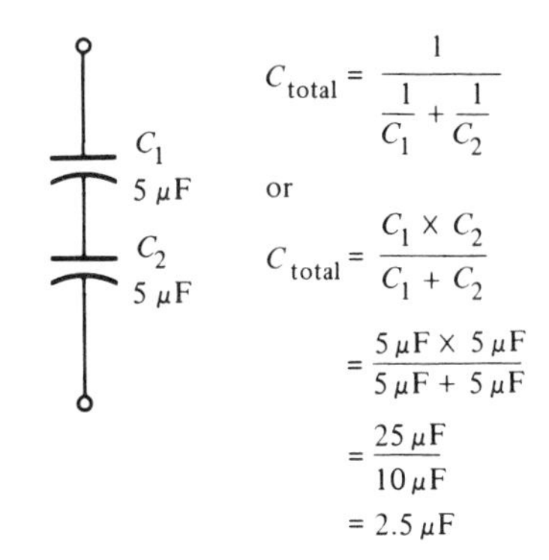

FIGURE 1–20
Finding Total Capacitance in Series

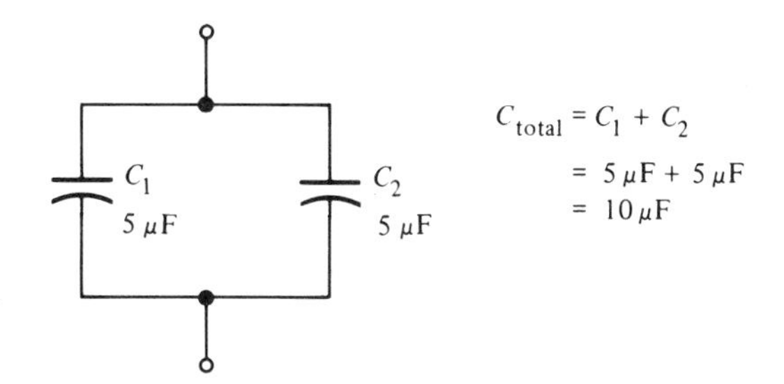

FIGURE 1–21
Finding Total Capacitance in Parallel

the top plate are pulled to the positive part of the supply. This pulling action leaves a deficiency of electrons on the positive plate and an abundance on the bottom plate. In a short time, the capacitor will become fully charged. (A special note here is that the capacitor will charge up only to the applied voltage.) Opening the switch causes the capacitor to remain charged. The capacitor can hold its charge until the two plates have been shorted out. This charging action of the capacitor causes a small current to flow in the circuit, but when the capacitor becomes fully charged, the current flow ceases. This charging rate is shown in Figure 1–22b. Based on this information, a general statement can be made: a capacitor will block all dc current. The ability of the capacitor to block dc current is very helpful in amplifier coupling.

RC Time Constant

The rate at which the capacitor charges and discharges can be controlled. The controlling device is the resistor. Because the resistor regulates current flow, it can be placed in the circuit to control the current flow to the capacitor. Figure 1–22a shows how this can be accomplished.

Not only can the charge be controlled, but a mathematical

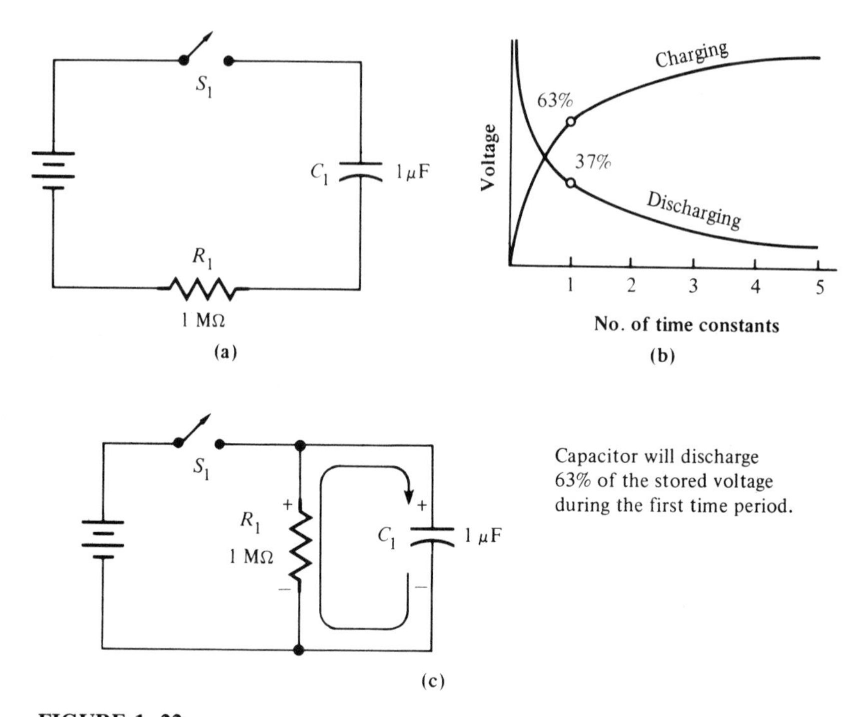

FIGURE 1–22
Charging Rate for a Capacitor: (a) Resistor Regulates Current Flow, (b) General Charging and Discharging Curves, (c) Discharging Path for C_1

equation can be used to predict the amount of time required to charge the capacitor. This formula is written as Equation 1–6.

$$T = R \times C \tag{1–6}$$

where:

T = time of one time constant
R = resistance (Ω)
C = capacitance (F)

This equation gives the time necessary for a capacitor to charge to approximately 63% of the applied voltage. During the charging process the capacitor will never become fully charged. The maximum value can only be 97% after the fifth time constant. At this point the capacitor is said to be "fully" charged. The graph in Figure 1–22b shows the general charging rate of the capacitor. This charging rate is very important in timing circuits.

When the switch is opened (Figure 1–22c), the discharging process begins. As in the charging cycle, the capacitor requires five time constants before it becomes fully discharged. Notice that during the first discharge time constant, the capacitor discharges about 63% of the applied voltage, leaving only around 37% of the voltage in the circuit.

The charging and discharging process is computed in time periods (time constants). Time periods later can be used to establish frequencies of oscillator circuits.

The ac Characteristics of a Capacitor

When a capacitor is connected to an ac source, it is constantly being charged and discharged because the current is constantly changing direction. This process gives the impression that the capacitor is passing the ac current. Thus, it is said that a capacitor will pass ac current.

When a capacitor charges and discharges with varying voltage applied, alternating current can flow. However, the capacitor's dielectric material will not allow current to flow. The charge and discharge of the plates produce current flow in the circuitry attached to the capacitor. The amount of current flowing in the circuit with a sine wave voltage applied depends upon the **capacitive reactance (X_C)** developed by the capacitor. Because capacitive reactance opposes current flow in a circuit, it is measured in ohms.

Two factors govern the reactance of the capacitor: the size of the capacitor and the frequency of the applied voltage. A mathematical expression can be used to determine the amount of reactance the capacitor will offer. This formula is given as Equation 1–7.

$$X_C = \frac{1}{2\pi f C} \tag{1–7}$$

where:

X_C = capacitive reactance (Ω)
π = 3.14
f = frequency (Hz)
C = capacitance (F)

As Equation 1–7 shows, both the size of the capacitor and the frequency applied to the capacitor will affect the reactance. Basically, as the size of the capacitor or the frequency is increased, the reactance will be decreased. The role of frequency in this action is shown diagrammatically by Figure 1–23. From the graph it can be seen that as the frequency increases, the reactance of the capacitor decreases, and as the frequency decreases, the reactance increases. This is true for any capacitor subjected to ac. With this knowledge, two general statements can be made about the capacitor in ac and dc circuits:

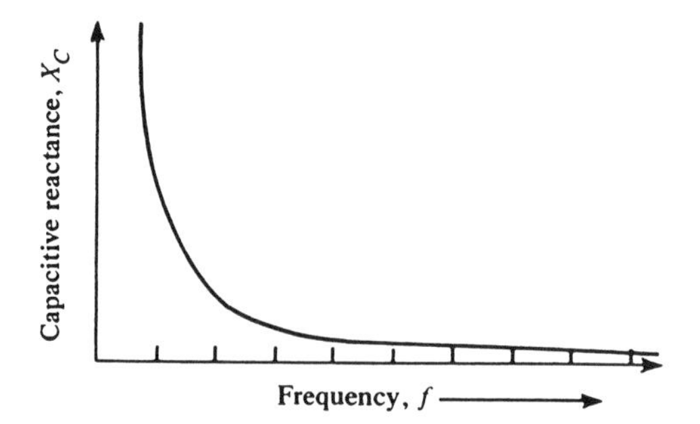

FIGURE 1–23
Effect of Frequency on Capacitive Reactance

1. A capacitor represents an open circuit to dc,
2. A capacitor represents less of an opposition to ac than dc.

INDUCTANCE

When a current is passed through a conductor, a small magnetic field is built up around the conductor. If this conductor is coiled, the magnetic fields aid one another, making a stronger magnetic field. To further strengthen the magnetic field, a core can be added. This principle of a conducting coil is called **inductance.** Inductors are made from coils of wire (conductors) with a core of material such as air, iron, or a ferrite material. Each of these substances, has certain characteristics that determine the value of inductance. The unit of measurement for inductance is the henry (H). The inductor is represented by L.

The dc Characteristics of an Inductor

One of the most basic principles of electricity and magnetism is that, whenever a conductor is in a moving or changing magnetic field, a voltage is induced in the conductor. This principle is called *electromagnetic induction.*

The circuit in Figure 1–24a shows what happens when dc current is passed through an inductor. As the current is passed, a magnetic field builds up around the coils of the inductor. As this current is building up, it develops a *bucking current,* which opposes the natural current flow. Once the magnetic field has reached maximum strength, the bucking ceases and maximum current begins to flow in the circuit once the switch S_1 is open (Figure 1–24b). Then the strength of the magnetic field collapses around the coil, and the magnetic field is induced into the coil, resulting in a higher current in the circuit. The induced current flows through the discharge path R_1. This higher current creates larger voltage across R_1. For a moment, current flows

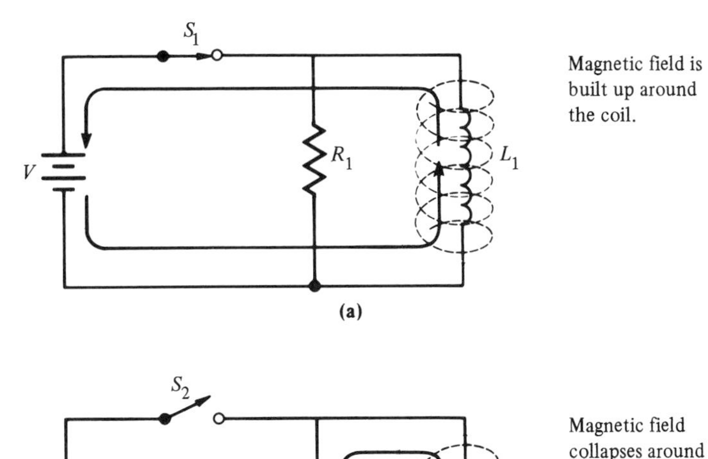

FIGURE 1–24
Passage of dc Current through an Inductor: (a) Switch Closed, (b) Switch Open

in the circuit (when bucking stops). When the field has collapsed totally, the current ceases to flow. Because of the inductor's ability to pass dc current once the field has been built up to maximum, the inductor is said to have low resistance to dc current.

The ac Characteristics of an Inductor

When an inductor is connected to an ac source, the current and the magnetic field are constantly changing. The different magnetic fields induced will try to oppose the changing current in the inductor. The opposition to current flow offered by the inductor is called **inductive reactance (X_L).** Inductive reactance is measured in ohms. A mathematical expression can be used to find the total ac opposition offered by the inductor. This formula is given in Equation 1–8.

$$X_L = 2\pi f L \qquad \textbf{(1–8)}$$

where:

X_L = inductive reactance (Ω)
π = 3.14
f = frequency (Hz)
L = inductance (H)

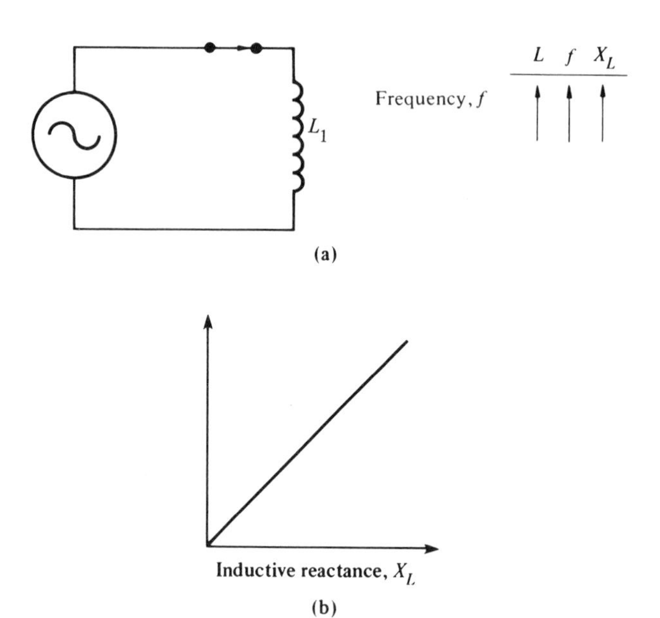

FIGURE 1–25
Effect of Frequency on Inductive Reactance

As can be seen in Equation 1–8, two factors will affect the amount of inductive reactance: the size of the inductor and the applied frequency. Figure 1–25a shows an inductor (L_1) connected to an ac signal source. Note that as the frequency across the inductor increases, the inductive reactance also increases. Figure 1–25b illustrates the graphical relationship between frequency increase and opposition offered by the inductor. Two important things should be remembered:

1. An inductor shows small resistance to dc,
2. An inductor shows large reactance to ac.

Transformer Operation

When ac current is passed through an inductor, a magnetic field is established around the inductor. If another coil is brought into the field of varying ac current, an ac current will be induced into that coil. This action leads the way to development of the transformer.

The coil in which the original magnetic field is set up is commonly called the **primary,** and the coil in which the current is induced is the **secondary.** This standard holds true for all transformers: the input is the primary, and the output is the secondary.

The amount of voltage induced from the primary to the secondary depends upon the **turns ratio.** The turns ratio is the ratio of the number of turns on the primary side to the number of turns on the secondary side. Transformers are divided into classes based on their turn ratios. A transformer may be a step-up, a step-down, or an isolation transformer.

The **step-up transformer** will increase the voltage between the primary and secondary sides, but will decrease the current between primary and secondary. The **step-down transformer** will decrease voltage at the secondary, but will increase current between primary and secondary. The **isolation transformer** will pass the same voltage from primary to secondary.

The action of the transformer of stepping up or down the voltage is accomplished through the turns ratio. In the step-up transformer, the secondary has more turns of wire than the primary. This increase in turns of wire in the secondary adds more resistance to that side. With more turns in the secondary, the magnetic field of the primary will be able to *saturate* (induce to the maximum) the secondary. This action causes an increase in voltage. However, the greater the number of turns, the greater the amount of resistance. This increase in resistance will decrease the current flow. The opposite actions occur in the step-down transformer. Equation 1–9 is a simple mathematical expression relating voltage-to-turns ratios of the primary and the secondary.

$$\frac{V_p}{N_p} = \frac{V_s}{N_s} \qquad \textbf{(1–9)}$$

where:

V_p = voltage in the primary
V_s = voltage in the secondary
N_p = number of turns in the primary
N_s = number of turns in the secondary

This expression also can be written as

$$V_p \times N_s = N_p \times V_s \text{ or } V_s = \frac{V_p \times N_s}{N_p}$$

A sample problem is given in Example 7.

EXAMPLE 7

A transformer has 2500 turns in the primary and 100 turns in the secondary. The voltage applied to the primary is 110 V(ac). What is the secondary voltage?

Solution:

$$V_s = \frac{V_p \times N_s}{N_p} = \frac{110 \times 100}{2500} = \frac{11{,}000}{2500}$$
$$= 4.4 \text{ V(ac)}$$

Therefore, this transformer is a step-down transformer, because the turns ratio reduces the voltage between the primary and secondary.

One important thing to remember about transformers is that the power always remains the same between the primary and secondary. If the primary has 110 V(ac) at 2 A, 220 W are developed at the primary. According to the maximum power transfer theorem, this means that 220 W also must be developed in the secondary side. Always remember that maximum power must be transferred between primary and secondary for the transformer to be operating at maximum efficiency.

The transformer can be an important component in consumer products. Its ability to control voltage and current between primary and secondary makes it a useful tool in power supplies. The transformer is able to tune the primary and secondary—to allow intermediate frequencies to pass from one amplifier to another. Because of its ability of maximum power transfer, a transformer can be used to couple the output of an amplifier to a speaker system, and thus deliver maximum power to that speaker.

IMPEDANCE

The total opposition to current flow found in an ac circuit is called the **impedance (Z).** This total opposition to current flow is the opposition found from resistors, capacitive reactance, and inductive reactance. The impedance of a circuit is the net ac resistance.

RESONANCE

In ac circuits that contain a capacitor and inductor, capacitive reactance and inductive reactance will at some frequency be equal, and will cancel out each other. The point at which $X_C = X_L$ is called **resonance.** The frequency at which these two components are equal is called the **resonant frequency.**

Series Resonance

When a coil and a capacitor are connected in series, as shown in Figure 1–26a, the circuit is called a **series resonant circuit.** At resonant frequency (Figure 1–26b), the capacitor and inductor's reactance cancel out each other. This cancellation brings impedance to the dc resistance of the circuit.

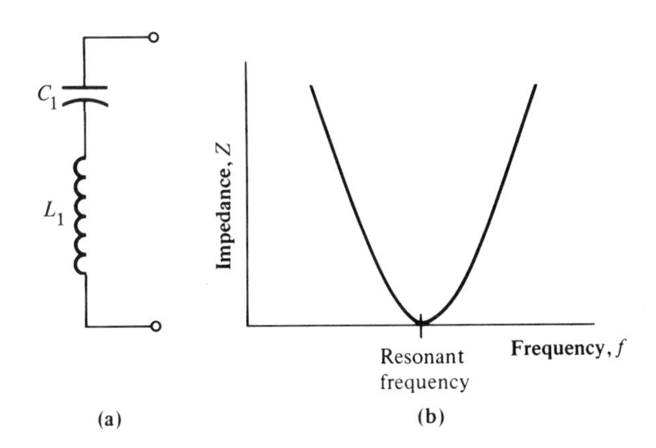

FIGURE 1–26
Series Resonant Characteristics

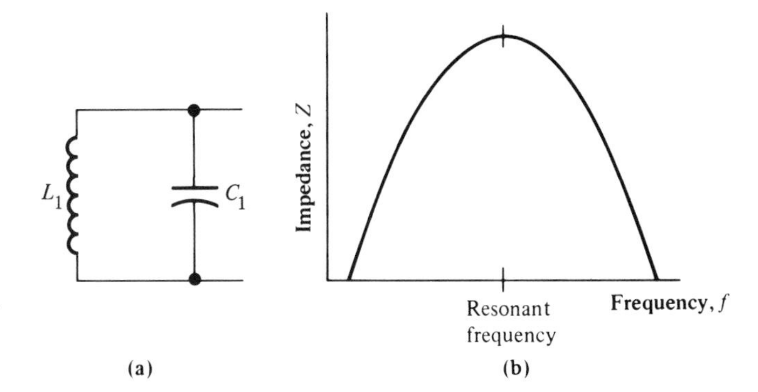

FIGURE 1–27
Parallel Resonant Characteristics of an *LC* Tank Circuit

Parallel Resonance

When the capacitor and inductor are placed in parallel, as shown in Figure 1–27a, the circuit is known as a **parallel resonant circuit,** or a **tank circuit.** This circuit differs from a series resonant circuit in that, at resonance, the impedance of the coil and capacitor, or *LC* tank circuit, is maximum and will not be cancelled out (Figure 1–27b). This means that current flow in the circuit is at minimum, and voltage is at maximum. Again, this is just opposite from the series resonant circuit. Because of the ability of the parallel resonant circuit to become resonant at a certain frequency, it is useful in oscillator circuits.

TUNED CIRCUITS

In consumer electronic devices, it often becomes necessary to select certain frequencies and reject others. To accomplish this task, a transformer and capacitor can be used. Because the transformer's primary

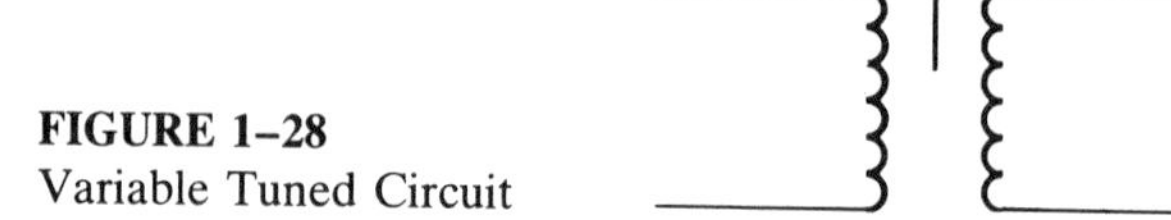

FIGURE 1–28
Variable Tuned Circuit

and secondary sides are inductors, a capacitor can be placed in parallel to make the circuit resonant.

If the primary and secondary of a transformer are made **tuned circuits,** frequency selective circuits, they become very selective in the frequencies they will pass. They will pass the resonant frequency and reject all others. A sample tuned circuit is shown in Figure 1–28. Notice that the transformer has an adjustable core (indicated by the arrow), which means that the technician can adjust the resonant frequency point. This ability makes such circuits useful in the IF (intermediate frequency) sections of radios, televisions, CB radios, and other consumer devices.

CERAMIC FILTERS

Devices that produce less noise than tuned circuits have recently been introduced. These devices are called **ceramic filters.** Ceramic filters are made of crystals such as quartz, Rochelle salt, and ceramic crystals. Cut to certain specifications, these materials respond to a specific frequency. In doing so, they will pass that certain frequency while rejecting all other frequencies.

The main purpose of the ceramic crystal is to narrow the bandwidth and make the circuit very selective. The ceramic crystal has also reduced the alignment time necessary in older consumer devices.

SUMMARY

Voltages exist in dc and ac forms. Direct current is steady and constant, whereas ac changes direction constantly.

Resistance is opposition to current flow. Resistance is used to restrict current flow, and to drop voltage. Variable resistors change either current or voltage, and can be linear tapered or nonlinear tapered.

Circuits are paths for current flow. A circuit must have a path for electrons, a power source, resistance, and a controlling device. Circuits come in series, parallel, and series-parallel forms.

Power is measured in watts. Power is the measurement of the amount of work being done in a circuit.

Ohm's law and Kirchhoff's voltage and current laws are important to every technician. With the knowledge of these basic laws, the values of voltage, current, resistance, and voltage drop can be found for any circuit.

Capacitance is found in almost every circuit. Because of the capacitor's ability to block dc and pass ac, the capacitor serves as a coupling device in amplifiers, and as a filtering component in power supplies. The important point to remember about a capacitor is that it offers reactance to the ac signal.

The inductor also has certain characteristics in regard to ac and dc. The inductor offers low resistance to dc, and high reactance to an ac circuit.

Because the inductor builds a magnetic field, its most common use is the transformer. Transformers have the ability to step up, step down, or isolate the primary and secondary voltages. The turns ratio between primary and secondary sides determines whether the transformer will step up or step down voltage. If the voltage is increased, the current will decrease on the secondary side. Because of this, all transformers have a power ratio. This means that the power primary will equal the power secondary.

Impedance is the total opposition to current flow found in an ac circuit. The opposition to current flow developed by capacitors, inductors, and resistors is very important to all ac circuits.

Capacitors and inductors are often combined to form resonant circuits. They can be combined in series or parallel to pass or reject certain frequencies. This combination causes the circuits to become tuned. Tuned circuits are often found in the IF sections of radios, televisions, and communication devices.

KEY TERMS

alternating current (ac)
average value
capacitance
capacitive reactance
capacitor
ceramic filter
current
dielectric
direct current (dc)
fixed resistor
impedance
inductance
inductive reactance
isolation transformer
Kirchhoff's current law
Kirchhoff's voltage law
linear tapered resistor
nonlinear tapered resistor
Ohm's law
parallel circuit
parallel resonant circuit
potentiometer
power
primary
reactance
resistance
resonance

resonant frequency
rheostat
root mean squared (RMS) value
secondary
series circuit
series-parallel circuit
series resonant circuit
step-down transformer
step-up transformer
tank circuit
thermistor
transformer
tuned circuit
turns ratio
variable resistor
varistor
voltage
voltage-dependent resistor (VDR)
voltage divider

REVIEW EXERCISES

1. What equipment is needed to measure ac voltage?
2. Give a simple definition of current. What is the basic unit of measurement for current?
3. Define resistance, and give the basic unit by which it is measured.
4. How is heat generated in a resistor measured?
5. What controls the resistance of a thermistor?
6. Name the four conditions necessary for a circuit.
7. Define series circuit.
8. Three resistors with values of 1 kΩ, 1.5 kΩ, and 1.2 kΩ are connected in series. What is the total resistance of the circuit?
9. Draw a parallel circuit having three separate loads of 1 kΩ, 3.3 kΩ, and 4.7 kΩ. Find the total resistance of the circuit.
10. Define Ohm's law, and give the letters representing voltage, current, and resistance.
11. State in your own words Kirchhoff's voltage law.
12. Write three formulas used to find power in a circuit.
13. Define power as it applies to electrical circuits.
14. Find the value of voltage drop developed across R_1 in the following figure:

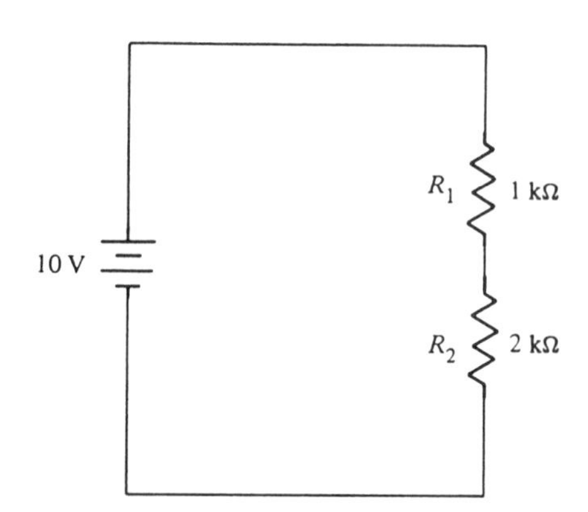

15. What does NPO mean in the rating of a capacitor?
16. How many time constants must a capacitor go through to charge to 99% of the applied voltage?
17. What is the opposition to current flow a capacitor offers to an ac circuit called?
18. What is the reactance to a 10 μF capacitor through which a 1 kHz frequency is applied?

19. What is the term used for the opposition an inductor offers to an ac signal?

20. What will happen to the amount of reactance as the frequency increases through an inductor?

21. A step-up transformer will (increase/decrease) voltage between primary and secondary, whereas the current will (increase/decrease) between input and output.

22. A certain step-up transformer develops 110 V at the primary and 330 V at the secondary. What is the turns ratio of this transformer?

23. Define impedance, and identify the circuit components that develop impedance.

24. Draw a schematic diagram of a series resonant circuit, and one of a parallel resonant circuit. Give the characteristics of each circuit.

25. What is the main purpose of ceramic filters in circuits? Where might these filters be found?

26. Give an example of a tuned circuit. Why should these circuits be made adjustable?

2 VACUUM TUBES AND SEMICONDUCTORS

OBJECTIVES

Upon completing this chapter, you should be familiar with:

—Vacuum tube characteristics
—Vacuum tube dc biasing
—Vacuum tube identification
—Cathode ray tube characteristics
—Troubleshooting vacuum tubes
—Semiconductor characteristics
—Semiconductor atomic structure
—Diode characteristics
—Special diode characteristics

INTRODUCTION

The real breakthrough in electronics came when the first amplifying device was invented. The first tube, called the *audion,* allowed signals to be amplified. Before the audion tube was developed, radio communication depended upon the strength of the signal. With the audion tube, the ability to transmit over great distances was realized. This first audion tube caused the explosion of the age of electronics. Technical development continued from the tube, to the semiconductor, to the integrated circuit.

The audion was an **active component.** Active components are circuit components that have the ability to change a signal. For example, an amplifier will increase the signal amplitude between input and output terminals. All active devices have three terminal leads: the source, the controller, and the output. Each lead relates to a specific job that must be accomplished by the active component. The *source* is used to supply the electrons necessary for amplifying. The *controlling device* is necessary for regulating the flow of electrons to the output circuit. The final element, the *output,* is used to deliver the signal to the load. Each of these elements will be covered briefly in this chapter, and discussed in greater detail in later chapters. Also considered in this chapter are the basics of rectifier devices. A rectifier is used to convert ac to pulsating dc voltage.

VACUUM TUBES

As mentioned, the pioneer in the electronics field was the audion, which enabled the input signal to be amplified. The basics of semiconductors, to be covered later, are more easily understood with a background in such vacuum tubes.

A **vacuum tube** consists of a glass or metal envelope encasing metal electrodes within a vacuum. The **electrodes** are elements used to receive or emit electrons. The electrode that emits the electrons is the **cathode.** Generally, the cathode must be heated by a **filament,** a wire material that generates heat when a current is passed through it. This heating action causes thermonic emission of electrons from the cathode; that is, the cathode, a solid body, gives off electrons. The electrode responsible for collecting the electrons emitted from the cathode is the **plate.** To develop this electron flow, a polarity must be developed between the cathode and the plate. In general, the plate is made positive and the cathode is made negative.

Making the plate positive with respect to the cathode results in plate current. An additional element, the **control grid,** can be placed between the cathode and plate. The control grid controls the number of electrons flowing between cathode and plate, and therefore controls the amount of plate current.

The development of vacuum tubes within the industry began with the diode vacuum tube, or simply **diode.** These tubes consisted of only two elements: a plate and a cathode. The next tube developed was the **triode,** which was composed of three elements. Following the triode were the **tetrode,** with four elements, and the **pentode,** with five elements.

All vacuum tubes used a cathode to emit the electrons and a plate to collect them. In order to supply vacuum tubes with the proper voltage polarity, three separate supplies were developed. An A source was developed to supply power to the filaments, a B source was developed for the plate-cathode polarity, and a C source was developed for the control grid source.

The Diode

As noted previously, a diode has only two elements: a cathode and a plate. The schematic symbol for this tube is shown in Figure 2–1a. The heater is not counted as an electrode because it is used simply to heat the cathode electrically.

Shown in Figure 2–1b is the physical placement of the electrodes in the tube. Note that the plate completely surrounds the cathode in order to capture the electrons emitted from the cathode so that plate current can be developed.

For the vacuum tube to function, the filament must transfer heat to the cathode. When the cathode is heated to a sufficient temperature, the electrons start to "boil-off" the cathode and are pulled toward the plate. This movement develops plate current.

In order to develop this flow of electrons from the cathode, the cathode can be constructed in one of two forms. These forms are shown in Figure 2–2. In Figure 2–2a the cathode and the filament are one and the same. When heated, the filament creates the thermonic emission in order to develop plate current. Figure 2–2b illustrates a vacuum tube with the filament and cathode separated. In this case the cathode is indirectly heated (by the **heater**) to create thermonic emission.

Power is necessary to deliver the proper voltage and current rating to each filament. These ratings would be found in a tube manual containing general specifications. For example, many tubes found within consumer products are rated at 6.3 V, 0.3 A for the heater. This means that 6.3 V will be applied to produce a heater current of 0.3 A. Most of the tubes used in consumer products are indirectly heated by an ac voltage operating at a 60 Hz frequency.

Vacuum tube operation is dependent upon the vacuum within

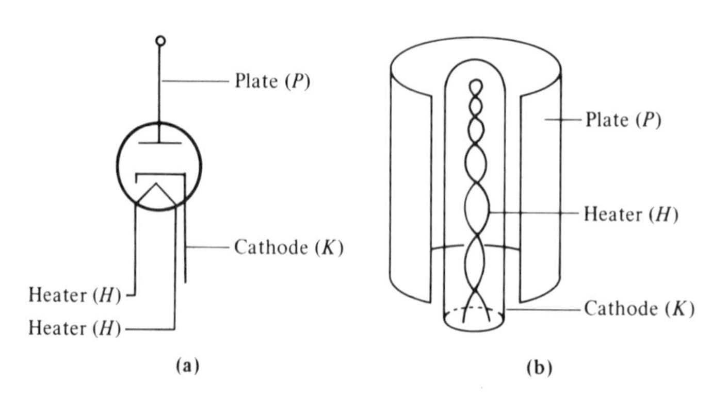

FIGURE 2–1
Diode Construction: (a) Schematic Diagram, (b) Physical Placement of Electrodes

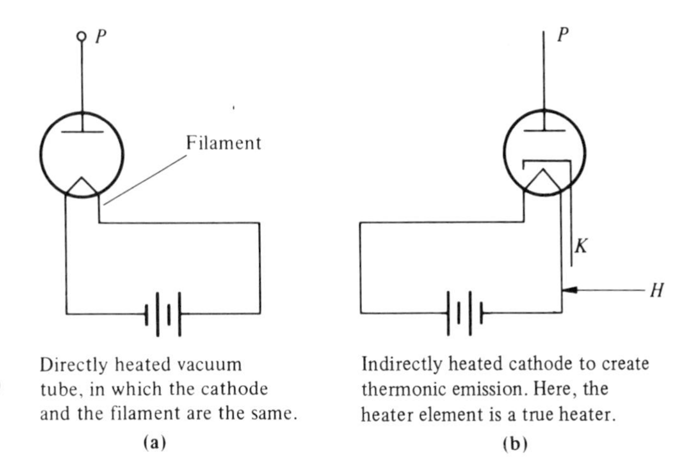

FIGURE 2–2
Cathode Construction: (a) Directly Heated, (b) Indirectly heated

the glass envelope. Without this vacuum, the filament would oxidize in the air and burn out. In addition, the cathode emits more electrons in a vacuum.

The vacuum created within the envelope is very important. When the vacuum tube is under construction, the glass envelope is sealed off. To improve the vacuum, the tube is heated. This heating process forces out any gases trapped within the electrodes. Mounted inside the tube is a **magnesium getter compound.** This getter compound will vaporize in the heating process and will insure that all air has been removed from within the envelope. When the tube cools, the vaporized getter condenses on the inside of the envelope, forming a silvery film, usually seen on the glass tube.

Diode Plate Current As stated, the plate is made positive with respect to the cathode. With this biasing arrangement, a plate current (I_P) is developed. Figure 2–3 shows an example of this connection. Notice that in the schematic diagram, a complete path for electrons is established between the cathode and the plate. As the power supply voltage is increased, the current in the plate also is increased. The current increases until it reaches a maximum point, called the **saturation** of plate current.

If the polarities of the cathode and plate are reversed, the vacuum tube will no longer produce plate current. Because of this action, the diode makes an excellent rectifier in power supplies. Figure 2–4 shows a simple half-wave rectifier circuit, a device used to convert ac to pulsating dc using only one-half of the input ac waveform. Notice that as the ac input voltage swings positive, it makes the plate positive with respect to the cathode. The diode conducts and develops a voltage

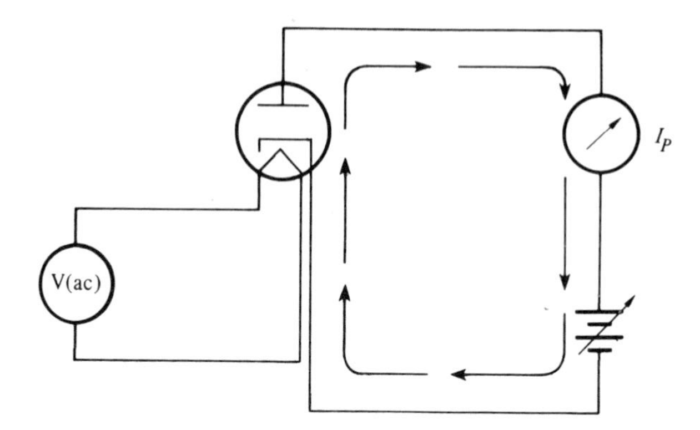

FIGURE 2–3
Biasing the Diode

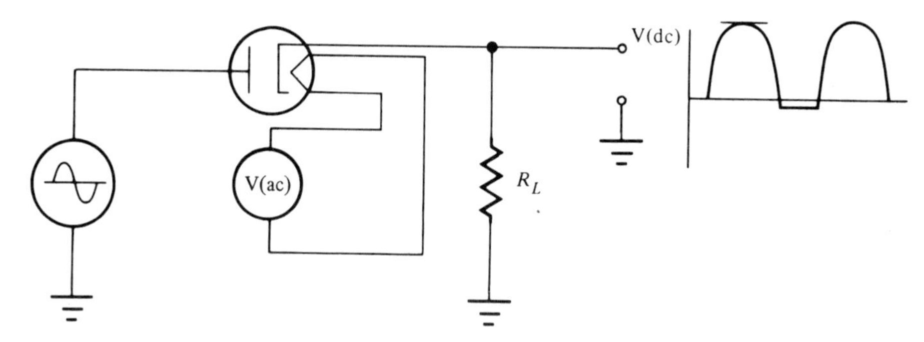

FIGURE 2–4
Half-Wave Diode Rectifier

across R_L. As the ac input voltage swings negative, it places a negative voltage on the plate. This negative voltage causes the tube to stop conducting.

The Triode

The triode, constructed basically like the diode, has in addition a thin mesh wire placed between the cathode and plate. This thin mesh wire is the control grid mentioned earlier. All electrons that travel between the cathode and the plate must pass through the control grid. The grid is attached to the base of the tube so that an external voltage can be connected to it. The voltage applied to the control grid regulates the plate current. The schematic diagram of the triode grid voltage is shown in Figure 2–5.

The grid voltage applied to the control grid is usually a small negative voltage with respect to the cathode. This connection is shown in Figure 2–6a. As shown in this diagram, the small negative voltage

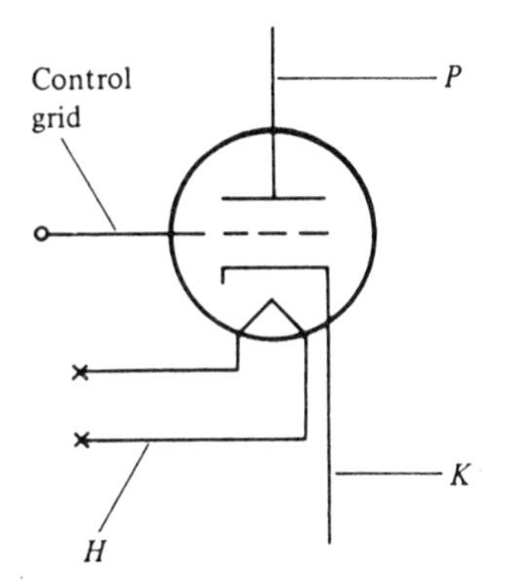

FIGURE 2–5
Triode

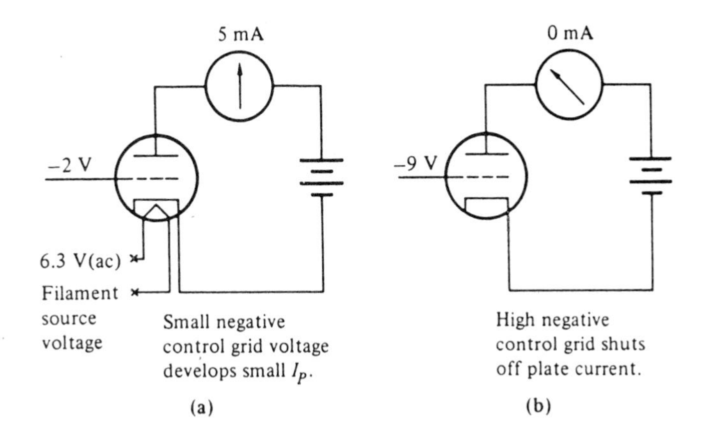

FIGURE 2–6
Biasing the Control Grid

is used to control plate current. The more negative the grid is made, the less plate current is developed, as shown in Figure 2–6b. The more positive the control grid, the more plate current developed.

The effects of the control grid on the plate current can be summarized as follows:

1. If the grid is made negative enough, there will be no plate current;
2. More negative grid voltage decreases plate current;
3. Less negative control grid voltage increases plate current.

Because of the ability of the control grid to control plate current, the triode will amplify an ac signal between the input and the output.

Biasing the Triode Figure 2–7 shows the proper dc biasing for the triode. Notice that control grid supply voltage is labeled C−. This dc

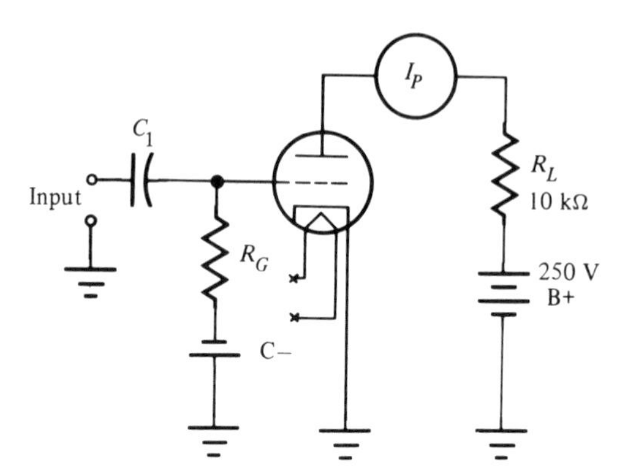

FIGURE 2–7
Biasing the Triode

power supply biases the control grid circuit. The control grid circuit consists of a current-limiting resistor, R_G, and the control grid. C_1 is placed in the circuit to block the C− supply from a low resistance path to ground through the signal source. The dc voltage used to bias the plate cathode circuit is developed from the dc power source, labeled B+. Components used to bias the plate cathode network are the plate cathode of the vacuum tube and the 10 kΩ load resistor, R_L.

In the external plate circuit, electrons flow through the 10 kΩ load resistor back to the B power source. Assuming that a 10 mA plate current is established, a voltage drop across the load resistor can be calculated as follows:

$$V_{RL} = I_P \times R_L = 0.01 \times 10{,}000$$
$$= 100 \text{ V}$$

Because the plate and cathode are in series with the load resistor, the voltage drop across the tube can be found. Since the source delivers 250 V minus the 100 V dropped across the load resistor, the voltage drop between plate and cathode is 150 V. A simple Kirchhoff's voltage equation can be written to solve the voltage drop in this circuit. This equation is given as Equation 2–1.

$$V_b = V_{bb} - V_L \qquad \textbf{(2–1)}$$

where:

V_b = voltage drop between the plate and the cathode

V_{bb} = B power source voltage

V_L = voltage drop across the load

This simple formula is used to determine the actual plate-to-cathode voltage.

EXAMPLE 1

Calculate V_b for 15 mA of plate current, a 2 kΩ R_L, and a V_{bb} equal to 300 V.

Solution:

$$\begin{aligned} V_b &= V_{bb} - V_L = V_{bb} - (I_P \times R_L) \\ &= 300 - (0.015 \times 2000) = 300 - 30 \\ &= 270 \text{ V} \end{aligned}$$

The Tetrode

The tetrode is basically the same tube as the triode, with the addition of a grid between the control grid and the plate. This second grid is called the **screen grid.** Figure 2–8a shows an example schematic diagram of the tetrode.

The typical biasing of the tetrode is illustrated in Figure 2–8b. As in the triode, the control grid is placed close to the cathode. This placement allows the control grid to regulate the electron flow between cathode and plate, and thus control plate current. The additional grid, the screen grid, is not used to control flow between cathode and plate, but rather to help accelerate the electrons toward the plate.

Because the screen grid is made positive, it will collect some of the electron flow, and therefore will develop a small current flow. To keep the screen grid operating at proper voltage, it generally will have a capacitor connected to ground (C_S in Figure 2–8b). This capacitor, C_S, is called the *screen grid bypass capacitor,* and is used to keep the voltage constant during operation. In Figure 2–8b, this capacitor should have a reactance equal to about one-tenth the resistance of the screen grid resistor, R_S. This will insure proper bypass action during the tube's operation. Because of its bypass action (the ac signal is passed to ground so that a signal develops in the load), this capacitor will show very low reactance to the signals being amplified by the tube.

One problem that occurs within the triode is the capacitance caused by the separation of the metal electrodes. The two electrodes—control grid and plate—act as plates of a capacitor, and the vacuum serves as a dielectric material. Such capacitance is generally in the low picofarad range. However, when a tube is amplifying a broad range of frequencies, this small capacitance could reduce the amplification factor. The screen grid is used to reduce this capacitance.

With the addition of a positive screen grid in the tetrode, the electrons are accelerated very fast toward the plate. Because of this action, the electrons bounce off the plate. This bouncing action inside the tube, called *secondary emission,* makes the tube very noisy during

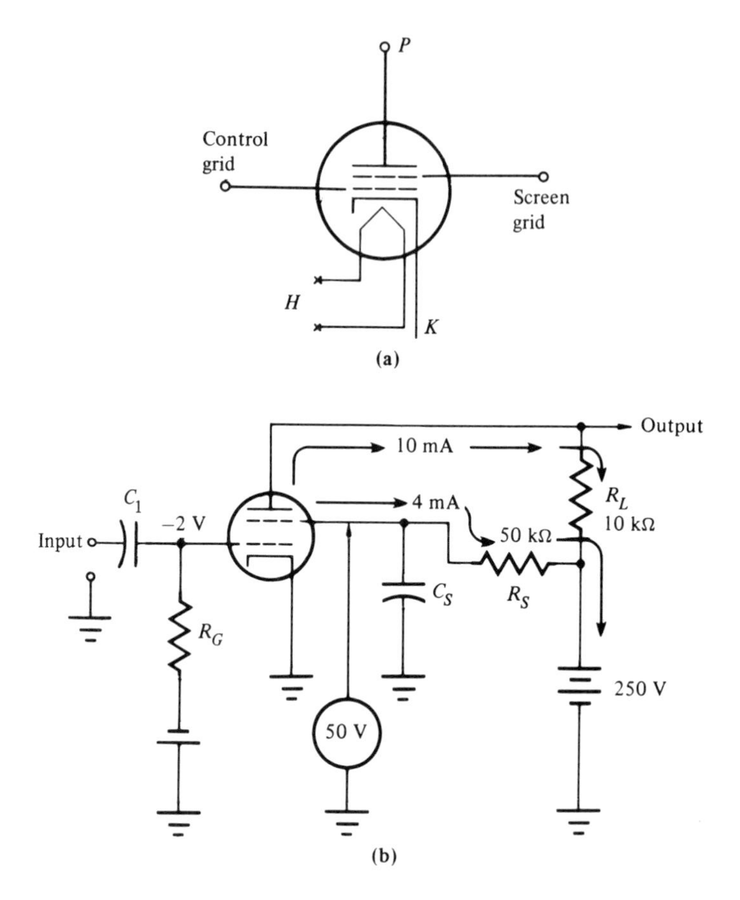

FIGURE 2–8
Tetrode: (a) Schematic Symbol, (b) Biasing the Tetrode

operation. To eliminate secondary emission, another grid is placed in the tube. This grid is called the **suppressor grid.** The addition of this grid changes a tetrode tube into a pentode tube.

The Pentode

The suppressor grid of the pentode is located between the screen grid and the plate. Figure 2–9a shows the schematic symbol for the pentode tube. The proper biasing of the pentode is illustrated in Figure 2–9b.

Notice that the biasing action of the cathode, plate, control grid, and screen grid are the same as for the tetrode. The additional component—the suppressor grid—is connected to a negative potential. Generally, in pentodes the suppressor grid is connected internally to the cathode. This connection will give the suppressor grid the necessary negative polarity. This negative polarity on the suppressor grid will repel the electrons back toward the plate, thus reducing secondary

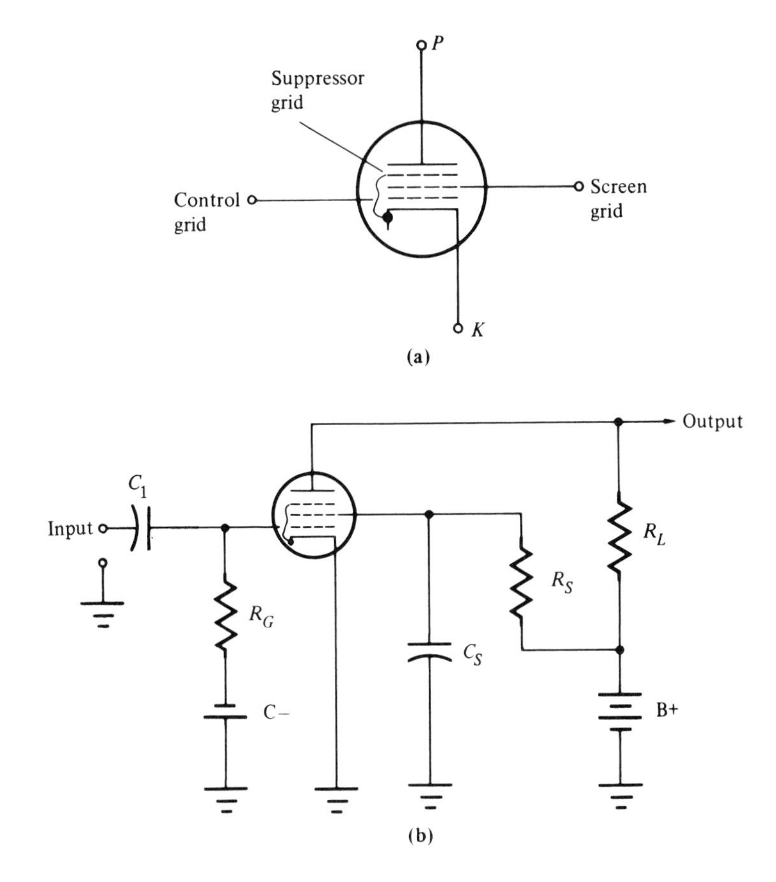

FIGURE 2–9
Pentode: (a) Schematic Symbol, (b) Biasing the Pentode

emission. With this action, the noise generated within the tube is greatly reduced.

VACUUM TUBE IDENTIFICATION AND PIN CONNECTIONS

The first digit of the tube number for amplifier tubes designates the heater voltage. For example, the 6GH8A has a heater voltage of 6.3 V, and the 36LR6 has a heater voltage of 36 V. The letters that follow the filament numbers identify tube use. A tube manual is needed to identify these letters.

It often becomes necessary to measure the pin voltage found on tubes. Figure 2–10 shows several diagrams of tube elements and their connection to external pins. When looking at the tube's socket from the bottom, the numbers are read clockwise. When looking from the

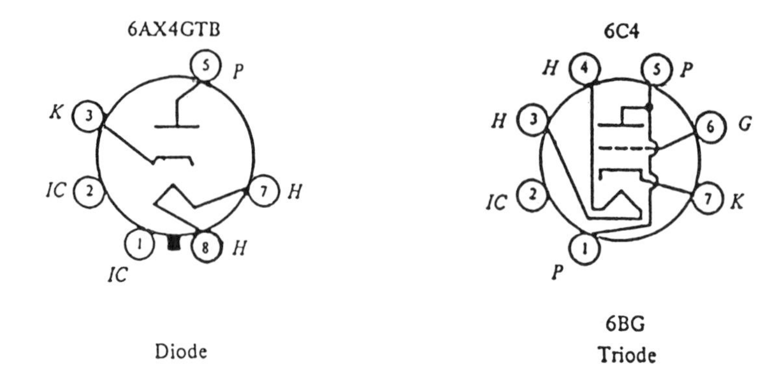

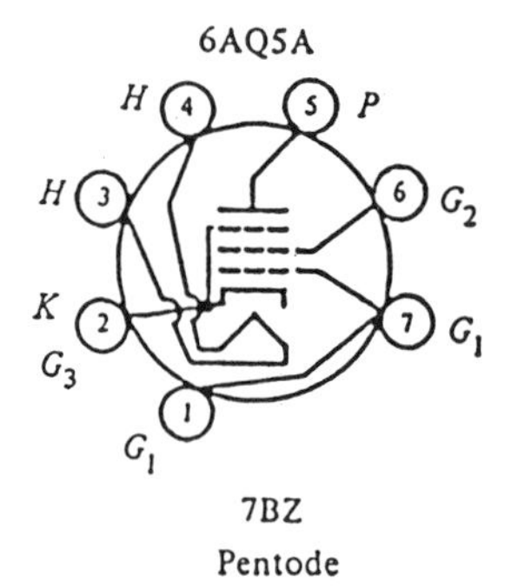

FIGURE 2–10
Tube Elements and External Pin Connections

top, the pins are numbered counterclockwise. Between two of the end pins, there will be a wide spacing. This allows the tube to fit into the socket in only one direction.

In most cases pins are arranged in either a 7-pin format or a 9-pin format. For some of the larger-power tubes, the pin arrangement is in a 12-pin, or duodecal, base.

CATHODE RAY TUBES

The vacuum tube used most often in test equipment, computers, and televisions today is the **cathode ray tube (CRT).** Figure 2–11 shows an example of a CRT found in many oscilloscopes. This tube consists of an electron gun assembly, deflection plates, and a fluorescent screen, all found within an evacuated glass envelope, that is, within the vacuum inside the tube.

In operation, the CRT has a heater that causes the cathode to develop thermonic emission. The control grid found within the gun

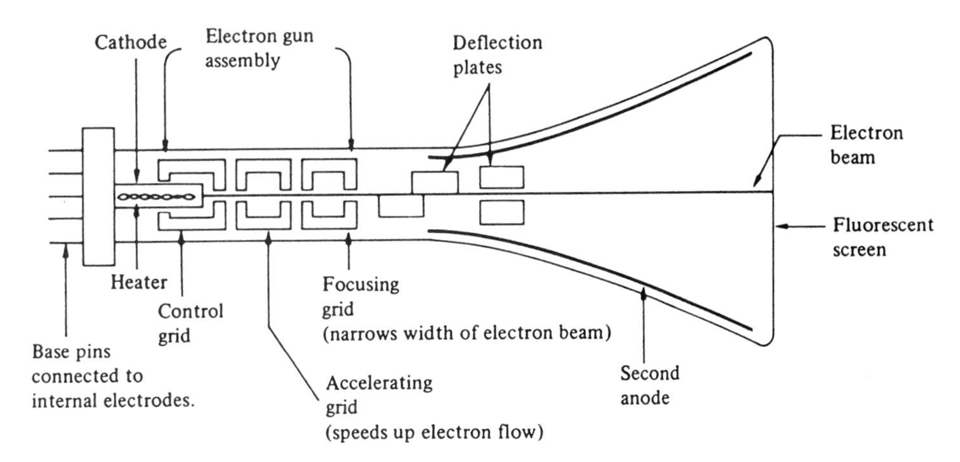

FIGURE 2–11
CRT Internal Electrodes

assembly controls the electron flow toward the screen (see Figure 2–11). (The CRT differs from other vacuum tubes in having as a plate a fluorescent screen instead of a metal electrode.) A positive voltage of 2 kV to 8 kV is applied to the fluorescent screen. This highly positive voltage will attract the electrons from the cathode. As the electrons are accelerated, they strike the fluorescent screen causing it to glow. A chemical reaction between the phosphorous screen and the electrons gives off light.

In order that this light beam can be moved from side to side and top to bottom to trace out patterns of waveforms, a set of deflection plates is placed inside the tube (see Figure 2–11). A pair of these plates is responsible for movement of the beam from top to bottom. The other pair is responsible for left-to-right movement.

Numbers on CRTs are different from those on vacuum tubes. The first digit denotes the diagonal measurement of the rectangular screen. The numbers at the end denote the type of phosphor found on the screen: P1 for green screen, P4 for white screen, and P22 for color phosphor of red, blue, and green.

SOLID-STATE DEVICES

Active devices in electronic circuits all have the same function: to control current flow. The vacuum tube accomplishes this by regulating the current flow between cathode and plate with a control grid. However, because the vacuum tube consumes power and space in an electronic circuit, newer, smaller devices have been developed and are now used extensively to control current. These new devices are called **semiconductors.** As the name implies, these devices operate between two states: the state of conductance and the state of nonconductance.

Conductors

All materials are made up of billions of atoms. Each atom has several parts. The center part is called the **nucleus** and is positively charged. Surrounding the nucleus are orbiting electrons. These electrons are negatively charged. Figure 2–12 shows an example of a conductor. Notice that this conductor is composed of many atoms, each containing a nucleus and orbiting electrons. The figure shows one orbiting electron in the outermost energy shell of each atom. (The other shells of orbiting electrons are not illustrated.) This electron, which has very little attraction to the atom's nucleus, is called the **valence electron.** The valence electron is very important because it is the current carrier for the atom, and therefore makes the material a conductor.

The valence electron is loosely attracted to the nucleus of the atom. Because the attraction is weak, these electrons are able to move easily when an electromotive force (voltage) is applied to the conductor.

With the application of voltage to the conductor, the valence electrons dislodge from their atoms and begin to drift through the conductor. Because the voltage has a positive source and the electrons are negatively charged, the electrons drift toward the positive end. Figure 2–13 shows a conductor with voltage connected and the drifting action of the electrons. The billions of valence electrons of the conductor allow electron flow to develop through the conductor.

Many materials can serve as conductors. The most popular conductors are copper, gold, and silver. Most conductors used in circuits today are made of copper. Wire, cable, and printed circuit trails are all made from this conductor. Copper is easy to work with,

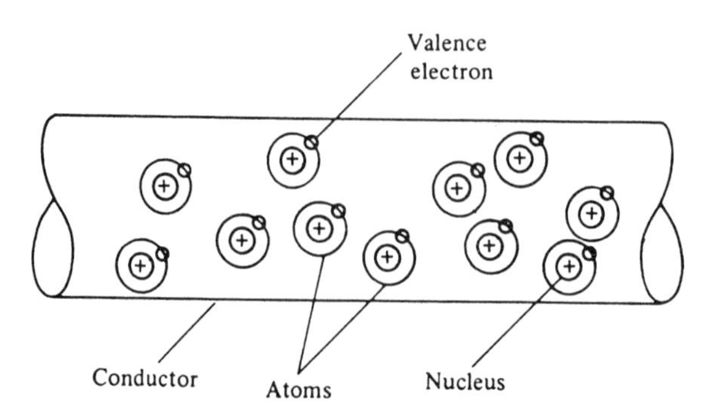

FIGURE 2–12
Conductor, Atoms, and Valence Electrons

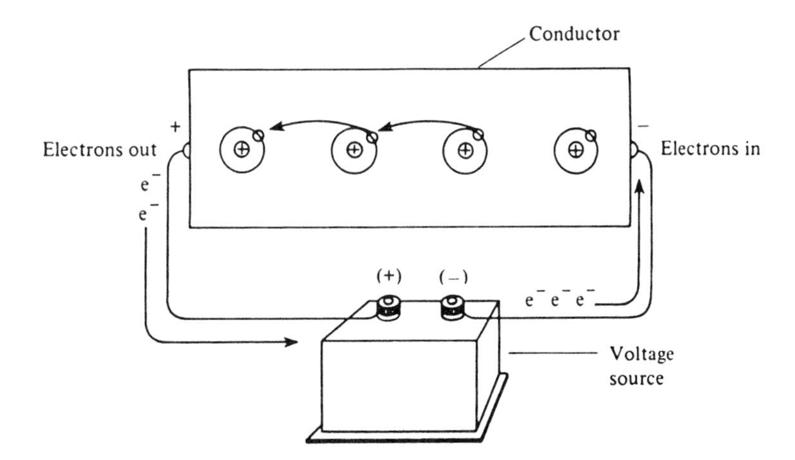

FIGURE 2–13
Electron Movement in a Conductor

and solder adheres to it well. Other conductors, such as silver and gold, are used only to a limited degree. Silver is a better conductor than gold, but because of high cost, neither is found often in electronic circuitry.

Heating a conductor will change its stability in conducting electrons. When the conductor is heated, the valence electrons become very active and begin to collide with each other. As the electrons drift toward the positive end, their collisions become more frequent. These frequent collisions resist the flow of electrons in the conductor. Therefore, the number of electrons reaching the positive terminal declines. When this happens, it is said that the resistance of the conductor has increased. This effect is called a *positive temperature coefficient*: as temperature increases, so does the resistance. Most conductors will react in this manner.

Semiconductors

Semiconductors behave differently than conductors. Unlike conductors, semiconductors do not allow current to flow easily. Under the proper conditions, the semiconductor will become an insulator. This ability to change resistance is a special feature of the semiconductor material.

In the manufacturing of semiconductors, two materials are used: **silicon** and **germanium.** These materials are used to make diodes, transistors, and integrated circuits. Silicon is very popular in the manufacturing of semiconductor devices.

To understand completely how silicon and germanium are used as semiconductors, their atomic structures should be studied. See Figure 2–14 for a diagram of the atomic structure of a silicon atom. Notice that there are three layers of orbiting electrons, each with a negative

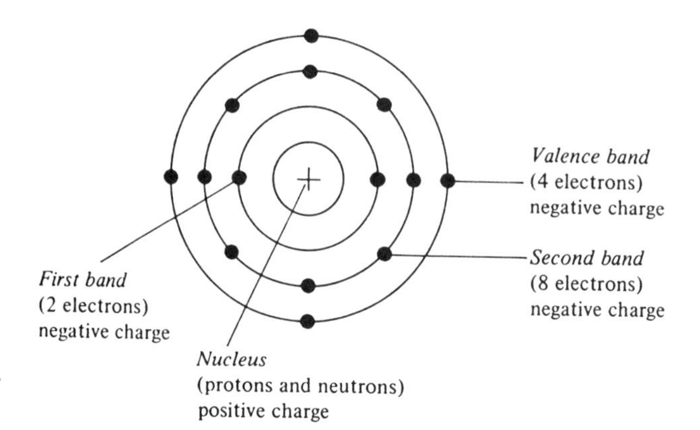

FIGURE 2–14
Atomic Structure of a Silicon Atom

charge. At the center is the nucleus, which contains *protons* and *neutrons.* This compact bundle of protons and neutrons has a positive charge. The positively charged center and the negatively charged orbiting electrons form the attraction necessary for the existence of the atom.

Notice in the figure that the first layer of orbiting electrons contains only two electrons. The second layer has eight electrons. The third and outermost layer has four electrons. This outer layer is called the **valence band.** Figure 2–15 shows the silicon atom with its nucleus and the four outermost valence electrons. Remember that the valence electrons are the current carriers.

Because in its natural state the silicon atom has only four electrons in the valence band, it is considered an unstable atom. This means that it will join with other atoms and share valence electrons. This sharing of valence electrons is called **covalent bonding.** For example, a silicon atom will combine with other atoms having four valence electrons, resulting in eight electrons in the valence band. Eight electrons in the valence band make the silicon atom very stable. To develop this stable state, silicon atoms are combined with other silicon atoms. When conditions are proper, the silicon atoms share their valence electrons. This covalent bonding causes the silicon to form a structure

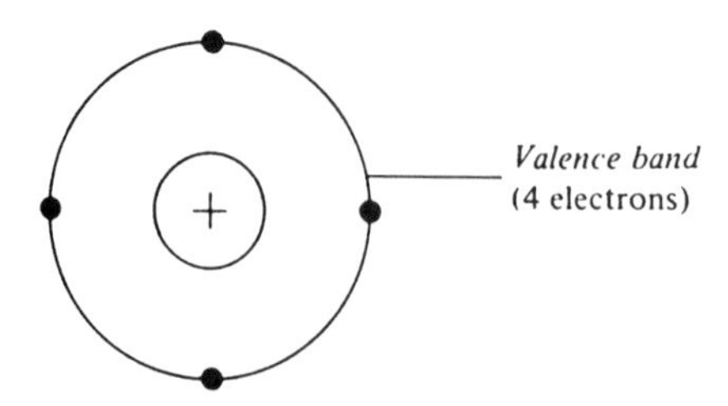

FIGURE 2–15
Silicon Atom's Four Valence Electrons

TROUBLESHOOTING

VACUUM TUBES

In all cases a defective tube must be replaced, either with an identical tube or a replacement brand.

Open Filament

An open heater circuit is a common problem in vacuum tubes. When there is no current flow through the heaters, the heaters remain cold. The tube that does not glow can be found. A quick check is to measure the filament's resistance. Generally, a good filament will develop about 2 Ω. Another check is to measure the voltage at each filament within the circuit.

Cathode-to-Heater Leakage

Leakage between the heater and the cathode results in hum within the tube. Generally, the heaters are powered by an ac voltage at 60 Hz. Because this frequency is audible, it will be passed from the cathode to the plate, and therefore develop a hum. To check this, a resistance check should be made between the heater pins and the cathode pin to insure that they have high resistance.

Tube Checker

The *tube checker* is an instrument that allows the technician to check if the tube is operational. The tube under test is placed in a socket. Conduction developed through this tube is measured on a meter. The meter will indicate if the tube is good, bad, or questionable.

called a **crystal.** Figure 2–16 shows the crystal atomic structure formed by the covalent bonding of silicon atoms. Notice that around each nucleus, there are eight electrons, making this a very stable atomic structure.

Because silicon is now stable, it behaves as an insulator at room temperature. If a small voltage is applied across this material, only small amounts of current will develop. One method of making the pure silicon crystal a conductor is to raise the temperature. As heat is applied, the electrons in the valence band become active and dislodge

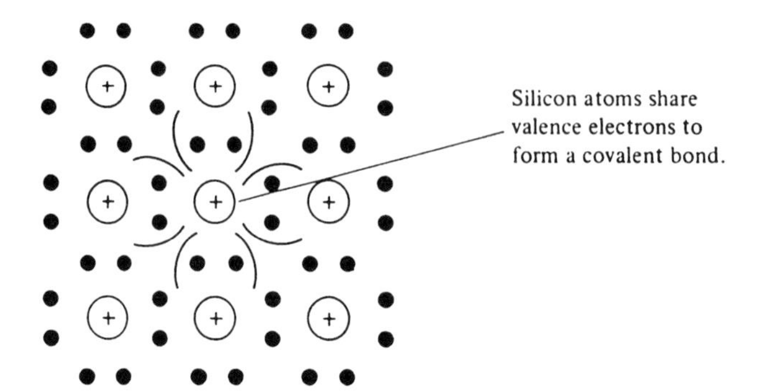

FIGURE 2–16
Covalent Bonding of Silicon

from the atomic structure, as shown in Figure 2–17. Then, when a small voltage is applied to the crystal structure, current flows.

Because of current flow developing when heat is applied, the silicon material is said to have a *negative temperature coefficient.* This means that as the temperature rises, the resistance of the material decreases. Silicon's resistance will decrease by one-half with a rise in temperature of 6°C.

As was mentioned earlier, germanium is also used for semiconductor devices. The germanium atom is similar to the silicon atom. Like silicon, the germanium atom has four valence electrons. It also combines with other atoms to have eight electrons in the outer orbit. In addition, germanium atoms can be combined with other germanium atoms to form covalent bonds and a crystal structure.

Although both germanium and silicon have a negative temperature coefficient, the silicon crystal can tolerate more heat before it becomes a conductor. Because of power requirements in consumer products, the majority of semiconductor materials are made from silicon.

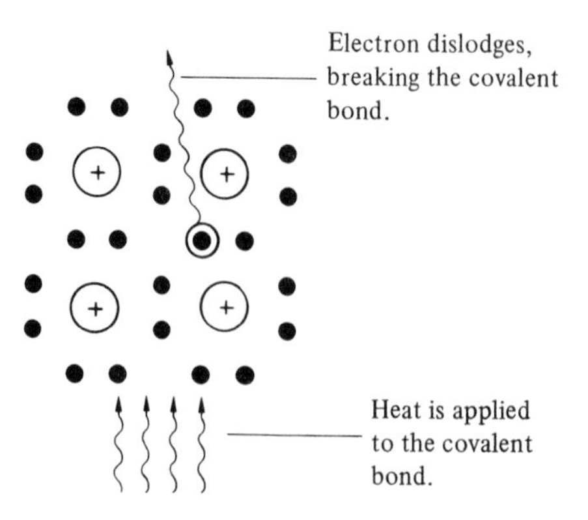

FIGURE 2–17
Dislodging a Silicon Electron by Applying Heat

P and N Semiconductor Material

By definition, for the silicon or germanium to operate as an active device, it must control current flow. As noted, at room temperature these materials are insulators. However, as temperature is increased, they become conductors. Raising and lowering temperature in pure semiconductor materials is impractical. To make these materials efficient conductors of current, **impurities** are added to the crystal structure. These impurities are either **trivalent atoms** with three valence electrons, or **pentavalent atoms** with five valence electrons.

The addition of impurity atoms (nonsilicon atoms) to a crystal to increase the number of conduction band electrons is identified as **doping.** When a crystal has been doped, it is called an **extrinsic semiconductor.** Atoms that have five valence electrons are added to the crystal. The covalent bond is formed. However, because the covalent bond requires only four additional electrons, there is an electron left over. This extra electron becomes the current carrier for that material. This doped material is called **N-type** (negative-type) **material** because it supplies an extra electron in the conduction band.

Atoms commonly used to form N material are arsenic, phosphorus, and antimony. Figure 2–18 shows the covalent bonding of silicon atoms and an arsenic atom. Note that the impurity has created one free electron. In the doping process, many atoms of the impurity are added to the silicon crystal to make the N material. The additional conduction band is generated by the addition of the pentavalent atom.

P-type (positive-type) **material** is formed in the same manner as N-type material. However, instead of using atoms with five valence electrons, the P material is formed using atoms with only three valence electrons. The impurity is added, the covalent bond is formed, but this time there is an electron missing. The lack of this one electron makes the bond weak. To make the bond stronger, an electron is taken from a nearby covalent bond. This stripping off of electrons leaves holes in the neighboring bonds. This constant moving of electrons from bond to bond creates a simulated movement of holes, giving the material a positive potential charge.

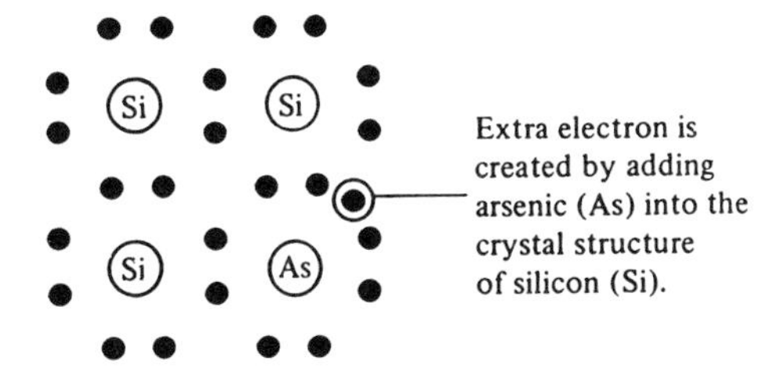

FIGURE 2–18
Doping the Silicon Crystal with an Additional Atom

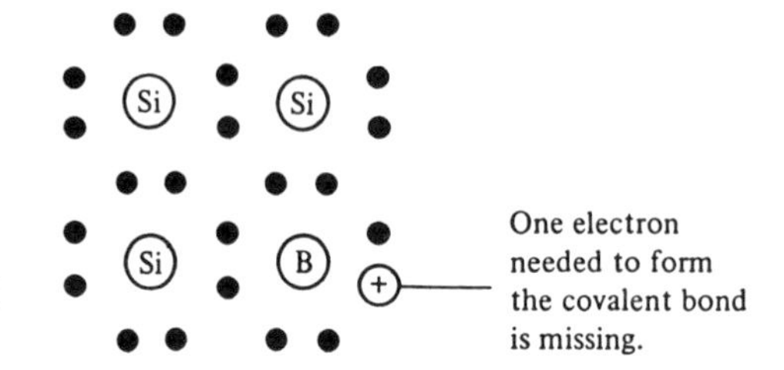

FIGURE 2–19
Doping P Silicon Material with Boron (B)

Atoms commonly used to dope the P material are aluminum, boron, gallium, and indium. Figure 2–19 shows a boron atom and its valence electrons being combined with several silicon atoms, and the bond that is developed. The addition of these atoms causes a **trivalent impurity,** which creates extra holes in the crystal.

In the P material there will be current flow. The current is carried by the simulated moving of the holes from bond to bond. In

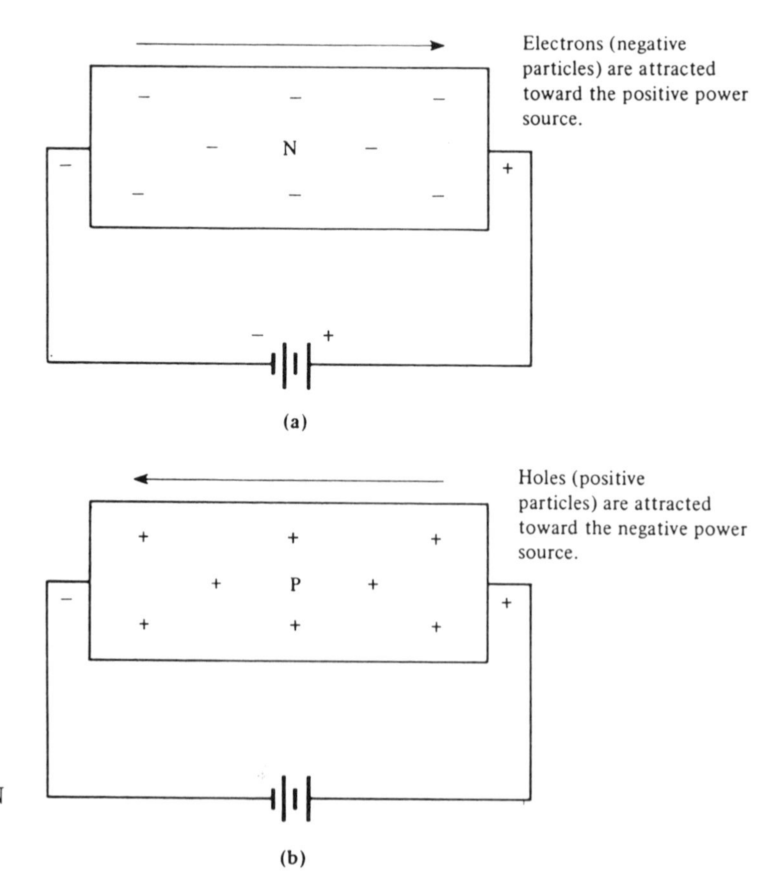

FIGURE 2–20
Current Carriers: (a) in N Material, (b) in P Material

contrast, the electrons are the current carriers in the N material. Figure 2–20 shows an example of the P- and N-type materials. Note in Figure 2–20a that when power is applied to the crystal material, the electrons move toward the positive part of the supply, whereas in Figure 2–20b the holes move toward the negative terminal of the battery.

PN Junction

The basic **PN junction** diode is shown in Figure 2–21. This diode contains a region of P-type material and a region of N-type material. This diode is formed by a solid piece of germanium or silicon crystal. The union, or junction, of the P and N material is also shown in Figure 2–21. The junction marks where the P material ends and the N material begins. This junction is not a mechanical junction. There is no physical union between the P and N material. Because of this solid structure, the electrons are allowed to move across the junction.

When the diode is formed, electrons drift across the junction. These drifting electrons fill in the holes that have been developed in the P material. Figure 2–22 illustrates this drifting of electrons and

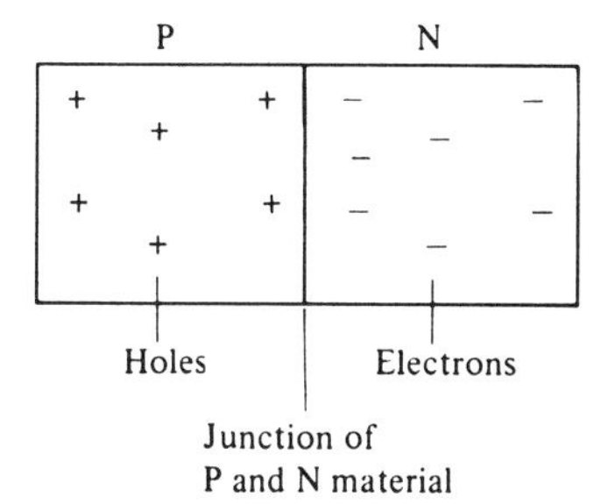

FIGURE 2–21
PN Junction Diode

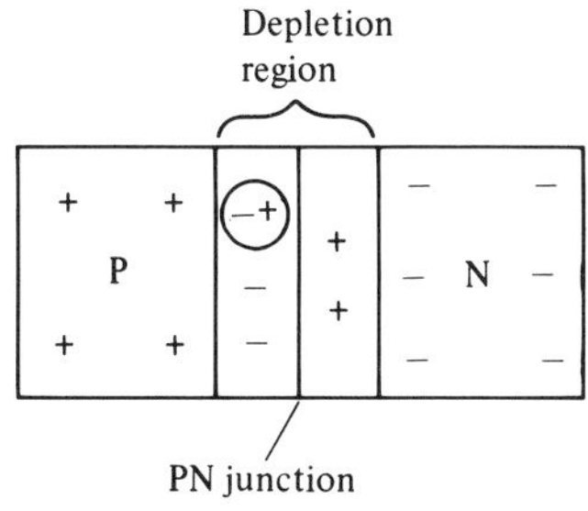

Hole movement across the junction (+ signs in the depletion region), as well as electron movement across the junction (– signs in the depletion region).

FIGURE 2–22
Depletion Region of the PN Junction

hole movement across the junction. The junction of the positive and negative ions is called a **dipole.** As the dipoles build up, the movement of these ions is reduced. This area of reduced movement is identified as the **depletion region.** An *ion* is an atom that has either gained or lost one or more valence electrons to become electrically charged. Positively charged ions have a deficiency of electrons, while negatively charged ions have a surplus of electrons.

The depletion region has no majority current carriers. Therefore, it would seem that the depletion region has formed an insulator between the P and N material, and therefore that the PN junction would make an excellent insulator. However, when an external voltage is applied to this junction, the depletion region becomes a semiconductor material.

Figure 2–23 shows an external voltage connected to the PN junction diode. Note that the positive pole of the battery is connected to the P material via a resistance. The negative pole of the battery is connected to the N material. This method of connection is called **forward bias.** In operation, the forward bias state has made the depletion region a very narrow area. The depletion region is reduced in size because of the *law of charges,* which states that like charges repel and unlike charges attract. Because the positive voltage is connected to the P material and the negative voltage is connected to the N material, the like charges repel. This repelling action sends the holes and electrons to the junction. This force reduces the width of the depletion region. This reduction of the depletion region allows current to flow through the semiconductor material.

Figure 2–24 shows the other condition of the PN junction. This connection is called **reverse bias.** Note the voltage from the battery.

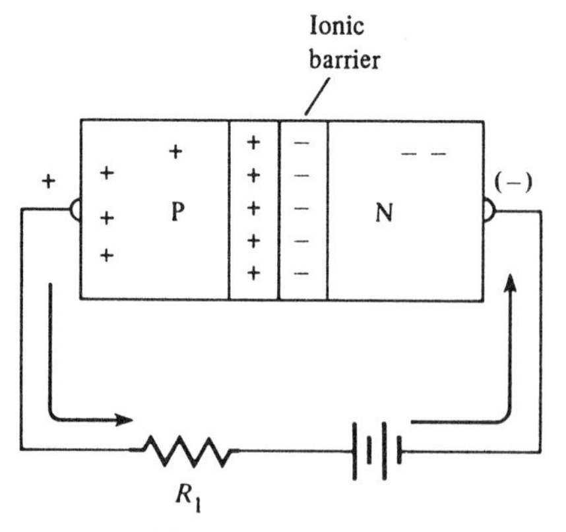

FIGURE 2–23
Forward Biasing the PN Junction

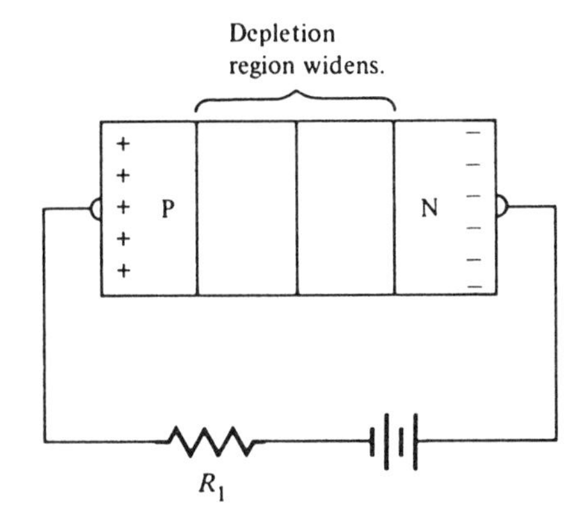

FIGURE 2–24
Reverse Biasing the PN Junction

The negative battery terminal is connected to the P material, and the positive battery terminal is connected to the N material. The law of charges holds here also. Opposites attract. This attraction makes the depletion region larger. The enlarged depletion region serves as a block to current flow. However, when the junction is reversed biased, a small amount of current will flow. This current is called **leakage current.** The leakage current is so small that it cannot perform any work in the circuit.

The temperature on the material will have a definite effect on leakage current. As was mentioned earlier, compared to germanium, silicon is a stable material when heat is applied. Therefore, the silicon junction diode will have lower leakage currents at higher temperatures than the germanium junction diode.

The PN junction diode is the basis of semiconductor study. The junction diode is made from either silicon or germanium. The material is covalent bonded to form a crystal. The crystal material is doped to produce a junction of P and N material. When an external voltage is applied, the junction diode will conduct or become an insulator. This action classifies the junction diode as an electronic *switch,* a component used throughout modern consumer electronic circuits.

Diode Characteristics

All manufactured diodes have certain characteristics important to the service technician. These characteristics are the schematic symbol, physical size, diode rating, and application.

Schematic Symbol Shown in Figure 2–25 is the schematic symbol for the diode. Notice that the diode can be divided into two sections: the triangular section and the bar section. These two sections have specific names. The triangular section is called the **anode.** The anode of the diode is made up of the P material. The bar section is called the

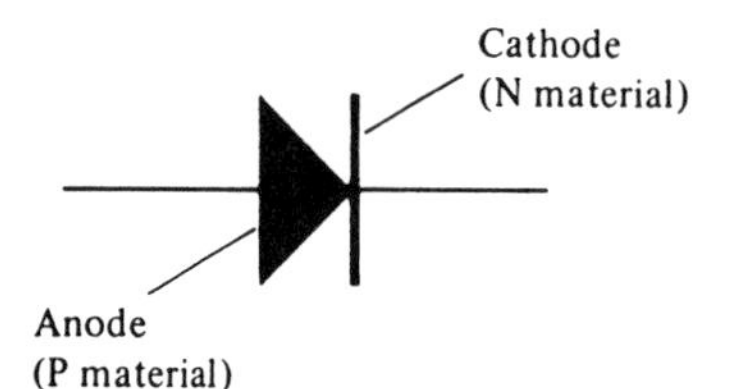

FIGURE 2–25
Diode Schematic Symbol

cathode. The cathode is made up of the N material. This information can be used to determine whether the diode is connected in forward or reverse bias.

Figure 2–26 shows the diode connected in forward bias. Notice that the cathode (N material) is connected to the negative part of the power source, while the anode (P material) is connected to the positive part of the power source. The arrow indicates the direction of current flow in the circuit. Notice that there is a resistance in the circuit. The only function of this resistance is to regulate the forward current flow in the diode. If the resistor were not present, the current flow might damage the diode.

Figure 2–27 shows the diode connected in reverse bias. Notice that the power supply connections are opposite those for the diode in forward bias. As was mentioned earlier about the semiconductor PN junction, the diode now blocks most of the current flow in the circuit. Only a very small leakage current will flow.

Another important fact must be understood about the action of the diode. The diode must have sufficient voltage applied in order to conduct. In general, the silicon diode requires a 0.7 V drop between anode and cathode before current starts to flow. The germanium diode requires only a 0.3 V drop. These voltage drops, once reached by the supply, will be maintained by the diode.

The graph in Figure 2–28 shows what happens when the diode begins to conduct. This graph is called a *volt-ampere characteristic*

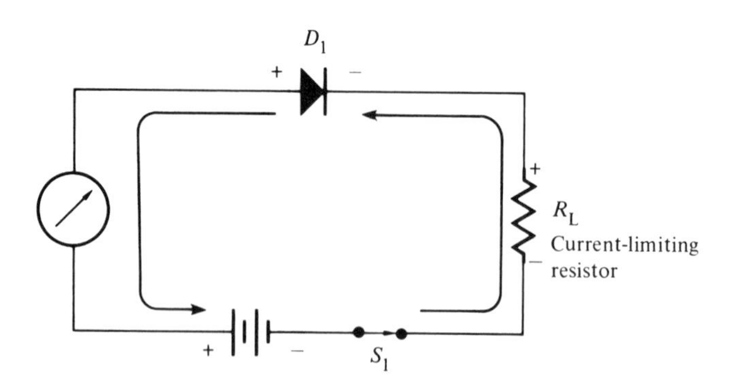

FIGURE 2–26
Diode in Forward Bias

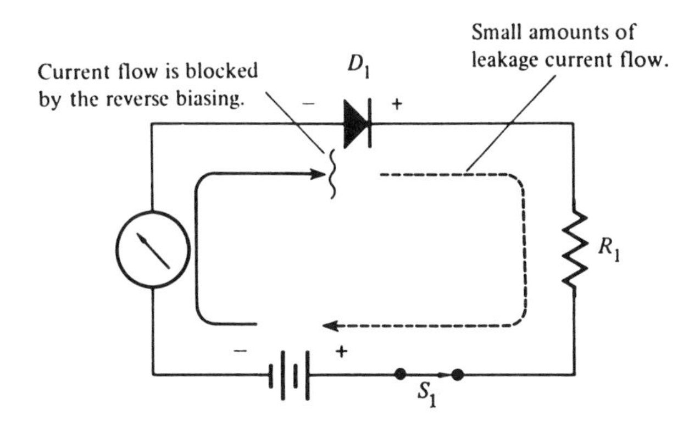

FIGURE 2–27
Diode in Reverse Bias

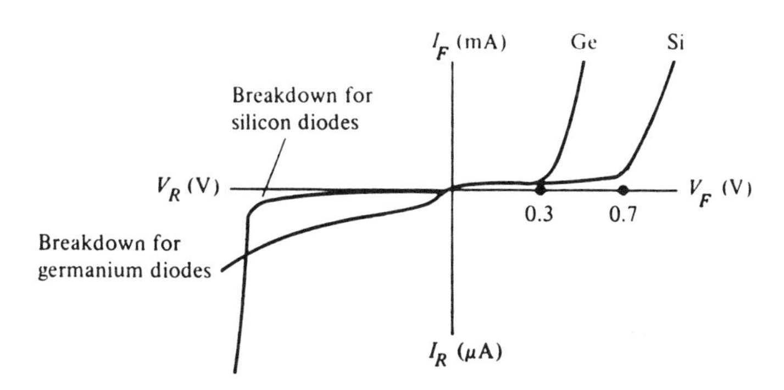

FIGURE 2–28
Volt-Ampere Characteristic Curve for Silicon and Germanium Diodes

curve. The graph shows that as a silicon diode approaches 0.7 V, very little current flows through the diode. Once the 0.7 drop is achieved, the diode has an almost instant rise in current flow. If the resistor were not placed in series with the diode, the diode would be damaged by an excessive amount of current flow developed in the circuit. This resistor prevents the current from going into avalanche effect, the fast increase in current flow. Once the 0.7 V level is achieved across the diode, the diode will keep itself regulated at that voltage. This regulated voltage drop is a very important characteristic of diodes. No other component develops a relatively constant voltage drop as voltage increases.

Figure 2–29a illustrates the connection of a silicon diode to a dc power supply. This simple circuit illustrates the dc characteristic of the silicon diode. The table in Figure 2–29b shows the voltage drops across the diode and the load resistor, R_L. Notice that when the power supply is set to 1 V, a 0.7 V drop appears across the diode, and a 0.3 V drop across the load. As the voltage is increased, the diode maintains its voltage drop and the resistor increases its voltage drop. This circuit

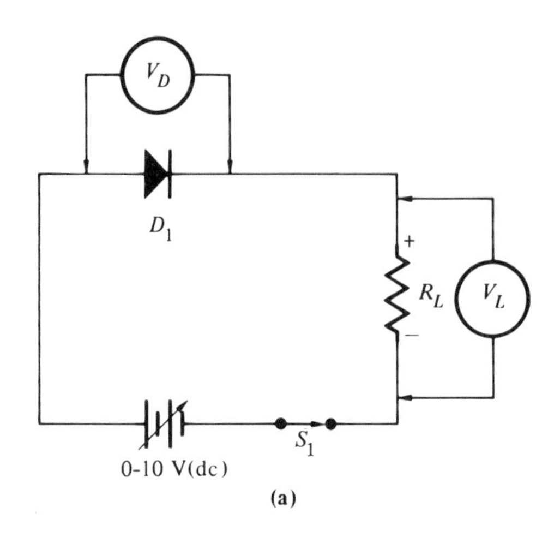

V_S	V_D	V_L
0 V	0 V	0 V
1 V	0.7 V	0.3 V
5 V	0.73 V	4.27 V
7 V	0.73 V	6.27 V
10 V	0.73 V	9.27 V

(b)

FIGURE 2–29
Maintaining a Constant Forward-Bias Voltage Drop

must hold true because of voltage drops developed in a series circuit as they relate to Kirchhoff's law.

Another important factor is the **reverse voltage** condition of the diode. When the diode goes into reverse bias, the depletion region increases in width, and this acts as an insulator. However, as with any insulator, when high enough voltages are applied, the insulation breaks down and becomes a conductor. If large reverse voltages are applied, the diode will break down and conduct. The graph in Figure 2–28 shows what happens in reverse biasing silicon and germanium diodes. Note that the silicon diode does not break down until a specific voltage is established. This voltage has a value of between 50 V and 1000 V, depending upon the rating of the diode. Once this voltage is reached, a sudden increase in current is present in the diode. If this reverse voltage continues for a long time, the diode will be destroyed. The germanium diode, on the other hand, begins to conduct current as soon as voltage is seen in the reverse direction. This is another reason silicon is preferred as a diode material.

Diodes come in different physical sizes and shapes. Figure 2–30 shows examples of some of the different diodes found in consumer products. Notice that each diode has a marking to indicate which end is the cathode. In general, a silver band or a plus sign is used to mark

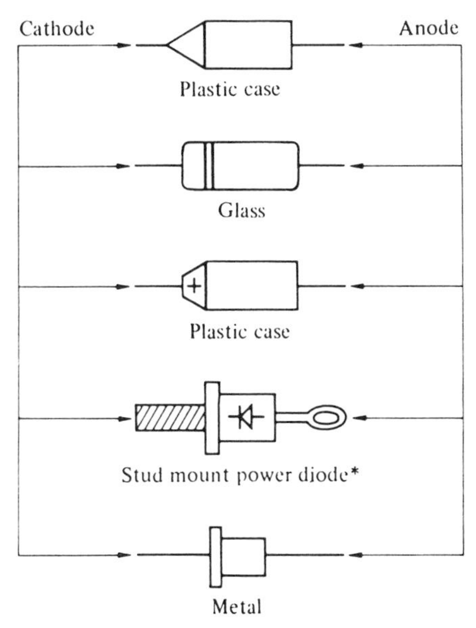

FIGURE 2–30
Physical Markings of Diodes

the cathode end. This marking enables the technician to replace the diode properly. If the technician fails to observe the proper diode placement, damage to the diode and surrounding components may result.

In some cases it may become necessary to identify the cathode and anode of the diode with the use of an ohmmeter. Because the ohmmeter uses only a small amount of voltage to measure resistance, the voltage will make an excellent reference to measure the forward and reverse resistance offered by the diode. Figure 2–31 shows an example of the resistance measuring process. Notice that in Figure 2–31a an analog ohmmeter is placed on the $R \times 100\ \Omega$ scale. The negative lead of the meter is placed on the cathode lead of the diode, and the positive meter lead is connected to the anode of the diode. This connection will result in a low resistance. In Figure 2–31b the leads of the ohmmeter are reversed. The positive is applied to the cathode, and the negative ohmmeter lead is connected to the anode. This connection will result in a high resistance reading. If these readings are found on the diode, then the diode is good.

The internal power sources of digital meters are not large enough to forward bias semiconductors. In order for these instruments to measure semiconductor resistance, a special range has been added. This range is called the *diode checker position.* Placing the meter into this position provides sufficient voltage to forward bias the diode's

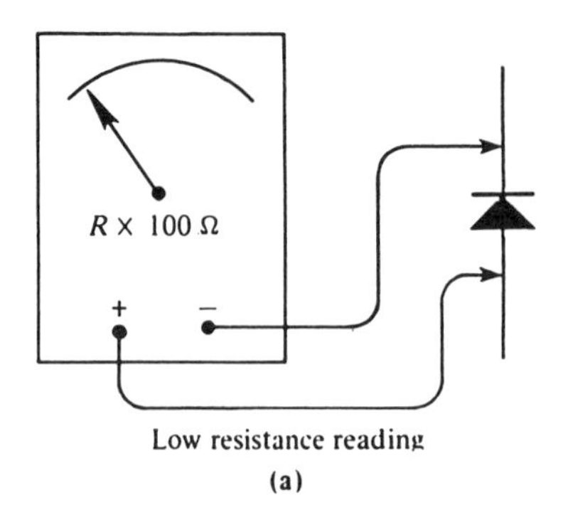

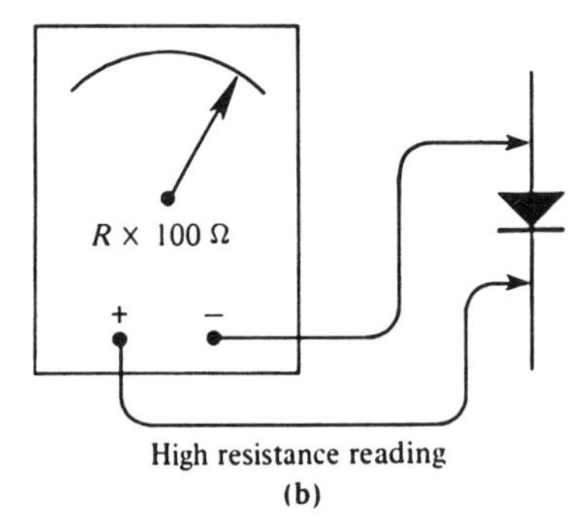

FIGURE 2–31
Measuring Diode Resistance: (a) Forward Low, (b) Reverse High

junction. Once the meter is placed in this position, the same procedure as detailed in Figure 2–31 is used.

Diode Rating Specific ratings have been established for each manufactured diode. These ratings, which give a written interpretation of the volt-ampere graph, are important for the service technician. Replacement information must be gathered from ratings on diode data sheets. The ratings are as follows:

—**Forward voltage (V_F)** is voltage that will be developed between the cathode and anode of the diode. At room temperature, the anode voltage should be 0.7 V for silicon and 0.3 V for germanium.
—**Reverse voltage (V_R)** is the difference in voltage between the cathode and the anode before the diode breaks down. This rating is also given in *peak inverse voltage* (PIV) or *peak reverse voltage* (PRV).
—**Forward current (I_F)** is the amount of current the diode can safely handle in forward bias.
—**Reverse current (I_R)** is the amount of current the diode can safely handle when a reverse voltage is applied.
—**Reverse breakdown voltage (V_{RB})** is reverse voltage beyond which the diode can no longer hold back reverse current.

—**Reverse recovery time** (t_{RR}) is the time necessary for a diode to recover from forward conduction and block reverse current.
—**Power dissipation** (P) is the maximum power the diode can handle in forward bias. Because P is made up of I and E, and E is fixed, P usually is the maximum current through a diode.

All ratings for a diode assume the diode is operating at room temperature (25°C). When the temperature changes, so will the current, voltage, and power figures on the diode.

Remember, when a diode needs replacement in a circuit, the specifications should be checked to make sure the new diode is an exact replacement for the defective diode.

Diode Application The diode has many uses in modern consumer products. One of the prime applications is in rectification. Because of its ability to function as a switch, the diode can be used to convert an ac waveform into pulsating dc. Another important use of the diode is as a regulator. The diode holds voltage constant in forward bias, and therefore keeps voltages from reaching high levels. Figure 2–32a shows

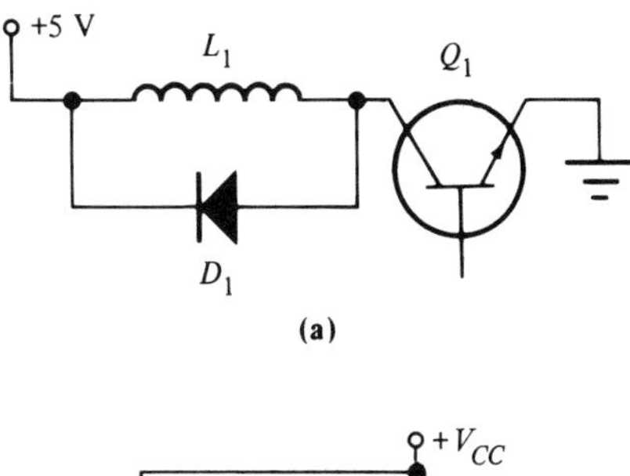

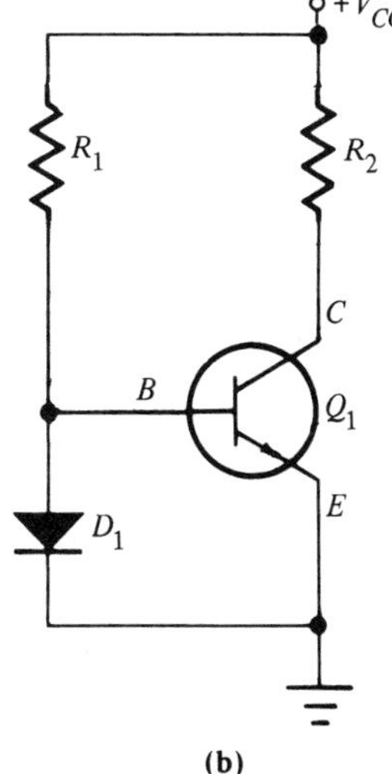

FIGURE 2–32
Diode Used as a Clamping Circuit: (a) to Keep Voltage Constant across L_1, (b) to Keep Voltage Constant between Base and Emitter

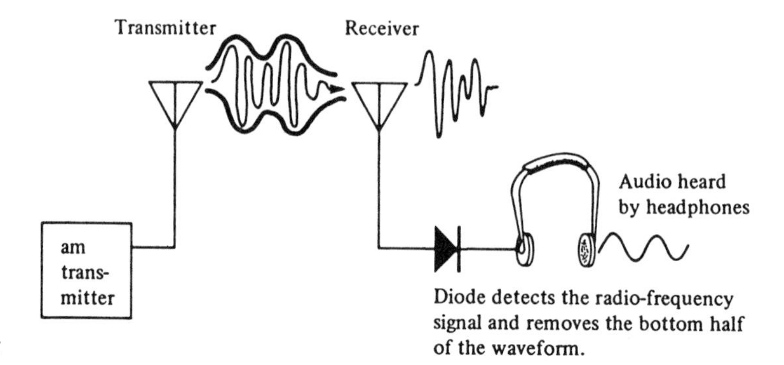

FIGURE 2–33
Diode Used as a Detector

an example of using a diode to hold the voltage across L_1 constant. This is also called *clamping.* Figure 2–32b illustrates another method to hold voltage constant with a diode. Here the diode is placed between the base and emitter of a transistor. If the diode is silicon and the transistor is silicon, then the forward bias voltage drop will be 0.7 V. Since the diode is in forward bias, it will maintain this 0.7 V drop. Also, the base emitter is in parallel with the diode, and it also will maintain the 0.7 V drop. This constant voltage drop on the transistor's base-emitter junction will keep constant current flowing through the collector-emitter's junction of the transistor. As illustrated, if the voltage tries to increase at the base emitter, the diode will hold this voltage constant.

The diode serves an important function in a radio. It is used to remove the transmitter's high-frequency modulation. Figure 2–33 shows a block diagram of this process and the waveforms produced at the output of the diode. Diodes used in this manner are called *detectors,* or *demodulators.*

Diode Ohmmeter Checks

A diode can be quickly determined as good or bad by the use of an ohmmeter. The internal battery of the ohmmeter is usually 1.5 V. This voltage is sufficient to forward or reverse bias the diode. If the positive (normally red) lead is connected to the anode, and the negative (normally black) lead is connected to the cathode, the diode will be in forward bias. The meter reading should indicate a low resistance. The $R \times 1$ kΩ or $R \times 10$ kΩ setting should be used when making this resistance check. If the leads are then reversed on the diode, the meter reading will indicate a high resistance. The general rule of thumb when measuring a diode is to look for these high and low resistance readings.

Their presence means the diode is good. Figure 2–31 shows an example of how to read this high-low resistance combination.

High-powered diodes often have to be checked by the service technician. Sometimes these high-powered diodes cannot be tested with the ohmmeter because they contain several diodes in series. This connection does not allow the diode to establish the high-low resistance reading.

Zener Diodes

Normal rectifying diodes are made to operate in the forward bias region. Special diodes can be made to operate in the reverse bias region. These diodes are called **zener diodes.** By increasing the doping process, an increase in the number of impurities is developed in the diode. Different levels of impurities will cause the diode to break down at different levels of reverse voltage. Zener diodes are available in voltage ratings from 2.4 V to 200 V, with power ratings of from 1/4 W to 50 W. Because of temperature considerations and current capabilities, silicon rather than germanium is usually used in the manufacturing of zener diodes.

Figure 2–34a shows a schematic diagram of the zener diode. Figure 2–34b presents a graphical analysis of the diode's operation. As shown by the graph, the zener diode operates in reverse bias. This

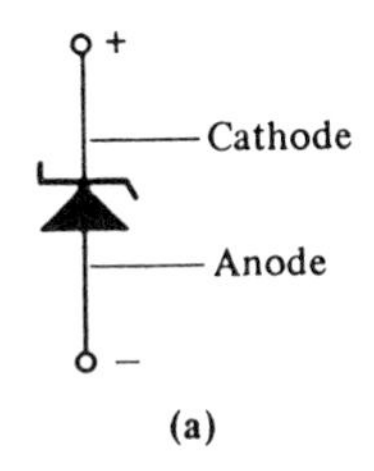

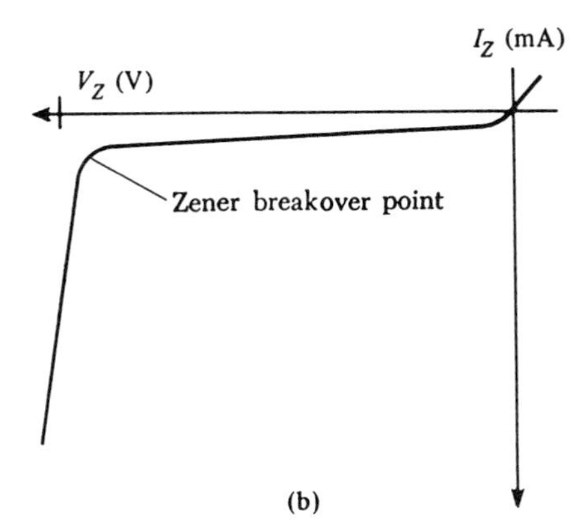

FIGURE 2–34
Zener Diode: (a) Schematic Symbol, (b) Volt-Ampere Characteristic Curve

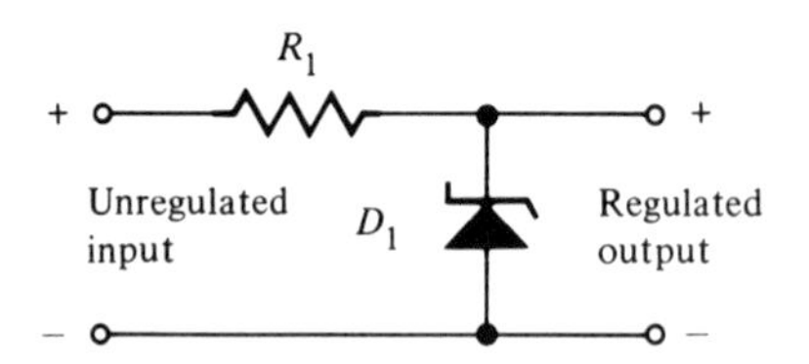

FIGURE 2–35
Zener Diode Used as a Voltage Regulator

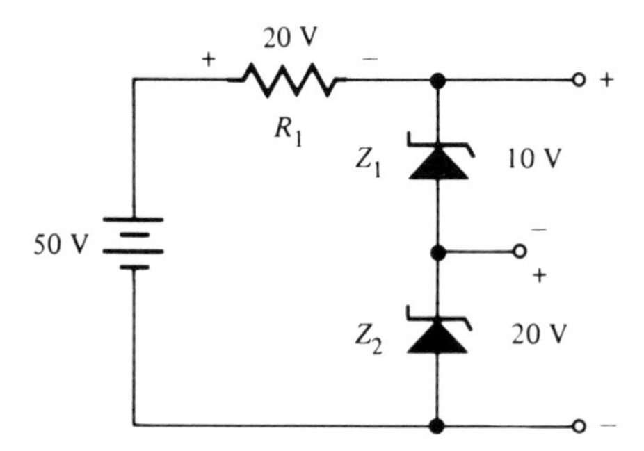

FIGURE 2–36
Developing Two Voltage-Regulating Circuits

means that the cathode is connected to positive voltage, and the anode to negative voltage. In operation the zener diode does not start to conduct until its reverse voltage is developed. In the graph this point is at 8 V. Once this voltage is reached, the voltage drop across the diode is held constant. If the voltage tries to increase across the zener diode, then a large current increase will be seen in the diode. Figure 2–35 shows an example of how the zener diode is used as a simple regulator.

Zener diodes can also be used to fix a reference point for several dc voltage outputs. Figure 2–36 shows two zeners connected in series to give two regulated output voltages. In addition, zener diodes can be used in ac circuits to clip waveforms to various peak-to-peak levels of voltage. Figure 2–37 shows an example of regulating ac voltage.

Zener diodes are numbered beginning with the designation 1N. For example, a 1N961 is a zener diode. To determine this zener diode's rating, a semiconductor data manual must be consulted.

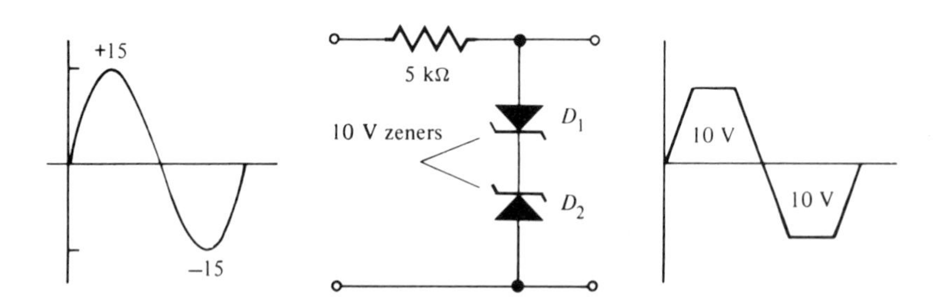

FIGURE 2–37
Using Zener Diodes to Regulate ac

Varicap Diodes

Varicap diodes, or **varactors,** are semiconductor diodes. They are voltage-dependent variable capacitors. Their mode of operation depends upon the amount of reverse bias voltage, because the depletion region's width is adjusted by this reverse bias voltage on the diode. Since the P and N junctions can be moved closer and farther apart, they act like plates of the capacitor. (It should be remembered that one determinant of the amount of capacitance is the distance between the plates.) Figure 2–38a shows the typical schematic symbols for varicap diodes. Figure 2–38b is a graphical representation of the relationship between the amount of reverse voltage and the amount of capacitance offered by the varicap diode. Notice that the capacitance drops off sharply as the reverse voltage is increased. Generally speaking, these diodes will offer a range of about 2 pF–100 pF, and up to 20 V. Because small amounts of leakage current are desired with these diodes, a silicon material is generally used.

Because their capacitances are in the picofarad range, varicap diodes are used for high-frequency application. They are found in automatic frequency control devices, adjustable bandpass filters, FM modulators, RF (radio frequency) circuits, and tuners.

Light-Emitting Diodes

As its name implies, the **light-emitting diode (LED)** gives off a visible light when biased properly. The schematic symbol for the LED is shown in Figure 2–39a. Notice that this diode has a lead identification of a cathode and an anode, and it must be forward biased in order to

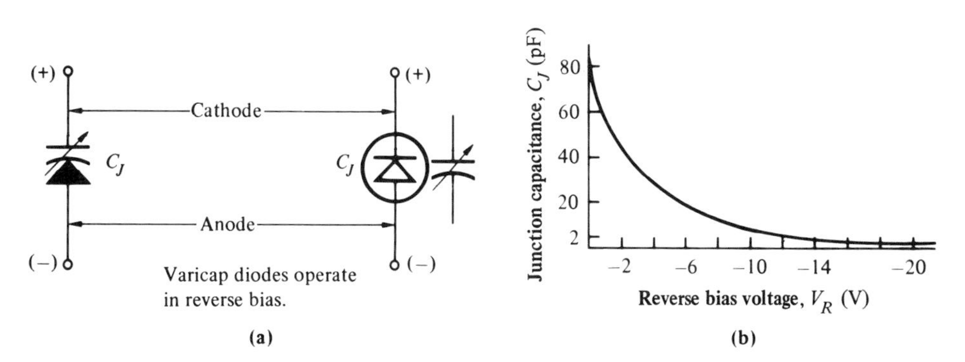

FIGURE 2–38
Varicap Diode: (a) Schematic Diagram, (b) Voltage versus Farad Rating

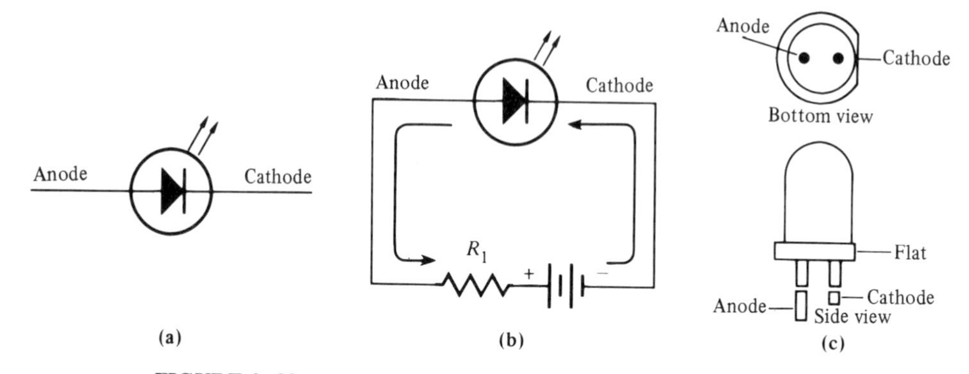

FIGURE 2–39
Light-Emitting Diode (LED): (a) Schematic Diagram, (b) Biasing, (c) Physical Shape

emit light. Also note that the LED uses a current-limiting resistor (see Figure 2–39b). This resistance will range between 150 Ω and 330 Ω. The larger the resistance, the dimmer the light. If no resistance is used, the current will destroy the LED. The technician should also be aware of the physical position of the anode and cathode of the LED. As shown in Figure 2–39c, the LED has a flat side. The lead closest to the flat side is the cathode.

The LED operates on the basis of movement of electrons in the N material into the holes of the P material. This movement causes energy to be given off within the diode. Some of this energy is in the form of heat, and some in the form of light. To insure that the light energy is given off, the LED is made with gallium arsenide phosphide or gallium phosphide. The photons of these materials give off the light energy. LEDs come in a variety of colors, including red, orange, yellow, and green.

LEDs can be combined into a **seven-segment display.** This display can create numbers between 0 and 9 when proper voltages are applied. Figure 2–40 shows an example of the seven-segment display. Applying a forward bias to the different segments of the LEDs will result in different numbers. In Figure 2–40, LEDs *a, b, c, d,* and *g* are emitting light to form the number 3. A specification sheet should be consulted for proper voltages on the LED display.

The typical voltages of the LED range from 1.7 V to 3.3 V. This voltage range for the LED makes it able to operate within the semiconductor range of voltage. If these LEDs are biased properly, they will operate for more than 100,000 hours.

Initially, LEDs and seven-segment displays were used primarily by the computer industry. However, because of the shift to solid-state

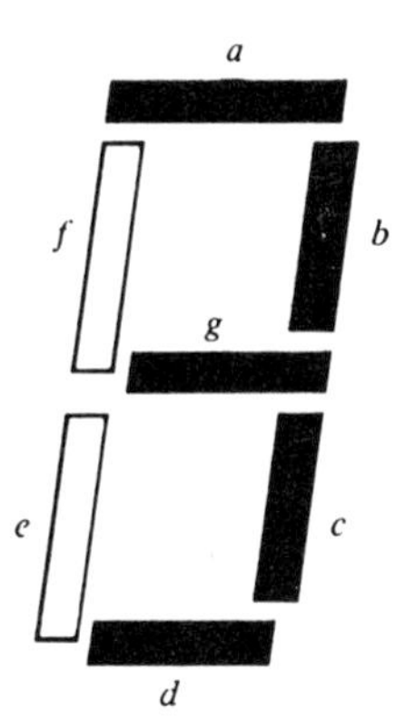

FIGURE 2–40
Seven-Segment LED Display

semiconductors in consumer products, they have found new applications. Seven-segment displays are used to display channel selection in television receivers, and the LEDs are used as meters to show levels of sound.

SUMMARY

Active devices are used to change signal amplitude. The first active devices were the vacuum tubes. Then, semiconductor devices such as transistors and integrated circuits became popular.

Vacuum tubes operate on the principle of current conduction from cathode to plate in a vacuum. Different types of vacuum tubes either rectify or amplify signals.

Semiconductors are made into crystal forms. Materials for the semiconductors are usually either germanium or silicon. Silicon is used most often.

For the semiconductor to have the properties of an insulator or a conductor, the material is doped. This doping process produces areas within the crystal of N-type material and P-type material. The joining point of the P and N materials is called the PN junction.

Forward bias applied to the PN junction will develop a conductive material. Reverse bias applied to the PN junction will develop an insulator.

The formation of the PN junction is widely used in a device called a diode. Diodes are used as rectifiers, regulators, and detectors. Diodes made from silicon require at least a 0.7 V difference between anode and cathode before conduction begins. Germanium requires at least a 0.3 V difference.

A rectifier diode is made to operate in forward bias. If placed in reverse bias, and its reverse voltage rating is exceeded, the diode will fail. Zener diodes are made to operate in reverse bias. They are also used to regulate dc and ac voltages.

Diodes that are placed in reverse bias and develop capacitance are called varicap diodes. Varicaps range from 2 pF to 100 pF. The amount of capacitance depends upon the reverse bias voltage.

Diodes that emit light are called light-emitting diodes (LEDs). LEDs can be arranged into a seven-segment display. Numbers on this display range from 0 to 9.

KEY TERMS

active component
anode
cathode
cathode ray tube (CRT)
control grid
covalent bonding
crystal
depletion region
diode
dipole
doping
electrode
extrinsic semiconductor
filament
forward bias
forward current
forward voltage
germanium
heater
impurities
interelectrode capacitance
leakage current
light-emitting diode (LED)
magnesium getter compound
N material
nucleus
PN junction
pentode
pentavalent atom
plate
P material
power dissipation
reverse bias
reverse breakdown voltage
reverse current
reverse recovery time
reverse voltage
saturation
screen grid
semiconductor
seven-segment display
silicon
suppressor grid
tetrode
triode
trivalent atom
trivalent impurity
vacuum tube
valence band
valence electron
varicap diode
zener diode

REVIEW EXERCISES

1. Name the three terminals common to all active devices.
2. Draw the schematic diagram of a diode vacuum tube, and label each of the terminals.
3. What is the polarity of the plate? of the cathode? Draw the vacuum tube diode in forward bias.
4. Draw a schematic diagram of a triode vacuum tube. Label each of the terminals,

and identify the proper polarity of each terminal.

5. Which element in the vacuum tube eliminates secondary emission?

6. Draw a schematic diagram of a pentode vacuum tube. Label the terminals, and show the proper biasing of each.

7. Which band of the atomic shell is loosely held to the atom's nucleus?

8. Explain the positive temperature coefficient of conductors.

9. Name the two materials commonly used to produce semiconductors. Which material is used most often?

10. What is the process of turning silicon into a crystal structure called?

11. What is the term given to the process of adding impurities to the crystal material?

12. Identify three substances used to dope P material.

13. Draw two examples of a PN junction, one in forward bias and one in reverse bias.

14. Draw a schematic of a diode, and label the cathode and anode.

15. What is the forward bias voltage for a silicon diode? for a germanium diode?

16. Describe the following characteristics of the diode:

a. Forward voltage
b. Reverse voltage
c. Forward current
d. Reverse current
e. Power dissipation

17. Describe the method of checking a diode with a digital ohmmeter.

18. Zener diodes are operational in (forward/reverse) bias.

19. Draw a circuit that will regulate a 15 V source and a 5 V source. (*Hint*: Use zener diodes as the regulating device.)

20. What are diodes called that offer capacitance to the circuit?

21. Draw the schematic symbol of the varicap diode, and place it in the proper bias mode.

22. Draw a schematic diagram of an LED, and label each of the terminals.

23. Why is a series-limiting resistor needed in LED circuits?

24. Draw a seven-segment display, and label each of the elements *a* through *g*.

25. List several basic uses for the general-purpose diode.

3 POWER SUPPLIES

OBJECTIVES

Upon completing this chapter, you should be familiar with:

—ac input
—Rectifier circuits
—Voltage waveform analysis
—Voltage multiplier circuit characteristics
—Filtering action
—dc power supply regulation
—The load
—Fuses and circuit breakers
—Troubleshooting power supplies

INTRODUCTION

In all consumer product devices, there is one central instrument: the *power supply*. The supply is appropriately named, because it delivers power to all other components within the unit. All consumer products containing oscillators, digital circuits, and amplifiers operate off dc. For the most part, utilities deliver ac. Therefore, ac must be changed into dc to activate all components in the circuit. This is done through the power supply.

The process of changing ac to dc can be illustrated by use of a block diagram. Figure 3–1 shows a typical block diagram of the dc power supply. Notice in the diagram that the ac voltage enters the block labeled *ac input*. Here, the voltage can be

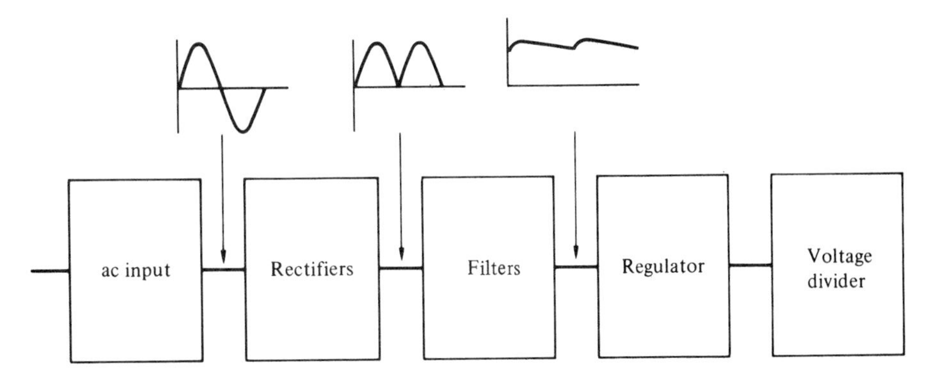

FIGURE 3–1
Block Diagram of dc Power Supply

stepped up, stepped down, or remain at the ac outlet level. The next block represents the *rectifiers,* which are used to convert the ac sine waveform into pulsating dc. The next block shows the *filters,* which are used to smooth out the pulsations to a constant level of dc voltage. The next block shows the *regulator,* an important component in modern consumer devices. The regulator delivers a constant voltage and current to the load. The final block shows the *voltage divider,* which drops the level of dc voltage down to different levels of voltages to the load. All these blocks fit together to perform an important task in the consumer product, to convert ac voltage into dc voltage.

AC INPUT

In the majority of power supplies used in modern consumer electronic products, the ac input is delivered to the rectifier block by a power transformer, which can be one of two types: a step-up power transformer or a step-down power transformer. Another way of delivering the ac input to the rectifier is called the direct connect or transformerless method.

Step-Up Power Transformer

The **step-up power transformer** will deliver an increased voltage at the secondary. While the secondary will show an increase in voltage, it will show a decrease in current flow.

In some older consumer devices whose active components are

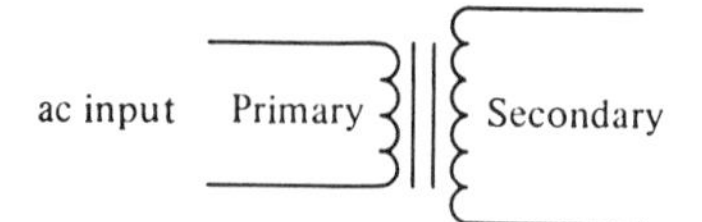

FIGURE 3–2
Schematic Symbol for Step-Up Power Transformer

vacuum tubes, a step-up transformer with a multitapped (more than one) secondary is used. The multitaps off the secondary are used to supply filament voltages for the vacuum tubes, and the necessary B+ voltage (dc operating voltage) for operating the active components. The symbol for the step-up transformer is shown in Figure 3–2.

Step-Down Power Transformer

The second type of transformer found in power supplies is the **step-down power transformer.** Recall that the voltage between primary and secondary depends upon the turns ratio. If the secondary has fewer turns than the primary, the voltage will be reduced. For example, if the number of turns on the secondary is one-half the number of turns on the primary, the voltage at the secondary will be one-half of the primary voltage.

If the voltage is stepped down, then an increase in current will be seen between the primary and secondary. As a general rule, the step-down power transformer will deliver high secondary currents at lower secondary voltages. Figure 3–3 shows the general schematic symbol for the step-down power transformer.

Transformerless Connection

Another type of ac input found in power supplies is the **direct connection,** also called the **transformerless connection.** Figure 3–4 shows an example of this type of connection. Notice that the ac is tied directly to the rectifier. The other lead is tied to the chassis for return paths to complete current flow. This arrangement is very popular in portable television receivers. The absence of the heavy, bulky, and expensive power transformer makes it possible to keep cost and weight down.

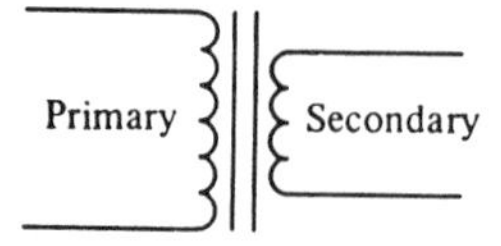

FIGURE 3–3
Schematic Symbol for Step-Down Power Transformer

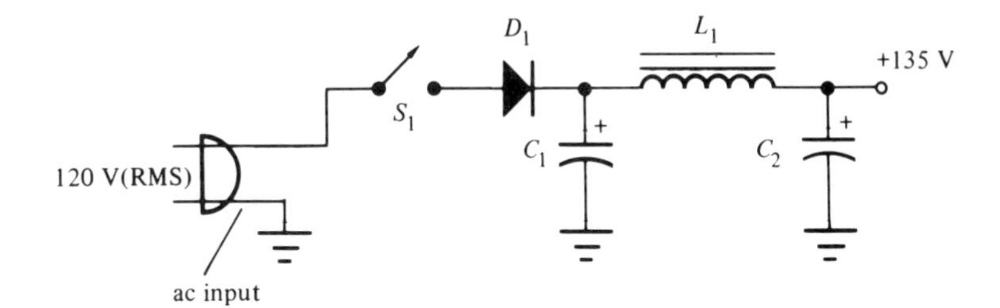

FIGURE 3–4
Direct ac Input Connection

RECTIFIERS

The **rectifier** circuit is the most important point in the power supply. The rectifier circuit is responsible for converting the ac waveform from the ac input into a pulsating dc waveform. One of several rectifier circuits can be used to accomplish this task. These circuits are the half-wave rectifier, the full-wave rectifier, the full-wave bridge rectifier, and the voltage doubler.

Half-Wave Rectifier

Figure 3–5 shows the schematic diagram for the **half-wave rectifier** circuit. Notice that the rectifier contains a diode and a load. The load is a resistance that will draw current from the power supply. The load is usually noted by the symbol R_L.

The circuit in Figure 3–6 shows that the transformer will reduce the voltage between primary and secondary. Point x of the secondary will connect to the anode of the diode, and point y will connect to the load. The other end of the load is then connected to the cathode. This completes the circuit path.

During the positive ac input cycle, a polarity is developed at the load. Figure 3–6 shows this polarity, which has reversed biased the diode. Because the diode has high resistance in reverse bias, most of the voltage is dropped across the diode. A small amount of leakage current developed in reverse bias will cause a small voltage drop at

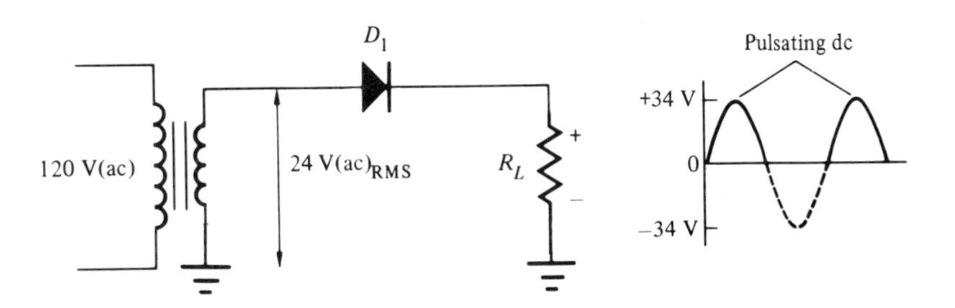

FIGURE 3–5
Half-Wave Rectifier

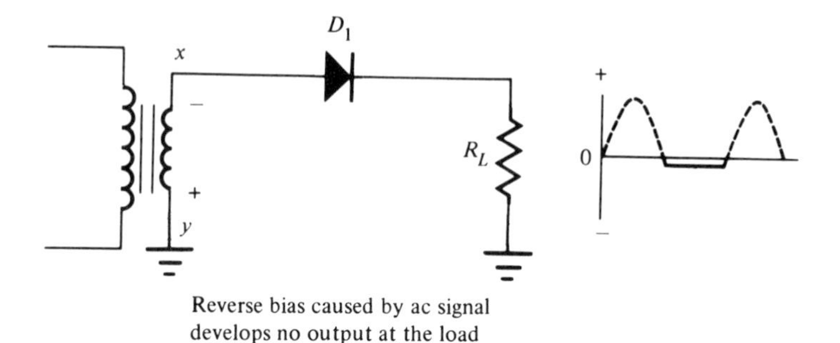

FIGURE 3–6
Reversed Biased Half-Wave Rectifier

the load. This small amount of current will develop a small voltage drop that will not be able to perform any work in the circuit.

As the ac input cycle starts into its negative half, the polarity at the load reverses. Figure 3–7 shows an example of this polarity, which has forward biased the diode. The load now has higher resistance than the diode, and all the voltage is dropped across the load.

The resulting waveform developed at the load is called a half-wave **pulsating dc.** The output waveform represents just the positive half of the cycle.

The technician sometimes may be requried to determine the polarity of dc voltage that will be developed at the load. In Figure 3–7, the cathode of the diode is pointing toward the load. This means that a positive voltage will be developed at the output resistor, R_L, as it relates to ground. In Figure 3–8, the diode has been reversed. Reversing the diode causes the anode to point toward the load. The dc output voltage has now switched polarity in relation to ground.

It is important that the technician pay attention to polarity when replacing diodes. If the diodes are replaced in the wrong polarity, a reversal in output voltage will occur. This sudden reversal of output voltage will cause damage to circuits connected in the load.

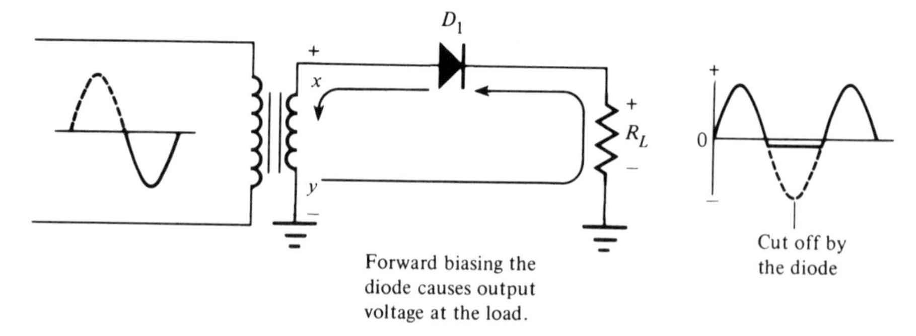

FIGURE 3–7
Positive Voltage Developed at the Load

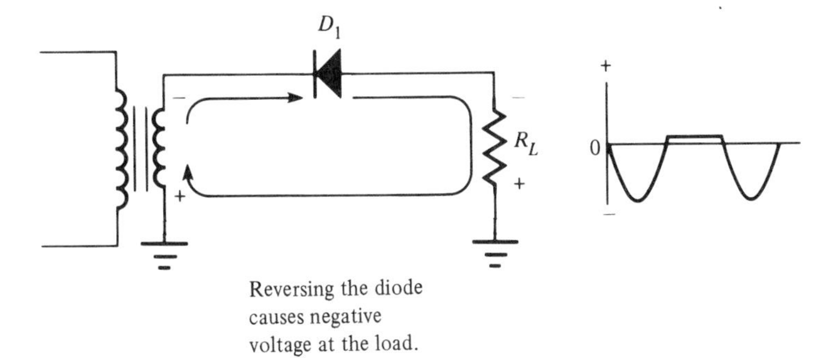

FIGURE 3–8
Negative Voltage Developed at the Load

The half-wave rectifier has one great disadvantage. During one-half of the input waveform, the voltage output is zero. To overcome this problem, a large filter circuit is used to convert the pulsating dc into smooth dc voltage.

Full-Wave Rectifier

Figure 3–9 shows a sample **full-wave rectifier** circuit. In this circuit, the center-tap power transformer must be used to split the phases of the ac input. When the phases are split, each diode will conduct during one-half of the input waveform.

The operation of this rectifier centers around the transformer. During one-half of the input cycle, point x is negative and point z is positive in relation to point y or ground in the circuit. (Figure 3–9a). Because point y is center-tapped, its potential rests at 0 V. The solid lines trace the path of current during the first half of the input cycle. Notice that the current path is through the load resistor and then through D_2. Diode D_2 will allow passage of current, because the anode is at a more positive potential than the cathode. At this point in time, point z is the most positive point in the circuit. This establishes a forward bias on D_2. Because point x has become the most negative part in the circuit, D_1 has become reversed biased. Notice that the reverse bias is developed from the negative voltage applied to the D_1 anode. Figure 3–9a shows an example of this.

During the next half of the input cycle, the polarities at points x and z reverse. Point x becomes positive, and point z becomes negative (Figure 3–9b). Point y remains at 0 V. The solid lines trace the current path during this half of the cycle. The current path is through the load again. Reaching the diode junction, the current path will be through D_1, because D_1 is forward biased and D_2 is reversed biased. Figure 3–9b shows an example of this action.

In this rectifier circuit the voltage drops to zero only for a few milliseconds, after which the voltage begins to rise again to a peak

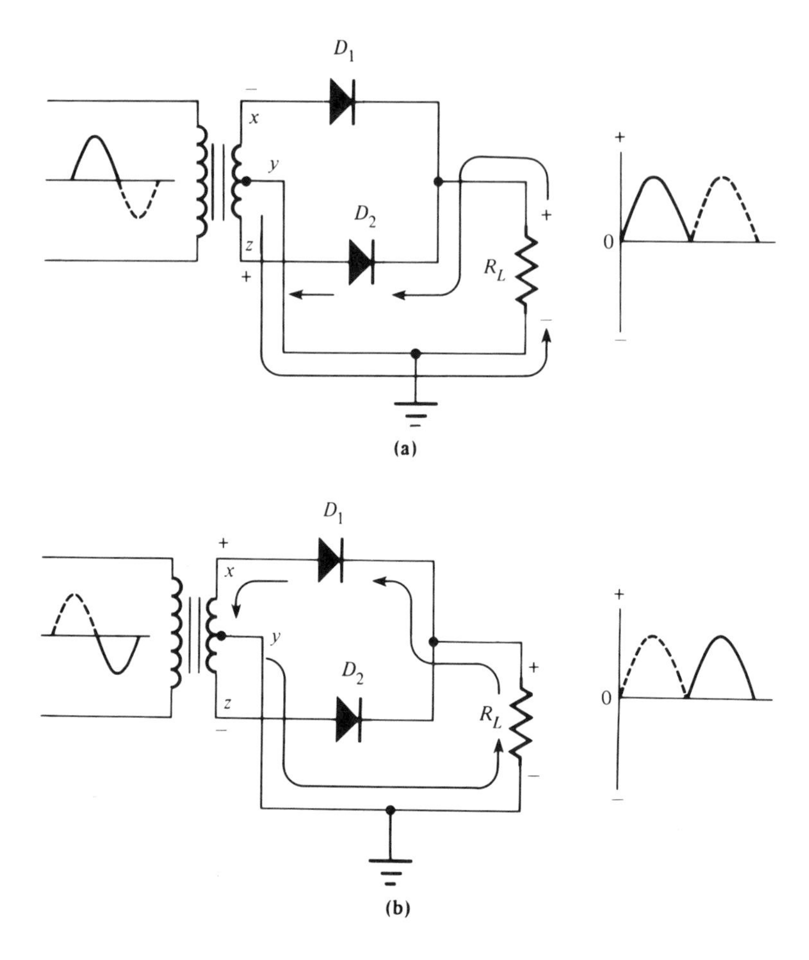

FIGURE 3–9
Full-Wave Rectifier: (a) Current Path during the Positive Input Cycle, (b) Current Path during the Negative Input Cycle

value. This means that there is output voltage for both halves of the input cycle. Notice that the output voltage fills in the gaps developed during half-wave rectification as shown in Figure 3–5.

Because of the action of both halves of the input cycle being developed, there are twice as many peaks in a full-wave rectifier than in a half-wave rectifier. This action develops an output cycle frequency of 120 Hz, as compared to the 60 Hz frequency of the half-wave rectifier with the same 60 Hz input.

Polarity of the diodes in the full-wave rectifier circuit is also important. The diodes must be placed in the right polarity, based upon the output voltage. Figure 3–9 shows a full-wave rectifier with positive dc output voltage. In Figure 3–10 the diodes have been reversed. This reversal causes a negative voltage to be developed at the load.

The full-wave rectifier has many advantages over the half-wave power supply. The major advantage is the delivery of twice as many peaks to the load. It also has its drawbacks. Because a center-tapped transformer is required, the voltage delivered to the load will be only

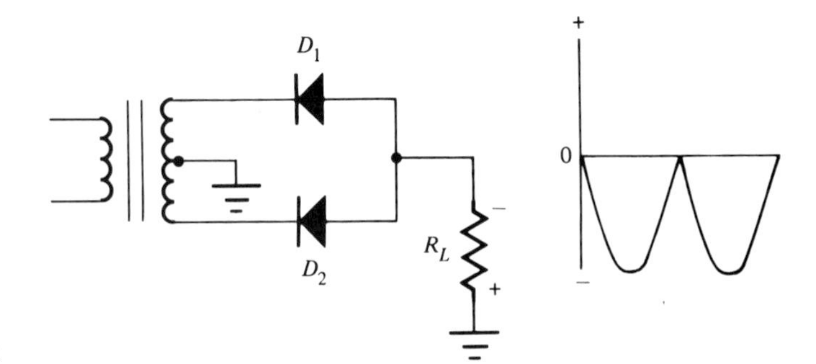

FIGURE 3–10
Full-Wave Rectifier Developing Negative Output

one-half the secondary voltage. In addition, the transformer adds cost, weight, and size to the consumer product.

Full-Wave Bridge Rectifier

An example of the **full-wave bridge rectifier** is given in Figure 3–11. This is another type of full-wave rectifier circuit. During the input ac

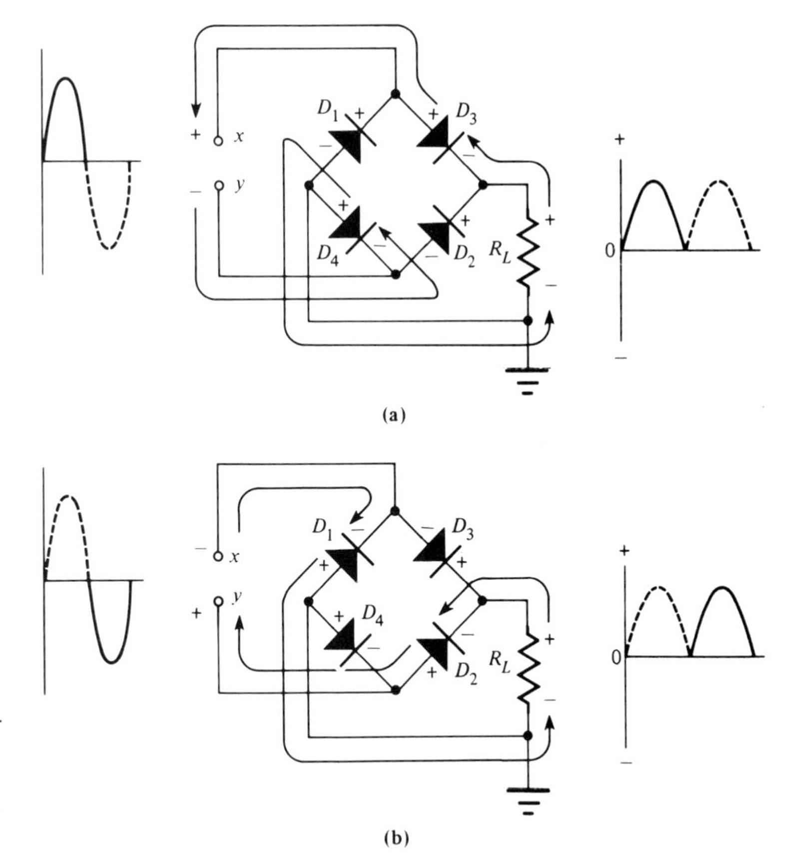

FIGURE 3–11
Full-Wave Bridge Rectifier: (a) Current Path during the Positive Input Cycle, (b) Current Path during the Negative Input Cycle

cycle, points x and y establish polarities. As shown in Figure 3–11a, during the first half of the cycle, point x becomes the most positive point in the circuit and point y becomes the most negative point in the circuit. The solid lines trace the current flow during this cycle. Here, D_3 and D_4 are forward biased, and D_1 and D_2 are reversed biased. The current's path is through D_4, then into the load. Through the load, the current flows to D_3, and then returns to the positive part of the transformer.

During the next half of the cycle, the polarities of x and y reverse. Point x becomes the most negative point in the circuit, and point y becomes the most positive point in the circuit. Again, the solid lines trace the current's flow through this half-cycle. Notice that again the current flow is through the load. Now D_1 and D_2 are passing the current flow in only one direction. The important point here is that during both halves of the input cycle, current flows through the load.

The bridge rectifier has advantages over the full-wave rectifier. First, it requires no transformer. However, if higher or lower voltages are needed, a transformer must be used. Second, the bridge rectifier circuit has a higher output voltage. This results because the bridge rectifier uses the whole secondary for each half-cycle of conduction, rather than splitting the voltage in half as does the full-wave rectifier.

Voltage Waveform Analysis

Because the output of the rectifiers delivers constant pulses to the load, different voltages can be obtained. These different voltages can be found by simple calculation. In the half-wave rectifier, the output voltage can be determined by multiplying 0.318 times the peak value of the input voltage. This mathematical formula is expressed in Equation 3–1.

$$V_{out} = 0.318 \times V_p \tag{3–1}$$

EXAMPLE 1

The half-wave rectifier circuit shown in Figure 3–12 develops 60 V(RMS) at the transformer secondary. What is the approximate output voltage for this circuit?

Solution:

First, the 60 V(RMS) must be converted into a peak value. Using Table 3–1, V_{RMS} can be converted, as follows:

$$V_p = 1.414 \times V_{RMS} = 1.414 \times 60 = 84.8 \text{ V}$$

Next, the output voltage can be found, using Equation 3–1:

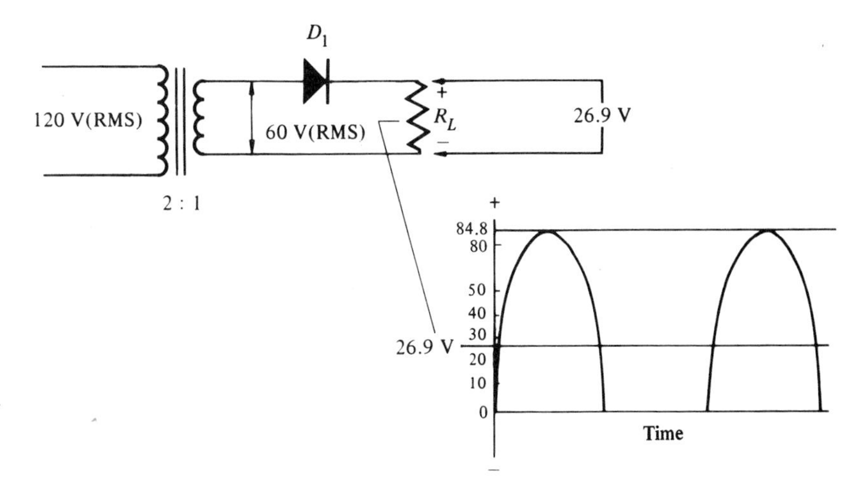

FIGURE 3–12
Output Pulsating dc Voltage developed from 84.8 V(p)

TABLE 3–1 Sine Wave Conversion Chart

From	To			
	RMS	**Average (av)**	**Peak (p)**	**Peak-to-Peak (p-p)**
RMS	—	0.900	1.414	2.828
Average	1.110	—	1.571	3.142
Peak	0.707	0.637	—	2.000
Peak-to-Peak	0.354	0.318	0.500	—

$$V_{out} = 0.318 \times V_p = 0.318 \times 84.8$$
$$= 26.97 \text{ V}$$

This example shows that approximately 27 V will be developed from the output of this half-wave rectifier.

Another equation can be used to solve the voltage output developed from the full-wave rectifier. This formula is shown in Equation 3–2.

$$V_{out} = 0.637 \times V_p \quad \textbf{(3–2)}$$

This conversion factor is also listed in Table 3–1.

EXAMPLE 2

Calculate the approximate output voltage of the rectifier circuit in Figure 3–13.

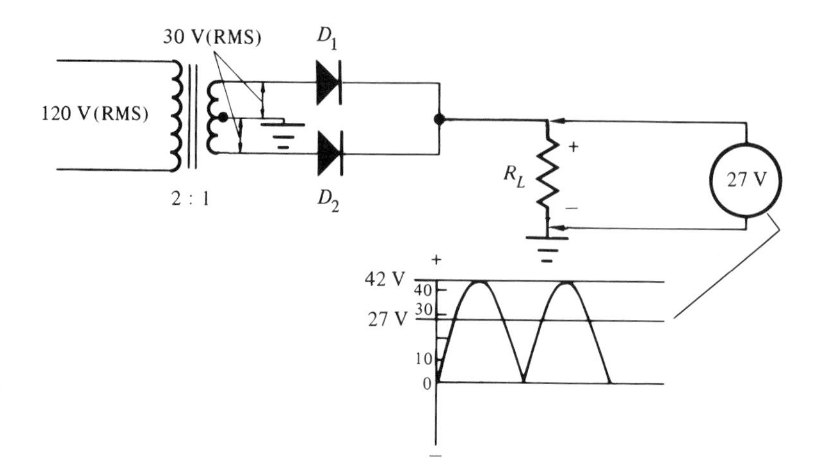

FIGURE 3–13
Output Pulsating dc Voltage Developed from a Full-Wave Rectifier

Solution:

Again, the first step is to solve for the peak voltage of the circuit:

$$V_p = 1.414 \times V_{RMS} = 1.414 \times 30 = 42.4\text{ V}$$

This value can now be placed into Equation 3–2:

$$V_{out} = 0.637 \times V_p = 0.637 \times 42.4 = 27.0\text{ V}$$

Therefore, if a meter were placed across the output load resistor, a voltage of approximately 27 V would be read.

EXAMPLE 3

A power transformer delivers 150 V(ac) at the secondary to a full-wave bridge circuit. What is the approximate output voltage developed from this circuit?

Solution:

Again, the V_{RMS} reading must be converted into a peak reading:

$$V_p = 1.414 \times V_{RMS} = 1.414 \times 150 = 212.1\text{ V}$$

Now, the output voltge can be calculated using Equation 3–2:

$$V_{out} = 0.637 \times V_p = 0.637 \times 212.1 = 135.1\text{ V}$$

Equations 3–1 and 3–2 are used only to approximate the output voltage developed by the rectifier circuit.

The conversion factors used in determining the various voltages are summarized in Table 3–1. For example:

$$60\text{ V(RMS)} \times 1.414 = 84.8\text{ V(p)}$$

This simple chart will aid the technician in making simple conversions.

Voltage Doubler

The circuit shown in Figure 3–14 is a dc **voltage doubler** circuit. Voltage doublers produce higher dc voltage than the normal half-wave and full-wave rectifier circuit. The doubler can be used as the power supply when low load currents are needed and where poor regulated power sources are used. As the name implies, the output dc level is approximately twice that of the input ac peak value. Two types of doublers are found in power supplies: the half-wave doubler and the full-wave doubler.

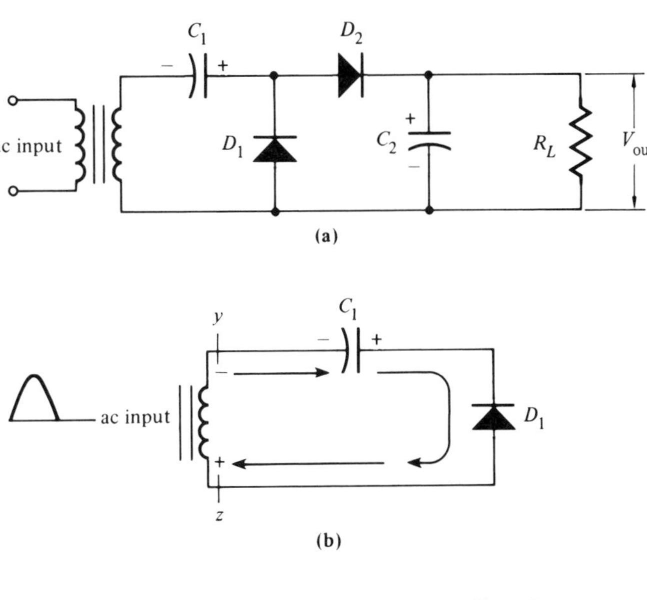

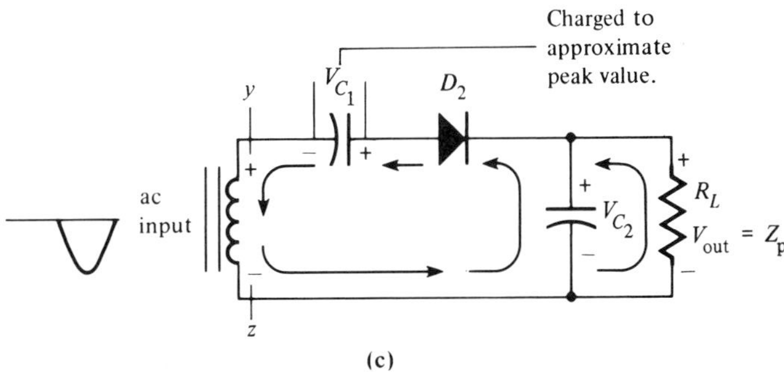

FIGURE 3–14
Half-Wave Voltage Doubler: (a) Schematic Diagram, (b) Current Path during the Positive Input Cycle, (c) Current Path during the Negative Input Cycle

Half-Wave Doubler The schematic diagram of a **half-wave doubler** circuit is shown in Figure 3–14a. During the input ac waveform, a polarity is established on the secondary of the transformer.

In the positive half-cycle of the ac input, point *y* becomes negative and point *z* becomes positive. Current flow during the positive cycle is indicated by the solid lines in Figure 3–14b. Current flow charges the negative side of C_1, and electrons are pulled away from the positive plate of C_1, Capacitor C_1 has now been charged to approximately the peak input voltage. This flow of electrons has forward biased D_1, which causes the current path to be completed to the positive part of the transformer. Because it has a peak value charge, C_1 acts as another voltage source.

When the input cycle goes into the negative swing, the polarities at the secondary of the transformer are reversed. As shown in Figure 3–14c, point *y* has become positive. This action places C_1 in series with the secondary of the transformer. The voltage stored in C_1 and the voltage at the secondary add together. This action doubles the peak voltage applied to the circuit. The current path is now through C_2, back through D_2, back to C_1, and then back to point *y* of the transformer. Because C_2 charges to the transformer's peak, and the peak stored in C_1 has flown to C_2, C_2 must be charged to double the peak voltage. Capacitor C_2 then discharges through the load at twice the input voltage.

If an ac voltmeter were to be used to measure the secondary of the transformer in a voltage doubler, it might be expected to multiply by two the voltage read. However, it should be remembered that the voltmeter reads only V_{RMS} values. Therefore, the meter reading must be multiplied by 1.414 to obtain the peak reading. Once this conversion has been made, the peak value can be doubled to find the total output voltage. An example will be helpful.

EXAMPLE 4

Using a voltmeter, the secondary of a transformer reads 50 V(ac). Approximately what should be the output voltage of the voltage doubler?

Solution:

$$V_{out} = 1.414 \times 50 = 70.7 \text{ V(p)}$$
$$= 70.7 \times 2 \text{ (because of the doubling action)}$$
$$= 141.4 \text{ V(ac)}$$

As the term *half-wave doubler* implies, the doubling action takes place only during one-half of the input cycle.

Full-Wave Doubler The half-wave doubler and **full-wave doubler** circuits are much alike. Figure 3–15a shows the schematic diagram of a full-wave doubler.

During the positive input ac cycle, points x and y establish polarity. Point x becomes positive, and point y becomes negative,

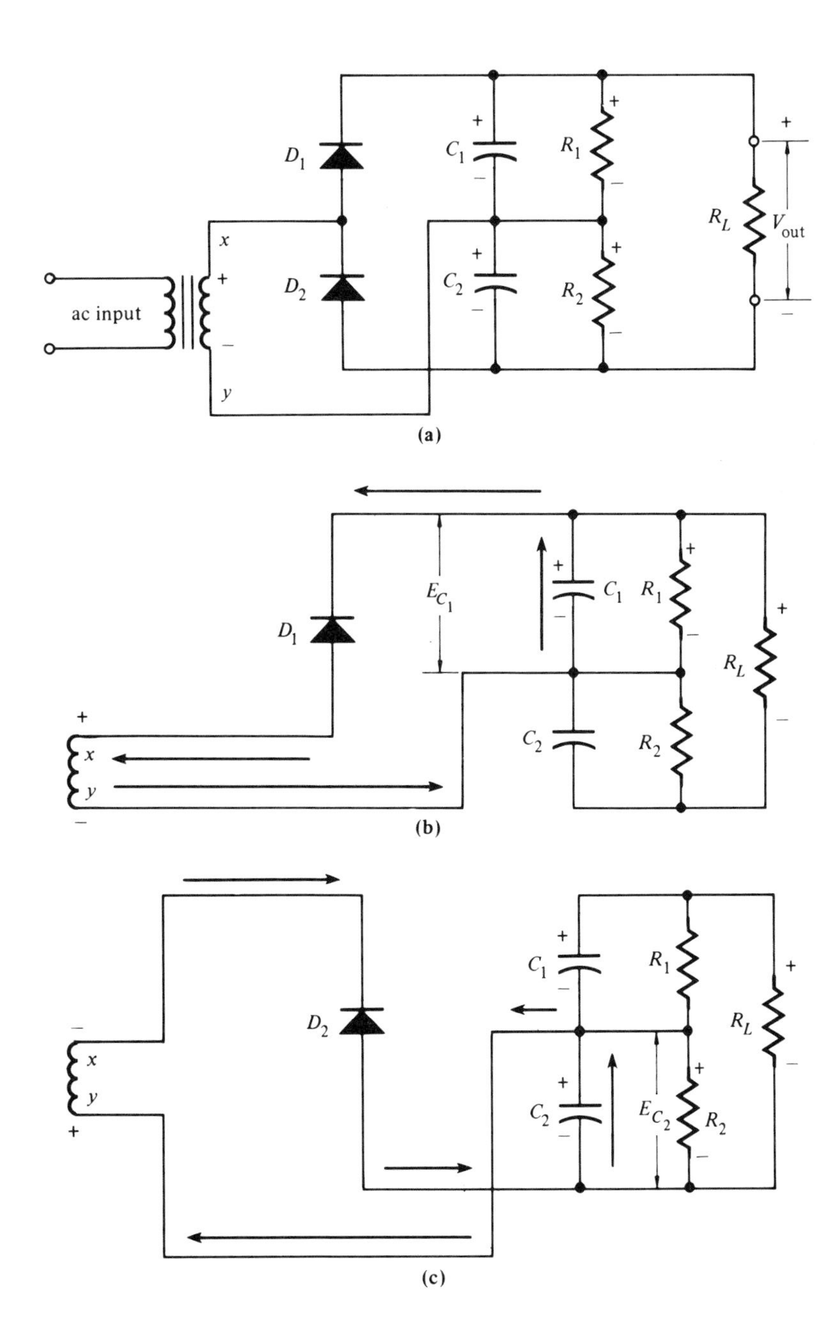

FIGURE 3–15
Full-Wave Doubler: (a) Schematic Diagram, (b) Current Path during the Positive Input Cycle, (c) Current Path during the Negative Input Cycle

causing a current to flow through C_1. This current charges C_1 to the peak value of applied voltage. The charging polarity on C_1 is shown in Figure 3–15b. Diode D_1 has been properly biased, and allows the current to flow back to the positive part of the transformer.

During the next half-cycle, the polarity at the secondary of the transformer changes. Points x and y switch polarities, which then properly biases D_2. This forward bias condition allows current flow to charge C_2 to the peak applied voltage. Figure 3–15c shows this action. Because C_1 and C_2 are connected in series, their voltages add up, producing twice the input voltage. The capacitors' charging up on both parts of the input cycle gives the full-wave doubler its name.

If a voltmeter is used to measure the secondary's voltage in a full-wave doubler, the same procedure would have to be followed as described in the half-wave doubler circuit. Remember that the voltmeter measures the V_{RMS} value.

The full-wave doubler has an advantage over the half-wave doubler in that it supplies a constant voltage to the load. This constant voltage is supplied by the charging and discharging of C_1 and C_2. However, these doubler circuits have poor regulation—the voltage from the output drops as the load current increases.

FILTERS

Rectifiers are used to convert the ac waveform into a pulsating dc form, but amplifiers, oscillators, and other load circuits require a smooth, steady flow of dc. Figure 3–16 includes the waveform that would be developed from a battery connected to the load. Notice that once the switch is closed, a steady flow of current is delivered by the power source. This is the ideal waveform into which the pulsating dc should be converted. To accomplish this task, capacitors and inductors are used to smooth out the pulsating dc to look like smooth dc voltage at the output. This is called **filtering.**

Some of the most common filters used in power supplies are

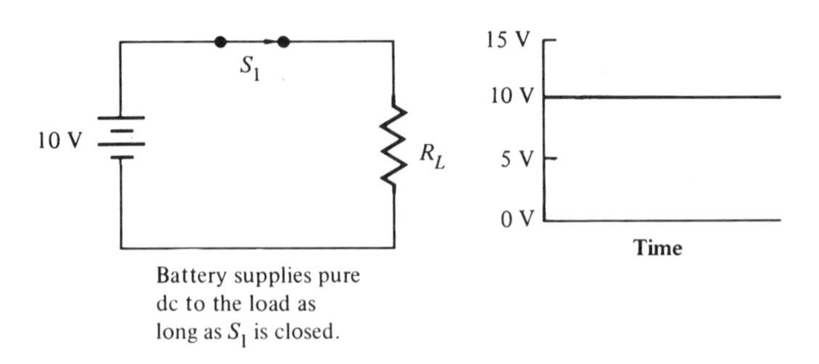

FIGURE 3–16
Waveform Developed by a Battery Connected at the Load

the capacitive filter, the choke input filter, the capacitive input filter, and the resistance-capacitance filter.

Capacitive Filter

The capacitor has the ability to store electrical energy, and when required can release this energy into the load. For this reason, the capacitor makes an excellent filtering component. Figure 3–17 shows an example of how a **capacitive filter** is connected to the rectifier circuits.

The capacitor operates in a simple fashion. During the input cycle, the capacitor charges to the peak applied voltage. Figure 3–18 shows an example of operation with the filter connected to a half-wave rectifier. At point A in Figure 3–18b, the capacitor has charged to the applied peak voltage value. Here, the waveform has reached its peak,

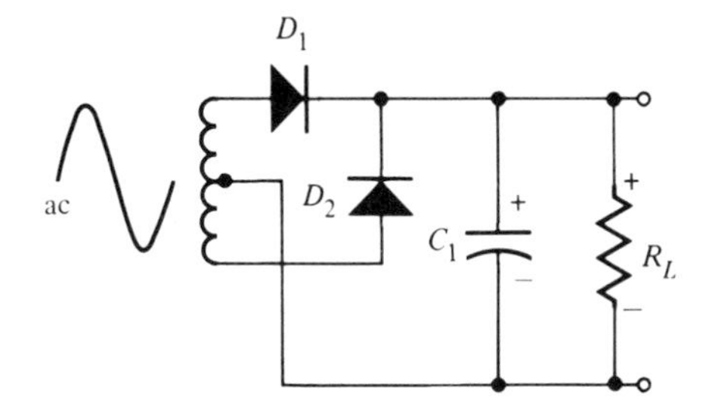

FIGURE 3–17
Capacitor Connected as a Filter

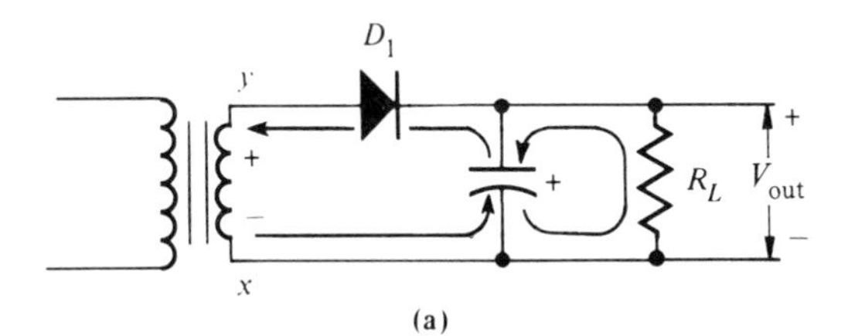

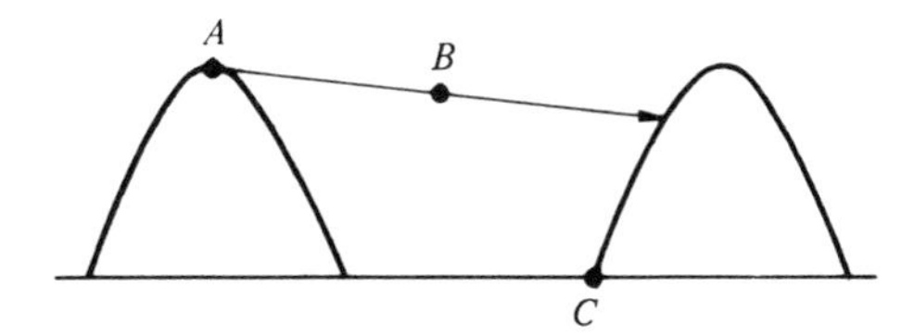

A – capacitor charges to the applied peak voltage
B – capacitor discharges through the load
C – new cycle starts to recharge the filter capacitor

(b)

FIGURE 3–18
Filtering of a Half-Wave Rectifier

and then begins to head toward negative current flow in the circuit. However, the capacitor sees this drop in current flow. The energy that has been stored on the capacitor is discharged through the load. (The discharge cycle is shown by the lines connected between the peaks.) This discharge process supplies a steady current to the load. If a steady current flow is seen at the load, a steady voltage is developed across the load.

Before the capacitor has a chance to discharge fully, a new input cycle starts. This new cycle begins to recharge the capacitor (see point *C* in Figure 3–18b). To be an effective filter, the capacitor must charge quickly, and then discharge slowly through the load. The capacitor cannot supply the load with a steady current flow for a long period of time. Therefore, the voltage begins to drop off in the load circuit. Figure 3–18 shows this decline in voltage. This variation of voltage is called **ripple,** and is developed in any filtering action.

The effectiveness of a given capacitive filter depends on several conditions:

—Time between pulses (frequency of power supply)
—Size of the capacitor
—Size of the load

In half-wave and full-wave circuits, there is a time span between the peaks of the waveform. Figure 3–19 shows examples of these frequencies. In the half-wave circuit, the time between pulses is 16.66 ms, which is equivalent to a 60 Hz frequency, as shown in Figure 3–19a. In the full-wave rectifier, the pulses arrive twice as often. Therefore,

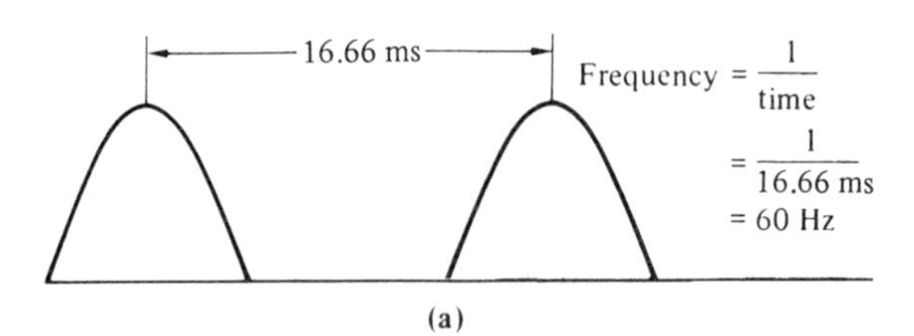

(a)

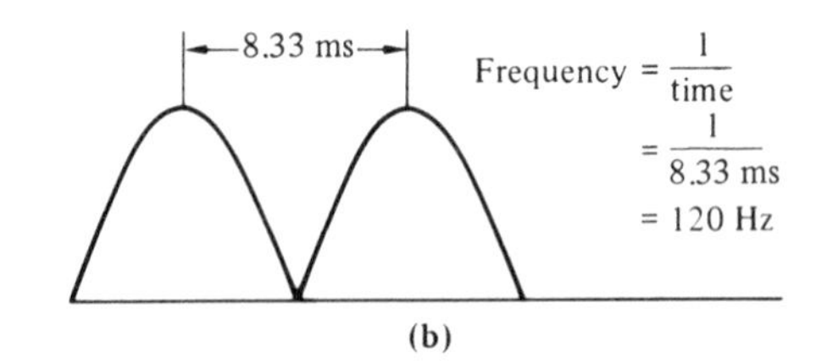

(b)

FIGURE 3–19
Pulsating dc Output Waveforms: (a) Half-Wave Frequency, (b) Full-Wave Frequency

the frequency for these circuits is 8.33 ms, or 120 Hz, as illustrated in Figure 3–19b. Because of this, the half-wave rectifier will have a larger ripple than the full-wave rectifier, since its capacitor has a longer discharge period.

The charging and discharging process of the capacitor can be controlled. To control the rate, a resistor can be placed in series with the capacitor. A formula can be used to calculate the rate of charging and discharging. The formula is shown in Equation 3–3.

$$T = R \times C \qquad \textbf{(3–3)}$$

where:

T = time (s)
R = size of resistor (Ω)
C = size of capacitor (F)

By combining the different sizes of resistors and capacitors, the technician can find the combination to match the time spans between the voltage pulses in the circuit. A general rule of thumb is, the larger the filter capacitor, the lower the ripple developed in the output of the power supply.

Because of the large capacitance needed for filtering, *electrolytic capacitors* are used. These capacitors will charge quickly when the diodes are forward biased, and discharge through the load slowly. Figure 3–20a shows the filtering capacitors and the 1 V(p-p) ripple developed at the output. However, if the load resistance decreases, then the discharge time of the filter capacitor is reduced. Figure 3–20b shows what happens when the load resistance is reduced to 200 Ω. Notice that the ripple voltage has increased to 4 V(p-p). This increase in ripple voltage will be developed in circuits supplied by the power supply and will cause these circuits to operate improperly.

The capacitive filter can lead to problems. If the filtering capacitor is too large, it may cause damage. This damage is the result of the large amounts of current required to charge the capacitor. Such large amounts of current could cause the transformer or the rectifiers to overheat, resulting in failure of these components.

The capacitive filter will always provide higher dc output voltages than an unfiltered supply. Figure 3–21 shows an example of these two voltage levels. Without the filter, the power supply will deliver about 108 V to the load:

$$V_{av} = 0.9 \times V_{RMS} = 0.9 \times 120 = 108 \text{ V}$$

When the switch is closed, the power supply will deliver about 169 V to the load:

$$V_{dc} = 1.414 \times V_{RMS} = 1.414 \times 120 = 169.6 \text{ V}$$

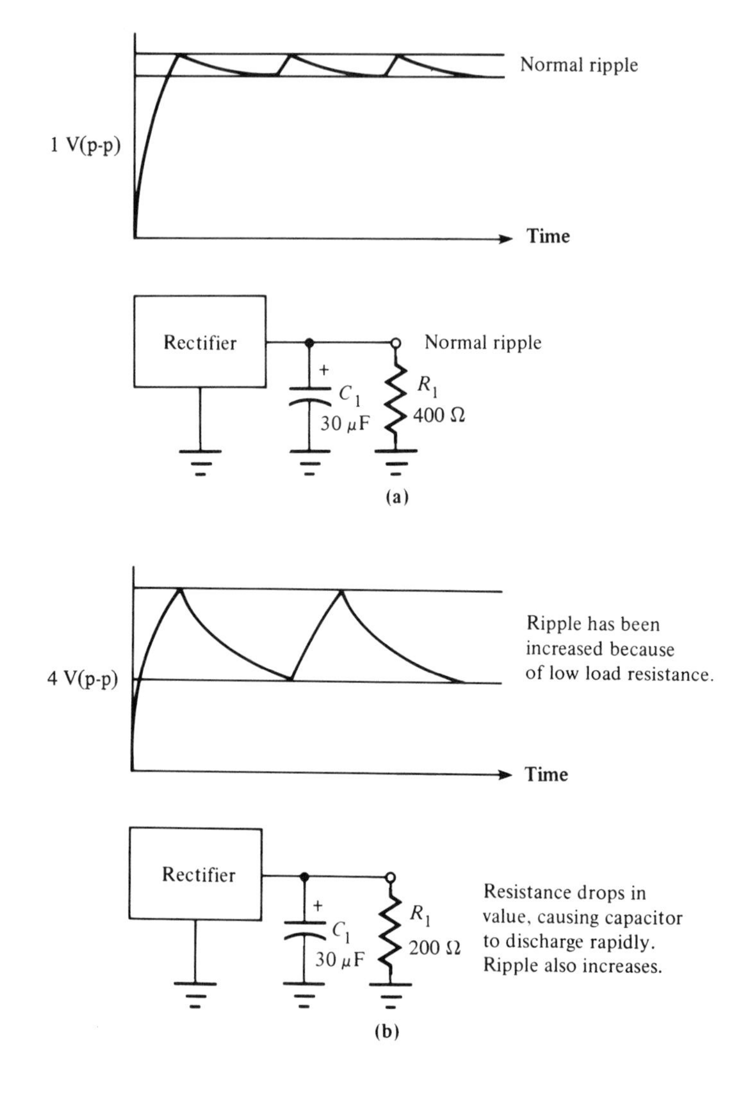

FIGURE 3–20
Ripple Increase Due to a Change in the Discharge Rate of a Filter

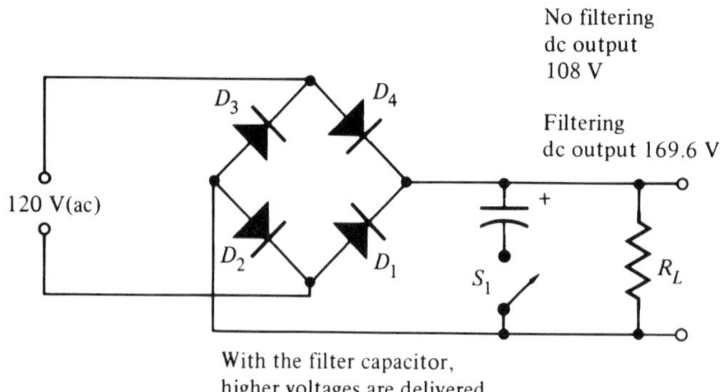

FIGURE 3–21
Increase of dc Output Voltage because of Filter

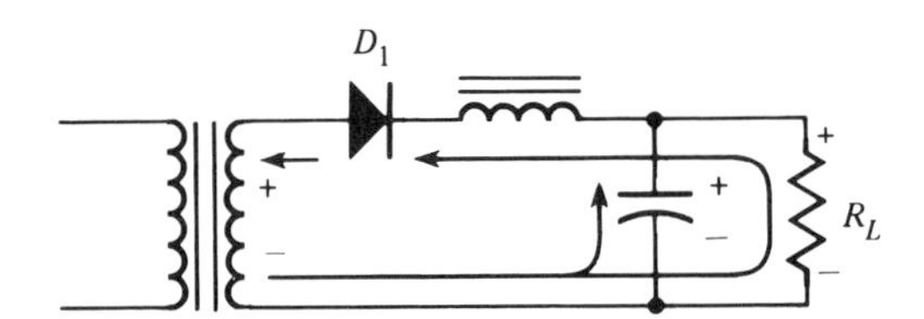

FIGURE 3–22
Choke Input Filtering

However, it should be remembered that if a load is applied to the output of the power supply, the voltage at the output will drop and the ripple will increase.

Choke Input Filter

Figure 3–22 is an example of the **choke input filter.** Because the inductor opposes any change in current flow, this filter can be used to smooth out any current variations. The choke input filter gets its name from the fact that the choke is the first component in the filter.

In operation, the choke input filter has basically the same action as the capacitive filter. When the diodes become forward biased, the current flow charges the capacitor. At the same time, a magnetic field builds up around the inductor. As the voltage starts to fall toward zero, the magnetic field collapses around the inductor. The magnetic field around the inductor can hold the current constant for only a short interval. Once the magnetic field has weakened to the point of being no help to the load, the capacitor discharges and aids the inductor. By this time, a new cycle appears and starts the charging process over again.

Because of the action of the inductor and the capacitor, the components of this circuit must have very large values. This means that the filtering section will drop a large portion of the dc output voltage. Therefore, this filter gives low ripple, but also gives lower dc output voltages.

Capacitive Input Filter

The **capacitive input filter** is so named because the capacitor is the first component in the filter. The action of this filter is very much like that for the choke input filter. Figure 3–23 shows an example of this circuit.

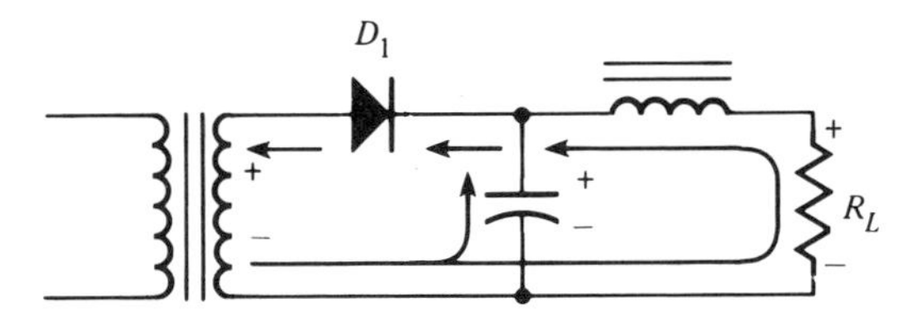

FIGURE 3–23
Capacitive Input Filtering

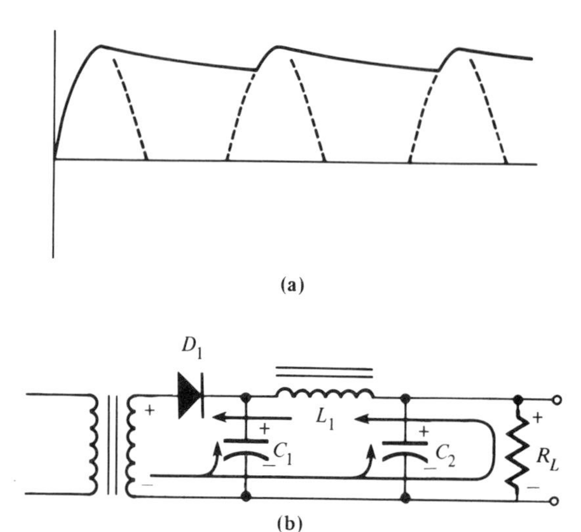

FIGURE 3–24
Pi Filtering

As the diode becomes forward biased, the capacitor charges to peak value. In addition, the magnetic field builds up around the coils of the inductor. Once the waveform begins to drop, the inductor and capacitor release their stored energy. This release causes the voltage to remain constant across the load.

For the capacitive input filter to work at maximum efficiency, the load must be extremely large. If the load is too small, the capacitor tends to discharge faster and causes variations in the dc output voltage.

A more elaborate capacitive input filter, called the **pi filter**, is illustrated in Figure 3–24. Figure 3–24a illustrates the pi filter connected to the output of a half-wave rectifier. Figure 3–24b illustrates the ripple waveform produced by the pi filter. This filter gets its name because it looks like the Greek letter π. The pi filter has two capacitors, which means it does a better filtering job than the regular capacitive input filter. The second capacitor has the same action as the first capacitor. It is used to help smooth out the pulsating dc voltages, and to reduce the ripple to a low level. The main disadvantage of this filter is that as the load switches from on to off states, the voltage from the power supply also fluctuates. This filter should be used where a constant load is seen by the power source.

Resistance-Capacitance Filter

Figure 3–25 shows an example of the **resistance-capacitance (*RC*) pi filter.** Because the resistor R_1 cannot store electrical energy or oppose a change in current, it is used only to lengthen the discharge time of capacitor C_1. If the time constant for discharge is lengthened, the

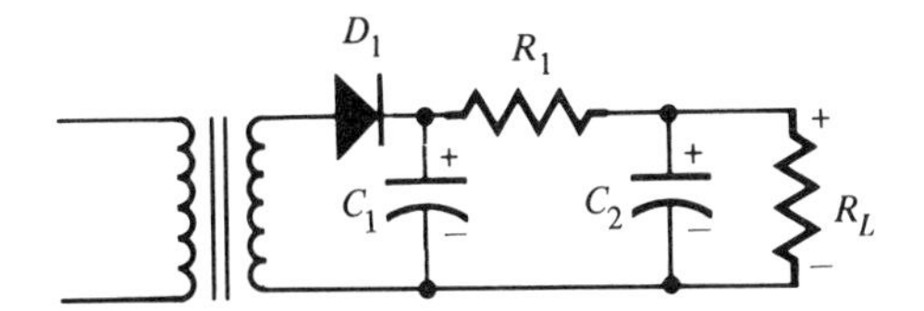

FIGURE 3–25
RC Pi Filter

voltage from the capacitor will fill in more of the valleys left by rectification. With the time increased for discharge, the voltage at the output is held fairly constant.

Calculating Ripple

The effectiveness of the filter can be checked by using a simple equation. The formula used to calculate the ripple is given in Equation 3–4.

$$\%\ \text{ripple} = \frac{\text{ac}}{\text{dc}} \times 100 \qquad \textbf{(3–4)}$$

where:

ac = measured value of the ripple RMS
dc = measured dc output voltage from the supply

Consider the following example.

EXAMPLE 5

A dc power supply develops a 2 V(p-p) ripple at the output. The dc voltage developed by this power supply is 40 V. What is the percentage of ripple?

Solution:

Substituting into Equation 3–4 gives:

$$\begin{aligned}\%\ \text{ripple} &= \frac{\text{ac}}{\text{dc}} \times 100 = \frac{0.707}{40} \times 100 \\ &= 0.0177 \times 100 \\ &= 1.77\%\end{aligned}$$

Note that from the example it was found that 1.77% of the dc voltage delivered to the load was ripple. Two points should be remembered when measuring the ripple. First, the ripple should always be measured under full load conditions. Any filter measured without a load will show low ripple, but when the load is applied, the ripple will increase. Second, an oscilloscope should be used when making the measurement. The oscilloscope can show the technician two important items about the ripple. It gives the amplitude of the ripple, and shows the general waveform of the ripple to be sawtooth in shape.

REGULATORS

In many modern consumer electronics devices, a well-regulated power supply is required. **Regulation** means that the power supply will deliver a constant current to the load. The dc output voltage can change for several reasons. The ac voltage might increase or decrease. If this happens, the output dc voltage increases or decreases proportionately to the input change. The type of filter used can also greatly affect the dc output. As seen with the choke input filter, large voltage drops occur across the inductor because active devices are turning on and off in the circuit, thus drawing more operational current from the power supply. Because of this action, regulation in the circuit becomes very important. Among the regulators found in consumer products are the zener diode regulator and the series regulator.

Basic Regulator

A basic regulator is shown in Figure 3–26. This type of regulator uses a variable resistance and a load resistance. Because these resistances are connected in series, their voltage drops must add up to the supply voltage. For example, if the dc output from the rectifier-filtered output were 10 V, that voltage would divide between the two resistances. The larger resistance would have the larger voltage drop. As the variable resistor was varied, the voltage drop across R_L would increase and

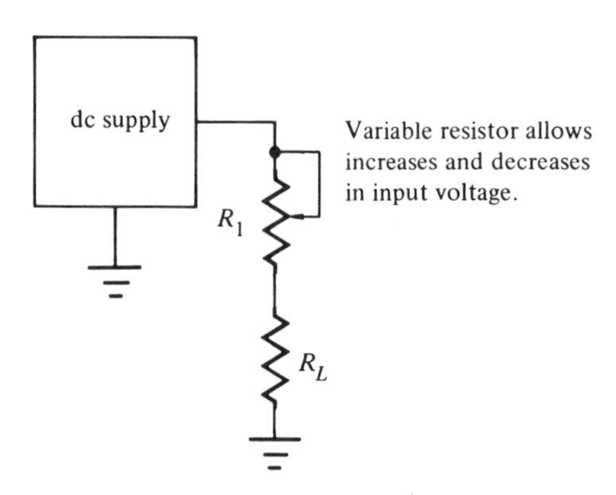

FIGURE 3–26
Basic Regulator

decrease. For example, assume that the input ac voltage increases. This will show up as an increase to a basic regulator circuit. Another condition is necessary: the load requires only 3 V to operate. Therefore, if the ac input increases by 2 V, the voltage across the load also increases. To keep R_L constant at 3 V, the variable resistor R_1 must be adjusted. Increasing R_1 resistance causes a greater voltage drop across the variable resistor. The greater voltage drop across R_1 means that R_L will drop back to 3 V.

The basic regulator is crude but operational. However, to keep a constant voltage across the load, this component must be monitored with a meter. If the voltage across the load increases or decreases, the variable resistor can be adjusted. With this regulator, a constant watch must be kept on the power supply.

Zener Diode Regulator

As stated, the basic regulator circuit requires constant monitoring of a meter to keep the output voltage at 3 V. To eliminate this need for continual adjustment, a zener diode can be used. As has been learned, the zener diode has the ability to keep voltage constant. Figure 3–27 shows an example of the **zener diode regulator** circuit.

A zener diode can be selected from a variety of voltage ratings. In Figure 3–27, a 3 V source is used. A zener diode is placed in the circuit as the regulator. When the voltage from the dc output increases or decreases, the zener diode keeps the voltage output constant. If the voltage from the supply rises 2 V, the zener diode maintains a 3 V output (see Figure 3–27b). Because R_1 and the zener diode are in series, their voltage drops must add up to the supply voltage from the filters. The zener diode in Figure 3–27b will hold the voltage constant at 3 V. Therefore, the additional 2 V drop will be seen across R_1. It must be pointed out that the zener diode requires at least 3 V before it starts to conduct. Thus, the dc output must be at least 3 V. The opposite action is shown in Figure 3–27c.

Resistor R_1 is not in the circuit simply to make up for excess voltage. It is there also to limit the current. The zener diode is a semiconductor, and semiconductors are very sensitive to current variations.

As can be seen, in this type of regulator circuit, the zener diode is connected in parallel with the load. Because of this connection, this regulator is also called a *shunt regulator*.

Series Regulator

Figure 3–28 shows an example of a basic **series regulator** circuit. As shown in the figure, the collector and emitter terminals are in series

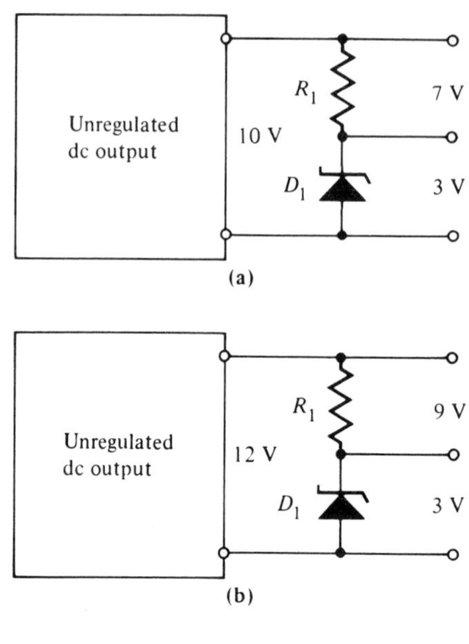

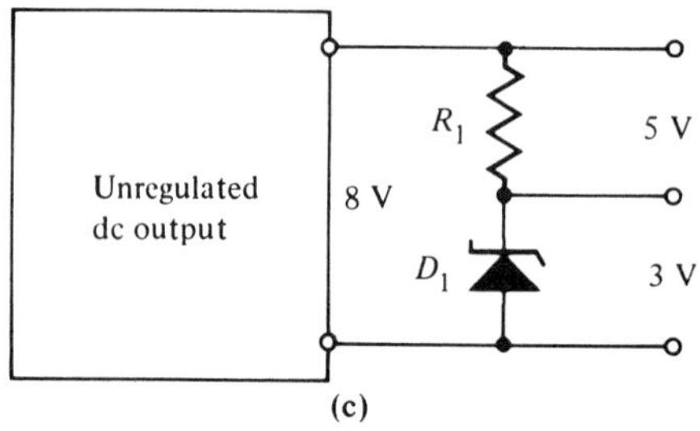

FIGURE 3–27
Basic Zener Diode Regulation: (a) 10 V Output Develops 3 V Output, (b) Input Voltage Increase Develops Constant Voltage at Output, (c) Decrease in Voltage to Regulator Keeps Output Constant

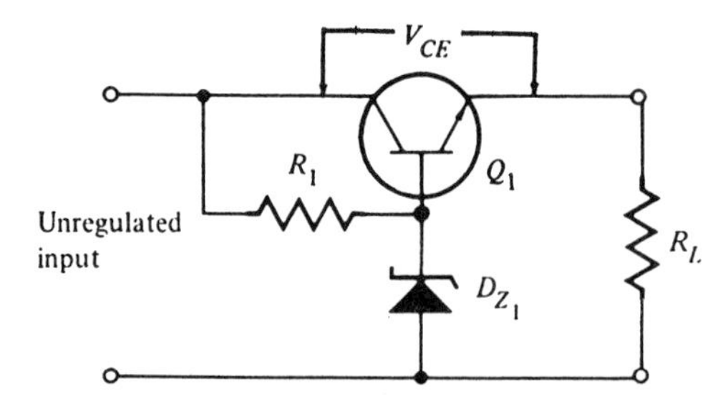

FIGURE 3–28
Series Regulator

with the load. The zener diode is connected to the base of Q_1 (the transistor), and acts as a reference voltage source for the transistor.

In basic operation, the zener diode is used to keep the voltage on the base constant. When the input voltage increases to the regulator, a constant current is delivered to the load. As the voltage increases, the zener diode voltage remains constant, but since R_1 and the zener diode are in series, the voltage drop increases across R_1. The zener diode has held the base voltage constant, causing constant current flow through the transistor and the load. If the output voltage decreases, the same action occurs. The zener diode holds the base voltage constant. The decrease in voltage appears across R_1. Because the base voltage remains constant, a constant current is delivered to the load. A constant current means constant output voltage.

Certain integrated circuits are used as regulating devices. The type of integrated circuit most often used for regulation is the *op amp,* which will be discussed later in this text. These regulators have all the necessary circuits for regulation contained in one housing. Their drawbacks are that they are limited in the amount of output current they can handle. To increase current output capabilities, a power transistor can be added.

Series regulators are the most popular type of regulator used in consumer products. They have the ability to keep current to the load constant. If current remains constant, the output voltage remains constant.

Calculating Regulation

A simple mathematical calculation can be used to determine the percentage of regulation found in the power supply. The formula for finding the percentage of regulation is given in Equation 3–5.

$$\% \text{ regulation} = \frac{V_{NL} - V_L}{V_L} \times 100 \qquad \textbf{(3–5)}$$

where:

V_{NL} = voltage with no load
V_L = voltage with load applied

The following is a simple example of calculation.

EXAMPLE 6

Without a load, a power supply delivers 20 V to the load. When the load is added, the output voltage drops to 15 V. What is the percentage of regulation for this power supply?

Solution:

$$\% \text{ regulation} = \frac{V_{NL} - V_L}{V_L} \times 100 = \frac{20 - 15}{15} \times 100$$

$$= \frac{5}{15} \times 100 = 0.333 \times 100 = 33.3\%$$

This means that when the load is applied, a voltage drop of 33.3% will be seen in the circuit.

To improve regulation in a filtered power supply, a **bleeder resistor** can be used. The bleeder resistor will always cause a current to be drawn from the filter, thus delivering a fixed voltage to the load. Not only does the bleeder resistor help to improve regulation, but it also will discharge the filter capacitor after the power supply has been shut off.

VOLTAGE DIVIDERS

Any circuit that draws current from the regulator of the power supply is called the **load.** Many circuits in consumer electronic devices require different voltage levels. These voltages would require a power supply for each circuit. For example, in a transistor amplifier, one supply would be needed for the base circuit, one for the emitter circuit, and one for the collector circuit. Multiple power supplies would be costly for the consumer. Instead of multiple power supplies, a voltage divider is established throughout the load to tap different voltage levels. With this process, a single power source can operate all the circuits found within a device.

In the voltage divider circuit of Figure 3–29, three separate voltages can be tapped. If a voltage were taken from point A to ground, a total of 20 V would appear. This results because R_1, R_2, and R_3 are connected in series. Kirchhoff's voltage law states that the voltage

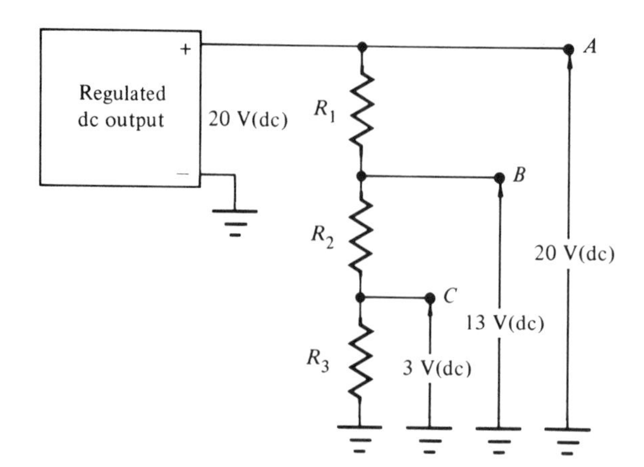

FIGURE 3–29
Voltage Divider Establishes Different Output Voltages

drops of the resistors must add up to the total supply voltage. This law holds for the voltage divider. If a voltage were measured from point B to ground, the power supply would deliver 13 V, and if a 3 V supply were needed, point C to ground could be used.

FUSES AND CIRCUIT BREAKERS

In most consumer products, some sort of protection is used to combat excessive current drains from the power supply. In many cases, these protection devices are **fuses** or **circuit breakers.**

The function of these devices can be seen in Figure 3–30. If the load resistance in the secondary decreases, an increase in secondary current will occur. Because the current in the secondary will reflect to the current of the primary (if one increases, the other increases), an overheating of the transformer will develop. This excessive amount of current passing through the transformer will cause damage to the transformer, and to other components connected to the secondary. For this reason, a fuse or circuit breaker can be placed in the primary circuit to protect against overloading.

Fuses and circuit breakers are not only found in transformer circuits. Often, they are found in transformerless connections. Figure 3–31 shows an example of such a connection.

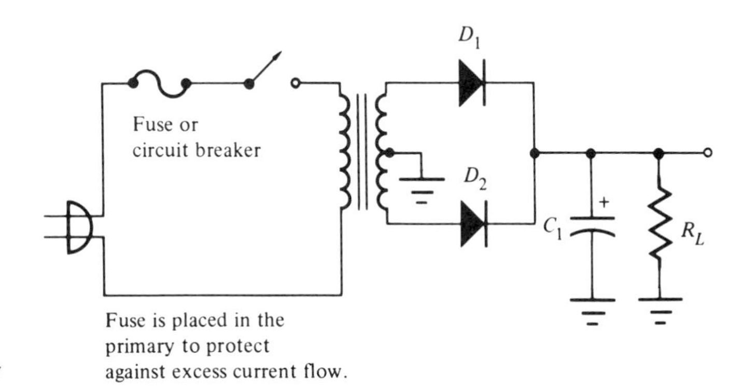

FIGURE 3–30
Protection for the Transformer from Excessive Amounts of Current Flow

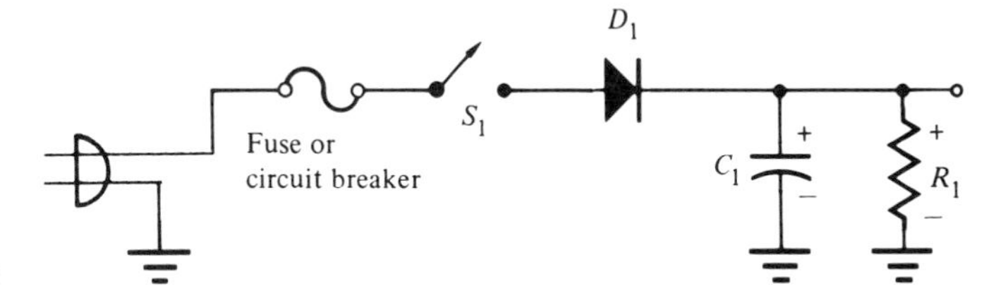

FIGURE 3–31
Fusing Power Supply to Protect from Currents Drawn from Power Supply

The method used to find the size of fuse needed to protect the circuit is fairly simple. A sample calculation follows in Example 7.

EXAMPLE 7

The transformer in Figure 3–32 develops 30 V(RMS) at the secondary, and the secondary is connected to a 30 Ω load. What size fuse should be placed in the primary to protect this circuit?

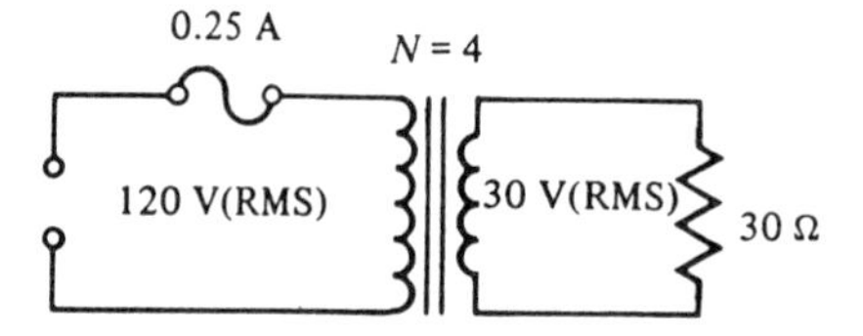

FIGURE 3–32
Finding Value of Fuse to Protect Transformer

Solution:
First, the secondary current must be found. Using Ohm's law, it is calculated as:

$$I_s = \frac{V_s}{R_L} = \frac{30}{30} = 1 \text{ A}$$

Next, the turns ratio of the transformer is determined:

$$N = \frac{V_p}{V_s} = \frac{120}{30} = 4$$

Finally, the current at the primary can be calculated:

$$I_p = \frac{I_s}{N} = \frac{1}{4} = 0.25 \text{ A}$$

This means that a fuse of 0.25 A at 125 V can be used to protect the power supply.

When a power supply is first turned on, a **surge current** often develops. Surge currents are momentary high currents. After these momentary high currents are seen, they reduce to normal currents. A fast-acting fuse placed in a circuit with a surge current would blow every time the supply was turned on. Therefore, a slow-blow fuse or a circuit breaker should be used. The use of a **surge resistor** will also limit the momentary rush current, as shown in Figure 3–35, where R_1 limits this current flow.

TROUBLESHOOTING

POWER SUPPLIES

The technician may often find the power supply to be the source of problems in the consumer product. The reason for this is obvious. The power supply delivers operating voltages or bias to all other stages within the device. If the power supply is nonoperational, then all the additional stages will cease to operate; and if a block of the power supply fails, distortion or failure will result in another section of the receiver. For example, assume that the technician is troubleshooting an AM-FM tape recorder. The first step in troubleshooting this device is to turn it on and identify the problem. Assume further that this unit, when turned on, has no sound output from either the radio or the tape recorder. In this condition, the first step is to check the power supply, because the power supply is responsible for voltage to operate both sections.

The technician should automatically think of the power supply in block diagram form, and begin to trace the problem with the block diagram in mind. A sample block diagram is shown in Figure 3–33. To begin, the technician measures the output dc voltage at TP_A (test point A), and verifies this reading with the schematic. If no voltage is present, a voltage reading is made at TP_B (test point B). If no voltage is found at this point, the technician works backward toward the ac input until the faulty block is found.

Once the faulty block has been located, the bad component in that block must be identified. First, several visual checks should be made. Has any component burned open? Has a capacitor blown apart? If these checks show nothing, then an ohmmeter is used to

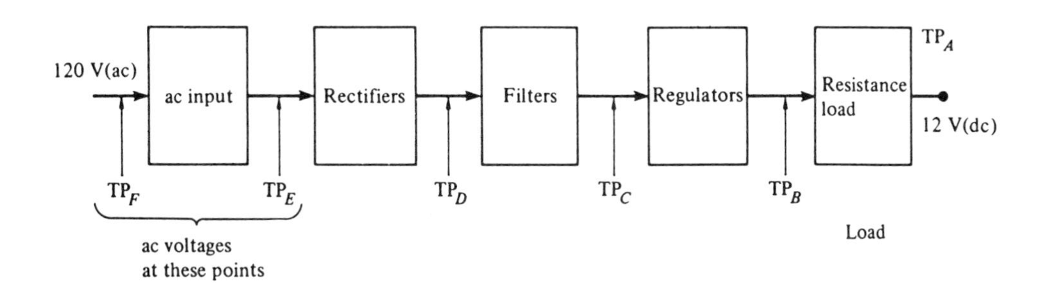

FIGURE 3–33
Block Diagram of dc Power Supply Showing Test Points for Scope or Meter Measurements

measure the resistance of each component. It is important that this check be done with the power off and disconnected. Once the faulty component has been found, it can be replaced. The power is then turned on to make sure the repair has restored output voltage. An oscilloscope should be used to insure that the ripple voltage is within specification of the power supply.

Power supplies have some very common problems:

—Lack of output voltage
—Low output voltage
—High output voltage
—High ripple voltage
—Continuous blowing of fusing device

We will discuss each of these problems individually.

Lack of Output Voltage

Lack of output voltage can be caused by several conditions. A voltage measurement of each block should be taken with a voltmeter. Some of the components that should be checked are given in the following list. Figure 3–34 shows these components in the schematic diagram of a full-wave rectifier.

1. *Defective Regulator Circuit*: Check R_1. If open, there will be no dc output.
2. *Defective Filter Choke*: Check. If open, there will be no dc output.
3. *Defective Diodes*: Check the resistances on the diodes. There should be high-low resistance ratio. Make sure that the power is off.

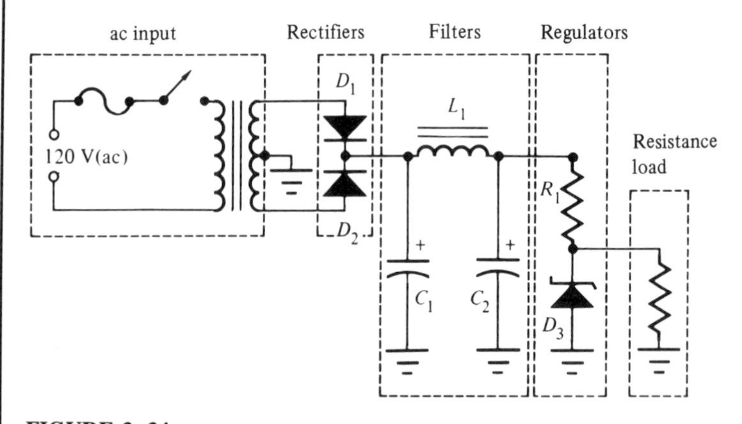

FIGURE 3–34
Block Diagram as It Relates to Components for a Full-Wave Rectifier

4. *Defective Transformer*: Measure the resistance of the primary and the secondary. Both should be low, but not zero.
5. *Defective ac Input*: Check the switch with an ohmmeter. Should show zero and infinite ohms resistance when turned on and off. Check the fuse or circuit breaker. Resistance should measure low. Power must be removed from the circuit for safety. Check the line cord. The voltage at the output should be at least 120 V(ac). Check the ac outlet. Be sure that the proper voltage is being delivered to the power supply.

Before making resistance checks, all power should be disconnected from the power supply, and the filter capacitors discharged. If power is not disconnected, it will cause damage to the test equipment and power supply. Another point to remember about measuring resistance of rectifier diodes is to make sure that one end of the diode is disconnected. If the diode is left in the circuit being checked, some false resistance readings may develop.

Low Output Voltage

Another problem associated with power supplies is low output voltage. Again, the problem can be traced to specific blocks on the power supply. See Figure 3–34. First, voltages at the output of the regulator should be measured. The low-voltage problem should be traced back toward the input as necessary.

Some of the conditions that cause low output voltage are given in the following list.

1. *Decreased Load Resistance*: If the load has a lowered resistance, the *RC* time constant has changed. The filter capacitors are discharging too quickly, and are not keeping the voltage to the load steady. Therefore, the load resistance must be checked. (*Hint*: First check all components in the load that are power amplifiers.)
2. *Leaky Filter Capacitors*: The dielectric materials in the filter capacitor may have become leaky, thus reducing the capacitance value. To check a filter capacitor, a capacitance tester can be used, or the ohmmeter can be used to measure the resistance of the capacitor. These filter capacitors should show 100 kΩ or more to be good.
3. *Open Diode*: One of the diodes in the rectifier could have opened. Two checks for this can be made. First, the resistance of the component should be measured. The diode should show a high-low resistance ratio. The next check is to use the oscilloscope. Figure 3–34 shows a full-wave rectifier. The waveform

for this rectifier should be the double-rectified pattern. If one of those rectified waveforms is missing, one of the diodes is open. This condition makes the rectifier a half-wave rectifier.

4. *Defective Transformer*: The secondary voltage will be reduced to 0 V by an open winding. The best check is to measure the ac secondary voltage and compare it to the schematic voltages. An ohmmeter can also be used to check the transformer's secondary resistance.
5. *Low ac Input Voltage*: Always make sure that 120 V(ac) is being delivered to the ac outlet.

High Output Voltage

The high output voltage found in some supplies is not the fault of the supply itself. The load is usually at fault. High output voltages mean that the draw of load current from the supply is decreased. Look for an open in the load circuits. If the supply has a transistor regulator circuit and the output voltage is high, a check should be made for bad regulator components.

High Ripple Voltage

High ripple voltage is usually connected with the filtering devices. If the ripple's peak-to-peak voltage has increased, the filter capacitors may be at fault. The symptom of high ripple voltage should be checked with an oscilloscope. The ripple measured on the oscilloscope should be checked against the schematic values. The high ripple value will usually also be associated with low dc output voltage. Ripple checks should be made when low output voltage is developed.

Continuous Blowing of Fusing Device

Direct shorts often develop in power supplies. A direct short causes the circuit protection to blow open constantly. In searching for this problem, the technician must use the ohmmeter.

Direct shorts can be caused by large amounts of current being drawn by the load, or by a short within the power supply. A quick check can be made by inserting a new fuse, or by resetting the circuit breaker. If the protection device blows a few moments after the supply is turned on, the trouble is in the load. Resistance checks of the load must be made. If, on the other hand, the device blows

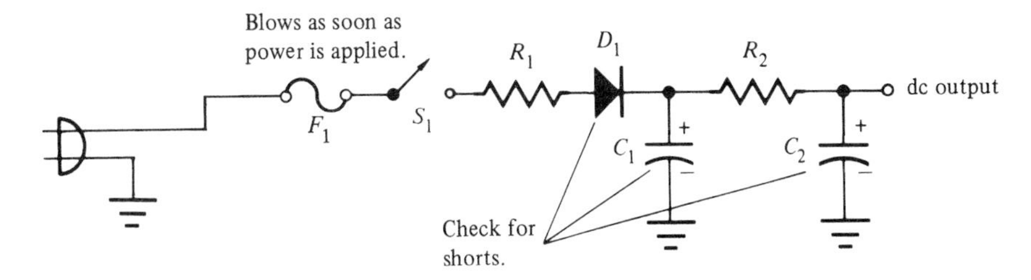

FIGURE 3–35
Components Likely to Cause Continuous Blowing of Fuses

as soon as power is applied, the short is probably in the power supply itself. Items to be checked in the power supply are capacitors, regulators, diodes, and the ac input. Figure 3–35 shows a half-wave power supply connected directly to the 120 V(ac) input. All components connected to ground should be checked.

These are only a few of the symptoms that might show up as dc power supply problems. Because the power supply is the central block in all consumer products, it should be checked for power output voltages.

SUMMARY

Power supplies are composed of five basic blocks: ac input, rectifiers, filters, regulators, and voltage divider networks.

The ac input could be a direct connection from the 110 V(ac) source, or a step-up or step-down transformer could be used. If a step-up transformer is used, an increase in voltage is seen at the secondary, along with reduced current output. If the step-down transformer is used, there will be a decrease in voltage, but an increase in current. All transformers used for power supplies are called power transformers.

Several types of rectifiers are found in power supplies. They are the full-wave rectifier, the half-wave rectifier, the full-wave bridge rectifier, and the voltage doubler circuit. All rectifiers give an output of pulsating dc voltage.

Filters are required to smooth out pulsating dc. All filtering systems are based on the capacitor's ability to charge and discharge. The capacitor charges to approximately the peak value of the pulsation, and then discharges its stored voltage to the load. All capacitors used in filters are electrolytic capacitors. The amount of pulsating dc remaining after filtering is called ripple. Ripple has a frequency of 60 Hz for half-wave circuits and 120 Hz for full-wave circuits if the ac input is 60 Hz.

The regulator is used to keep the voltage in the load constant. A zener diode or a transistor can be used to keep a constant level of voltage applied to the load. Most regulator circuits contain either a series regulator transistor or a shunt regulator (zener diode).

The last block in the power supply is the voltage divider. To make the dc output more versatile, a voltage divider is employed to drop the voltage to usable levels.

The technician may find it necessary to troubleshoot the power supply system. Some common problems associated with power supplies are lack of output voltage, low output voltage, high output voltage, high ripple voltage, and continuous blowing of the fusing device.

The power supply is the "heart" of consumer electronic devices. Its "pumping" of constant voltage and current keeps the consumer product working to its maximum specifications.

KEY TERMS

bleeder resistor
capacitive filter
capacitive input filter
choke input filter
circuit breaker
direct connection
filtering
full-wave bridge rectifier
full-wave doubler
full-wave rectifier
fuse
half-wave doubler
half-wave rectifer
load
pi filter
pulsating dc
RC pi filter
rectifier
regulation
ripple
series regulator
step-down power transformer
step-up power transformer
surge current
surge resistor
transformerless connection
voltage doubler
zener diode (shunt) regulator

REVIEW EXERCISES

1. At which block does the ac voltage enter the power supply?
2. Draw a block diagram of a typical dc power supply, and label each of the blocks.
3. Which factor determines whether a transformer is a step-up or a step-down transformer?
4. Where is a transformerless power supply used?
5. Name the three basic rectifier circuits found in dc power supplies.

6. Identify the output frequency for the half-wave and full-wave rectifier circuits, assuming a 60 Hz input frequency.

7. In the figure below, match each circuit with its waveform.

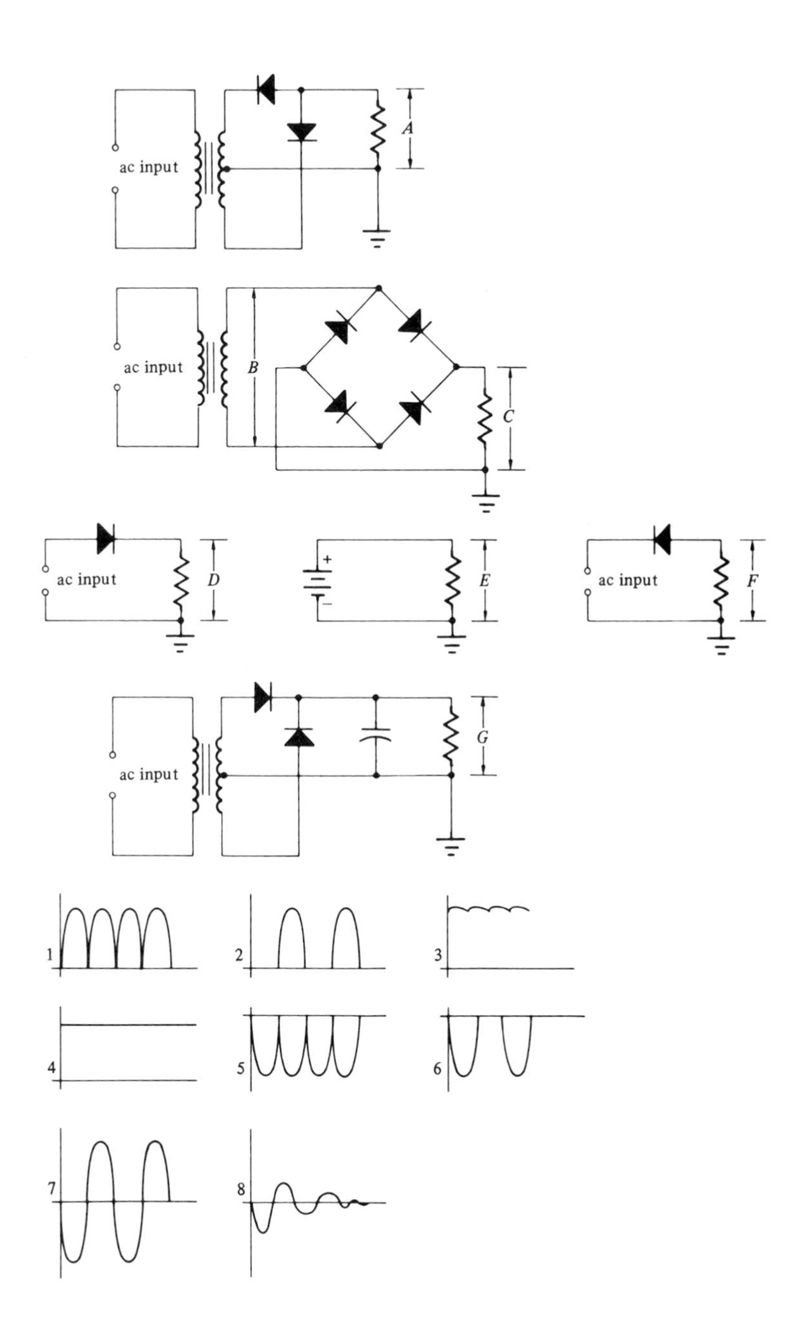

8. Find the approximate output voltage for a half-wave rectifier that develops 30 V(ac) at the transformer. What would the output voltage be for a full-wave rectifier?

9. What is the approximate output voltage from a voltage doubler circuit that has 75 V(p) applied to the input?

10. Perform the following conversions:

    a. 60 V(p-p) = ______V(p)
    b. 70 V(RMS) = ______V(p-p)
    c. 120 V(RMS) = ______V(p-p)
    d. 38 V(p-p) = ______V(RMS)

11. Name three basic filtering circuits found in dc power supply circuits. Draw a schematic diagram of each.

12. Why do large filter capacitors reduce the amount of ripple in a circuit?

13. A power supply delivers 10 V(dc) to the load, and has a 20% ripple. Is this power supply effective?

14. Draw the general waveshape of the filtering block. What is this waveform called?

15. What is the major function of the regulator in a dc power supply?

16. Draw a schematic diagram of a basic series regulator. Label all parts of this circuit.

17. Define load.

18. What are two types of circuit protection devices found in consumer products?

19. List some of the common troubles found in power supply circuits.

20. List some of the areas to check for low output voltage.

21. List some of the areas to check for lack of output voltage.

22. Which stage should be checked for higher than normal ripple voltage?

23. What causes a circuit protection device to blow over and over again?

4 BASIC BIPOLAR TRANSISTOR

OBJECTIVES

Upon completing this chapter, you should be familiar with:

—Transistor construction
—Transistor resistance
—Base, emitter, and collector current relationships
—Transistor current control
—Base-emitter forward biasing
—Transistor operation between cutoff and saturation
—Transistor operating characteristics
—Substitution transistors
—Transistor packaging
—Troubleshooting transistors

INTRODUCTION

The first radio consisted of an antenna, a crystal detector, and a headphone set. Only one person could listen to the broadcast at a time. With the advent of the vacuum tube, that radio signal could be amplified. This amplified signal was then sent through a speaker so that more than one person could listen. In 1945 in the Bell Laboratories, a signal was amplified in the first **transistor.**

This transistor had three elements, and allowed the signal to be increased between the input circuit and the output circuit. It did not require the heaters, high voltages, and long warm-up time necessary to the vacuum tube. When power was applied, the transistor became an amplifier. No longer did the consumer have to wait for the tube to warm up to hear music from a radio. This breakthrough of the transistor was a boon for the consumer products industry. Radios, televisions, and communications devices all became lightweight and portable because of this semiconductor amplifying device.

TRANSISTOR CONSTRUCTION

The transistor contains three parts of doped **semiconductor** material: either two N parts and one P part, or two P parts and one N part. In other words, a transistor is either an **NPN transistor** or a **PNP transistor.** Figure 4–1a shows the three layers of the PNP transistor; Figure 4–1b shows the three layers of the NPN transistor. These layers are called the **base (*B*),** the **emitter (*E*),** and the **collector (*C*).** Notice that leads are connected to each layer.

The transistor often is called a **bipolar junction transistor (BJT).** The word *bipolar* describes the fact that within the transistor, there are two majority current carriers: holes and electrons.

The bipolar **junction** transistor is basically a junction diode, plus an additional junction. The junction is the point at which the N and P materials of the transistor are joined. The transistor is made of one of the two materials used for semiconductors: **silicon** and **germanium.** The pure silicon or germanium is doped with impurities, which develops the N and P regions. This doping process was described in Chapter 2.

Each of the three sections of the transistor has a specific function. The emitter is heavily doped with current carriers, which are emitted. The collector is moderately doped with current carriers because it conducts almost as much current as the emitter. The base is lightly doped with current carriers because of the small amount of current it must handle. The current carriers pass through the base region, and are collected in the collector section. Because the current carriers are allowed to pass through the base region, the base is an excellent place to control the current flow. Just as the control grid in a vacuum tube regulates the current, so also does the base of the

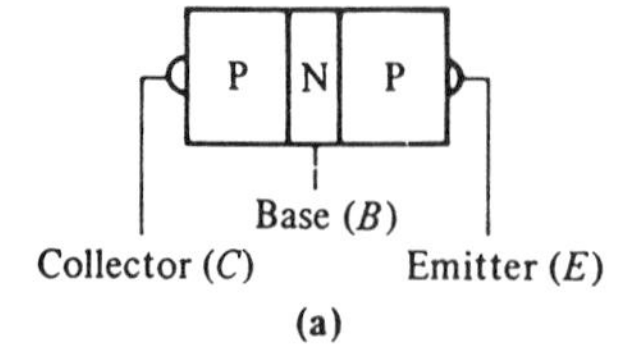

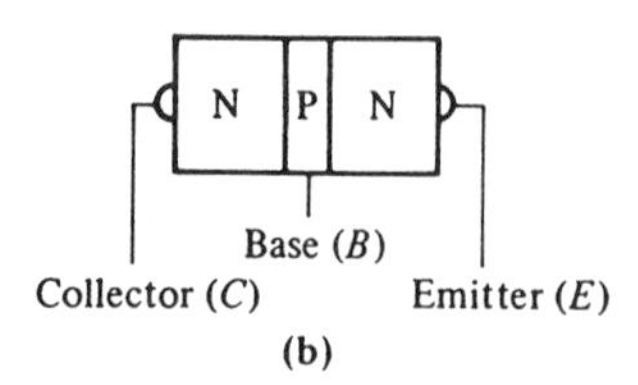

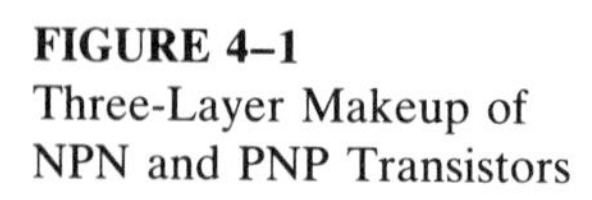
FIGURE 4–1
Three-Layer Makeup of NPN and PNP Transistors

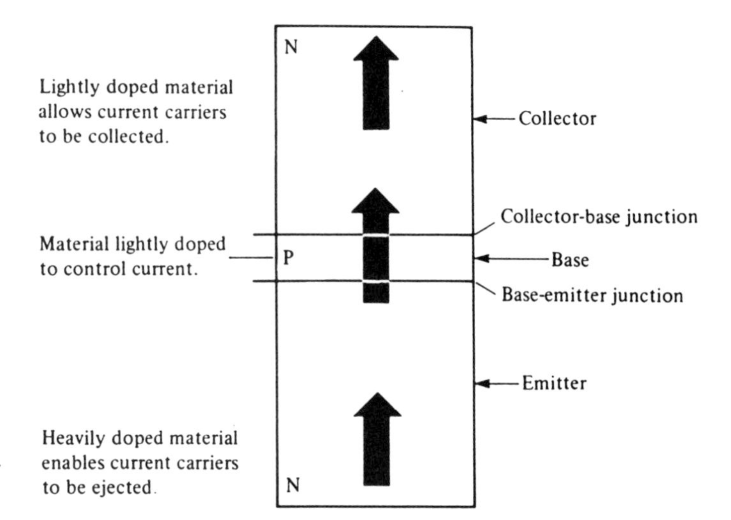

FIGURE 4–2
Current Path Developed through the NPN Transistor

transistor. Figure 4–2 shows the block diagram of the transistor and its current path.

Block diagrams of the transistor are not used in schematics. Figure 4–3 shows the schematic symbols for the transistor. In each symbol the emitter is identified by the arrow. The direction of the arrow indicates the type of transistor. For example, on the NPN type (Figure 4–3a), the arrow points away from the base. The arrow always points toward the negative material. Since the arrow is pointing at the emitter, the emitter is N-type material. Because emitter and collector are made from the same material, the collector also is N-type material. This leaves the base. The material of the base is always the opposite type material from that of the collector and emitter. Thus, the base is P-type material. The PNP type has the arrow pointing toward the base (Figure 4–3b). Therefore, the base is N material. This means that the collector and emitter are P material. The polarities on the transistor and the lead identification are important items, and should be remembered.

As with the semiconductor diode, the transistor must have ap-

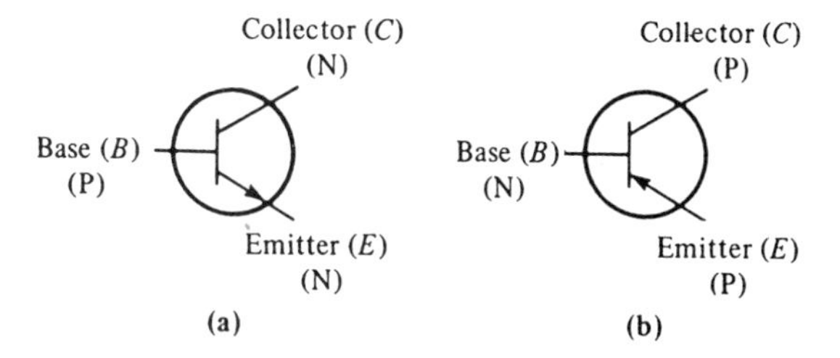

FIGURE 4–3
Transistor Schematic Symbols: (a) NPN Type, (b) PNP Type

plied voltage to its terminals in order to function. To illustrate this biasing, both the NPN and the PNP transistor lead biasing will be shown. Figure 4–4 illustrates the biasing action on the NPN transistor. First, the base emitter junction of the transistor must be biased. The base emitter will always be forward biased if the transistor is to operate as an amplifier. The emitter of this transistor is made from N material. Therefore, in order to be forward biased, it must have a negative voltage applied. This negative voltage is applied from the dc power supply, labeled V_{BE}. The base is made of P material and it also must be forward biased. The positive terminal of the V_{BE} supply is connected to the base. Figure 4–4 confirms this connection. The next junction that must be biased is the collector base. This junction must be reversed biased in order to operate as an amplifier. Figure 4–4 illustrates this connection also. Notice that a separate power supply is used. This source is called V_{CB}, or, in other words, the voltage for the collector base junction. The N material of the collector has a positive polarity supplied by the V_{CB} source. And the base P material has negative voltage applied. This applied voltage has the collector base junction in reverse bias. The NPN transistor is properly biased and will conduct current through its junctions.

Electrons in the NPN transistor are the majority current carriers. The electrons are injected into the emitter. They are attracted to the thin base region and pass across it. Once across the base region, the electrons come under the force of the highly positive collector. This force causes the electrons to be pulled through the collector to the positive V_{CB} source. Thus a current flow developed in the emitter causes a current flow through the collector. This action demonstrates why the transistor is identified as a current operating device.

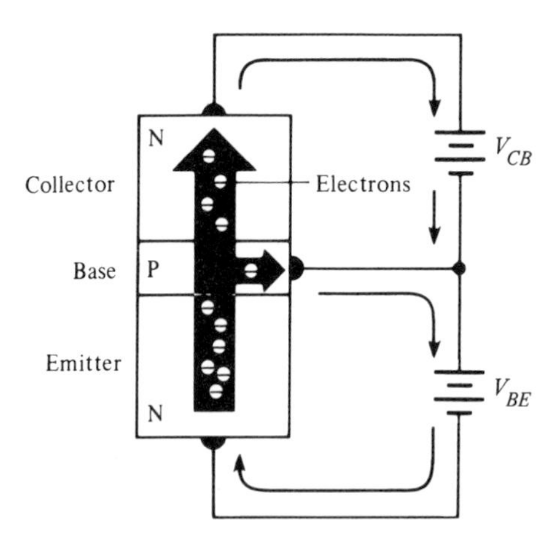

FIGURE 4–4
Biasing the NPN Transistor

Figure 4–5 illustrates the block diagram of the PNP transistor. Notice in this transistor the connection is reversed from the NPN transistor. The emitter is now positive material, and the base is now negative material. The power source V_{BE} is used to bias the base emitter junction. The emitter has the positive part of V_{BE} connected to its terminal. The negative base material has a negative source connected to its terminal. This connection has the base emitter in the forward bias mode. The collector base junction must be reversed biased. Since the collector is P material, the V_{CB} source must have the negative terminal connected to the collector lead. The base region will have positive voltage connected to its terminal. This connection will reverse bias the collector base junction.

Current flow through this transistor is identified as hole movement. The holes are injected into the emitter and drift toward the base. Since the base emitter is forward biased, the junction develops low resistance and the holes drift toward the collector. The negative polarity on the collector will attract the holes. Therefore, as in the NPN transistor, current flow was developed through the transistor by the injection of current carriers into the emitter. From this point forward in the book, the reader will only identify current flow being from negative to positive. The hole flow discussion will only relate to the flow in the PNP transistor.

Because the biasing methods for the NPN and PNP are completely different, these two types of transistors cannot be interchanged in a circuit. Even though both transistors perform the same functions as amplifiers, oscillators, and regulators, these two transistors cannot replace one another unless the biasing of the circuit is changed. This process is very involved; therefore, only identical replacement transistors should be used.

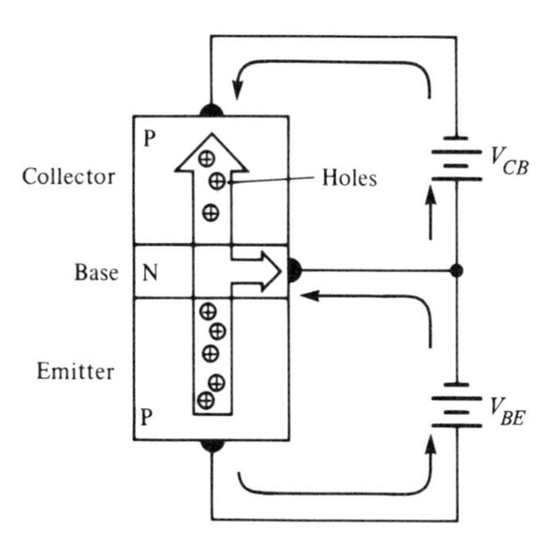

FIGURE 4–5
Biasing the PNP Transistor

TRANSISTOR RESISTANCE

Because of the different biasings found on the transistor, the junctions are at different resistance levels. The base-emitter junction is forward biased. A forward biased semiconductor will always have low resistance. Since the collector-base junction is reversed biased, it will have extremely high resistance. It is because of this large resistance difference that the transistor is able to control current and thus amplify.

BASE, EMITTER, AND COLLECTOR CURRENTS

Figure 4–6 shows a properly biased NPN transistor. Resistor R_1 serves as a current-limiting resistor for the emitter. The biasing supplies have been renamed. Voltage for the emitter is V_{EE}. The collector supply is V_{CC}.

The application of an emitter voltage causes an **emitter current,** which is labeled I_E. The **base current** is designated as I_B, and the **collector current** is identified as I_C.

An important relationship develops between these three current flows. Because the emitter is the injection point for current carriers, 100% of all current flow will be developed in the emitter. Since the base is a thin doped layer, little current is developed there. Generally, about 3–5% of the total current is developed in the base. The remaining 95–97% is seen at the collector. The general relationship between base, emitter, and collector currents is given in Equation 4–1.

$$I_E = I_B + I_C \tag{4–1}$$

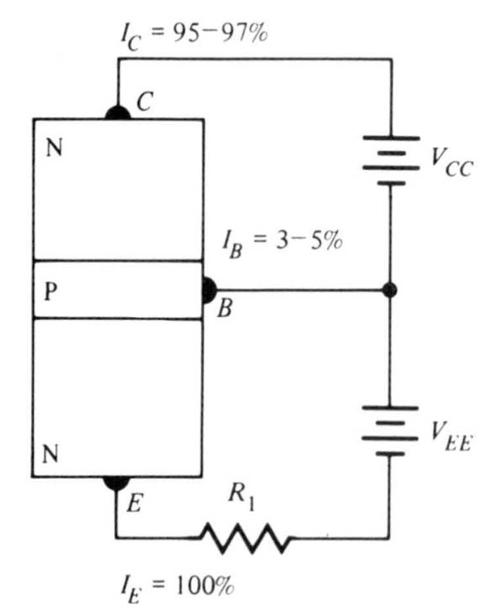

FIGURE 4–6
Current Flow Developed in the NPN Transistor

where:

I_E = emitter current
I_B = base current
I_C = collector current

Using this equation with any two known current values, the third current can be found. A sample problem follows in Example 1.

EXAMPLE 1

A transistor develops 1 mA of base current and 50 mA of collector current. What is the emitter current for this transistor?

Solution:

The known current values are placed into Equation 4–1:

$$I_E = I_B + I_C = 1 \text{ mA} + 50 \text{ mA} = 51 \text{ mA}$$

This current relationship also holds for the PNP transistor.

CURRENT CONTROL IN A TRANSISTOR

One of the most important functions of a transistor is to control current. As stated earlier, the emitter will see all of the current flow, and the base and the collector will share in the emitter's current, with the base taking considerably less.

One of the fundamental concepts of amplification in a transistor is that of a small current controlling a larger current. In the transistor the base current controls the current flow to the collector. The base behaves like the control grid in a vacuum tube, which controls plate current. Control of the collector current by the base current in a transistor is illustrated diagrammatically in Figure 4–7.

Figure 4–7 shows an example of an NPN transistor. Here, when the current in the base is increased from 1 mA to 1.5 mA, the collector current increases from 40 mA (Figure 4–7a) to 60 mA (Figure 4–7b). When the current is decreased from 1.5 mA to 0.5 mA, the collector current decreases (Figure 4–7c).

Although the base current is quite small, it is very important to transistor operation. An example of what happens when the base current is removed is given in Figure 4–8. Here, the base lead has been removed from the power source. This means that no base current will flow. The two power sources, V_{CC} and V_{EE}, add together. This

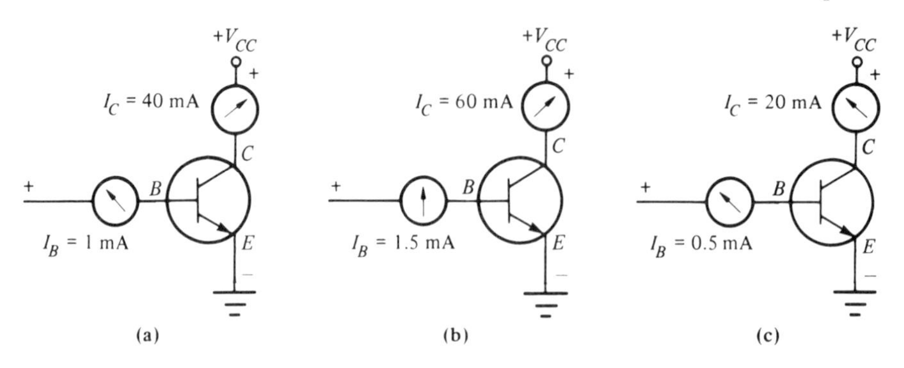

FIGURE 4–7
Transistor Base Current Controlling Collector Current

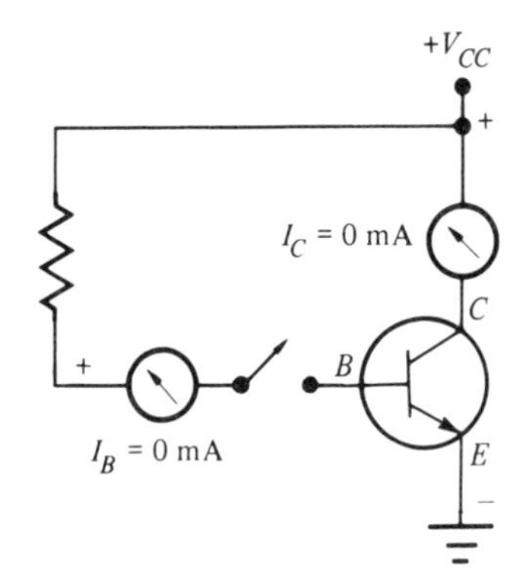

FIGURE 4–8
Removing Base Current Turns Off the Transistor

makes the collector more positive than the emitter. A difference of potential exists, and current should flow. However, there is only a small amount of current flow, identified as leakage current. In order to have conduction through the transistor, the base-emitter junction must be forward biased. Opening the base lead has removed the forward bias. Therefore, even though the base current is small, it controls the larger collector current.

Because of this relationship of base current controlling collector current, the transistor is capable of good **current gain.** This relationship is commonly expressed as the ratio of collector current to base current. This ratio, one of the most important transistor characteristics, is called **beta** (β), or $\boldsymbol{h_{FE}}$. This beta relationship is given mathematically in Equation 4–2.

$$\beta \text{ (or } h_{FE}) = \frac{I_C}{I_B} \tag{4–2}$$

The following is an example of finding the beta of a transistor.

EXAMPLE 2

A transistor has a collector current of 40 mA, and a base current of 1 mA. What is the beta of this transistor?

Solution:

Using Equation 4–2, the problem is solved as follows:

$$\beta = \frac{I_C}{I_B} = \frac{40 \text{ mA}}{1 \text{ mA}} = 40$$

Because milliamperes occur in the denominator and the numerator, they cancel out each other. This cancellation leaves beta as a number without a unit of measurement. The beta value calculated here means that, under these biasing conditions, the transistor will produce a current gain of 40.

Beta values of transistors vary greatly. In small-signal transistors, beta can be as low as 20, and as high as 400. In some larger-power transistors, beta is as little as 20. A typical beta gain for a medium-range amplifier will be between 90 and 100.

TRANSISTOR BASE-EMITTER BIASING

In order for the emitter to inject electrons into the transistor, an external voltage must be applied. Figure 4–9 shows an example of the dc biasing required for an NPN transistor. As always, the base-emitter junction must be forward biased. The circuit shows positive voltage

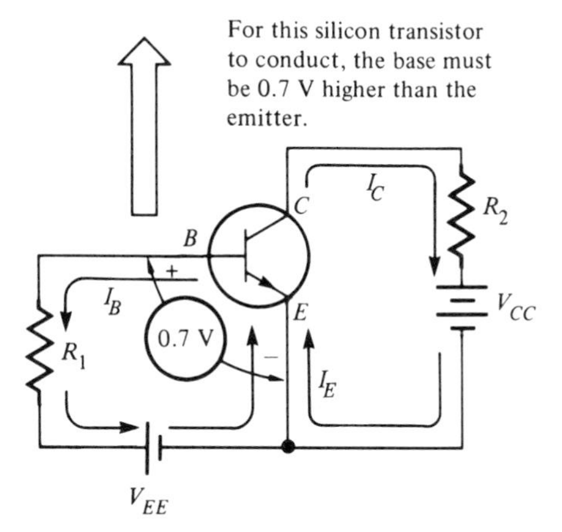

FIGURE 4–9
Properly Biased Base-Emitter Junction of an NPN Transistor

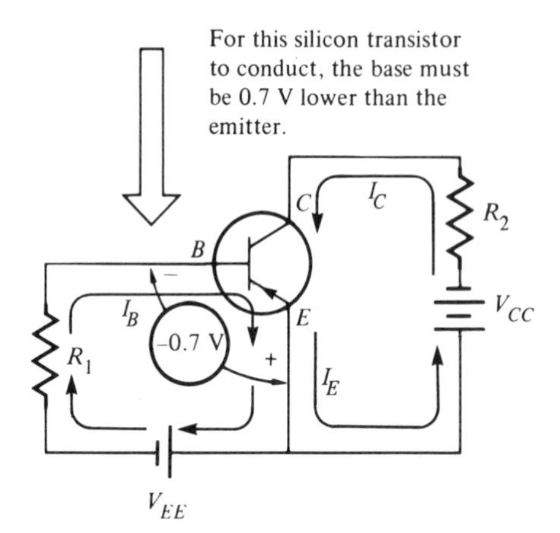

FIGURE 4–10
Properly Biased Base-Emitter Junction of a PNP Transistor

connected to the base through R_1, and negative voltage applied to the emitter. The collector of this transistor is N material. The collector must be reversed biased. This means that positive voltage should be connected, which it is. This proper biasing has the transistor conducting.

The arrows in Figure 4–9 trace the path of current flow. The emitter current flows from V_{EE} into the base-emitter junction. A small portion then flows to the base, and the remainder flows to the collector. This current flow between the base and the emitter has developed a voltage drop because of the base-emitter resistance. The voltage drop between these terminals is 0.7 V for silicon, and 0.3 V for germanium. Because the base is positive, for silicon the base voltage must be at least 0.7 V higher than the emitter voltage for conduction to occur. For germanium it must be at least 0.3 V greater. These voltage drops are very important for the operation of the transistor.

Figure 4–10 shows an example of a PNP transistor and its proper biasing. Again, a silicon transistor is used. Here, the base must be more negative than the emitter. This negative voltage is developed through R_1 and V_{EE}. The important fact about this transistor's voltage is that the base voltage is 0.7 V less than the emitter voltage.

TRANSISTOR CUTOFF TO SATURATION

As was mentioned before, the emitter is a current carrier, or injector, into the transistor. This means that the transistor is a current-operating device. If so, why are the voltages around the transistor so important? The answer to this question is simple. In actual circuit conditions, it is easier to measure voltage than to cut into the printed circuit board and measure current. Therefore, the technician must be able to analyze voltages around the transistor. These voltages will show if the transistor is conducting current.

The transistor can be in one of three states of operation: cutoff, linear operating range, and saturation. **Cutoff** in a transistor means that the forward bias has been removed from the base-emitter junction. As was mentioned before, when the base-emitter bias is removed, current ceases to flow. Figure 4–11 shows an example of what happens when the base-emitter bias falls below the 0.7 V point on a silicon NPN transistor.

In the **linear operating range,** the forward bias is established. Normal collector current flows, and beta falls into the general range of 50–100. It is within this range of the transistor that most consumer amplifiers operate.

Hypothetically, if the base bias is increased, the collector current should increase by a factor of beta. However, in actual transistor operation, base bias can change only so much, until no real increase in collector current is seen. This maximum level of increase, which is different from transistor to transistor, is called **saturation.** Figure 4–12 shows an example of how this level of saturation operates. Notice that in Figure 4–12a, a current flow of 1 mA on the base establishes 50 mA on the collector. A simple calculation for beta shows that a gain of 50 will be seen. Hypothetically, with this biasing network any change in base current will result in a 50-fold increase in collector current. Figure 4–12b illustrates such an increase. When the base current increases to 5 mA, 250 mA would be expected to be seen on the collector. Instead, only 101 mA are seen at the collector. At this point, the transistor has become saturated, as shown in Figure 4–12c. Saturation will be discussed in more detail in Chapter 5.

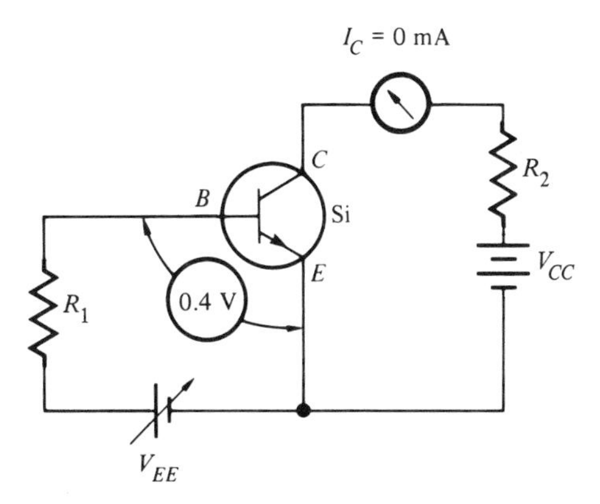

FIGURE 4–11
Reducing Forward Bias below 0.7 V

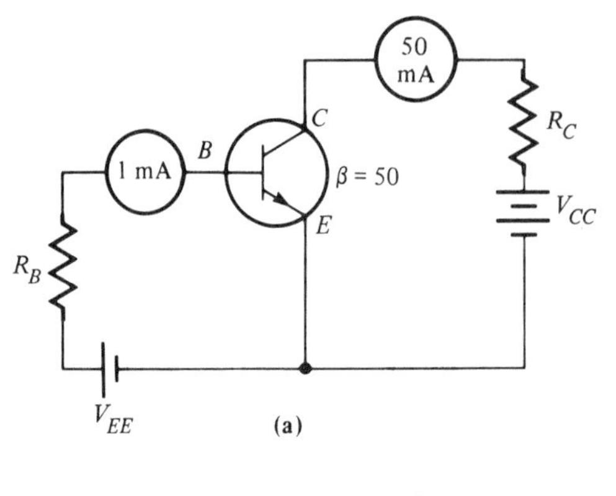

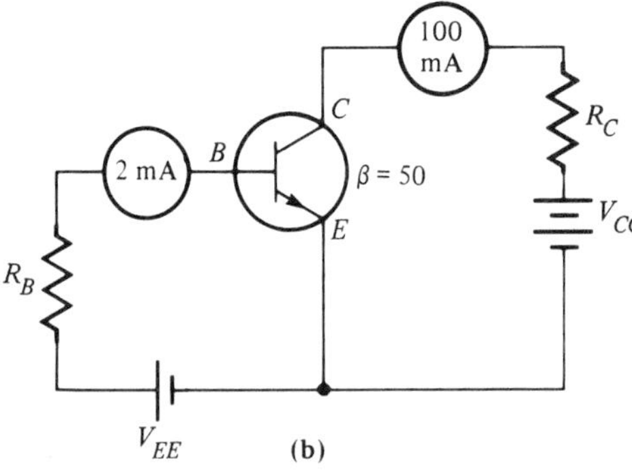

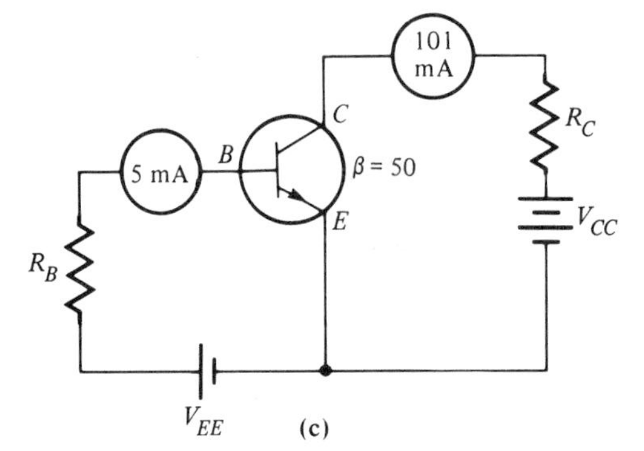

FIGURE 4–12
Biasing the NPN Transistor: (a) Establishing the Operating Point, (b) Increasing Base Current Increases Collector Current, (c) Placing the Transistor into Saturation

TRANSISTOR CHARACTERISTICS

As has been noted, the transistor is basically a current-operating device. It is used in many home entertainment products as an amplifier. It also has been stated that PNP and NPN transistors have different characteristics, such as amount of current gain, maximum terminal voltages, temperature range, and power dissipation. The technician should understand the characteristics of each transistor type, and how they relate to transistor application.

As stated previously, all bipolar transistors have two junctions. One junction is between the base and the emitter, and the other is between the base and the collector. From this knowledge the numbering system is identified for the transistors. Each bipolar transistor made in the United States has a prefix of 2N, the number 2 referring to the number of junctions. Following the 2N prefix is a series of numbers. These digits identify the particular transistor. An example is the 2N2222 junction transistor. All numbering sequences of transistors made in the United States, including replacement transistors, are registered with the Electronics Industrial Association (EIA). Foreign transistor types do not conform to this numbering system. Japanese transistors all begin with 2S. Generally, PNP transistors are designated 2SA, and NPN types are designated 2SC. Table 4–1 is a general table of codes for the Japanese transistor types.

Figure 4–13 is a typical data sheet for a 2N2222 junction transistor. All data sheets are basically the same, whether for U.S. or foreign transistors. The service technician must be able to understand the various characteristics listed on the data sheet in case transistor replacement is necessary.

Notice in Figure 4–13 that a short statement identifying the use of the transistor is given. In the case shown, the transistor is an NPN silicon transistor. It is suitable to use in high-speed switching circuits and in general-purpose amplifiers. The data sheet also lists the type of case in which the transistor is housed, and the transistor's high-frequency range. Also shown are the maximum terminal voltages found on this transistor. It is important to note that all these measurements on the transistor were taken at 25°C (room temperature). If for any reason the temperature surrounding this transistor were to rise, the transistor characteristics would change. Such a change in ratings would mean drastically reduced performance of the transistor.

TABLE 4–1 Lettering Code for Japanese Transistors

Letter*	Polarity and Application
A	PNP transistor, high-frequency
B	PNP transistor, low-frequency
C	NPN transistor, high-frequency
D	NPN transistor, low-frequency

*Each letter here is preceded by the 2S designation.

2N2222
NPN EPITAXIAL PLANAR SILICON TRANSISTOR

High-reliability silicon transistor suitable for high-speed, medium-power switching and general-purpose amplifier applications

- TO-18 case
- High transistion frequency (f_T), 350 MHz typical

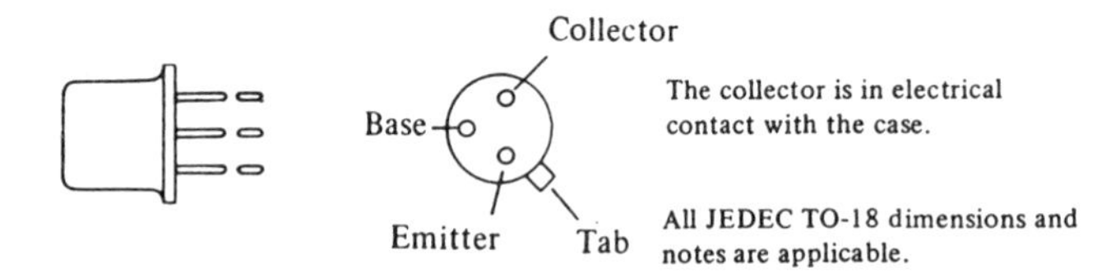

Absolute Maximum Ratings at 25°C Free Air

Collector-base voltage	60 V
Collector-emitter voltage	30 V
Base-emitter voltage	5 V
Collector current	800 mA
Maximum dissipation	500 mW
Maximum dissipation at 25°C case temp	1.8 W

Derate linearly to 175°C case temperature at rate of 12.0 mW/°C

Typical Electrical Characteristics:

I_{CBO}			10 nA (max) 10 μA (max at 150°C)
I_{EBO}			10 nA (max)
h_{FE}	V_{CE} = 10 V	I_C = 100 μA	35 (min)
		I_C = 1 mA	50 (min)
		I_C = 10 mA	75 (min)
		I_C = 150 mA (pulse)	100-300 (min-max)
		I_C = 500 mA (pulse)	30
f_T (transistion frequency)			350 MHz (typical)

Switching Characteristics:

Turn-on time	25 ns
Turn-off time	200 ns

FIGURE 4–13
Data Sheet for the 2N2222 Transistor

SUBSTITUTION TRANSISTORS

Many electronic product manufacturers manufacture their own transistors. These transistors have their own numbering systems. For example, RCA manufactures transistors identified as the SK series, Sylvania identifies its transistors by the ECG series, and General Electric uses the GE series. All of these transistors are direct replacements for many of the transistors found within consumer electronic products. The service technician should be aware of these various replacement transistors. Often the original transistor cannot be located. If a general replacement can be found, it will function properly within the unit.

TRANSISTOR PACKAGING

Once the junctions of the transistor have been formed, the transistor is placed into a metal or plastic package. The service technician should become familiar with the various packaging styles. Some typical examples of four common packaging styles are shown in Figure 4–14.

A TO-5 package is illustrated in Figure 4–14a. The drawing contains side and bottom views of the transistor. Of importance is the identification of the three leads. Note the tab on the bottom view. The closest lead to this tab is the emitter lead, the next lead is the base lead, and the lead farthest from the tab is the collector lead. Without the tab, the lead configuration of the TO-5 package is generally the

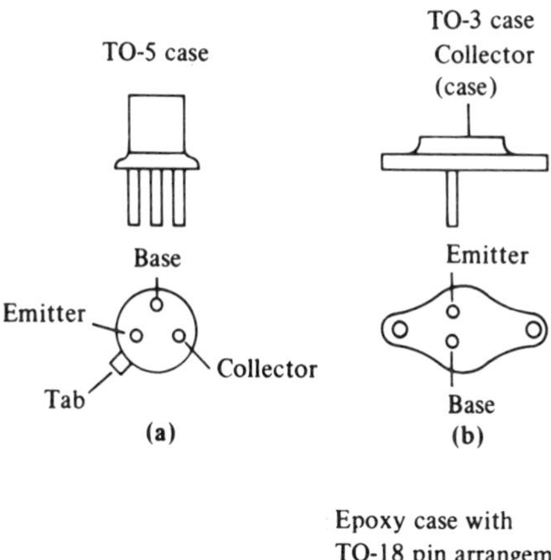

FIGURE 4–14
Common Transistor Packaging Styles

same, but checking a transistor data sheet will verify the lead identification.

Figure 4–14b shows a TO-3 case. The TO-3 package is generally used in high-power applications. Notice that on this packaging style, only two leads—the base and the emitter—are identified. The collector is always tied to the outside case. Therefore, the collector is made part of the transistor case's body.

Figure 4–14c shows another type of case found in housing high-power transistors: the TO-220 case. As a general rule, the two outside leads are identified as the base and the emitter. The center lead is identified as the collector. Note that the tab on this transistor also connects to the collector lead.

Because the TO-220 case and the TO-3 case usually are used for high-power applications, they should be heat-sinked properly to dissipate the heat. Figure 4–15 shows an example of a typical heat

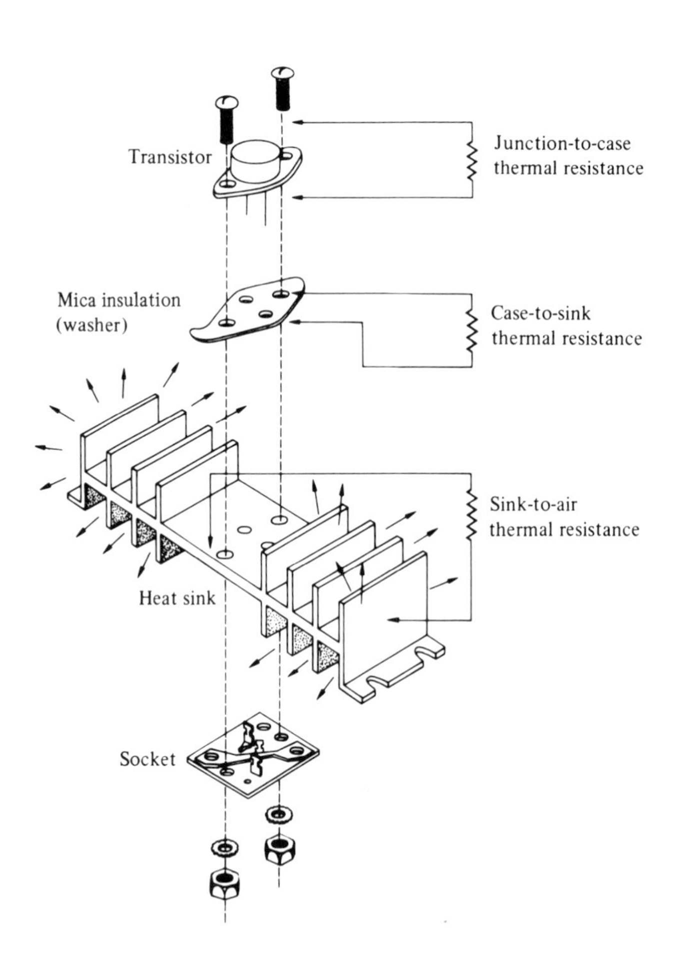

FIGURE 4–15
Heat Sink Used for the TO-3 Power Transistor Package

sink used for the TO-3 style of transistor. This type of heat sink is found in the audio output section of AM-FM stereo receivers. Notice that at each point of the heat-sinking process, contact is made between the transistor and the heat sink, keeping the transistor operating within its temperature characteristics. Because many of the transistors used with heat sinks are TO-3 and TO-220 cases, and the collector is tied to the tab, the transistor must be insulated from the heat sink. Insulating from the heat sink can be accomplished by placing a plastic insulator such as a mica washer between the transistor and the heat sink. Figure 4–15 illustrates the use of a mica washer. A silicon grease can be used between the transistor and the heat sink. This grease is used to transfer heat between the transistor and the heat sink.

The fourth type of case common in consumer products is the epoxy case style with TO-18 pin arrangement. This popular style of case is shown in Figure 4–14d. As a general rule, the base lead is the center lead, and the emitter and collector leads are on either side of the base lead.

The four housing types just described are the most common transistor housings found within consumer products. Additional package styles are too numerous to list. If a question arises about the case style or lead identification, a transistor data book should be consulted. This book should contain all the important information about transistors.

TROUBLESHOOTING

TRANSISTORS

In many applications the transistor must be checked to insure that it is operating properly. The check on the transistor can be accomplished by using expensive test equipment, or by making a simple ohmmeter check. All tests should be conducted out of circuit.

First, before using an ohmmeter, it is important to identify the polarity of the ohmmeter leads. In some ohmmeters, the red lead is not always positive. Therefore, to insure the proper polarity, the output of the ohmmeter leads should be measured with a voltmeter. Once the polarity has been established, the transistor's junction resistance can be measured.

Using an NPN transistor, the negative ohmmeter lead should be placed at the emitter, and the positive lead at the base. This should result in a low resistance reading, somewhere between about 200 Ω and 500 Ω. Then, the leads of the ohmmeter on the base and emitter should be reversed. The reading should be high. A high-low resistance reading means that the base-emitter junction is good.

If the resistance readings for this junction show any deviation, such as low-low or high-high, the junction is faulty, and the transistor should be replaced. Figure 4–16 shows an example of this test.

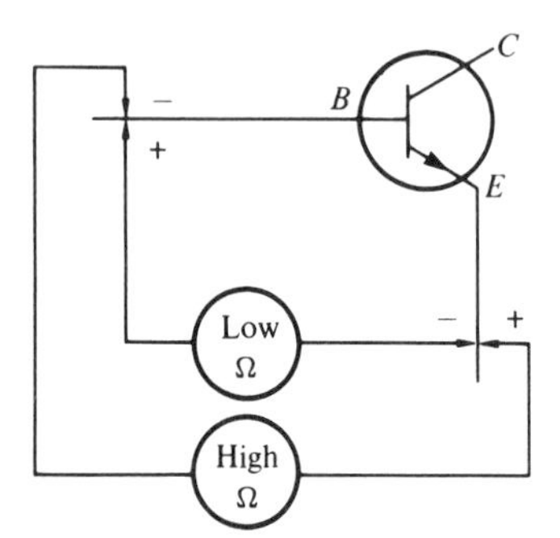

FIGURE 4–16
Resistance Check of the Base-Emitter Junction

Next, the terminals should be checked. In this measurement, the emitter has the negative ohmmeter lead, and the collector has the positive ohmmeter lead. The resistance reading for this junction should be between about 10 kΩ and 100 kΩ. Temporarily short the collector to the base. This now forward biases the transistor, and will reduce the emitter-collector resistance. If these readings are achieved, then the transistor is considered to be in operational condition. Figure 4–17 shows an example of testing the emitter-collector resistance on the NPN transistor.

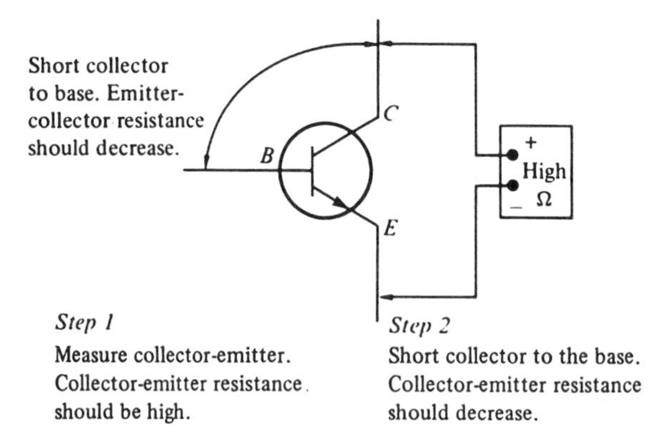

FIGURE 4–17
Resistance Check of the Emitter-Collector Junction

If the transistor were a PNP transistor, the testing procedure would be reversed. The important point in testing the transistor is the high-low resistance relationship.

Another transistor that might have to be tested is the high-powered transistor. In this transistor the resistance found between the base and the emitter will be about 30–50 Ω. The emitter-collector resistance will measure several hundred ohms. Shorting the collector to base, jumping a wire between the two leads, will reduce the emitter-collector resistance reading. Again, when testing the PNP or the NPN power transistor, the resistance relationship between junctions should be good. The power transistor should be checked carefully because of the low resistance readings, which might lead the technician to think that the transistor is shorted.

The ranges of resistance readings given for the transistor tests are approximate. If a given reading falls within the correct range, the transistor is assumed to be good.

An in-circuit test can be performed on the transistor by reading the voltage drop between the base and the emitter. If the reading for the transistor is 0.7 V for silicon or 0.3 V for germanium, the transistor is operational.

SUMMARY

Transistors are constructed in two styles: NPN type and PNP type. Each type is composed of three layers. Attached to each layer is a terminal. The three terminals are the emitter, the base, and the collector.

The transistor is often called a bipolar junction transistor, or BJT.

The transistor is made of either silicon or germanium. A doping process is used to construct the various N and P layers.

The emitter section of an NPN transistor emits electrons, the base is the controlling terminal, and the collector collects the current carriers.

The PNP and NPN each have schematic symbols. In both symbols the emitter is identified with an arrow. If the arrow points away from the base, the transistor is NPN type. If the arrow points toward the base, the transistor is PNP type.

In order for the transistor to operate, the base-emitter junction must be forward biased, and the base-collector junction must be reversed biased. The forward biased base-emitter junction will have low resistance, whereas the base-collector junction will have high resistance. This resistance ratio enables the transistor to be an excellent current amplifier.

The following mathematical equation can be used to calculate a transistor terminal's current flow: $I_E = I_B + I_C$. One hundred percent of the current will be developed in the emitter, 3–5% in the base, and 95–97% in the collector.

The relationship between collector current and base current is called beta. Because of this relationship, the base current controls collector current.

The biasing between base and emitter will be approximately a 0.7 V difference for silicon, and a 0.3 V difference for germanium.

The transistor has three ranges of operation: the cutoff range, the linear range, and saturation. At cutoff, no collector current flows. At saturation, maximum collector current flows. The linear range operates between cutoff and saturation.

Most transistors are in the 2N or the 2S series. The 2N designation is American, and the 2S denotes Japanese transistors.

If the ratings of two different transistors are exactly the same, the transistors can replace each other in circuit operation. Many manufacturers make their own transistors, and have their own numbering sequences.

The lead identification of transistors differs among the different transistor housings. Many styles of housing are available. If the technician is not sure which housing to use, a reference data sheet should be consulted. Each transistor has a data sheet listing the type of transistor, application, temperature rating, and terminal maximum voltages.

Often, the transistor can be checked out of circuit to determine if it is good or bad. If the transistor is good, it will have proper resistance readings between its terminals. Resistance readings out of the given range identify which transistor needs to be replaced.

KEY TERMS

base
base current
beta
bipolar junction transistor
collector
collector current
current gain
cutoff
emitter
emitter current
germanium
h_{FE}
junction
linear operating range
NPN transistor
PNP transistor
saturation
semiconductor
silicon
transistor

REVIEW EXERCISES

1. Define bipolar as it refers to transistors.
2. Transistors are made of what two basic materials?
3. Draw the schematic diagrams of an NPN transistor and of a PNP transistor. Identify each of the leads.
4. What bias should be applied to the base-emitter junction of the transistor? to the collector-base junction?
5. Draw schematic diagrams of the NPN and PNP transistors applying proper biasings.
6. What resistance will be developed across the forward biased base-emitter junction?
7. What percentage of current flow is found in the base lead? in the emitter lead? in the collector lead?
8. If 1 mA of current is developed in the emitter, and 0.9 mA in the collector, what is the current flow in the base?
9. With which element of the vacuum tube can the base of a transistor be compared? Why?
10. Explain why the transistor is capable of good current gain.
11. What term is used to identify the collector current to the base current?
12. What does the term h_{FE} mean?
13. What is the typical beta range for a small-signal transistor?
14. A transistor develops 10 μA of base current, and 1 mA of collector current. What is the beta for this transistor?
15. What is the voltage drop on a silicon base-emitter junction? What is the voltage drop on a germanium base-emitter junction?
16. Name the three stages of transistor operation.
17. Define saturation.
18. Identify the applications of the following transistor prefixes:
    a. 2SA
    b. 2SB
    c. 2SC
    d. 2SD
19. Give the use of each of the following packaging styles:
    a. TO-3 case
    b. TO-220 case
    c. TO-5 case
20. Should a high or low resistance be developed between the following transistor terminals of a forward-biased transistor?
    a. Base-emitter
    b. Base-collector
    c. Emitter-collector
21. What is the main function of a heat sink?

5 BIASING THE TRANSISTOR

OBJECTIVES

Upon completing this chapter, you should be familiar with:

—Characteristic curve application
—Transistor dc biasing
—Bias stabilization components
—Troubleshooting transistor dc biasing

INTRODUCTION

Transistors are widely used in consumer electronic products as amplifiers and switches. To better understand how transistors operate in these two areas, biasing methods should be explored further.

Biasing of the transistor is very important. The technician must understand the different biasing methods developed in the three common transistor configurations (the common base, the common emitter, and the common collector circuits). In addition, knowledge of the characteristic curve development for each transistor will help the technician understand dc biasing methods.

CHARACTERISTIC CURVES

Basically, each transistor operates between two points in a circuit: cutoff and saturation. If a point is picked somewhere between cutoff and saturation, the transistor will be used as an amplifier. If the transistor is to be used as an amplifier, an operating point should be selected in the region approximately halfway between the two points. However, if the transistor is going to be used as a switch, an operating point at or beyond cutoff should be chosen.

A set of **characteristic curves** is used to select the operating point for a transistor. A set of characteristic curves identifies several important parameters of a given transistor. For example, the set for a certain transistor might indicate maximum emitter-collector voltage, maximum collector current, and maximum base currents. Figure 5–1 shows a typical example of a set of characteristic curves for a bipolar transistor. Notice that the vertical axis is labeled $I_{C(\max)}$, and that the horizontal axis is labeled V_{CE}. The horizontal lines above the horizontal axis are the various base currents at which this transistor can operate.

On the graph in Figure 5–1, several **operating points** have been selected. At point *A*, the transistor is in cutoff, since point *A* shows no base current, no collector current, and no emitter-collector voltage drop. Point *B* is an operating point that gives several clues as to the operation of this transistor. Note that point *B* is located on the I_B = 60 μA line. This means that for this transistor, 60 μA of dc current will flow through the base. In addition, axes values for point *B* show that 8 mA of collector current will flow, and that a voltage drop of 10 V between collector and emitter will be developed. Point *C* on the graph identifies another area in which this transistor could operate. Note that if dc biasing were used to operate the transistor in this

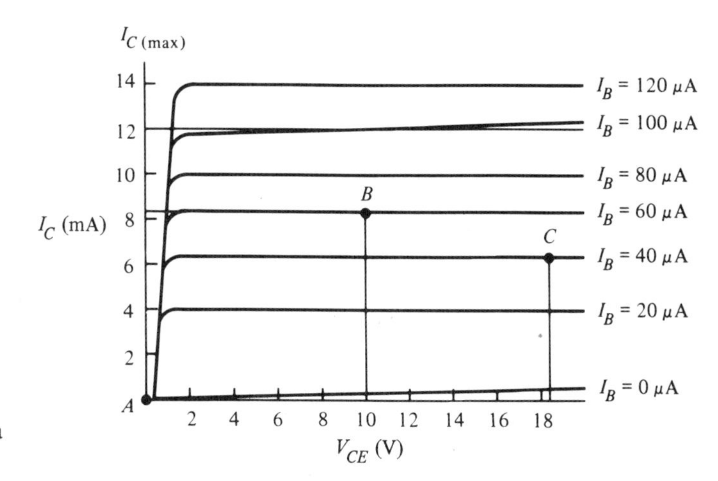

FIGURE 5–1
Characteristic Curves for a Transistor

region, a base current of 40 μA would be required. A collector current of about 6 mA would be developed, along with a collector-to-emitter voltage drop of about 18 V. Points *A, B,* and *C* are operating points that can be developed for this transistor. If the proper passive components were used to establish these operating points, the transistor could be used as an amplifier or a switch. If the transistor were biased so that it would operate outside of these limits, the usefulness of the transistor would be much less, and the life of the transistor would be shortened drastically. Limiting the transistor to operate within the set of characteristic curves will allow maximum response and life. This is one reason that exact replacement components must be used when repairing transistor circuits.

DC BIASING THE TRANSISTOR

As noted at the beginning of this chapter, there are three methods of applying the dc bias to the transistor. Each of these three configurations—common base, common emitter, and common collector—needs to be fully understood by the technician.

Common Base Biasing

Figure 5–2a shows an example of the **common base (*CB*) circuit.** The term *common base* means that the base is common to the input dc biasing loop and to the output dc biasing loop.

Note that two supplies are used for **common base biasing.** One supply is for the **emitter loop** (the emitter-base circuit). This supply is labeled V_{EE}. The other supply, labeled V_{CC}, is biasing the **collector loop** (the collector-base circuit). In both the emitter and the collector loops, resistors are used to develop necessary current flows and voltage drops for the operation of this transistor.

Figure 5–2b shows an example of a set of characteristic curves for this transistor. From this set of curves, a suitable operating point can be selected. According to the set of curves, the emitter can handle only 4 mA. The collector can handle only 4 mA also. The maximum voltage that can be developed between collector and base is only about 20 V. This graph has set limitations for this transistor.

Figure 5–3 shows an example of the components that make up the emitter loop, which is the first loop to begin the development of biasing. Note that this emitter loop is nothing more than a series circuit. This circuit consists of a power supply, V_{EE}; a series emitter resistor, R_E; and the base-emitter junction of the transistor. Because of the series circuit, Kirchhoff's law for voltage drop will apply. The formula for the emitter loop is given in Equation 5–1.

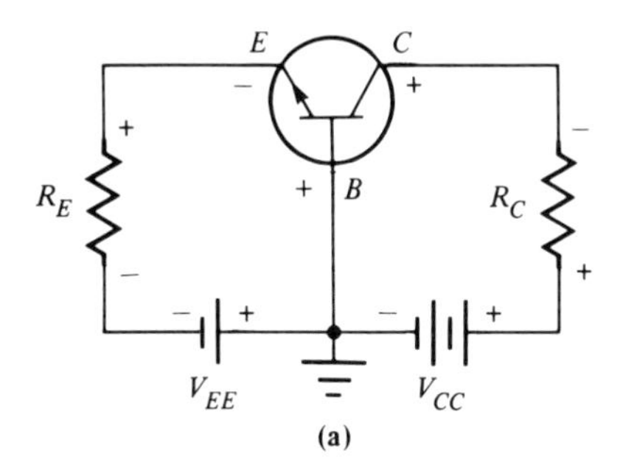

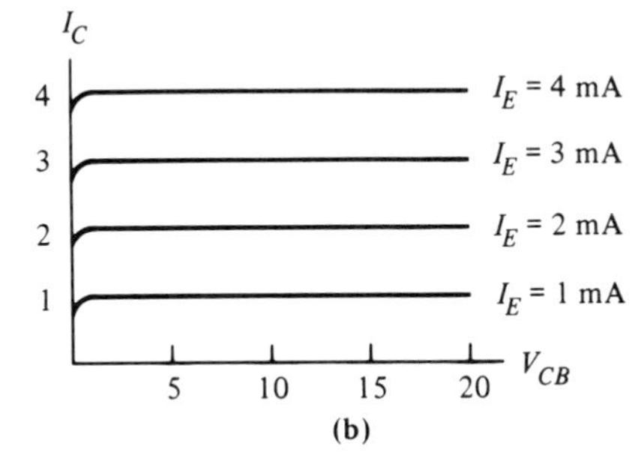

FIGURE 5–2
Common Base Biasing: (a) Schematic Connection of the Common Base, (b) Characteristic Curves for the Common Base

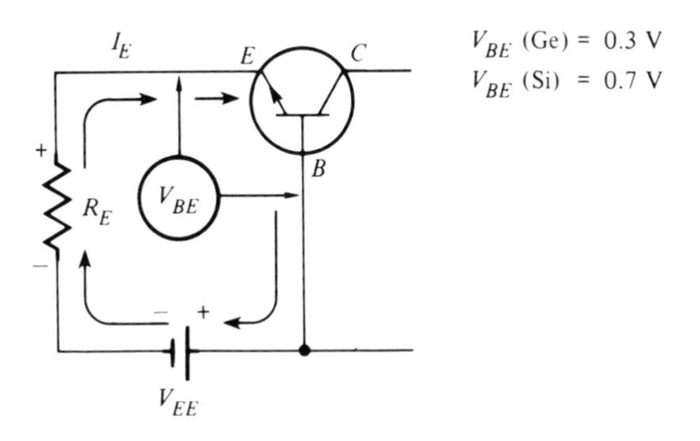

FIGURE 5–3
Base-Emitter Loop Biasing

$$0 = V_{EE} - V_{R_E} - V_{BE} \quad \textbf{(5–1)}$$

where:

V_{EE} = emitter voltage supply
V_{R_E} = voltage developed across emitter resistor
V_{BE} = voltage developed between base and emitter

To solve this equation, the voltage drop across the emitter resistor must be found. The base-emitter voltage used is the standard 0.7 V for silicon, or 0.3 V for germanium. A sample calculation will make use of this equation more understandable.

EXAMPLE 1

A silicon transistor has an emitter supply (V_{EE}) of 10 V. What is the voltage dropped across the emitter resistor?

Solution:

The known factors are placed into Kirchhoff's law, as follows:

$$0 = V_{EE} - V_{R_E} - V_{BE} = 10\text{ V} - V_{R_E} - 0.7\text{ V}$$
$$V_{R_E} = 10 - 0.7$$
$$= 9.3\text{ V}$$

Therefore, the voltage drop across R_E is 9.3 V.

Of course, Ohm's law could also be used to find the voltage drop across R_E:

$$V_{RE} = I_E \times R_E$$

The collector loop is the next step in this configuration. Notice in Figure 5–4 that the collector loop consists of a power source, V_{CC}; a load resistor for the collector, R_C; and the collector-base junction. Because these components of the circuit are connected in series, the collector loop voltage drop equation can be written. This Kirchhoff's law equation is given as Equation 5–2.

$$0 = V_{CC} - V_{R_C} - V_{CB} \quad \textbf{(5–2)}$$

where:

V_{CC} = collector voltage supply
V_{R_C} = voltage developed across collector load
V_{CB} = voltage across the collector-base junction

Solving this part of the circuit is not as simple as solving for the emitter loop. The only voltage source known is the supply. The voltage drops

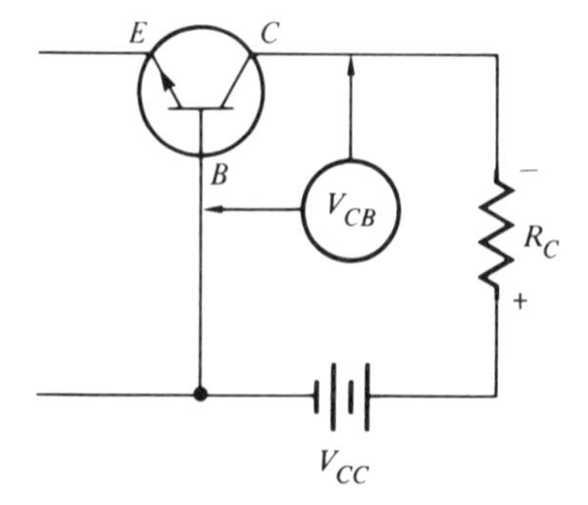

FIGURE 5–4
Collector Loop Voltage Drop across the Collector-Emitter Junction

across R_C and the collector-base junction are not known. Two methods can be used to find the unknowns in this circuit. One way would be to use a set of characteristic curves. The other way would be to use Ohm's law (the more practical approach).

The first step in solving for the unknowns in this collector loop using Ohm's law would be to identify the emitter current in the emitter loop. By identifying the emitter loop current, the collector current can be approximated. As has been stated, 100% of all current flows through the emitter, 95–97% is found in the collector, and the remaining 3–5% is found in the base. With this knowledge, the following approximation of current flow in the collector can be used:

$$I_C \approx I_E$$

Using this approximation, voltage drops around the collector loop can be found. Once these voltages are determined, they can be substituted into Kirchhoff's law and the equation can be solved. A sample problem will help illustrate this procedure.

EXAMPLE 2

See the common base circuit in Figure 5–5. Find the following dc biasing conditions:

1. Base-emitter voltage,
2. Emitter current,
3. Collector current,
4. Collector-base voltage.

Solution:

The following methods are used to find these dc biasing conditions.

1. The base-emitter voltage is equal to 0.7 V. This solution is simple: 0.7 V is the forward bias voltage drop for a silicon transistor.

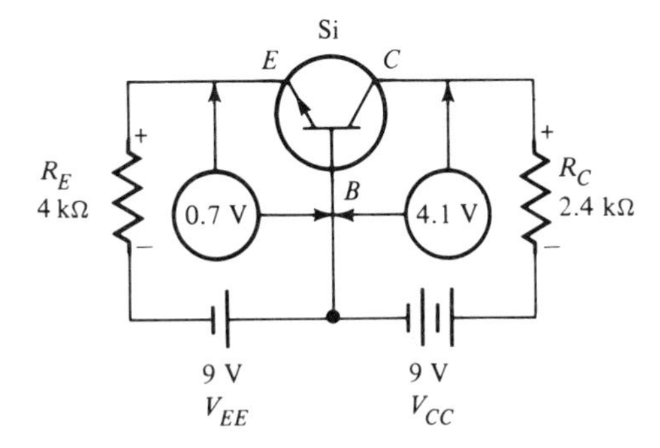

FIGURE 5–5
Common Base Circuit to Find Emitter and Collector Loop Voltages

2. The emitter current is found as follows:

$$I_E = \frac{V_{EE} - V_{BE}}{R_E} = \frac{9\text{ V} - 0.7\text{ V}}{4000\ \Omega} = \frac{8.3}{4000}$$
$$= 2.07\text{ mA}$$

Since the emitter loop is a series circuit, the current is constant throughout the loop.

3. $I_C \approx I_E$. Therefore, $I_C = 2.07$ mA.
4. The voltage drop across the collector-base junction can be found by the equation

$$V_{CB} = V_{CC} - V_{RC}$$

and since $V_{RC} = I_C \times R_C$, the equation can be written as follows:

$$V_{CB} = V_{CC} - I_C \times R_C = 9\text{ V} - (2.07\text{ mA} \times 2.4\text{ k}\Omega)$$
$$= 9 - 4.9$$
$$= 4.1\text{ V}$$

This simple equation should be used to find the voltage drops around the input and output loops of the common base circuit. The technician may never have to calculate the operating conditions, but to fully service the circuit, a complete understanding of biasing is a must.

Common Emitter Biasing

One of the most popular transistor configurations is called the **common emitter (*CE*) circuit.** Figure 5–6 shows an example of this type of

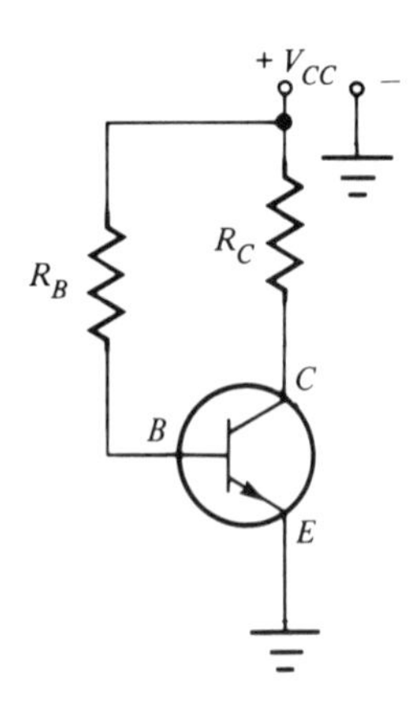

FIGURE 5–6
Biasing the Common Emitter Circuit

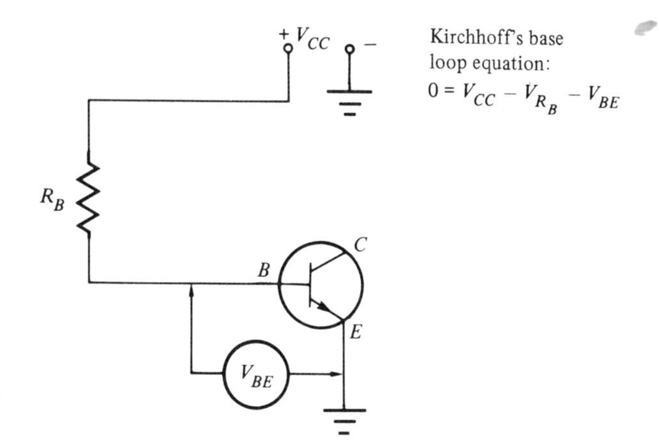

FIGURE 5–7
Base Loop Developed for the Common Emitter

circuit. Notice that in the common emitter, only one dc power supply is needed. With this one power source, the necessary forward bias for the base emitter junction, and the reverse bias for the collector-emitter junction, can be established.

As with the common base circuit, the two dc current loops are found in the common emitter circuit. Breaking down the circuit into these two basic circuits will make the **common emitter biasing** calculation simple, and will help illustrate where the voltages on the transistor leads are developed.

To develop the voltage drops on the **base loop,** a series circuit can be established. Figure 5–7 shows an example of the base loop. This series circuit consists of resistor in the base, R_B; a voltage drop between base and emitter, V_{BE}; and the voltage source, V_{CC}. Since these three components make up a series circuit, a form of Kirchhoff's law will apply. This equation can be identified as Kirchhoff's base loop equation, which is given as Equation 5–3.

$$0 = V_{CC} - V_{RB} - V_{BE} \qquad \textbf{(5–3)}$$

With knowledge of any two of the three values in this equation, the third can be found. Again, the standard voltage drops for the silicon and germanium transistors must be used. A sample calculation using this equation follows in Example 3.

EXAMPLE 3

A base dc biasing loop has a V_{CC} of 10 V. A resistor is placed in the base circuit and a silicon transistor is used. What is the voltage drop found across the base resistor?

Solution:

Solving this problem requires the use of Kirchhoff's base loop equation, as follows:

$$0 = V_{CC} - V_{RB} - V_{BE} = 10\text{ V} - V_{R_B} = 0.7\text{ V}$$
$$V_{R_B} = 10 - 0.7$$
$$= 9.3\text{ V}$$

Therefore, the voltage developed across the resistor in the base is 9.3 V.

Another method of finding the voltage drop in the base is by using the Ohm's law equation:

$$V_{R_B} = I_B \times R_B$$

However, to solve this equation the current in the base must be determined. The base current can be calculated from another Ohm's law equation:

$$I_B = \frac{V_{CC} - V_{BE}}{R_B}$$

This equation can be narrowed down because of the forward bias developed across the base-emitter junction. Because this junction is forward biased, it has very low resistance, and it develops only a small voltage drop. Therefore, the V_{BE} can be eliminated, leaving

$$I_B = \frac{V_{CC}}{R_B}$$

This new equation gives an approximation of base current. This approximation then is used to calculate the voltage drop across the base resistor. All such approximations can be verified with actual circuit voltage readings. Trouble in the circuit will show as a great difference between the calculated and measured values.

The common emitter has a collector loop that also must develop a dc bias. Figure 5–8 shows the collector loop of a common emitter circuit. This circuit is comprised of the power supply for the collector, V_{CC}; the load resistor for the collector, R_C; and the voltage drop between collector and emitter terminals, V_{CE}. With these components used to bias the collector loop, the following Kirchhoff's collector loop equation can be used:

$$0 = V_{CC} - V_{RC} - V_{CE}$$

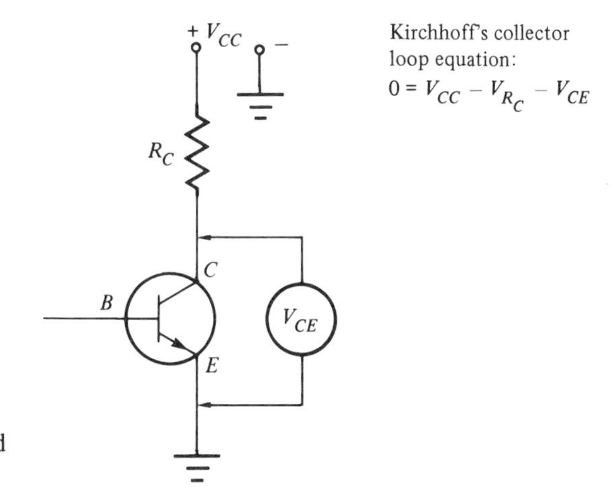

FIGURE 5–8
Collector Loop Developed for the Common Emitter

Again, with the knowledge of any two of the factors, the voltage in the loop can be found.

In biasing the circuit, it may become necessary to use Ohm's law equations to find voltage drops. For example, a voltage drop across the collector resistor can be found by

$$V_{RC} = I_C \times R_C$$

By substitution of known factors within this collector loop circuit, the voltage drop at R_C can be determined. Once this voltage drop is calculated, it can be placed into Kirchhoff's voltage drop law.

There is one drawback to using Ohm's law. To find the voltage drop across R_C, the current in the collector must be known. The current in the collector can be found in several ways. One method is to use the characteristic curves. Another is to measure the value. The third method is to remember that collector current relates to base current. If base current increases, so does collector current. As noted in Chapter 4, this ratio between the two currents is called beta. The relationship of collector current to beta is given as Equation 5–4.

$$I_C = \beta \times I_B \qquad \textbf{(5–4)}$$

where:

I_C = collector current
β = current gain
I_B = base current

The collector current determined by one of these three methods can be used to calculate the voltage drops around the collector loop.

The equations for finding voltage drops around the loops are only guidelines to insure that proper voltage drops are being developed. One important relationship must be remembered: the voltage drops around the loops must add up to the supply voltage. If they do not add up to the supply, there is trouble in the circuit.

Use of the equations for biasing the collector loop of the common emitter circuit is illustrated in Example 4.

EXAMPLE 4

Find the following for the common emitter circuit in Figure 5–9:

1. Base current,
2. Collector current,
3. Collector-emitter voltage.

Solution:

The calculations are as follows:

1. $I_B = \dfrac{V_{CC} - V_{BE}}{R_B} = \dfrac{V_{CC}}{R_B} = \dfrac{12 \text{ V}}{200 \text{ k}\Omega} = 60\ \mu\text{A}$
2. $I_C = \beta \times I_B = 50 \times 60\ \mu\text{A} = 3 \text{ mA}$
3. $V_{CE} = V_{CC} - I_C \times R_C = 12 \text{ V} - (3 \text{ mA} \times 1.5 \text{ k}\Omega)$
   $= 12 - 4.5$
   $= 7.5 \text{ V}$

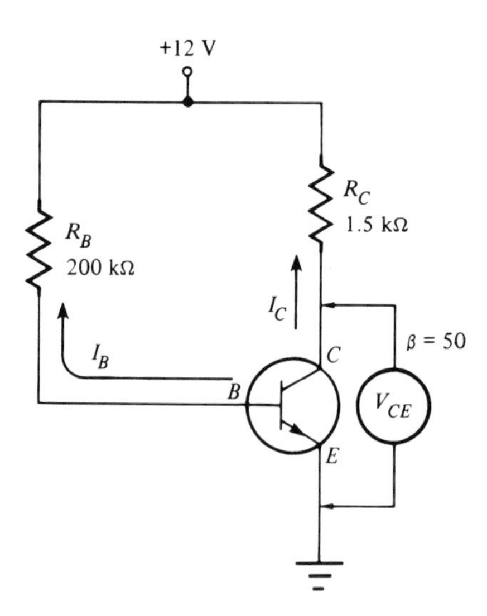

FIGURE 5–9
Common Emitter Biasing Solution

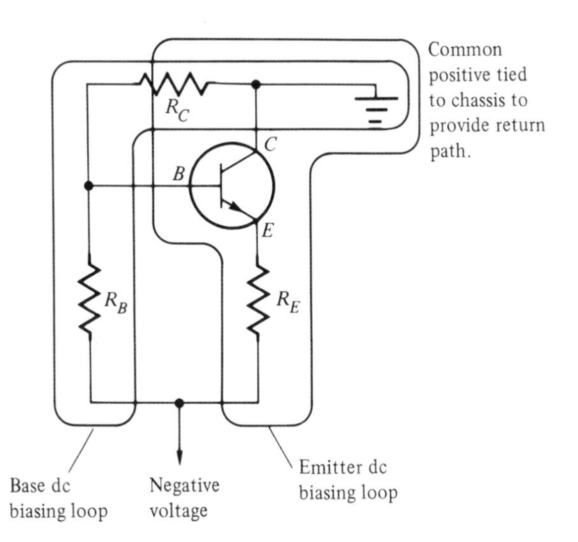

FIGURE 5–10
Biasing the Common Collector Circuit

Common Collector Biasing

The final type of biasing circuitry found in consumer products is the **common collector (*CC*) circuit.** One of the differences between the common collector and the common emitter circuits is the terminals of the transistor where the input signal is applied and where the output signal is developed. Note that the input terminals are the base and collector, while the output terminals are the emitter and collector. A sample of **common collector biasing** is shown in Figure 5–10.

Notice that in this drawing, the base loop consists of a base resistor, R_B, and the voltage drop between the collector and the base. The emitter loop is comprised of an emitter resistor, R_E, and the voltage drop between the collector and the emitter. The voltages around these two loops can be found in the same manner used for the common emitter and common base circuits. Generally, the same biasing rules apply to all three circuits.

BIAS STABILIZATION

Once the transistor has been biased to operate at a certain point, it is the function of the surrounding circuits to maintain this biasing point. In any transistor the operating point selected might shift. (The shifting might be the result of a temperature increase; with any semiconductor, heat will cause the operating point to change.) Three different circuits can be used to aid in **bias stabilization:** the emitter resistor, the voltage divider, and the collector feedback.

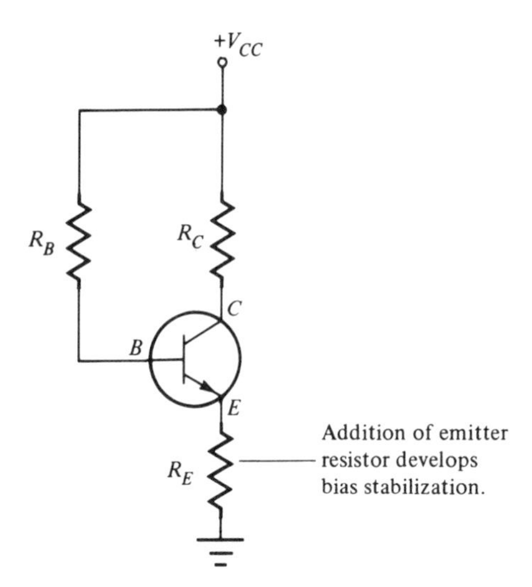

FIGURE 5–11
Emitter Resistor Used for Bias Stabilization

Emitter Resistor

Figure 5–11 shows an example of the first type of bias stabilization method. An **emitter resistor** has been added to the circuit. This resistor is used to maintain a constant current flow through the emitter. Since the emitter develops 100% of the current flow, placing a resistor in

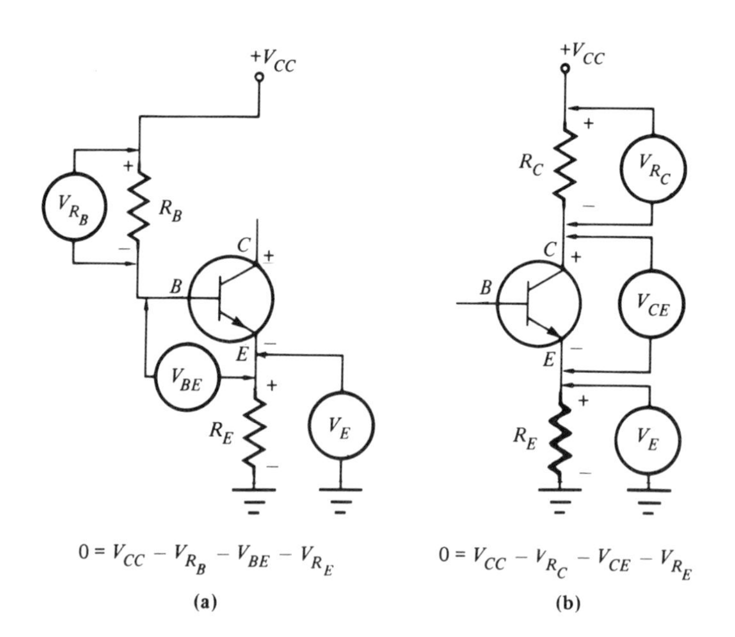

FIGURE 5–12
Common Emitter Configuration: (a) Base Loop Biasing, (b) Collector Loop Biasing and Voltage Drops

this location will keep the transistor stable under varying operating conditions.

Because an emitter resistor has been placed in the emitter leg, a voltage drop will be seen at the emitter. Since the emitter is part of the dc bias loops, this voltage drop must be taken into consideration. Figure 5–12a includes a Kirchhoff's dc bias loop equation for the base loop. Figure 5–12b contains the collector loop equation. The voltage drop across R_E figures into both loop equations.

The resistance of the emitter lead is an important part of keeping the transistor stabilized. A sample problem will illustrate how to calculate the value of this resistor.

EXAMPLE 5

Refer to Figure 5–13. This circuit provides the following known values:

$$V_{CE} = 10\text{ V}$$
$$\beta = 200$$
$$I_C = 2\text{ mA}$$

Solution:

With these known factors, the following Kirchhoff's collector loop equation for the collector loop can be used:

$$0 = V_{CC} - V_{R_C} - V_{CE} - V_{R_E}$$

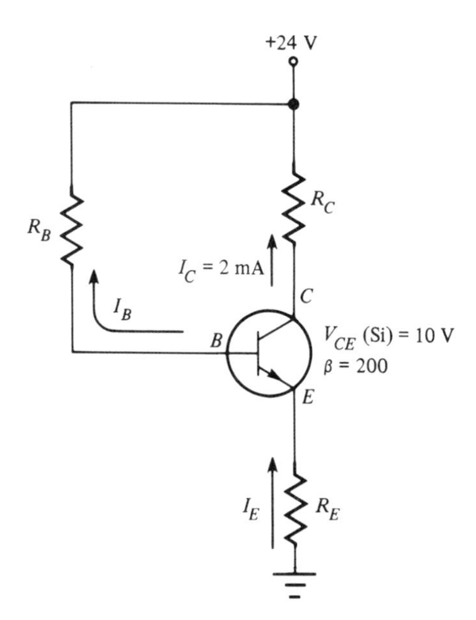

FIGURE 5–13
Calculating the Size of an Emitter Resistor

Since the voltage drop across the emitter resistor is to be found, this equation is rearranged to read:

$$V_{R_E} = V_{CC} - V_{R_C} - V_{CE}$$

Before this equation can be solved, a value is needed for V_{R_C}. An approximation of V_{R_C} is possible. It is known that a 10 V drop occurs across the collector-emitter terminals. To keep this transistor stable, it is assumed that the voltage drop across R_C equals V_{CE}. Since V_{CE} equals 10 V, then V_{R_C} is assumed to equal 10 V. Now, quantities can be substituted into the equation:

$$V_{R_E} = V_{CC} - V_{CE} - V_{R_C} = 24 \text{ V} - 10 \text{ V} - 10 \text{ V} = 24 - 20$$
$$= 4 \text{ V}$$

Therefore, the value of resistance in the collector can also be found using Ohm's law:

$$R_C = \frac{V_{R_C}}{I_C} = \frac{10 \text{ V}}{2 \text{ mA}}$$
$$= 5 \text{ k}\Omega$$

The resistance in the emitter can also be found. Again, Ohm's law can be applied. The emitter current is not known, but an approximation can be made. Because I_E and I_C are almost equal, the value for I_E can be approximated as 2 mA. The solution, therefore, is as follows:

$$R_E = \frac{V_{R_E}}{I_E} = \frac{4 \text{ V}}{2 \text{ mA}}$$
$$= 2 \text{ k}\Omega$$

This calculation means that a 2 kΩ resistor is placed in the emitter lead.

Voltage Divider

The second type of bias stabilization circuit is called the **voltage divider.** Figure 5–14 shows an example of this circuit. Notice that this circuit is very similar to the emitter-resistor stabilizing circuit. To increase the stability of this circuit, a resistor, R_S, has been added in parallel with the base-emitter and emitter-resistor circuit.

The connection of resistors R_B and R_S in Figure 5–14 develops a voltage divider network for the base. The voltage developed across

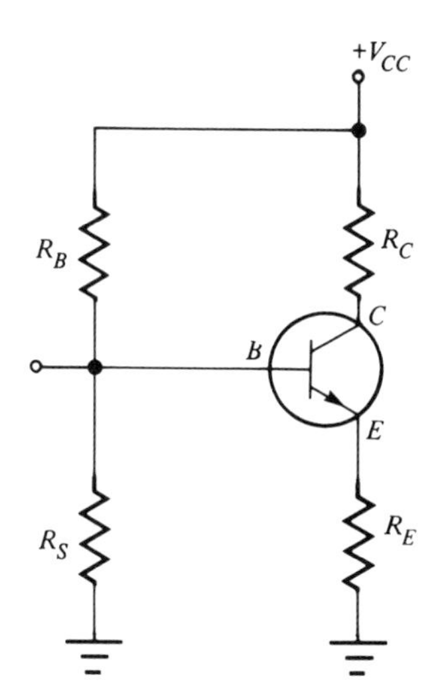

FIGURE 5–14
Voltage Divider Used as Bias Stabilization

R_S will be the voltage developed for the base. To calculate this base voltage (V_B), a simple voltage divider network formula can be used:

$$V_B = \frac{R_S}{R_B + R_S} \times V_{CC}$$

Once the base voltage has been found, the following simple equation can be used to find the emitter voltage:

$$V_E = V_B - V_{BE}$$

Therefore, the voltage at the emitter (V_E) is found by subtracting the standard base-emitter voltage drop from the base voltage. Again, the standard voltage is 0.7 V for silicon, and 0.3 V for germanium.

The amount of emitter current can be calculated using the following formula:

$$I_E = \frac{V_E}{R_E}$$

The collector current can be approximated by the emitter current, since:

$$I_C \approx I_E$$

Once the collector current is known, the voltage drop across the collector resistor can be calculated with the following form of Ohm's law:

$$V_{RC} = I_C \times R_C$$

Going one step further, the voltage at the collector can be found by using the output loop equation:

$$V_C = V_{CC} - V_{RC}$$

And, finally, the voltage between the collector and the emitter (V_{CE}) can be determined by

$$V_{CE} = V_C - V_E$$

Collector Feedback

The third type of stabilizing circuit operation is called the **collector feedback.** The action developed in this circuit is to couple a part of the collector voltage back to the base. An example of collector feedback biasing is given in Figure 5–15.

One of the main reasons for transistor instability is an increase in temperature. If the temperature in the semiconductor increases, there will be an increase in current through the transistor. If current increases, the voltage at the collector terminal will also drop. In the collector feedback circuit, this decrease in collector voltage is fed back to the base. Any decrease in voltage at the base decreases the forward bias on the transistor. Reducing forward bias action decreases collector current, therefore increasing the collector voltage. This action tends to keep the transistor voltage at a very stable operating point.

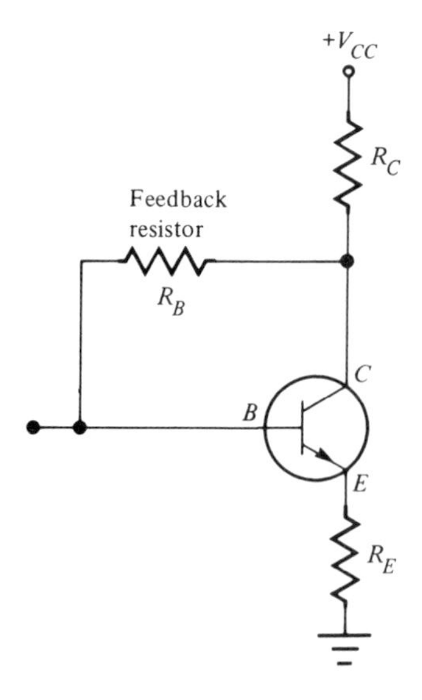

FIGURE 5–15
Collector Feedback Developing Bias Stabilization

TROUBLESHOOTING

DC BIASING

The service technician will deal with many different types of transistor biasing circuits. However, in general, any biasing operation can be broken down into one of the three basic biasing circuits covered in this chapter. When trouble appears in the dc biasing network, voltage measurements must be taken to find the defective component in the circuit.

The circuit that the technician will deal with most often when servicing transistor circuits is the common emitter. For this reason, the operation of the common emitter will be stressed here. The troubleshooting procedures presented will also apply to the common collector and common base circuits.

Correct Voltages—NPN Type

The NPN transistor requires the proper bias voltage at the collector, base, and emitter terminals in order to operate. The transistor base-emitter junction must have a forward bias voltage applied. This means that the base must be more positive than the emitter. A reverse bias voltage must be applied to the collector. When these voltages are present, and are separated by the proper voltage levels, the transistor is ready for operation.

Figure 5–16 shows an example of an NPN silicon transistor with

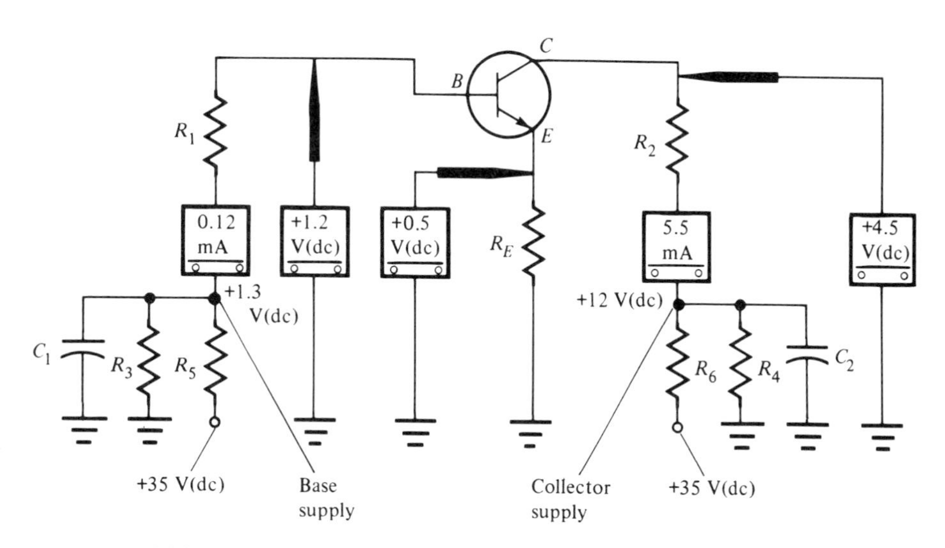

FIGURE 5–16
Common Emitter Configuration Using an NPN Transistor

the proper dc biasing. Notice that the base voltage is 0.7 V higher than the emitter voltage, and that the collector is more positive than the emitter. Also note that R_3, R_4, R_5, and R_6 are used as voltage divider networks to drop the 35 V source to usable base and collector voltages. In addition, R_1 and R_2 are used to bias the base terminal and the collector terminal, respectively. These voltages found around the NPN transistor are established to keep this transistor operating at a stable point, and operating within its linear region between cutoff and saturation. Any voltage measurement that deviates from the normal reading means problems within the biasing circuit.

Correct Voltages—PNP Type

Figure 5–17 shows a PNP transistor with the proper dc biasing applied. Notice that each of the supply voltages is of negative polarity. As in the NPN transistor, R_3, R_4, R_5, and R_6 are used as voltage dividers to drop the dc biasing source to a usable operating level. Resistors R_1 and R_2 are again used to develop the necessary base and collector voltages. Most important, however, is the fact that the terminal voltages of the transistor are of negative polarity. The negative polarity is a result of the correct biasing of this transistor. Because the base is negative, it must have at least a 0.7 V difference for silicon, and a 0.3 V difference for germanium, in order to operate. If these voltages are found around the transistor,

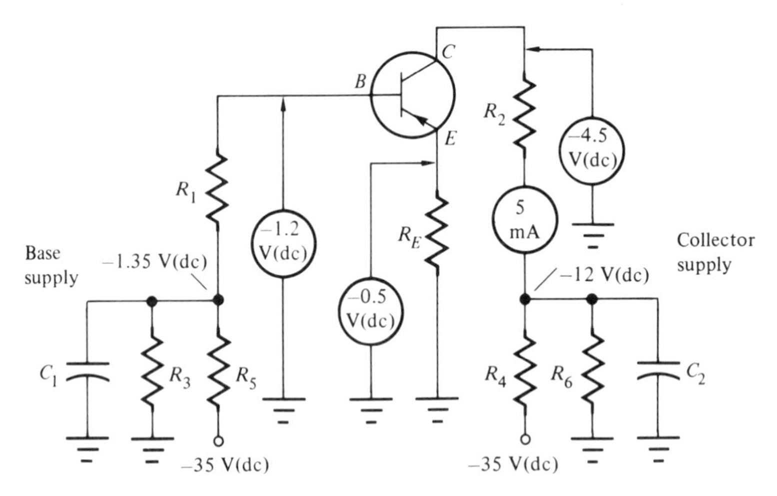

FIGURE 5–17
Common Emitter Configuration Using a PNP Transistor

the device is operating within its linear region, and under stable conditions.

An example of biasing for the PNP transistor with a positive dc supply is shown in Figure 5–18. Note that proper biasing is achieved for the base, emitter, and collector, but that these voltages are higher than those found when using the negative power source.

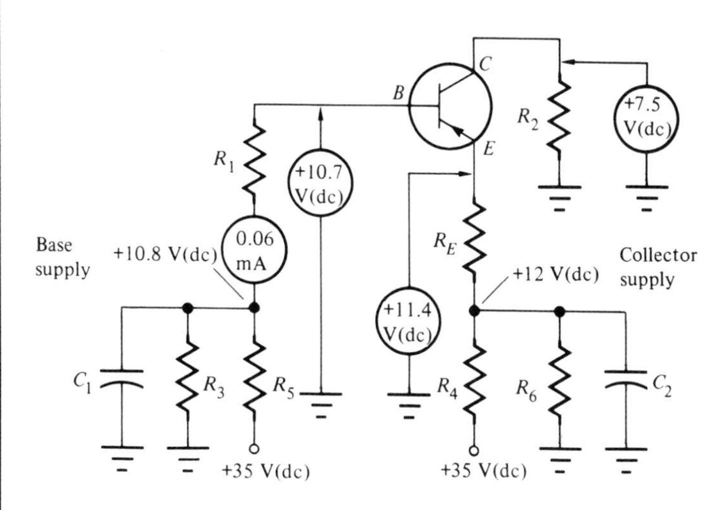

FIGURE 5–18
Biasing a PNP Transistor Using a Positive dc Supply

The troubleshooting information in the sections that follow pertains to the NPN transistor, a popular component of consumer electronic products. Because the NPN and PNP transistors are biased by similar methods, the troubles noted and the tests proposed also apply to PNP transistor operation.

Incorrect Voltages

Figure 5–19 shows the NPN transistor connected in the common emitter biasing mode with incorrect voltages. The key to troubleshooting the transistor circuit is to read the base-emitter voltage drop. If the voltage drop is present, and the base and emitter are separated by their 0.7 V or 0.3 V difference, then the transistor is operational. In Figure 5–19, however, both base and emitter read zero. This means that forward bias has been removed. Removing this forward bias has placed the transistor in cutoff. Whenever the

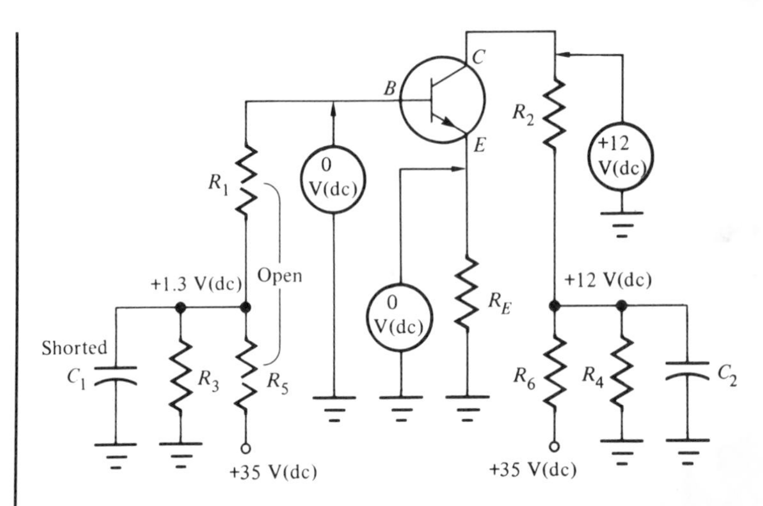

FIGURE 5–19
Troubles in an NPN Transistor Circuit

transistor is placed in cutoff, the collector voltage will increase to the collector supply voltage.

Under these conditions, the technician should make checks within the base biasing line. If R_5 or R_1 were to open, there would be no base conduction path. Also, if C_1 were shorted, the 35 V supply would have a direct path to ground, causing the base voltage to drop to zero.

This problem is easy to diagnose. No voltage on the base and emitter and higher voltage on the collector means that the transistor is in cutoff. A good rule to remember when troubleshooting these circuits is as follows: *Abnormal base voltage with respect to ground means trouble in the base circuit.*

Open Emitter

Figure 5–20 shows the emitter circuit with the emitter resistor opened. When the base voltage is measured to ground, it is found to be normal. Thus, there is no problem in the voltage divider circuitry of R_1, R_3, and R_5. Measuring the emitter voltage might cause the technician to assume that the stage is conducting. However, the clue here is that the emitter voltage is the same as the base voltage. The internal resistance of the base-emitter junction has caused this emitter voltage. Some transistors will cause a lower reading on the emitter. Lower voltage on the emitter might lead the technician to think that the transistor has a light forward bias. However, if the voltmeter is placed between base and emitter, no

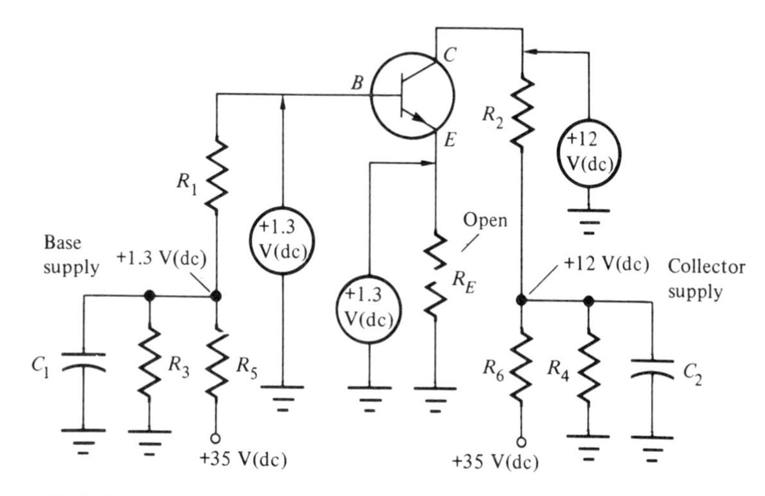

FIGURE 5–20
Open Emitter Resistor

difference will be seen. This means that the problem is in the emitter leg.

The final measurement should be made at the collector. The collector, when measured, shows 12 V. Note that the dc voltage on the collector terminal has risen to the collector supply voltage. This indicates that the transistor is cut off. Therefore, no current conduction is developed through the transistor. Also, note the base voltage. This 1.3 V value indicates that there is no conduction through the base-emitter junction. Therefore, there must be an open in the emitter since no current is flowing through the base or the collector terminals. The problem is an open emitter resistor, R_E.

Collector Circuit Defects

Figure 5–21 shows an example of a defective collector circuit. Measuring the dc voltages gives some interesting values. Base voltage is about normal, and the base-emitter voltage has about the correct forward bias. These measurements have ruled out the base or the emitter as the problem. Remember, there is still a current path between the emitter and the base.

The 0.06 V on the collector is the clue. When the voltage on the collector drops below the base voltage, the base-collector junction becomes forward biased. This forward bias action develops the proper voltage drop. The collector has dropped to the emitter voltage value.

An open collector circuit can be suspected when all the voltages

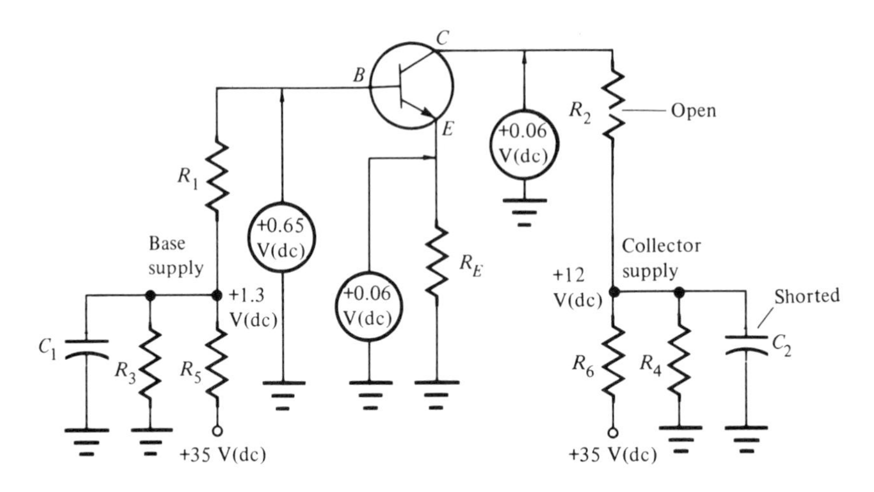

FIGURE 5–21
Defective Collector Circuit

on the transistor's terminals are about the same, and the base reading is normal.

Open Transistors

Often the problem is within the transistor itself. Measurements around the transistor might lead to troubleshooting resistors and capacitors that develop supply voltage. An open base lead inside the transistor will cut off the transistor. No current will flow to develop a voltage at the emitter resistor. Figure 5–22 shows the voltages measured at the transistor leads. Since the transistor is cut off, the collector voltage will increase to the collector supply value.

The key to troubleshooting this problem is that the voltage between the base and emitter is good. There is no drop in either base or collector supply, and no voltage drop signifies an open on the external base lead of the transistor. The fact that the emitter has no voltage means that there is an open in the base-emitter junction resistance. If the emitter resistor were open, the emitter would assume the base voltage, but the reading at the emitter here is zero. The following is a good rule of thumb for such problems: *When the transistor has good forward bias and still does not conduct, the open is inside the transistor.*

An open collector lead inside the transistor will produce normal base voltages when read from ground. Forward bias will be close to

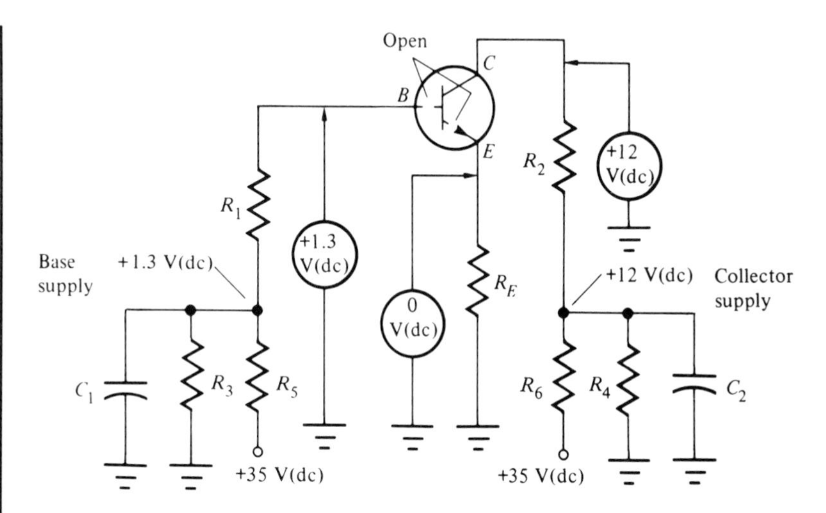

FIGURE 5–22
Open Lead Either in the Emitter Circuit or the Base Circuit

normal. The key is that the collector voltage has risen to the supply voltage value. This should not happen if the transistor is conducting. A small voltage drop will appear across the emitter resistor. Again, the base-emitter junction is conducting current. Figure 5–23 shows a schematic diagram with the open collector.

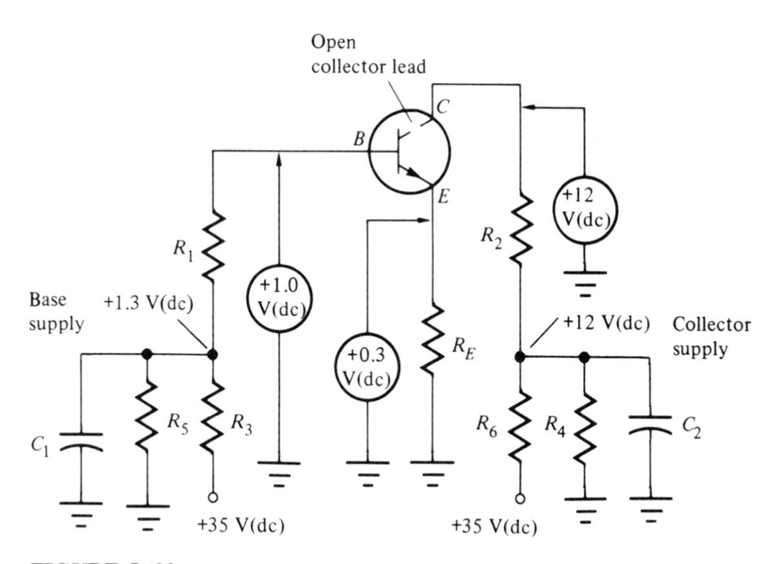

FIGURE 5–23
Open Internal Collector Lead

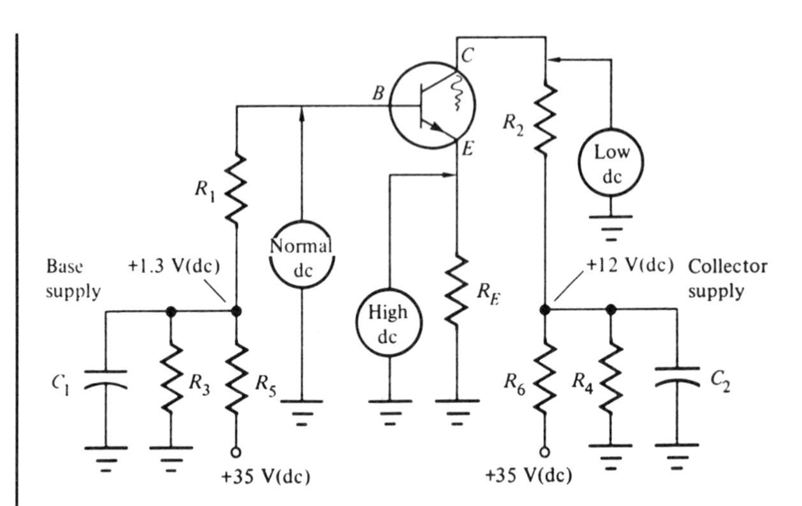

FIGURE 5–24
Leaky Collector-Emitter Junction

Leaky or Internally Shorted Transistor

The transistor's base-collector junction is reversed biased. Recall that when the transistor is reversed biased, the junction should block a good portion of current flow. Only a small amount of leakage current flows. Oftentimes, however, this leakage current increases beyond normal limits for the transistor. This condition characterizes a leaky transistor.

Figure 5–24 shows the typical voltage readings that might be found on a leaky transistor. Notice that the base voltage remains fairly constant, but that the emitter voltage has risen. This means that more current is being conducted through the transistor.

The key to the leaky condition is the low dc voltage reading at the collector. When the transistor conducts heavy collector current, the emitter voltage will rise. If this happens, the emitter resistor may exceed its wattage rating and burn open.

Biasing Multitransistor Circuits

Figure 5–25 shows an example of two transistors connected together to form a biasing network. Both transistors Q_1 and Q_2 are arranged in a common emitter form, and are directly coupled to each other.

The interesting feature of this circuit is that an NPN and a PNP transistor are being used. They operate from a negative ground source (refer to Figure 5–18). The base-emitter forward bias for Q_2

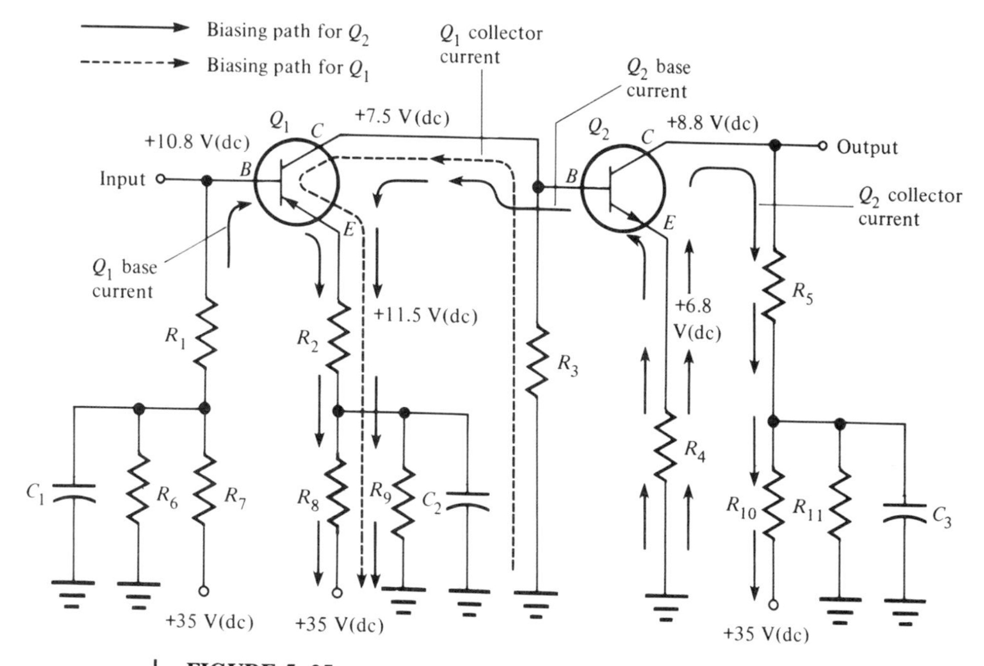

FIGURE 5–25
Biasing Two Transistors from a Single Supply

is derived from proper conduction of Q_1. What happens in Q_1 biasing will directly affect Q_2 operation. This two-stage amplifier operates as a single unit.

A quick check can be made to trace the current paths of the circuit. The following three points will make tracing the circuit easy:

1. Electron flow is always in a loop (a circuit),
2. Electrons always rush to the most positive point in the loop,
3. The point at which the electrons enter a component is considered to be negative, and the point at which they leave the component is considered to be positive.

Because electron flow (negative to positive) applies in Figure 5–25, the action in this circuit can be traced by using these three circuit-tracing rules. Current flow starts in the base of Q_1. The electrons are attracted to the emitter because of the higher dc voltage. The current flows through R_2 and R_8, then to the 35 V supply. The loop is completed from voltage source to ground, and then back to R_1 through the R_6 connection.

Collector current for Q_1 is established through R_3. The current then flows through the collector-emitter junction to R_2, and then completes its path to the 35 V source through R_8. The loop is completed back to ground through the source. The dotted lines trace this path.

Transistor Q_2 has no base voltage-divider network. Instead, a current path is established from ground side of R_4. The current moves through R_4, and forward biases the base-emitter junction. The current then flows through Q_2's base-emitter junction, and becomes part of the collector current flow of Q_1. It then passes through Q_1's collector-emitter junction, and back to the 35 V source. The arrows designate this current flow. Collector current for Q_2 flows from ground in R_4 through the emitter-collector junction of Q_2 and back to the 35 V source via R_5 and R_{10}. The circuit is completed through the ground side of the power source.

Being able to trace the biasing paths in a multisection transistor network is a must for the technician. This skill will lead to quick, effective troubleshooting.

A final, important point to remember is that *current flow is always to the most positive connection.* All these positives and negatives in a circuit might become confusing, but measuring between elements will show the correct polarity. All voltage measurements in an electronic circuit are referenced against ground.

SUMMARY

One of the most important factors related to the operation of the transistor is its proper biasing network. A correctly biased transistor is in its static state. Static operation, which will be discussed in Chapter 6, means that the transistor has a fixed operating point.

Biasing the transistor sets the operating point. The operating point is selected from the set of characteristic curves. The characteristic curves detail the maximum ratings of the transistor: the maximum collector currents, base currents, and collector-emitter voltage drops.

Transistors can be connected in three general biasing conditions: common base, common emitter, and common collector. Biasing for the common base condition requires two power supplies, one for forward biasing the base-emitter junction, and one for reverse biasing the collector-base junction. The common emitter and common collector circuits require only one power source.

A biasing network can be divided into two loop circuits. These loop paths are series circuits. Because of this, Kirchhoff's voltage law applies. The transistor requires two loops: the base (base-emitter) loop, and the collector (collector-emitter) loop.

Variations of Ohm's law can be applied to the transistor circuit to find values of voltage, current, and resistance.

To keep the transistor in a stable condition, three methods of dc stabilization are used. These are the emitter resistor, the voltage divider, and the collector feedback.

Troubleshooting biasing problems requires only quick voltage measurements of the base circuit, the emitter circuit, and the collector circuit. Voltage measurements on these three elements will identify an open transistor lead or a leaky junction.

Sometimes transistors are directly coupled. The biasing is accomplished by current flow from negative to positive.

Tracing a circuit shows that the point at which electrons enter a component is negative, and the point at which electrons leave the component is positive. Electrons always flow to the most positive point.

KEY TERMS

base biasing loop
bias(ing)
bias stabilization
characteristic curve
collector biasing loop
collector feedback
common base biasing
common base circuit
common collector biasing
common collector circuit
common emitter biasing
common emitter circuit
emitter biasing loop
emitter resistor
operating point
voltage divider

REVIEW EXERCISES

1. Where should the operating point of a transistor be selected for linear operation?

2. What important information does a set of characteristic curves contain?

3. Refer to Figure 5–1. If the collector-emitter voltage were reduced from 10 V to 6 V, and base current were held constant, what would be the collector current?

4. What are the two biasing loops found on a common base configuration?

5. Using the circuit shown, locate and identify the V_{EE} supply voltage.

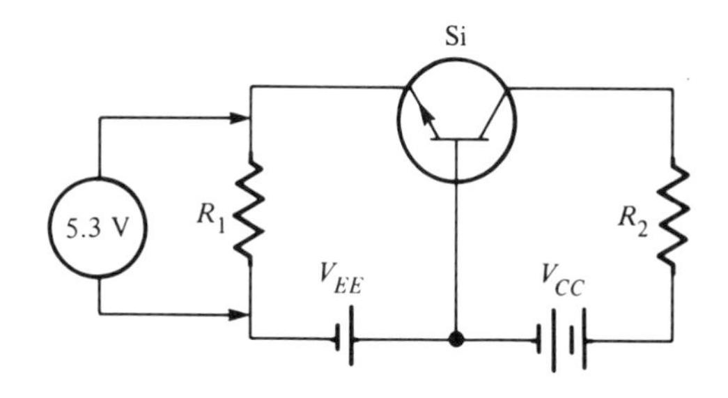

6. A common base biasing circuit develops 6 mA of emitter current. What is the approximate collector current?

7. What is the name of the most popular transistor biasing configuration?

8. Why does Kirchhoff's voltage law apply to the voltage drops in biasing loops?

9. Write the emitter loop equation for the common base circuit in Exercise 5.

10. Write the base and collector loop equations for the common emitter circuit in the accompanying figure.

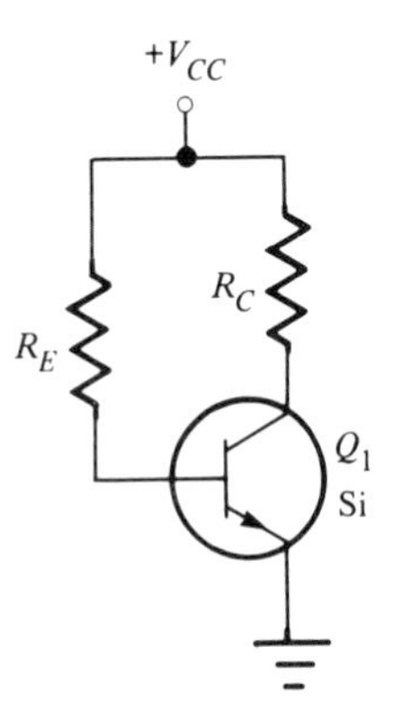

11. A transistor has a beta of 50, and a base current of 10 μA. What is the collector current?

12. Why might the operating point of a transistor shift?

13. Draw schematic diagrams of several methods of bias stabilization.

14. A transistor develops 2 V at the emitter, and an emitter current of 2 mA is established. What size emitter resistor is placed in the circuit?

15. If voltages developed on the base circuit are incorrect, which circuit requires troubleshooting?

16. Voltages on the collector and emitter read close to each other, and the base voltage reads normal. Which circuit is at fault?

17. If a transistor develops good forward bias and still does not conduct, where is the problem?

18. Base voltage reads close to normal, and the emitter voltage is increased. What is the likely problem?

19. Which polarity is established on components where the electrons enter?

20. What is the importance of an emitter resistor?

21. Describe the action of collector feedback used to stabilize transistor operation.

6 AMPLIFIER PRINCIPLES

OBJECTIVES

Upon completing this chapter, you should be familiar with:

—Basic amplifier gain principles
—Decibel application to gain
—Amplifier characteristics
—Amplifier operation classes

INTRODUCTION

The use of the **amplifier** in electronic products is very important. The main function of the amplifier is to increase signal strength. In other words, the amplifier makes the output signal larger in amplitude than the input signal. This increase in amplitude is called the **gain** of the amplifier.

To provide gain within the circuit, active devices, or components, are used. These components could be vacuum tubes, transistors, or integrated circuits. The latter two are used most often in consumer products.

The consumer electronics technician must be able to do simple calculations to understand the relationship of gain in amplifiers, and also must be able to relate this gain to the standard decibel scale. All these concepts are presented in this chapter.

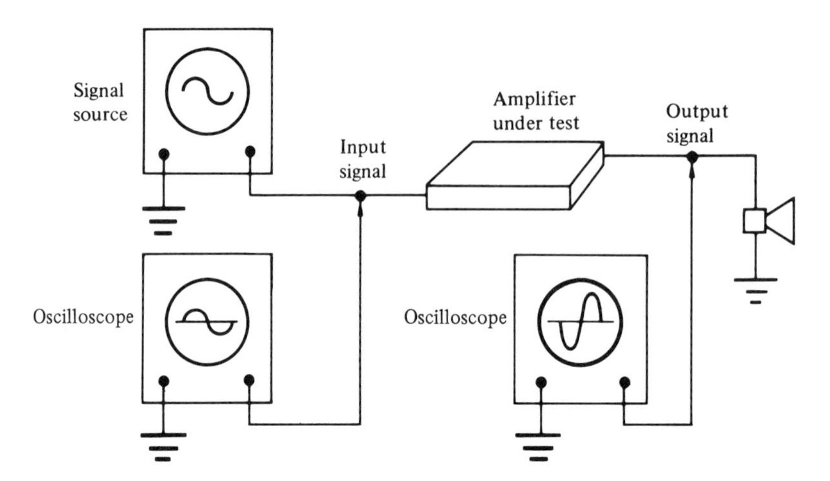

FIGURE 6–1
Measuring Gain of an Amplifier with Signal Phase Shift

AMPLIFIER GAIN

As stated, the basic function of amplifier circuits is to produce gain. Gain is the comparison of the amount of signal entering the active component to the amount of signal leaving that component. Because of the active component's ability to increase signal, the output signal will be larger.

To insure that the amplifier is supplying the proper gain, the gain must be measured. An example of measuring the gain of an amplifier is given in Figure 6–1. Note that the amplifier under test has a fixed-amplitude signal injected into the input. An oscilloscope is used to measure this amplitude. The output of the amplifier is connected into a load, in this case, the speaker. An oscilloscope is used to measure the output of the amplifier. Notice that the amplifier has caused a 180° phase shift of the signal. The gain of this amplifier can be calculated by using Equation 6–1.

$$\text{gain} = \frac{\text{output signal}}{\text{input signal}} \qquad \textbf{(6–1)}$$

Example 1 illustrates the application of this formula to find gain.

EXAMPLE 1

An amplifier has an input signal of 0.5 V(p-p), and the output signal measures 10 V(p-p). What is the gain of this amplifier?

Solution:

The values can be placed into Equation 6–1, as follows:

$$\text{gain} = \frac{\text{output signal}}{\text{input signal}} = \frac{10\ \text{V(p-p)}}{0.5\ \text{V(p-p)}}$$
$$= 20$$

Therefore, the gain of this amplifier is 20. Notice that in the equation, the peak-to-peak units of measurement cancel out each other. That is, it is not correct to describe this amplifier as having a gain of 20 V.

Amplifiers can be divided into two general operational classifications: small-signal amplifiers and large-signal amplifiers. Generally, the *small-signal* amplifier produces a voltage gain, whereas the *large-signal amplifier* generates a power gain. If the amplifier in Figure 6–1 were a large-signal amplifier, the gain of that stage (block, or gain block) could be calculated by the same formula given in Equation 6–1. The only difference would be the replacement of voltage with wattage, as shown in Example 2.

EXAMPLE 2

A large-scale amplifier has a 1 W input, and a 50 W output. What is the gain here?

Solution:

The gain for this stage is calculated as follows:

$$\text{power gain} = \frac{\text{output power}}{\text{input power}} = \frac{50\ \text{W}}{1\ \text{W}}$$
$$= 50$$

Again, the units cancel out each other, and the gain is given as 50 for this amplifier.

The Decibel

In the early years of the telephone, a unit of measurement was developed to show the loss of signal in two miles of telephone wire. This basic unit was called the *bel.* Because the bel was a large unit, a subunit was defined to accommodate the small workings of electronics. This subunit is the **decibel (dB).** Decibels are based on logarithms.

Decibels are used in electronics to describe a change between two measurements. This change may be in voltage, power, sound loudness, or sound pressure. In the case of the amplifier, the decibel can show gain. For example, an amplifier might have a +20 dB change.

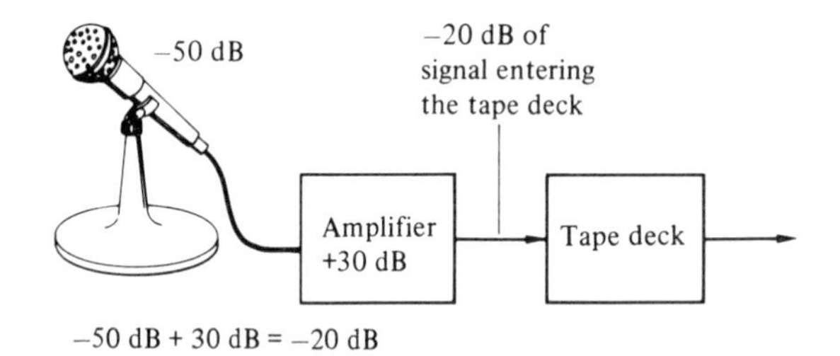

FIGURE 6–2
Signal Measured in Decibel Format

This value tells the service technician of an increase between input and output. On the other hand, a microphone might show a −10 dB change. This value means a reduction in signal between input and output. Therefore, the decibel is a handy tool for identifying amplifier gains and different component losses. Figure 6–2 shows what type of signal might be expected at the input of a tape deck.

Decibels of Voltage and Power As stated before, the decibel is used to describe a relationship between an input and an output. This relationship could result in an increase between input and output, stated as a plus dB change, or a reduction between the input and output signal, stated as a minus dB change.

Because the decibel relates a change, it can be used to describe the input and output of various components found in consumer products. For example, the change developed between the input signal and the output signal of an amplifier could be described as a decibel change, and the signal difference between the input and output of an antenna could be described as a decibel change. The standard used throughout the consumer product industry for signal measurement is 1 V = 0 dB. The term used in many audio equipment measurements is *dBV*. The dBV is a measurement of the increase or decrease in voltage that is independent of impedance levels. The decibel unit can be added or subtracted to develop a total change throughout the entire amplifier stages.

Power change between various stages in amplifiers can also be related to the decibel. In the communications industry, the standard unit for the decibel is the amount of change developed in a 600 Ω line when 1 mW is applied. This standard is the *volume unit* (*VU*) used in tape recorders and other audio equipment.

Often for audio equipment, the ratings of different stages are given in decibels. To insure that the stage is functioning properly, the decibel value might have to be converted to a voltage or power unit. Table 6–1 is a chart that can be used to convert decibel values into voltage and power gains. These simple calculations will aid the technician in proper servicing and adjustment of the consumer product. Examples 3 and 4 serve as practice problems for these conversions.

TABLE 6–1 Voltage and Power dB Conversion

Power		Voltage	
Loss	**Gain**	**Loss**	**Gain**
−1 dB = 0.79	1 dB = 1.26	−1 dB = 0.89	1 dB = 1.12
−3 dB = 0.50	3 dB = 2.00	−3 dB = 0.71	3 dB = 1.41
−10 dB = 0.10	10 dB = 10.00	−10 dB = 0.32	10 dB = 3.16
		−20 dB = 0.10	20 dB = 10.00

EXAMPLE 3

An amplifier has a 12 dB gain. What is the power gain for the unit?

Solution:

Reference to the power chart in Table 6–1 shows no entry for 12 dB. Therefore, 12 dB must be broken down into several parts, the values of which can be found in the chart. In this case, the following breakdown is used:

$$12 \text{ dB} = 10 \text{ dB} + 1 \text{ dB} + 1 \text{ dB}$$

From Table 6–1:

$$\begin{aligned} 10 \text{ dB} &= 10.00 \\ 1 \text{ dB} &= 1.26 \\ 1 \text{ dB} &= 1.26 \end{aligned}$$

By multiplying all these factors together, a total power gain can be calculated for the amplifier:

$$10.00 \times 1.26 \times 1.26 = 15.876$$

Therefore, the amplifier will increase the power between input and output by a factor of about 16.

EXAMPLE 4

A filter causes a −9 dB change at 5 kHz. Measurement of the output at 1 kHz shows the output voltage to be 1 V. When the frequency is increased to 5 kHz, the signal drops to 0.35 V. Is the filter working properly?

Solution:

To solve this problem, the −9 dB must be broken down:

$$-9 \text{ dB} = -(3 \text{ dB}) + (-3 \text{ dB}) + (-3 \text{ dB})$$

Then, using the voltage chart in Table 6–1, the solution is as follows:

$$-9 \text{ dB} = 0.71 \times 0.71 \times 0.71 = 0.35$$

Multiplying 1 V times the −9 dB change gives the output voltage:

$$1 \times 0.35 = 0.35 \text{ V}$$

According to the specifications stated in this example, the filter has properly reduced the signal's voltage level.

The technician should become familiar with the voltage and power chart shown in Table 6–1. Knowledge of this chart will help the technician determine whether the consumer product under test is operating within its specifications.

Current Change

One of the most important characteristics of an active device is the ability to regulate current flow. The base current of a transistor can control the amount of current flow through the collector. The capacity of the smaller current flow to control the larger current flow is the basic concept of amplification.

In Chapter 4, beta was identified as the relationship of current change in the collector to current change in the base. With the proper biasing applied, the transistor will operate within this range. However, in an amplifier the goal is to take a small ac waveform at the input of the transistor and change it to a large waveform at the output. This action through the active device is illustrated in Figure 6–3. The amplifier is developing a 180° phase shift between the input and the output.

To insure proper amplification in an active device, a set of characteristic curves is used to identify the operating range of the transistor. As was mentioned in Chapter 4, this set of characteristic

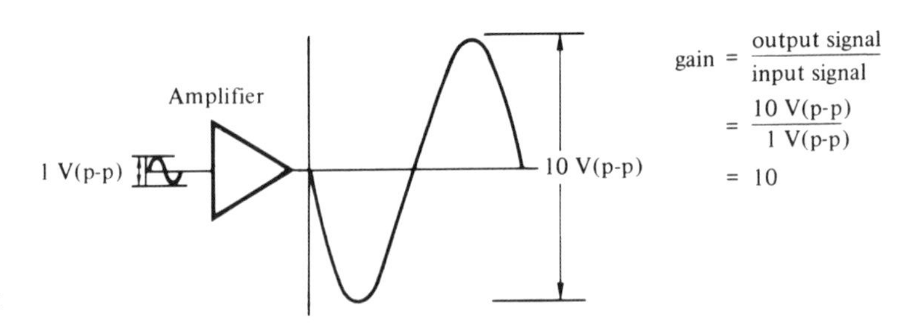

FIGURE 6–3
Active Device Used to Change Signal Strength between Input and Output

curves gives the maximum values for that particular transistor. At some point on the set of curves, an operating point is chosen for the device, and biasing is established. The operating point must fall within linear limits, or the amplifier will develop distortion.

Load Line Development

To insure against distortion of the signal, the amplifier is designed to operate within a set of characteristic curves. Figure 6–4 shows a set of characteristic curves for a typical silicon transistor. To operate the amplifier in the linear section, a **load line** is drawn on the characteristic curves. The development of the load line is based on two assumptions. The first is that the transistor is saturated. The second is that the transistor is in cutoff. The first assumption means that maximum current is flowing in the collector. This maximum current can be found by Ohm's law, as follows:

$$I_{C(\max)} = \frac{V_{CC}}{R_2} = \frac{12\text{ V}}{1\text{ k}\Omega} = 12\text{ mA}$$

Since the transistor is saturated, the voltage drop across the collector-emitter junction is assumed to be zero. Therefore, the first point is established at $I_C = 12$ mA, $V_{CE} = 0$. This point, point X, is plotted on the graph in Figure 6–4.

The second assumption means that there is no flow of collector current, and all the V_{CC} is dropped across the collector-emitter junction.

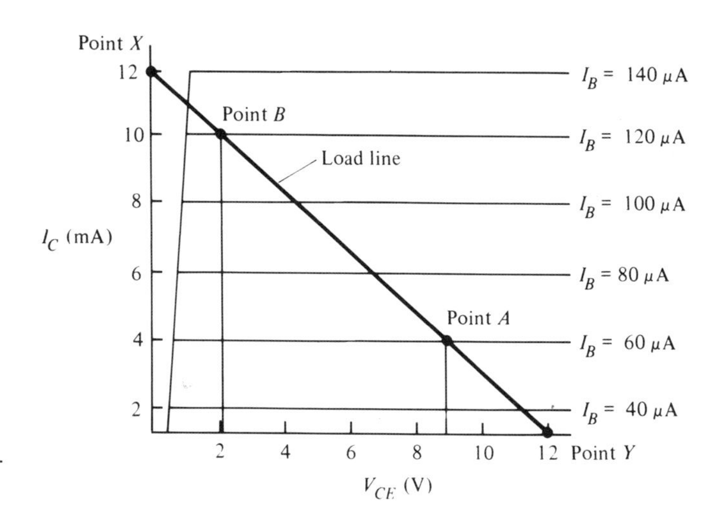

FIGURE 6–4
Set of Characteristic Curves for a Silicon Transistor

Therefore, $I_C = 0$ mA, and $V_{CE} = 12$ V. This point, point Y, is also plotted on the graph in Figure 6–4.

Points X and Y are then connected. The line formed is the load line. The amplifier can be operated at any point along this line. For example, if the amplifier base current is set at 120 μA, values for V_{CE} and I_C can be read from the graph. Point B in Figure 6–4 designates this operating point. From the graph, $V_{CE} \cong 2$ V and $I_C = 10$ mA. The operating point can be changed. Assume that base current is set at 60 μA. The operating point has moved to point A. This means that $V_{CE} \cong 9$ V, and $I_C = 4$ mA.

The cutoff and saturation regions are also shown in Figure 6–4. The cutoff area is where the base current falls below 40 μA, and the saturation level is the point at which $I_C = 12$ mA and $V_{CE} = 0$ V. These two areas must be avoided if the amplifier is to operate in the linear region. The best location for amplifier operation is the center region.

Figure 6–5 shows the movement of the operating point up and down the load line, and the resultant effect on the output signal. In Figure 6–5a, note that the operating point for the amplifier has been placed at the center of the load line. The input signal is centered at the operating point. The positive and negative swing of the input signal causes current to flow in the collector, and thus produces a voltage between the collector and emitter. The input signal keeps the transistor amplifier operating between cutoff and saturation. The output signal is not distorted.

Figure 6–5b identifies what happens to the output signal when the operating point is moved toward saturation point on the load line. The input signal causes the biasing on the transistor to change. This change causes the amplifier to be overdriven and results in a clipped, distorted waveform.

Figure 6–5c identifies another condition that can develop on the transistor amplifier. This state shows an operating point at the center of the load line. This location of the operating point should keep the transistor amplifier operating in the linear region, but the amplitude of the input signal is too large for the amplifier to handle. This means that the signal causes the transistor amplifier to be biased into saturation and cutoff. This will result in those peaks that overdrive the amplifier to be cut off. This action develops an output waveform that is clipped and, therefore, distorted.

Figure 6–5d shows another type of movement of the operating point on the load line. This time the operating point is moved toward the cutoff region of the amplifier's operating region. When the input signal is developed around this operating point, it forces the transistor into cutoff. This will then cause the transistor amplifier to cut off a part of the input signal. Note that in the drawing, the bottom part of

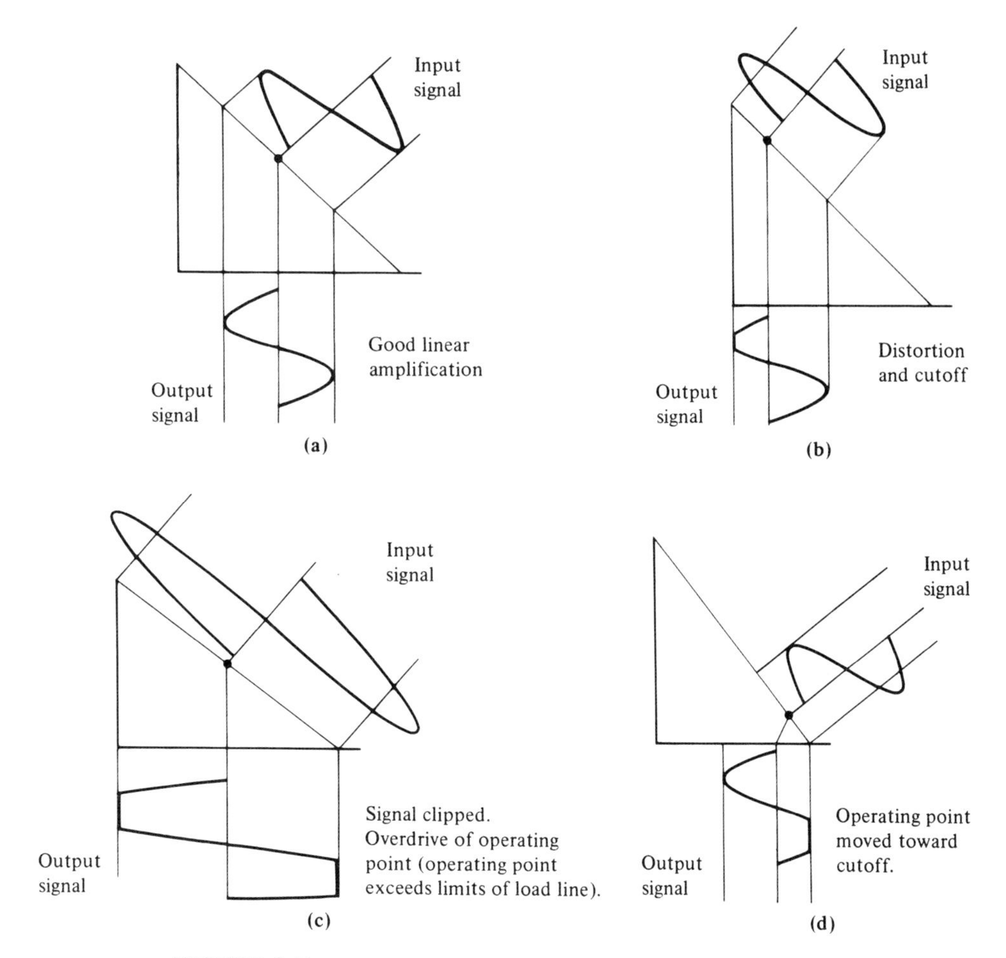

FIGURE 6–5
Movement of Operating Point: (a) Linear Region Develops Linear Output Signal, (b) Changing Operating Point Develops Distortion, (c) Increasing Input Signal Amplitude Develops Distortion, (d) Operating Point Moved Toward Cutoff Develops Distortion

the input signal is cut off in the output. In some cases the operation of the amplifier will be biased at this location near cutoff. This point is called operating the amplifier in the class B region.

The load line provides information that the technician can use when troubleshooting amplifier sections of the consumer product. The intent here is not to make the service technician the design engineer. However, understanding how the amplifier is designed aids in the repair of that device.

AMPLIFIER CONFIGURATIONS

Common Emitter Amplifier

Figure 6–6 shows a typical schematic diagram of the **common emitter amplifier.** This amplifier is so named because the emitter is common to both the input circuit and the output circuit. The input signal is applied across the base-emitter leads of the transistor, while the output signal is taken from the collector and the emitter. Resistors R_1 and R_2 are dc biasing resistors used to operate the active device. Capacitor C_1 is the input coupling device, and capacitor C_2 is the output coupling device. Capacitor C_1 is used to block any dc voltage from the signal source, and to block a current flow from the generator to the V_{CC} source. If the resistance in the signal source is lower than the resistance of the base-emitter junction, current will stop flowing through the transistor, placing it in cutoff. Current through the signal source is shown in Figure 6–7.

The transistor must be biased properly before it can become an amplifier. This biasing is called the **static state** of the transistor. Figure 6–8 shows a typical common emitter amplifier. Enough information is given about the amplifier to find base current, collector current, and voltage drops in the collector. Simple Ohm's law calculations given in Chapter 5 will apply here.

The current through the base is calculated as follows:

$$I_B = \frac{V_{CC} - V_{BE}}{R_1} = \frac{12\text{ V} - 0.7\text{ V}}{250\text{ k}\Omega}$$
$$= 45.2\ \mu\text{A}$$

With this information, the current in the collector can be determined.

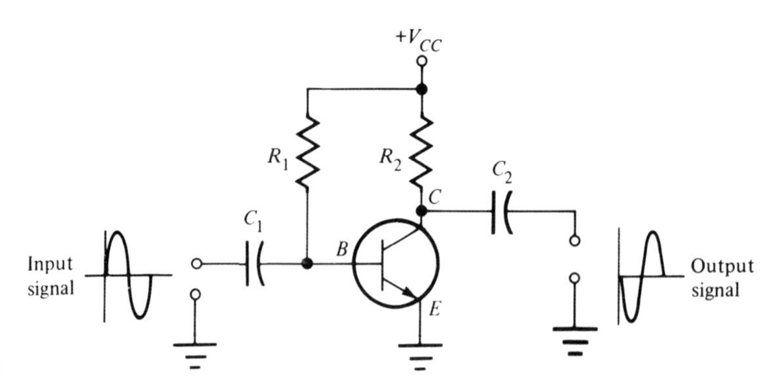

FIGURE 6–6
Common Emitter Amplifier

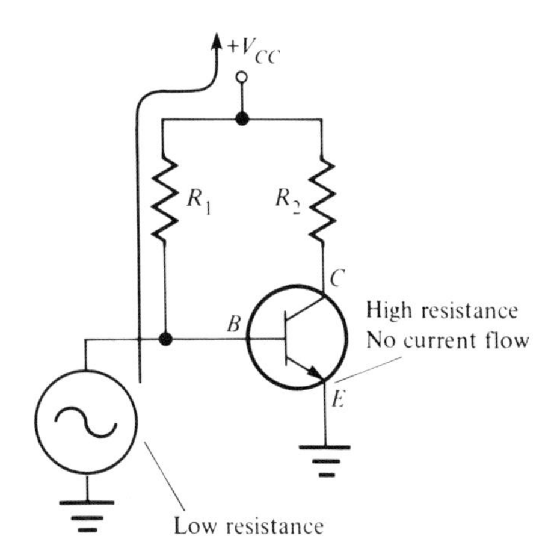

FIGURE 6–7
Low-Resistance Path for a Signal Source

Note that the transistor in Figure 6–8 has a beta of 50. Therefore,

$$I_C = \beta \times I_B = 50 \times 45.2\ \mu\text{A}$$
$$= 2.26\ \text{mA}$$

The voltage drop across R_2 can then be calculated from the following Ohm's law expression:

$$V_{R_2} = I_C \times R_2 = 2.26\ \text{mA} \times 2.1\ \text{k}\Omega$$
$$= 4.74\ \text{V}$$

The voltage between collector and emitter can be found as follows:

$$V_{CE} = V_{CC} - V_{R_2} = 12.0\ \text{V} - 4.74\ \text{V}$$
$$= 7.26\ \text{V}$$

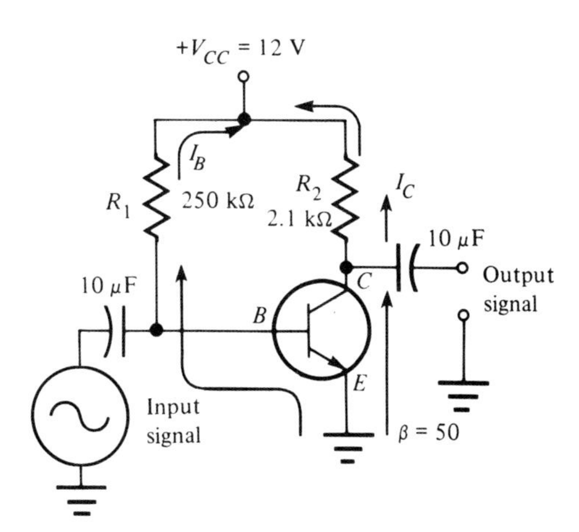

FIGURE 6–8
Biasing for Static Operation of a Transistor Amplifier

Voltage drops at the transistor leads are found by using Ohm's law's basic formulas and Kirchhoff's voltage drop law. The transistor in Figure 6–8 is in its static state. Now, an ac signal can be applied to the amplifier.

In Figure 6–8 the signal is applied to the base of the transistor. Since the signal is a varying ac voltage, it causes a change in base current. This change in base current then causes the collector current

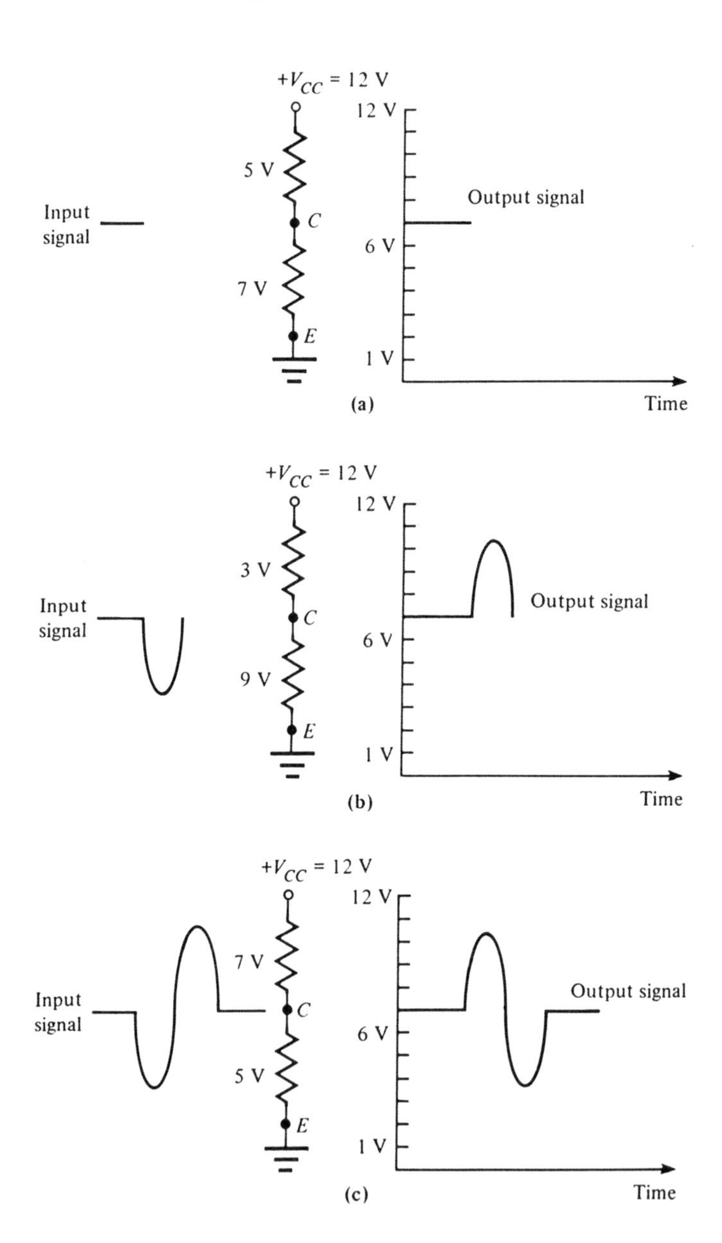

FIGURE 6–9
Input Signal Effect on Voltage Drops in Collector: (a) No Input Signal, (b) Negative Input Signal, (c) Positive Input Signal

to vary. This results because the base current controls the collector current. As the collector current changes, the voltage drops across R_2 and across V_{CE} are constantly changing. This presence of varying current in the collector characterizes the transistor as a variable resistor.

Figure 6–9a shows the two resistances in the output loop circuit. During the first time period, the voltage drops across R_1 and V_{CE} are shown. In Figure 6–9b, the input signal swings negative. This negative swing causes the base current to change, pushing the transistor toward cutoff. With the decrease of base current, a decrease in collector current results. Therefore, it is said that the resistance of the collector circuit has increased; and when the resistance increases, the voltage drop will also increase. Notice that the voltage drop across R_2 has decreased to 3 V, and that the voltage at the collector has increased to 9 V. As the base voltage starts to swing positive, the transistor state begins to change. The positive base voltage swing causes the transistor to conduct more current. This heavy conduction through the collector-emitter junction means that the resistance of the junction has decreased. This lower resistance signifies a lower voltage at the collector terminal. This action is shown in Figure 6–9c.

As indicated in Figure 6–9, there has been a **phase shift** between the input and the output. Between these two points, the signal has shifted 180°. This 180° phase shift is a very important characteristic of the common emitter amplifier circuit, as will be shown later in the chapter.

When the signal on the output is an amplified replica of the input signal, the amplifier is said to be a **linear amplifier.** A linear amplifier is illustrated in Figure 6–10a.

A **nonlinear amplifier** is given in Figure 6–10b. The nonlinear amplifier is characterized by an output signal that is not an amplified replica of the input signal. When a nonlinear waveform appears in audio amplifiers, it causes distortion in the signal, which ends up at the speaker. This distortion may be caused by one of two conditions within the amplifier. First, the volume may be set too high. If so, the

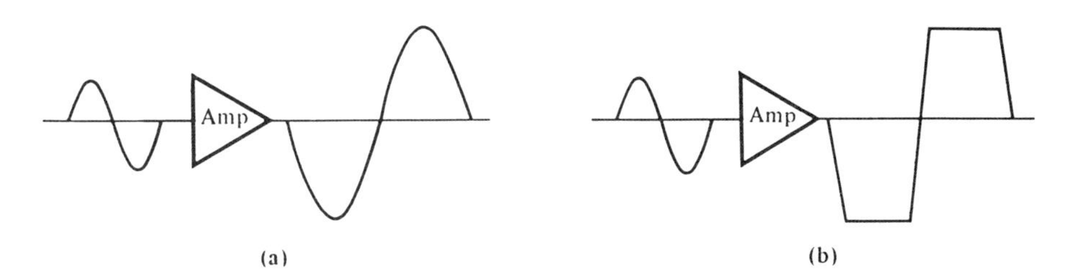

FIGURE 6–10
Amplifier Conditions: (a) Linear, (b) Nonlinear

input signal is reaching the limits of the amplifier, causing it to go into cutoff or saturation. Second, the biasing network may have caused the transistor to reach cutoff or saturation. The waveform in Figure 6–10b illustrates that the positive and negative alterations have been clipped, cut off at the top and the bottom.

Common Collector Amplifier

Another type of amplifier is the **common collector amplifier.** The schematic diagram for this device is given in Figure 6–11. At first glance, this amplifier looks almost like the common emitter amplifier. In fact, the dc current paths are the same as the current paths for the common emitter amplifier. However, there are some important differences. The first is that the input signal of the common collector is developed between the base and the collector, while the output is taken from the collector-emitter terminals. Also, notice that the load is in the emitter lead, and that a bypass capacitor is placed off the collector to ground. This capacitor places the collector at signal ground. Because the output signal is taken from the emitter, there is no phase inversion between input and output signals. This amplifier also gives no voltage gain. That is, the input signal will always be larger than the output signal. However, since the output signal is taken from the emitter, this amplifier provides good current gain. There is also a moderate power gain in this amplifier.

Because of the lack of phase inversion between base and emitter, the common collector amplifier is sometimes called an **emitter follower.** As the base voltage goes positive, so does the emitter voltage. In other words, the emitter voltage *follows* the base voltage.

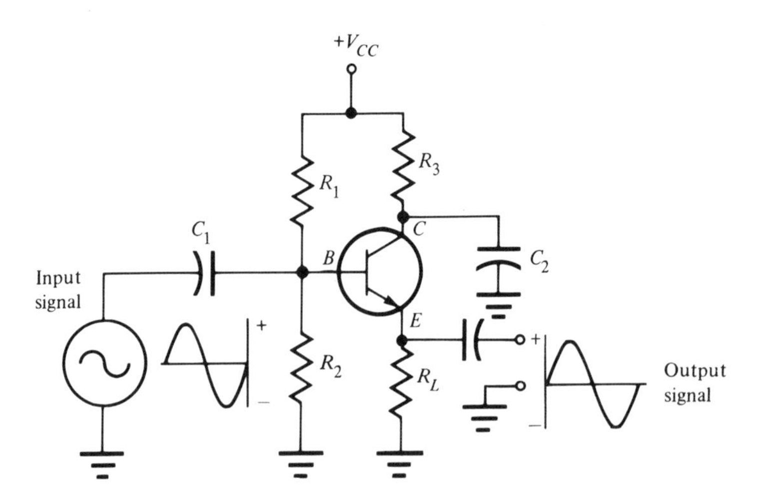

FIGURE 6–11
Schematic Diagram of Common Collector Amplifier

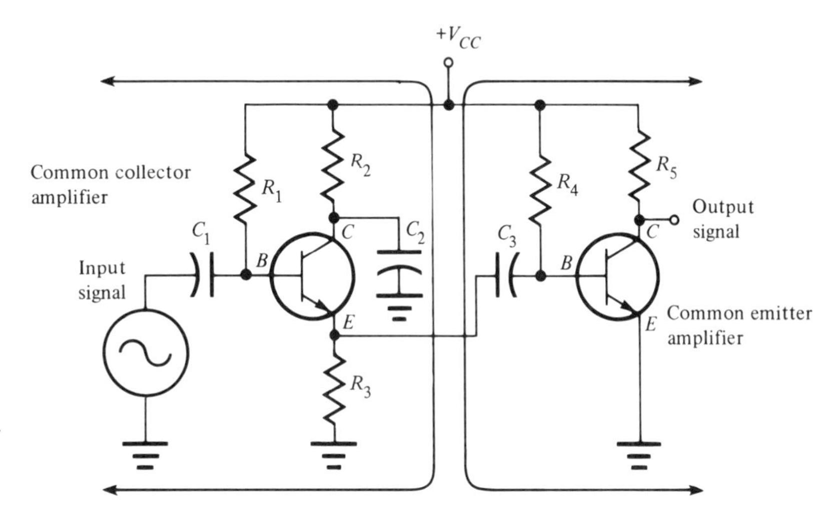

FIGURE 6–12
Common Collector Amplifier Used as an Isolation Amplifier

Another important characteristic of the common collector amplifier is its high input impedance. The input develops several thousand ohms to the input signal. If the signal is very high in amplitude and can withstand a large input impedance, the common collector might be a good selection. However, if the signal is low in amplitude, all of the signal will be dropped at the input, and therefore none will occur at the output. Because of this, the common collector is often used as an **isolation amplifier.** Its high impedance loads the input signal and causes very little current to flow. This then causes the signal to be isolated from the other stages. Generally, when a common collector is used in this way, a common emitter amplifier follows to provide good gain. Figure 6–12 shows an example of this connection.

Common Base Amplifier

The final configuration for the amplifier is the **common base amplifier.** Figure 6–13 shows a typical schematic diagram for this amplifier type. Again, it should be noted that the dc current paths are the same as for the common emitter and common collector amplifiers. However, several characteristics separate this amplifier from the common emitter and common collector amplifiers. One is that it has a low input impedance, but a high output impedance. The input terminals of the forward biased base-emitter junction develop the low input impedance, while the reversed biased output of the collector-base junction causes the high output impedance. This amplifier performs well in the high-frequency range. Its voltage gain is high, but its current gain is less than one.

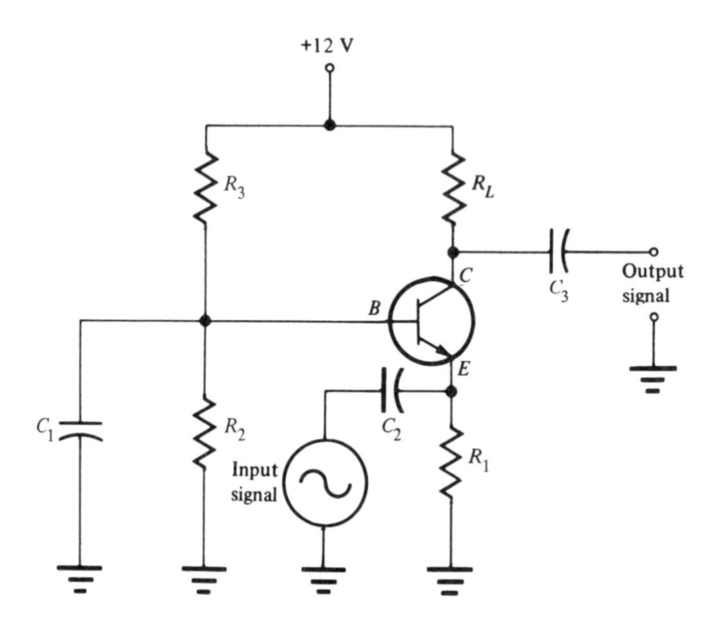

FIGURE 6–13
Schematic Diagram of a Common Base Amplifier

Because of its high-frequency operation and low input impedance characteristics, the common base amplifier makes an excellent preamplifier for televisions and FM tuners.

The chart in Table 6–2 is a summary of the characteristics of the common emitter, common collector, and common base amplifiers.

TABLE 6–2 Chart Comparison of Three Types of Amplifiers

Comparison Characteristic	Amplifier Type		
	Common Emitter	**Common Collector**	**Common Base**
Power gain	Good (highest)	Good	Good
Voltage gain	Good	No (less than 1)	Good
Current gain	Good	Good	No (less than 1)
Input impedance	Medium (about 1 kΩ)	Highest (about 300 kΩ)	Lowest (about 50 Ω)
Output impedance	Medium (about 50 kΩ)	Lowest (about 300 Ω)	Highest (about 1 MΩ)
Phase shift	Yes	No	No
Application	Universal; used in many consumer circuits	Used mainly as an isolation circuit	Used mainly in high-frequency applications

AMPLIFIER IMPEDANCE CHARACTERISTICS

Amplifiers have an important characteristic: impedance. In Chapter 1 *impedance* was defined as the total opposition offered to an ac signal. **Input impedance** is the loading effect the amplifier will present to a signal source. Figure 6–14 is an example of a signal source connected to an amplifier. When the input signal source is connected to the amplifier, the signal source sees the amplifier as a load, not as an amplifier. The load seen by the signal source is the input impedance of the amplifier.

As was mentioned earlier, different amplifier designs will offer different impedances to the signal source. Therefore, just as amplifiers offer different input impedances, so also do signal sources. For example, a microphone might have an impedance of 100 kΩ, whereas an antenna might offer an impedance of only 50 Ω.

In Chapter 1 the theorem of **maximum power transfer** was introduced. This theorem holds for amplifiers. If the maximum signal power is to be transferred from the output of the signal source to the input of the amplifier, the impedance of the signal source and the amplifier must be made equal. This process of making the impedances equal is called **impedance matching.**

The importance of having equal impedances for the load and the signal source is illustrated in the three parts of Figure 6–15. In Figure 6–15a, Z_G is identified as the impedance offered by the signal source, and Z_L represents the impedance offered by the amplifier circuit. Also, note the signal source produces an output signal equal to 10 V. The two impedances are connected in series and therefore can be added. This will result in a total impedance of 10 Ω for the

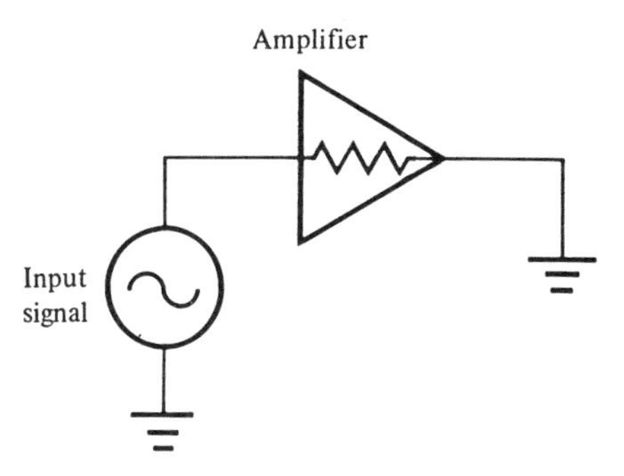

FIGURE 6–14
Load Delivered to Signal Source

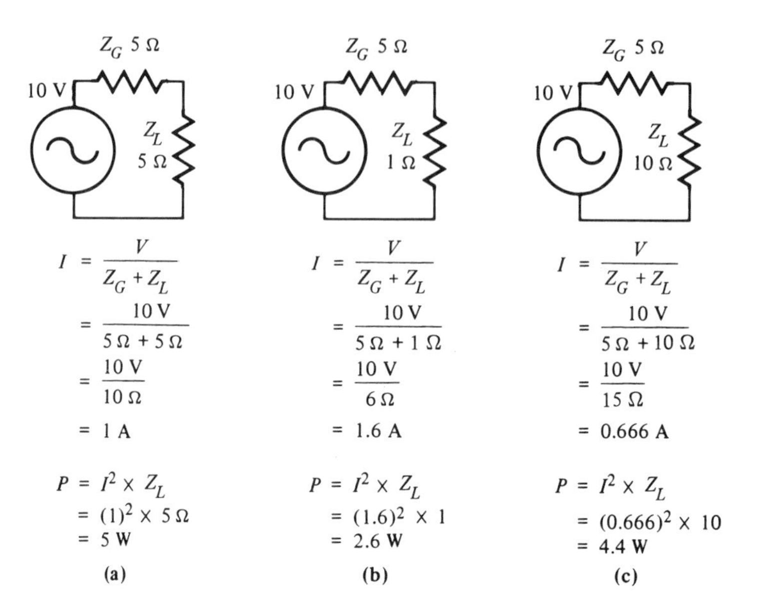

FIGURE 6–15
Impedance Matching of Load to Signal Source: (a) Load Impedance Matches Generator Impedance, (b) Load Impedance Is Less than Generator Impedance, (c) Load Impedance Is Greater than Generator Impedance

circuit. The simple Ohm's law calculation shows that a total current of 1 A is developed in this circuit. Power developed in the load can now be calculated, using the equation $P = I^2 \times Z_L$. Solving this equation, the load power is found to be 5 W. In Figure 6–15b, note that the load impedance has decreased. The current for this circuit has increased to 1.6 A. The power for the load, 2.56 W, has decreased. In Figure 6–15c, note that the load impedance has increased. This results in a

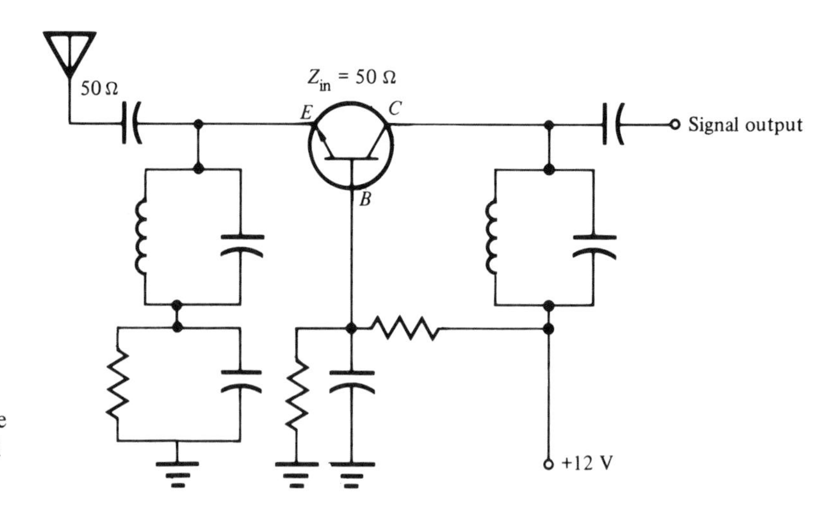

FIGURE 6–16
Common Base Amplifier Used to Match Impedance between the Antenna and Input of a Radio

decrease in current, as shown. The power for the load has decreased from the 5 W shown in Figure 6–15a.

The only part of this example where the power is equally divided between the load and the signal source impedance is the point when the signal source impedance and the amplifier's resistance are equal. At this point the maximum power is developed at the amplifier.

To accommodate the maximum power developed between the signal source and the load, the proper input impedance should be used. These proper impedances can be obtained by using one of the three types of amplifier designs. For example, Figure 6–16 shows a common base amplifier connected to the antenna of a radio. Because the antenna develops 50 Ω, and the input impedance of the common base matches that impedance, this amplifier configuration is a good selection.

AMPLIFIER CLASSES OF OPERATION

The transistor amplifier can be made to operate in several different areas of the load line. The biasing of the amplifier can be selected so that the amplifier will operate in the linear region, or the amplifier can be biased to operate near cutoff or saturation. Figure 6–15 shows several examples of moving the biasing point to change the output waveform. Often, because portions of the waveform are missing, the technician might think that the amplifier is not operating properly. However, it may be that the amplifier is biased so that it will produce only a fraction of the input waveform at the collector. This dc biasing selects the operating point of the amplifier, which in turn selects the **class of operation.**

The different classes of operation are designated by letters, the first being class-A. The **class-A amplifier** is biased so that the collector current flows for the entire input signal. In other words, the class-A amplifier operates at the linear part of the load line. The **class-B amplifier** amplifies only 180° of the input signal. Therefore, collector current flows for only one-half of the input waveform. The **class-C amplifier** amplifies less than one-half of the input cycle.

Class of operation is determined by the bias current or voltage, because the bias determines the operating point on the load line. Transistors connected to operate in the three different classes are shown in Figure 6–17. Table 6–3 summarizes the advantages, disadvantages, and uses of the different amplifier classes. Each class of operation has specific applications. For example, the class-A amplifier is used for audio amplifiers, where reproduction of the entire waveform is needed. Class-B amplifiers are useful as power amplifiers. Class-C amplifiers are useful for high-power amplifiers, transmitters, and oscillators because of their short burst of power.

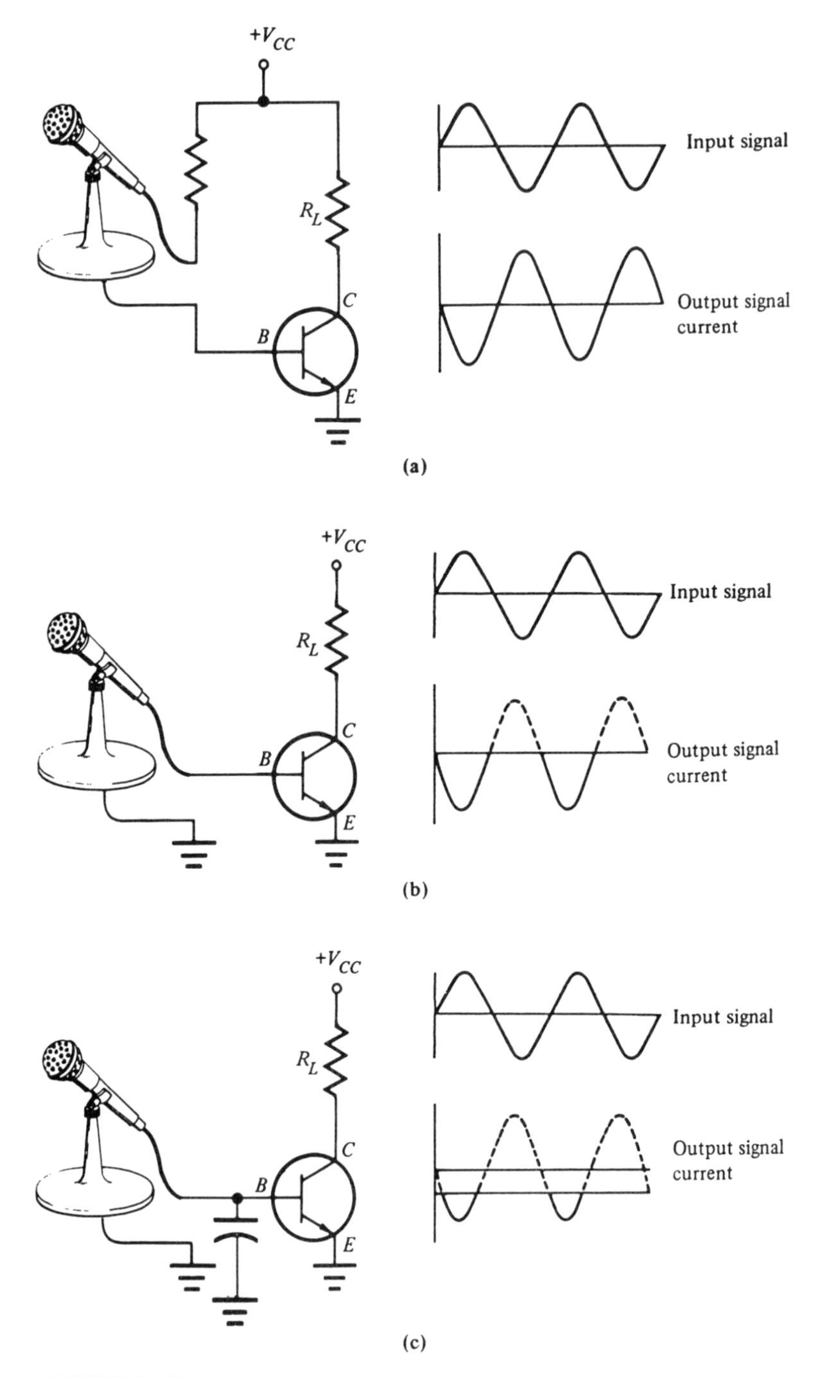

FIGURE 6–17
Biasing of an Amplifier to Operate at Different Classes: (a) Class-A Operation, (b) Class-B Operation, (c) Class-C Operation

TABLE 6–3 Comparison of Class-A, Class-B, and Class-C Operation

Class of Amplifier	Advantages	Disadvantages	Applications
A	Lowest waveform distortion	Lowest efficiency; used only for small-signal application	Preamplifier low-power stages
B	Fairly good efficiency, since no power used unless signal is present	Only half of waveform present; used mostly with pair of transistors to reproduce both halves	Power amplifiers, audio outputs
C	Highest efficiency, because current flows in bursts during part of waveform	Severe distortion; only 30–45% of each waveform amplified	Used in transmitters, oscillators

SUMMARY

Most amplifiers used in consumer products are designed to increase the signal between the input and the output. This increase in signal is called gain.

Amplifiers are divided into two general classifications: small-signal amplifiers and large-signal amplifiers. Generally, the small-signal amplifiers are voltage amplifiers and operate class-A in common emitter or common base connection. Large-signal amplifiers are identified as power amplifiers. These power amplifiers are biased to operate at class-B or class-A, and are arranged in either a common emitter, common collector, or common base configuration.

A change in voltage, current, or power in a circuit can be measured in decibels. A positive decibel value means an increase, and a negative decibel value shows a decrease, when comparing the input and the output.

The most popular amplifying components in consumer products are transistors and integrated circuits. The transistor causes a change in current between the input and the output. This change is expressed as a beta value.

The transistor amplifier is designed to operate within certain limits. Having the amplifier operate outside these limits will result in distortion. To insure proper operation of the transistor as an amplifier, a set of characteristic curves is used.

There are three different classifications of amplifiers: the common emitter, the common collector, and the common base. The common emitter is a linear amplifier, because it reproduces the full input at the output. The signal is applied to the base-emitter leads, and the output is taken from the collector-emitter leads. Simple forms of Ohm's law and Kirchhoff's law can be applied to solve amplifier biasing problems.

Common collector amplifiers are designed so the signal input is found on the base and collector terminals, and the output is taken from the collector-emitter terminals. Because output is taken from the emitter, there is no phase shift between input and output. This amplifier is sometimes called the emitter follower. It is commonly used as an isolation amplifier, and is often followed by a common emitter.

The common base amplifier is the third type of amplifier found in consumer products. The input of this amplifier is between the base and emitter, and the output is taken from the base and collector. This amplifier is generally used in high-frequency areas such as RF amplifiers.

Each amplifier has a set of impedance characteristics. To achieve maximum power transfer from the signal source to the amplifier, the impedances of the generator and load must be matched. If these impedances are not matched, the signal will be reduced. Each of the three configurations of amplifiers has specific input impedance characteristics important to service technicians.

Amplifiers can be biased to generate different degrees of waveform at the output. Class-A amplifiers generate a full 360° waveform at the output. Class-B amplifiers generate 180° of waveform at the output, and class-C amplifiers generate less than 180°.

KEY TERMS

amplifier
class A amplifier
class B amplifier
class C amplifier
class of operation
common base amplifier
common collector amplifier
common emitter amplifier
decibel
emitter follower
gain
impedance matching
input impedance
isolation amplifier
linear amplifier
load line
maximum power transfer
nonlinear amplifier
phase shift
static state

REVIEW EXERCISES

1. Write the formula used to compute gain.
2. An amplifier develops a 0.05 V(p-p) input signal, and a 0.5 V(p-p) output signal. What is the gain of this amplifier?
3. What do small-signal amplifiers generally amplify? large-signal amplifiers?
4. What is the term applied to the 0 dB reference level used for tape recorders and other audio equipment?
5. An amplifier has a 10 dB change between input and output. What kind of change should a technician expect between the input and output?
6. A filter develops a −12 dB change in signal. What reduction will be seen at the output of the filter?
7. Describe the main function of amplification.
8. If an amplifier operates outside its set of characteristic curves, what will happen at the output signal?
9. In a common emitter amplifier, to which terminals is the input signal applied?
10. What does biasing a transistor in the static state mean?
11. Why is a coupling capacitor used at the input and the output of a circuit?
12. Describe how a phase shift is developed in a common emitter amplifier. What is the degree of this phase shift?
13. Define linear as it applies to transistor operation. Draw an example.
14. Draw a schematic of a nonlinear amplifier.
15. In order to develop a load line, what two conditions are assumed on the transistor?
16. Draw a schematic diagram of a common collector amplifier, and describe several characteristics of this amplifier type.
17. List several characteristics of the common base amplifier configuration.
18. Why is it so important that the output impedance of the antenna in a circuit match the input impedance of the amplifier in the circuit?
19. Give a brief description of each of the different classes of amplifier operation.
20. At what location in the circuit would an emitter follower be used? Give reasons for its use.

7 SMALL-SIGNAL AMPLIFIERS

OBJECTIVES

Upon completing this chapter, you should be familiar with:

—Amplifier coupling principles
—Amplifier feedback principles
—Amplifier frequency response principles
—Amplifier distortion characteristics

INTRODUCTION

In all consumer products, increases in signal strength are accomplished through amplifiers. As stated in Chapter 6, amplifiers come in two basic forms: large-signal and small-signal. In this chapter the small-signal amplifier is discussed. The small-signal amplifier is usually found in RF (radio frequency), IF (intermediate frequency), and audio amplifier sections.

Small-signal amplifiers generally give medium signal gain. They also give the best frequency response to the signal as compared to large-signal amplifiers. Because these amplifiers provide only a medium gain, it may be necessary to couple amplifiers together. Several methods of coupling found in consumer products will be discussed.

Many small-scale amplifiers also include a gain control method. The amplifier will develop a negative or a positive feedback in order to control the gain.

Finally, distortion in the amplifier system is considered.

Coupling, distortion, and frequency response will relate to any type of amplifier but will show up more often in the small-signal class of amplifier or circuit configuration.

AMPLIFIER COUPLING

It often becomes necessary to feed the output of one amplifier into another amplifier. This process is called **cascading,** or creating a *chain* of amplifiers. The most common types of amplifier coupling found in consumer products are the directly coupled amplifier, the transformer-coupled amplifier, and the capacitor-coupled amplifier.

Direct Coupling

Direct coupling is feeding the output of one amplifier directly into the input of the next. This type of coupling is generally found in amplifier circuits operating under low frequencies. In this type of coupled amplifier, there are no reactive or resistive components between the two stages. Therefore, the directly coupled amplifier is able to pass very low frequencies.

Figure 7–1 shows an example of the directly coupled amplifier. Notice that the collector of Q_1 is connected directly into the base of Q_2. The base biasing established for Q_1 is from R_1 and R_2. The base biasing network for Q_2 is made up of R_3, the collector-emitter junction of Q_1, and R_4. If R_3 is made large in value, any change in collector current of Q_1 will develop large base voltage changes, resulting in large changes in base current for Q_2. This is the one disadvantage of the

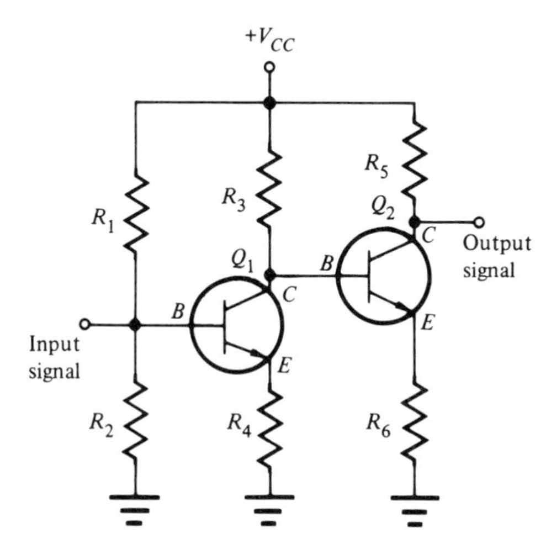

FIGURE 7–1
Directly Coupled Amplifier

directly coupled amplifier: any changes in one amplifier will affect the operation of the other amplifier.

The gain developed from the amplifier in Figure 7–1 is very small. Because the emitter resistors of Q_1 and Q_2 are not bypassed with a capacitor, the gain is reduced. These unbypassed emitter resistors cause negative feedback in the amplifier, which will be identified later in this chapter. If a capacitor were placed across the emitter resistor, the gain of the amplifier would be improved, but the ability of this amplifier to amplify low frequencies would be reduced.

Amplifier stability is very important. Stability in the directly coupled amplifier is poor. Any change in collector current of Q_1 due to a temperature change will show up in Q_2 base current. Thus, any thermal change would be amplified in Q_2. This thermal change would shift the operating point of Q_2, and develop distortion. The action of this circuit change can be seen by looking at Figure 7–1. As stated earlier, if a temperature change develops in Q_1, the leakage current and beta value increase. When this happens, Q_1 starts to conduct harder. More current flow through the collector-emitter circuit means that the voltage from collector to ground will be reduced. This reduction at the collector will show up as a lower voltage at the base of Q_2. This lower voltage at the base of Q_2 will shift the operating point of Q_2. This shift will move the entire amplifier system out of the linear range of operation, and cause distortion.

Some amplifiers do have directly coupled stages. Directly coupled amplifiers are usually found in the audio amplifier sections of radio receivers.

A popular type of directly coupled amplifier is the **Darlington amplifier.** This connection is shown in Figure 7–2a. A Darlington amplifier consists of two or more common collector circuits. As shown

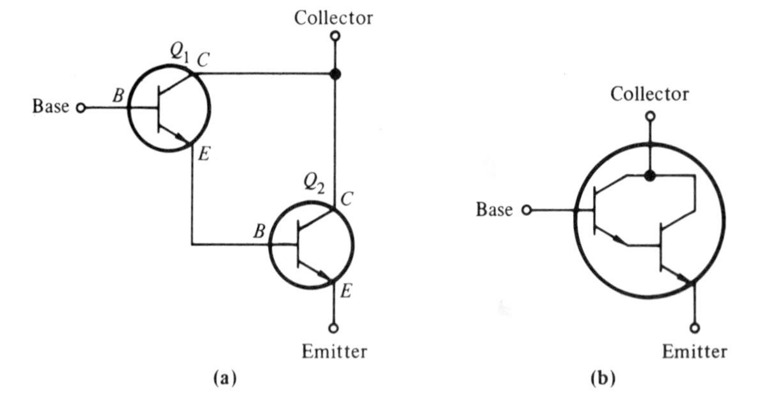

FIGURE 7–2
Darlington Amplifier Connection: (a) Transistor Connection, (b) Darlington in Single Package

in Figure 7–2a, the emitter of one amplifier is fed directly into the base of the next amplifier. Therefore, the emitter current of Q_1 is the same as the base current for Q_2. Generally, Darlington amplifiers are found in a single case with only three external leads: the base, the emitter, and the collector. Figure 7–2b shows this type of connection. The gain for the Darlington amplifier is extremely large and can be calculated by Equation 7–1.

$$\text{current gain} = A_i = \beta_1 \times \beta_2 \qquad \textbf{(7–1)}$$

Example 1 is a sample calculation of this gain.

EXAMPLE 1

Find the total current gain of a Darlington amplifier if each transistor has a gain of 50.

Solution:

Using Equation 7–1, the gain of this stage is as follows:

$$A_i = 50 \times 50 = 2500$$

The Darlington connection is a good amplifier to use when large current gains are needed, or when the signal at the input requires a large input impedance. Because of the high input impedance, the amplifier requires a low input signal to operate. This makes the Darlington amplifier an excellent circuit for the audio output section in many radio receivers.

Transformer Coupling

Another type of coupling method is called **transformer coupling.** Figure 7–3 illustrates this type of connection. The transformer is an ideal component to use because of its ability to match the impedance of the output of one amplifier to that of the input of another amplifier or to an output device such as a speaker. It should be remembered that when impedances are matched, the maximum amount of signal power can be transferred. The transformer is the only passive component that will allow impedance matching and maximum signal power transfer.

In Figure 7–3 the output impedance of Q_1 is about 30 kΩ, and the input of Q_2 is about 1 kΩ. To transfer the maximum amount of signal power between stages, the resistances must be matched. This matching is accomplished by using a transformer with the proper turns ratio. Equation 7–2 can be used to determine this ratio.

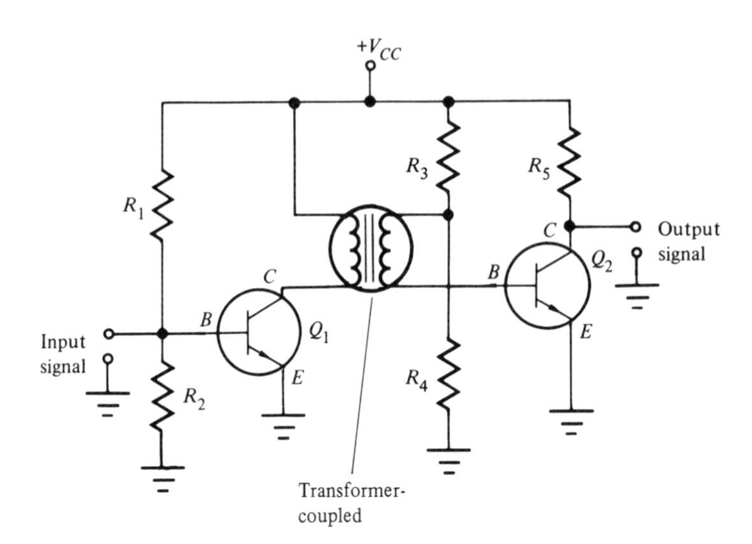

FIGURE 7–3
Two-Stage Transformer-Coupled Amplifier

$$N = \sqrt{\frac{Z_p}{Z_s}} \tag{7–2}$$

where:

N = turns ratio of the transformer
Z_p = impedance of the primary
Z_s = impedance of the secondary

EXAMPLE 2

An example calculation will prove useful here. Refer to Figure 7–3. Assume that the output impedance of Q_1 is 30 kΩ, and the input impedance of Q_2 is 1 kΩ. If the amplifier is transformer coupled, what turns ratio should be used?

Solution:

$$N = \sqrt{\frac{Z_p}{Z_s}} = \sqrt{\frac{30{,}000}{1000}} = \sqrt{30} = 5.48$$

Thus, in order to match the impedances of the primary and secondary, the transformer should have a 5.48 : 1 turns ratio. Generally, the transformers used for coupling are step-down transformers. Because of this action, transistor Q_1 sees a load impedance of 30 kΩ instead of 1 kΩ.

Another important characteristic of the transformer-coupled amplifier is that the output stage is isolated from the next stage. Therefore, the dc operating voltages found in the amplifier of Q_1 will not show up in the next stage. This means that a shift in the operating point of Q_1 will not affect the operating point of Q_2.

The transformer-coupled amplifier has a couple of advantages over the directly coupled amplifier. One is that the transformer isolates one amplifier from another. The other is that the transformer will match the impedances between stages.

The main disadvantages of the transformer-coupled amplifier are the low-frequency application and the cost. Many transformer-coupled amplifiers will not pass all the frequencies developed in amplifier stages. Generally, the transformer-coupled amplifier is found in the IF stages of radio receivers, and in the final power output stage of audio sections.

The transformer-coupled circuit can be used to select the IF frequency of radios and televisions. If a capacitor is placed across the secondary or primary, these circuits become tuned to a certain frequency. These tuned circuits are then able to select and reject frequencies. This circuit action is illustrated in Figure 7–4.

As shown in Figure 7–4a, the secondary of the transformer and the capacitor form an LC tank circuit. When the reactance of the capacitor and the reactance of the inductor are equal and opposite, the circuit will become resonant. This resonant circuit will then pass one frequency, which is identified as the **resonant frequency.** Equation 7–3 is the formula used to calculate the resonant frequency.

$$f_r = \frac{1}{2\pi\sqrt{L \times C}} \tag{7–3}$$

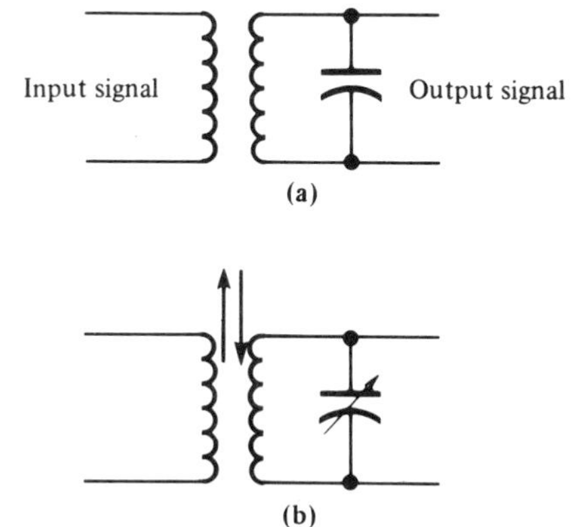

FIGURE 7–4
Tuned Transformer-coupled Circuits: (a) Fixed Tuned Circuit, (b) Variable Tuned Circuit

At resonant frequency, the maximum output voltage will be developed at the secondary. At frequencies above and below resonant frequency, the output voltage will be greatly reduced. Therefore, maximum signal voltage will occur across the capacitor at resonant frequency.

It may become necessary to adjust the tuned circuit to the proper frequency. To accomplish this task, the capacitor or transformer used is variable. In the IF sections of televisions and radios, the transformer usually can be adjusted by changing the position of the ferrite core of the transformer. This then allows the circuit to be tuned over a wide range of frequencies. An example of this variable tuned circuit is shown in Figure 7–4b.

Figure 7–5a shows another type of tuned circuit used to couple the signal from one stage to the next. Note that in the circuit shown in this part of the figure, the secondaries of the transformers have a capacitor placed in parallel. This parallel capacitor will tune the secondary, and thus pass only the resonant frequency of the tuned circuit

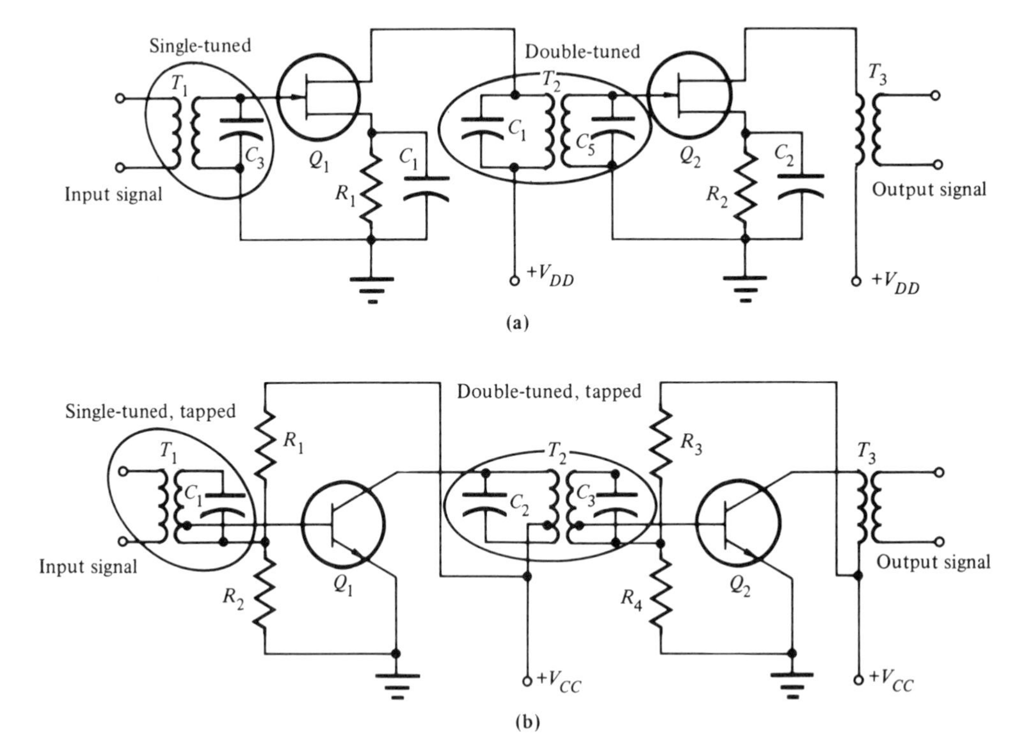

FIGURE 7–5
Tuned Circuits Used as Coupling Devices: (a) Single- and Double-Tuned Circuits Used in an FET Amplifier, (b) Signal Taken from the Taps of the Transformer

established by the secondary of the transformer and the capacitor. This arrangement is identified as a **single-tuned transformer** coupled circuit. Also, note that in the second stage of the amplifier, the primary and the secondary are tuned with the parallel capacitor. This type of connection is known as a **double-tuned transformer** connection.

In Figure 7–5b another type of single- and double-tuned transformer connection is shown. Note that the secondaries of both the transformers are tapped. This allows the secondaries to match the impedances between the low impedance of the transistor and the high impedance of the tuned circuit at resonance. If a transformer were placed across this circuit, it would load the circuit down, and the necessary signal would not develop. This is the reason the signal is developed from the taps of the transformer.

Capacitive Coupling

Another method of coupling the signal between two stages is to place a capacitor between the output of the first stage and the input to the second stage. This type of coupling, called **capacitive coupling,** is illustrated in Figure 7–6.

The input for the amplifier in Figure 7–6 is developed across two components: C_1 and R_2. Because these two components are connected in series, Kirchhoff's voltage law will apply. Therefore, there will be a voltage division between the capacitor and the resistor. This means that because this circuit will handle different frequencies, the

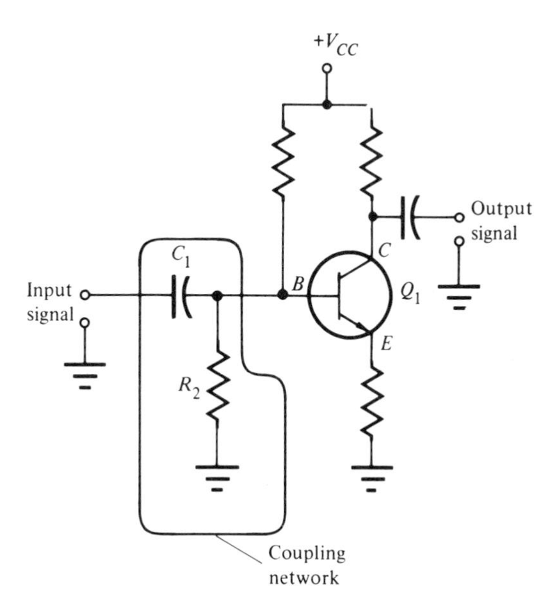

FIGURE 7–6
Capacitive-Coupled Amplifier

capacitor must have the lowest possible impedance for the frequencies developed.

Keeping the impedance of the capacitor low depends on two circuit characteristics: the signal current and the reactance of the capacitor. If the reactance of the capacitor is too large, all of the input signal will be developed across the capacitor, and none across the resistor. Therefore, there will be no voltage developed across the input of the amplifier.

Remember that the capacitive reactance depends upon the size of the capacitor and the applied frequency. For a given value of capacitance, the reactance will drop off as frequency is increased. Thus, at higher frequencies the capacitor will develop a small voltage drop, and the input of the amplifier will develop a large signal. Also of concern is that the capacitor will isolate the dc voltage of one stage from the other stage. The only transfer between the two amplifiers is that of the signal.

The major disadvantage of capacitive coupling is the mismatch of impedances for the two amplifier stages. Because of this, some of the signal will be lost in the transfer. To overcome this loss, additional amplifier stages can be included in the amplifier section.

The capacitive-coupled amplifier is the most common type of amplifier connection. It generally is found in audio sections, since low frequencies pass through these stages. The *RC*-coupled amplifiers are not found in IF and RF sections because of the mismatch of impedance created by the coupling circuit.

AMPLIFIER FEEDBACK

To create an oscillator, an amplifier is equipped with a feedback signal circuit. (Oscillators will be discussed in detail in Chapter 11.) This feedback signal then is amplified, and the process is repeated, thus creating an output signal. The feedback signal is in the same phase as the output signal. This signal is called **positive feedback,** or *regenerative feedback.* Another type of feedback is called **negative feedback,** or *degenerative feedback.* With negative feedback, the signal fed back to the input is 180° out of phase from the input signal. Examples of positive and negative feedback in amplifiers are shown in Figure 7–7.

Positive feedback is illustrated in Figure 7–7a. Note here that the signal is fed back through a feedback network, and is combined at the input with the signal to be amplified. The two signals at the input are said to be in phase because they are both going in the same direction.

Negative feedback is illustrated in Figure 7–7b. Again, part of the input signal is fed back through the feedback network. However,

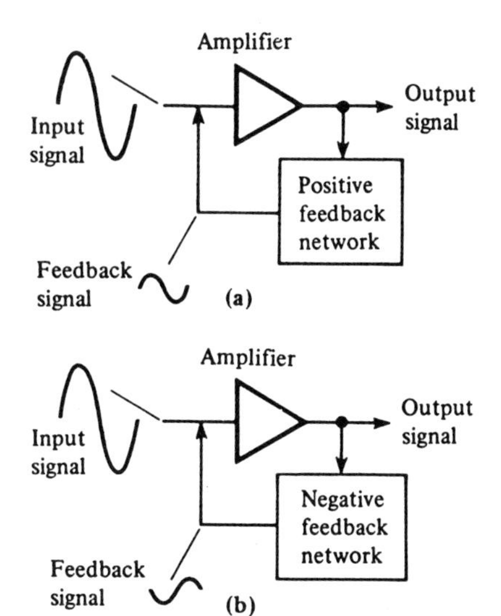

FIGURE 7–7
Amplifier Feedback:
(a) Positive Feedback,
(b) Negative Feedback

the feedback signal is now shifted 180° out of phase from the input. Thus, the feedback signal will reduce the amplitude of the input signal, and therefore reduce the gain of the amplifier.

Figure 7–8 shows how the two types of feedback signals are combined with the input signal. The positive feedback signal is given

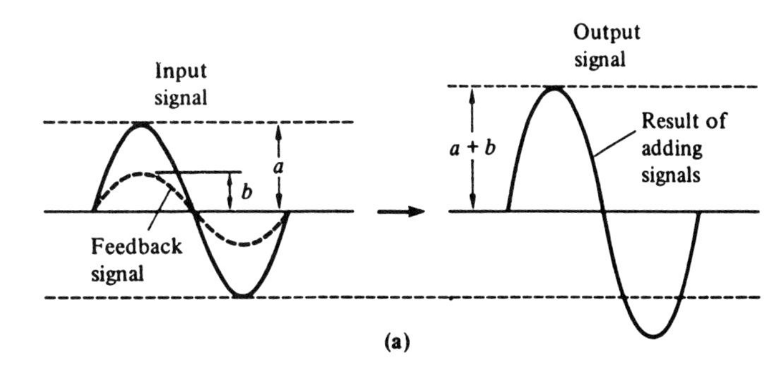

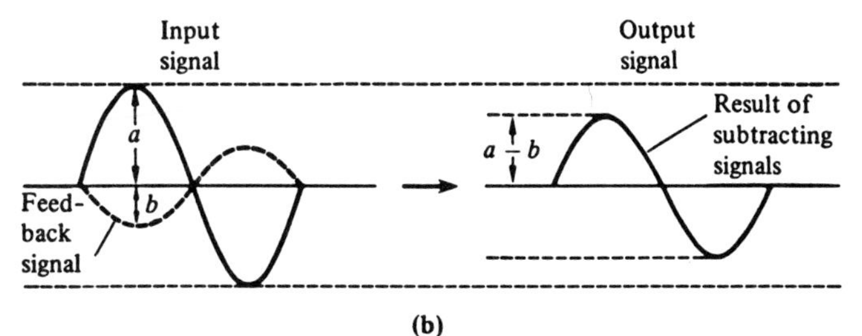

FIGURE 7–8
Waveforms Showing Feedback Signal Amplitude:
(a) Positive Feedback Results in Addition of Waveform Amplitudes,
(b) Negative Feedback Results in Subtraction of Waveform Amplitudes

in Figure 7–8a. The feedback signal is indicated by the broken line, and the input signal is shown by the solid line. These two signals are added together at the input. The waveform resulting from this additive process is shown at the right.

The negative feedback signal is shown in Figure 7–8b. Note that the two signals are 180° out of phase. Again, the feedback signal is traced by the broken line, the input signal by the solid line. When the solid line is going positive, the broken line is going negative. The feedback signal will reduce the amplitude of the input signal. Figure 7–8b shows the resulting output signal waveform. Notice that it is smaller than that for the original input signal.

Positive Feedback

Positive feedback is usually found in oscillator circuits. Because of the feedback signal, the amplifier is able to sustain oscillation. Positive feedback is also found in other applications. In some high-frequency amplifiers, it is used to increase the amount of gain in the amplifier. Often the transistors used in these circuits will develop signal loss because of the capacitance developed in the component. To increase this gain, positive feedback is developed. The positive feedback signal adds to the input, and thus gives higher amplification. The circuit in Figure 7–9 illustrates this type of configuration.

In Figure 7–9, Q_1 serves as an RF amplifier. A separate power supply is used to bias the collector, and a second power supply to bias the base-emitter circuit. The input signal is coupled via C_1, and the

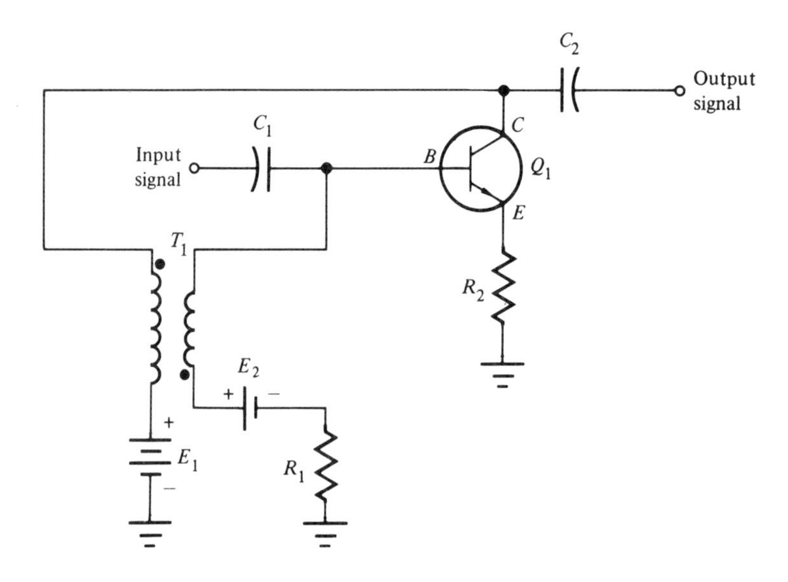

FIGURE 7–9
Signal Path Developing Positive Feedback

output signal is coupled via C_2. Positive feedback is taken from the collector of Q_1 and fed back to the primary of T_1. Notice that the schematic symbol of the transformer contains two dots. These dots denote at which points the primary and secondary signals will be in phase. The dots found on the component primary and secondary are important to the service technician. If the transformer is installed incorrectly, the wrong type of feedback will be developed in the circuit.

The polarity of the feedback signal comes from the collector of Q_1. The signal passes through T_1 to the base of Q_1. The feedback signal and the input signal are then combined at the base of Q_1. The signal polarities mean that the two signals are added together. This causes an increase in the input signal. Since Q_1 is an NPN transistor, an increase of base voltage in the positive direction will increase the gain of the amplifier stage.

There may be a problem with this type of feedback. If the feedback signal is too large in amplitude, the amplifier may become an oscillator. To prevent this, a step-down transformer is used for T_1. The step-down transformer will reduce the amplitude of the signal. Placing R_2 in the emitter will reduce the gain of the stage. Just enough signal is fed back to the input to develop the necessary gain in the amplifier.

The service technician should understand the operation of positive feedback. The most important factor is the amplitude of the signal fed back. Therefore, the feedback signal will increase the input signal in amplitude. This change in amplitude can develop oscillations in the RF amplifier, and the input signal will be destroyed.

Negative Feedback

Negative feedback is developed by taking the input signal, changing the phase, and feeding the signal back to the input. This procedure reduces the amplitude of the input signal. An example of negative feedback is shown in Figure 7–10.

The configuration in Figure 7–10 is that of a common emitter amplifier circuit. The input signal is developed on the base of Q_1, and is developed at the output on C_2. Resistors R_1 and R_2 are used to establish the base biasing network. Resistor R_3 is the load resistor, and resistor R_4 is an emitter stabilization resistor.

Because of the stabilization resistor in the emitter circuit, a voltage drop is developed. The voltage drop is a signal and a dc voltage. Of importance is the signal voltage. As the voltage on the base increases because of the input signal, the voltage on the emitter also increases. This means that the input signal makes the base more positive. The emitter is also made more positive because of the signal. When the signal on the base becomes negative, the emitter becomes more neg-

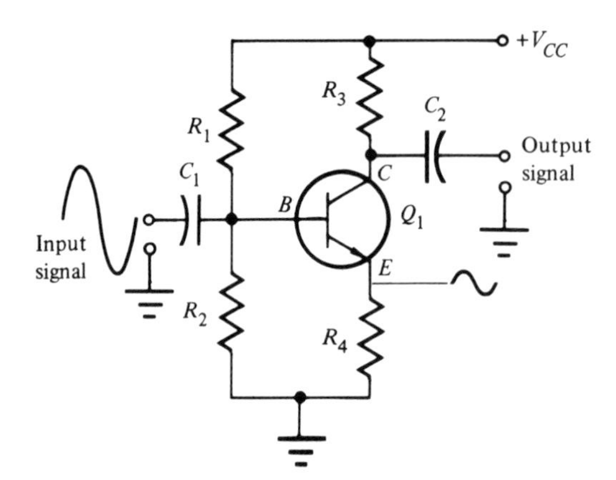

FIGURE 7–10
Current Feedback Path Developing Negative Feedback

ative. The emitter follows the base by becoming more negative. This is the standard action of the amplifier.

Because the emitter follows the base voltage, and the base-emitter forward bias controls the amplifier action, the gain of the amplifier can be controlled. The action of this control is shown in Figure 7–10. As the base voltage becomes more positive, the forward bias of the transistor should increase. However, because the emitter is also going positive, the gain of the amplifier is reduced. This action of increasing the base-emitter voltages is negative feedback. This type of feedback can be developed in two ways: by current feedback, and by voltage feedback.

Current Feedback Figure 7–10 shows the overall result of the presence of the emitter resistor (R_4). The main purpose of this resistor is to stabilize the dc bias in the circuit. It will also reduce the gain of the amplifier. This type of control is known as **current feedback.** It is called current feedback because the current flow in the emitter will develop a signal voltage in phase with the base, thus reducing the gain of the circuit.

Negative feedback serves an important function in amplifier circuits. The main purpose of gain reduction is to make the amplifier able to amplify a given set of frequencies. Remember that as the gain of an amplifier is increased, the range of frequencies it can amplify is reduced. Therefore, if the gain is reduced, the range of frequencies that can be amplified is increased. If negative feedback were not used in an amplifier, all the audio frequencies might not be amplified equally. There is a tradeoff in amplifier circuits between gain and bandwidth.

In Figure 7–10, R_4 serves the purpose of introducing current feedback, and thus developing negative feedback. If the circuit is working properly, the amplifier will not break into oscillation. Negative feedback is used in audio amplifier circuits in radios and television receivers.

Voltage Feedback Negative feedback can also be obtained by **voltage feedback.** Figure 7–11 can be used to explain voltage feedback. A two-stage *RC*-coupled amplifier circuit is shown in this figure. Here, R_1, R_2, R_5 and R_6 provide the biasing network for the base circuit. Resistors R_3 and R_7 are the load resistors for the two circuits. Capacitor C_1 is a coupling capacitor. Resistors R_4 and R_8 are used as stabilizing resistors. As pointed out earlier, R_4 and R_8 will not only provide a stable operating point, but will also introduce negative feedback.

The waveforms developed at the base of Q_1 are also indicated—the solid line for the positive polarity, and the broken line for the negative alternation. Because Q_1 is a common emitter amplifier, the output signal is shifted 180° from the input signal. Waveform *b* shows this shift. Since the coupling capacitor passes this signal, waveform *c* is the same polarity as waveform *b*.

The second amplifier stage is also a common emitter amplifier. This stage will also develop a phase shift of 180° between the input and the output. This phase shift is shown by waveform *d*. Note that waveform *a* and waveform *d* have the same polarity.

Resistor R_9 and capacitor C_2 return part of the output signal to the emitter of Q_1. An isolating capacitor, C_2, is used to block the higher dc voltage at the collector of Q_2 from the lower emitter voltage of Q_1. Resistor R_9 is used as a voltage-dropping resistor. This resistor

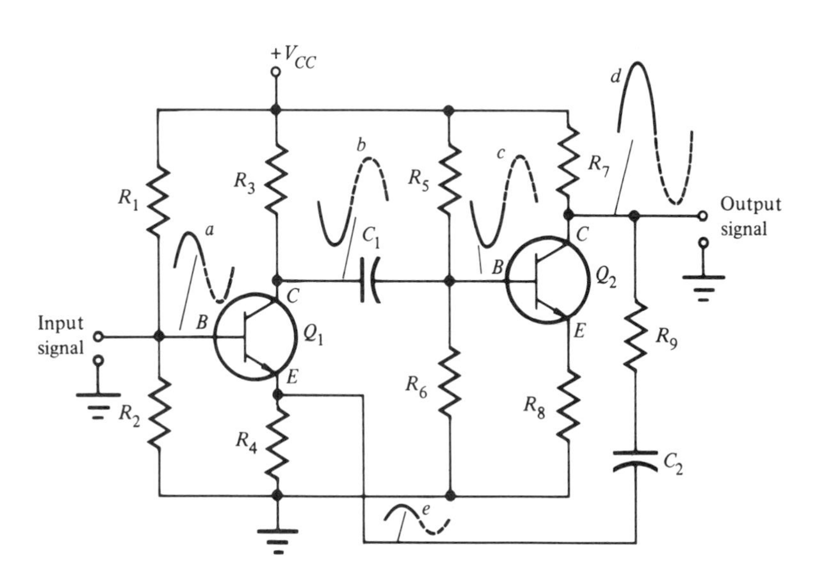

FIGURE 7–11
Two-Stage Amplifier Developing Voltage Feedback through R_9 and C_2

will lower the amplitude of the feedback signal. The feedback signal is shown by the letter e. Because this positive-going signal appears at the emitter of Q_1, it causes the emitter to become more positive. Making the emitter of an NPN transistor more positive means that the gain of the transistor will be reduced.

Negative Feedback Avoided

In semiconductor operation it is important that the component operate within a stable range. To accomplish this an emitter resistor is placed in the emitter circuit. This resistor will stabilize the operation of the transistor. However, this resistor will also introduce negative feedback into the amplifier. As stated earlier, negative feedback in an amplifier will reduce gain, and improve the bandwidth of the amplifier. In some amplifiers this is the desired result. In other cases a large gain and a narrower bandwidth are desired. To accomplish this a capacitor is

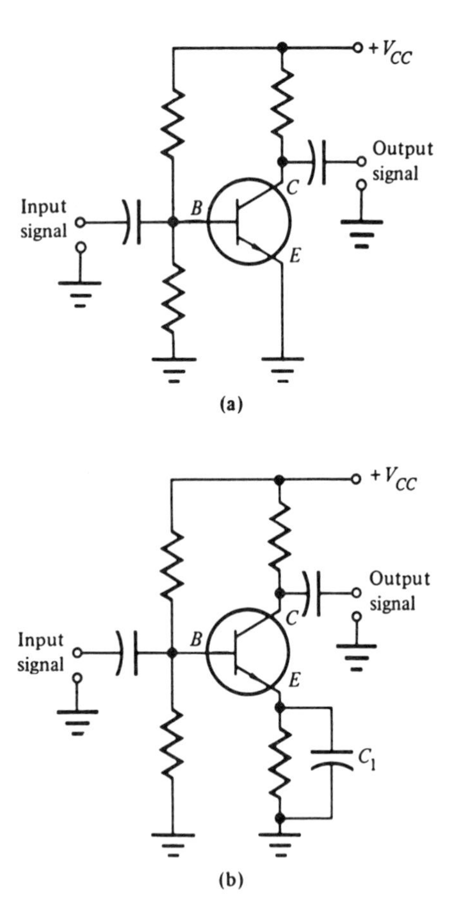

FIGURE 7–12
Reducing Negative Feedback Effects: (a) Circuit That Is Temperature Unstable, (b) Bypass Capacitor Used to Reduce Negative Feedback

placed across the emitter resistor. This capacitor is called an **emitter bypass capacitor.**

A schematic diagram of a circuit that will eliminate negative feedback is given in Figure 7–12a. Note that the emitter resistor has been removed. This will eliminate the negative feedback, but will also place the transistor in an unstable operating mode. This connection will result in higher gains within the amplifier, but that gain improvement will be affected by the raising or lowering of temperature around the transistor. In other words, the temperature will affect the amount of gain. If the transistor is hot, there will be more gain, and if the transistor is cold, there will be less gain.

To work around this problem, a bypass capacitor is added. This configuration is shown in Figure 7–12b. The main function of the bypass capacitor is to keep the voltage at the emitter constant. Its action is much like that of the filter capacitor in a power supply. Figure 7–13 shows how this capacitor will charge and discharge.

As the current begins to flow in this circuit, capacitor C_1 begins to charge, as shown in Figure 7–13a. This action is shown by the

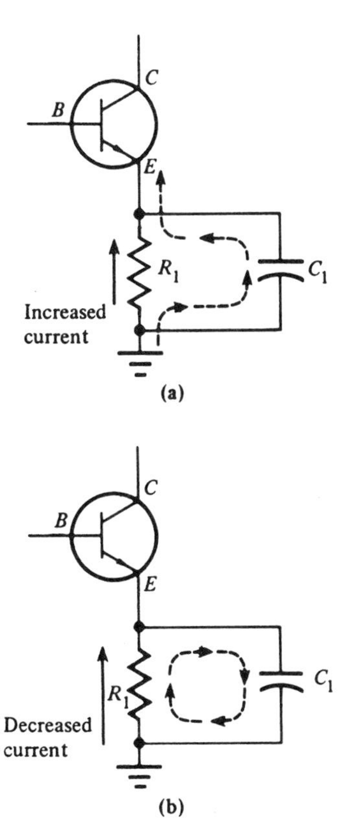

FIGURE 7–13
Bypass Capacitor Operation: (a) Charging Path of the Bypass Capacitor, (b) Discharging of the Capacitor through the Load

broken line. When the current begins to decrease in the emitter circuit, the voltage drop across the emitter resistor also begins to fall. As this voltage begins to decrease across the emitter resistor, the capacitor will discharge, as shown in Figure 7–13b. This discharging capacitor current will add to the falling emitter current and keep it constant. The constant current through the emitter resistor will keep the voltage constant at the emitter. If the voltage remains constant, the gain of the stage will remain the same.

As stated previously, the reactance of a capacitor is low at the higher frequencies, and high at the lower frequencies. Therefore, if the proper size of emitter bypass capacitor is used, the capacitor will behave as a short to the audio frequencies. This short will develop a low impedance path for the audio frequencies around the emitter resistor. Thus, the emitter resistor will not develop any voltage drop for these audio frequencies. The resistor then can develop a steady flow of current, and not develop any negative feedback in the amplifier.

The size of the bypass capacitor is very important. Generally, the reactance developed by the capacitor at the lowest frequency to be amplified should be one-tenth to one-half the value of the emitter resistor. With the selection of an appropriate bypass capacitor, the signal variation will not affect the gain of the amplifier.

The process of selecting an appropriate bypass capacitor is illustrated in Example 3. Equation 7–4 is the formula used to calculate the size of a capacitor.

$$X_C = \frac{1}{2\pi fC} \qquad \textbf{(7–4)}$$

EXAMPLE 3

What size capacitor is needed to bypass a 200 Ω emitter resistor found in the audio amplifier section? This amplifier will operate in the 20 Hz to 20 kHz range.

Solution:

The capacitor must be able to bypass the lowest audio frequency developed in the amplifier, which is 20 Hz. If the capacitor is able to bypass this frequency, then all the higher frequencies will also be bypassed. The resistance of the emitter resistor is 200 Ω. Assume that the capacitive reactance should equal about one-tenth the value of the emitter resistor. Therefore, the value of reactance should be approximately 20 Ω.

The value of the capacitor can be found by taking the capacitive reactance formula given in Equation 7–4,

$$X_C = \frac{1}{2\pi fC}$$

and rearranging it to read as follows:

$$C = \frac{0.159}{fX_C} = \frac{0.159}{20\text{ Hz}(20\ \Omega)} = \frac{0.159}{400}$$
$$= 397\ \mu\text{F}$$

Therefore, a 397 μF capacitor will give the necessary bypass action in the circuit. A 397 μF capacitor might be hard to find, but a capacitor with a rating of between 300 μF and 400 μF would work in this circuit.

The typical values for bypass capacitors are in the range of 20–1000 μF. These capacitors must be electrolytic because of the high capacitances required. Use of the new, small tantalum capacitors has reduced the physical size of the circuit board.

AMPLIFIER FREQUENCY RESPONSE

An amplifier should amplify each frequency that appears at the input. Ideally, the small-signal amplifier will give equal gain to each of these frequencies. However, because in operation the amplifier is not ideal, it will not give equal amplification to each frequency. The set of frequencies the amplifier will amplify is identified as the **frequency response** of the amplifier.

In any solid-state amplifier circuit, a change in input current will develop a change in output current. If these changes are at a slow rate and the circuit cannot respond, the amplifier would have poor low-frequency response. If the amplifier were not able to respond to the rapid rate of change of high frequencies, the amplifier would be characterized as having poor high-frequency response. Generally, silicon-type semiconductors have poorer frequency response than germanium materials. Because the majority of active devices found in amplifiers are made of silicon, the majority of amplifier circuits have poor frequency response.

Each amplifier will have a high- and a low-frequency cutoff point. The **low-frequency cutoff** point is usually the result of the type of coupling used in the amplifier. As has been pointed out, if inductors or transformers are used as coupling devices, they will have lower impedance to the lower frequencies. The value of the coupling capacitor will also have an effect on the transfer of signal from one stage to another. The **high-frequency cutoff** point is generally determined by the type of active component used. Because of the junctions found within a transistor, the reverse bias develops stray capacitance. This capacitance will cause the signal to be diverted around the amplifier and not be given the proper amplification.

Figure 7–14 shows an example of a typical response curve for an *RC*-coupled amplifier. As with all amplifier response curves, it shows the relationship between gain and frequency. Note that as the frequency is increased, the gain of the amplifier increases and then remains relatively constant. Then as the amplifier reaches higher frequencies, the gain drops off. However, look at the low frequencies. Here, the gain is very low, because of the reactance offered by the capacitor. The broken line shows the ideal level of gain for the signal.

The coupling capacitor found between the stages is the reason for the poor low-frequency response of the amplifier. As stated earlier, as the frequency increases, the reactance of the capacitor decreases.

In order to offset this drop in gain at low frequencies, an ***RC* compensating circuit** is added. This type of circuit is shown in Figure 7–15 for a two-stage common emitter amplifier. To make up for the lost gain by the coupling circuit, additional gain will be developed in the first stage of the amplifier.

Recall that the gain of the stage is determined by the amount of load impedance seen in the output circuit. In order to have an increase in gain at lower frequencies, an *RC* compensating network is placed in the collector circuit. This network will develop higher impedance at the lower frequencies, and lower impedance at the higher frequencies.

Note in Figure 7–15 that the load impedance in the collector circuit of Q_1 is developed by resistors R_1 and R_2 in parallel with the Z_{in} of the next stage. Z_{in} is the input impedance developed by the

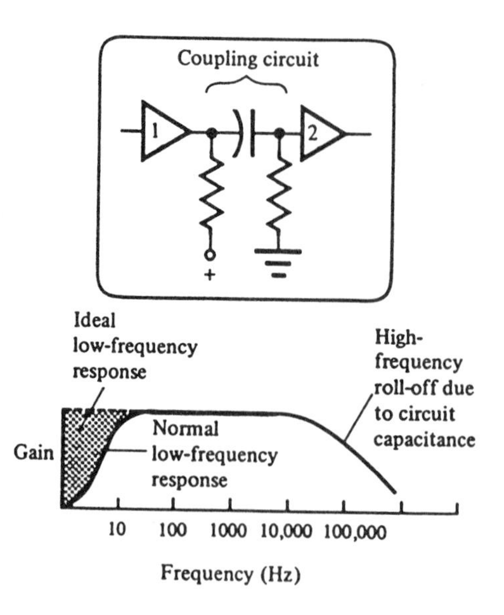

FIGURE 7–14
Response Curve for the *RC*-Coupled Amplifier

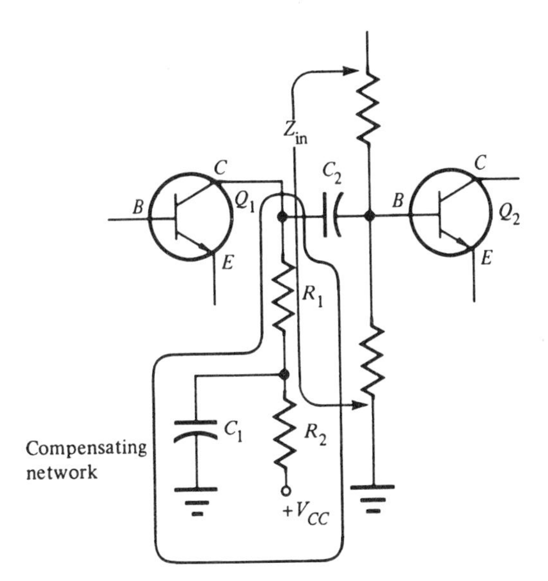

FIGURE 7–15
Two-Stage Amplifier Using Compensating Network to Improve Frequency Response

combination of C_2 and Q_2 input impedance. Capacitor C_1 is tapped off between R_1 and R_2. This capacitor will act as the compensating component. At the lower frequencies, C_1 will develop high reactance. It will behave as an open to the low frequencies. Therefore, the total load impedance will be a combination of R_1 and R_2 in parallel with the Z_{in} of the next stage. At the higher frequencies, the capacitor will act as a short. This will decrease the load impedance. The only impedance left will be R_1 in parallel with the Z_{in} of the next stage. Resistor R_2 will be shorted for high frequencies.

The result here is that the amplifier has improved the gain of the lower frequencies. This then will offset the loss created by the coupling components. The *RC* compensating network will give higher gain to low frequencies, and lower gain to high frequencies.

Also of importance is that in some amplifiers, the high-frequency component is reduced. This is generally the case in phonographs and tape players. The simplest type of compensating network is shown in Figure 7–16. Notice that a capacitor and resistor are placed in parallel. At the lower frequencies, the low-frequency component of the input signal is reduced, because of the reactance offered by the capacitor.

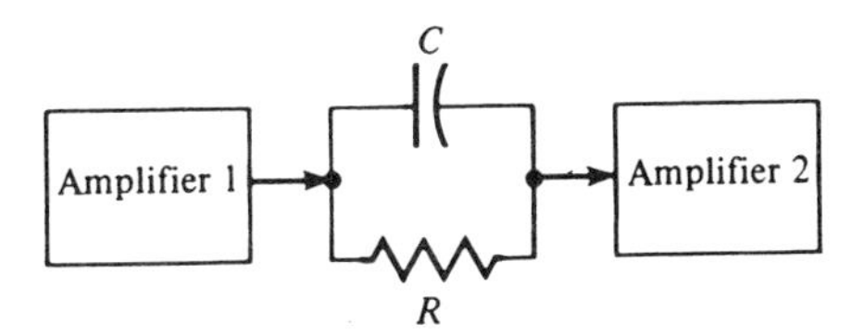

FIGURE 7–16
Coupling Components Used to Improve High-Frequency Component

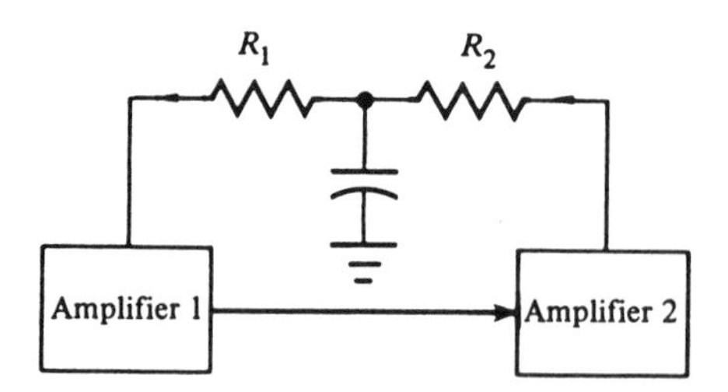

FIGURE 7–17
Feedback Path Developing Negative Feedback and High-Frequency Boost

The signals are passed by the resistor. At the higher frequencies, the capacitor becomes the low-impedance path. These frequencies are then passed by the capacitor. Thus, the capacitor will allow the frequencies to pass, and the lower frequencies will be reduced by the resistor. This type of network will give a boost to the higher frequencies. Generally, this type of compensating network is found in the preamplifier sections of tape recorders and phonographs. These circuits are required because of the poor high-frequency response of the input transducers.

Another type of circuit that provides negative feedback, and a high-frequency boost for the signal, is shown in Figure 7–17. This network will provide compensation as well as reduce the gain of the amplifier.

AMPLIFIER DISTORTION

The function of most amplifiers is to raise the level of the signal with a minimum amount of distortion. **Distortion** is defined as any change from the input signal to the output signal. For example, a class-B amplifier develops only one-half of the output signal. This is cause for distortion in the amplifier. The input signal is not reproduced at the output. In all consumer product amplifiers, there will be some amount of distortion. It will be found that practical amplifiers, amplifiers found in actual circuit operation, do not give constant gain to all input signals, nor can they withstand all sorts of different input amplitudes without distortion.

The distortion generally found within consumer product amplifiers can be divided into several basic classes. These classes of distortion are harmonic distortion, intermodulation distortion, crossover distortion, clipping distortion, and transient distortion.

Harmonic Distortion

One of the common types of distortion found in consumer products is called **harmonic distortion (HD).** Harmonic distortion results when an

FIGURE 7–18
Development of Harmonic Distortion: (a) Fundamental Frequency, (b) Upper Harmonic, (c) Combined Waveforms Showing Harmonic Distortion

original frequency recreates itself at even or odd multiples. Each time one of the multiples of the frequency is recreated, the amplitude of the signal is changed.

Harmonic distortion is created within an amplifier when feedback is created within the amplifier. Harmonic distortion is very unpleasant to hear. The squeals or harsh sounds developed are amplified, and eventually reproduced at the output of the system.

Figure 7–18a gives the waveform of the fundamental frequency of 2 kHz. Figure 7–18b shows the first even harmonic of 4 kHz. When these two signals are combined within the amplifier, the output waveform is developed. This waveform is shown in Figure 7–18c.

If harmonic distortion is left unchecked, the system can go into total harmonic distortion. In total harmonic distortion, all the harmonics of a certain frequency are developed. The additive process completely distorts the signal developed at the output.

Harmonic distortion is generally found in audio amplifier sections of consumer products. The industry has set a standard for the manufacturers of these products. This standard is that harmonic distortion should be less than 1% when the signal is fed into an 8 Ω load. Distortion measurements are made using a signal generator that produces a sine wave at 100 Hz. The signal is fed into the input of the amplifier, and is measured at the output with a distortion analyzer.

Intermodulation Distortion

Intermodulation distortion (IM) is another type of distortion developed within audio amplifiers. This distortion is developed when signals are added to and subtracted from one another. For example, if a 200 Hz tone and a 600 Hz tone are present at the amplifier at the same time, the signals will add together, producing an 800 Hz signal. They will also subtract from each other, and produce a 400 Hz tone. Because these two signals fall within the audio range, they are amplified by the amplifier along with the original 600 Hz and 200 Hz tones, and reproduced at the output. Intermodulation distortion is found in all amplifiers, generally at the lower end of the frequency range.

To test an audio amplifier for intermodulation distortion, two test signals are injected into the input of the amplifier, for example, a 60 Hz signal and a 7 kHz signal. The 7 kHz signal is about four times greater in amplitude than the 60 Hz signal. The output signal is measured with a distortion analyzer.

Crossover Distortion

Crossover distortion is generally found within the (push-pull) class-B amplifier. Figure 7–19 illustrates this type of distortion. In Chapter 8 a detailed explanation of crossover distortion is given as it relates to the power amplifier.

Clipping Distortion

When the input signal is too large, it drives the amplifier into distortion. This type of distortion is called **clipping distortion.** When saturation or cutoff is reached, the rounded peaks of the amplifier waveform are cut off. Figure 7–20 shows an example of this type of distortion.

Clipping distortion is found in almost all types of amplifiers, although primarily in small-signal amplifiers. Clipping causes the large-amplitude signals to become distorted in their reproduction. The addition of components within the amplifier will reduce the clipping distortion.

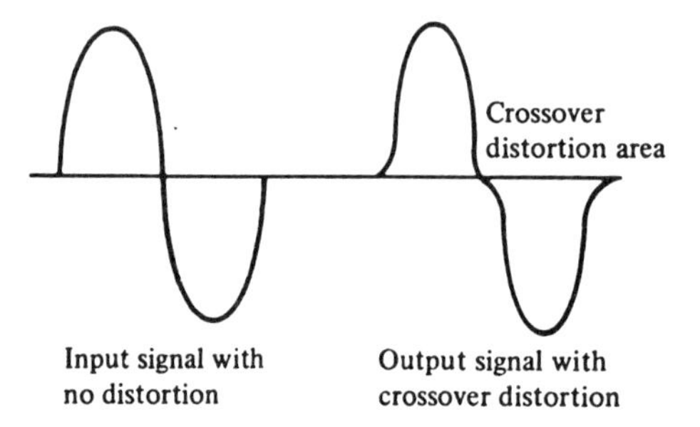

FIGURE 7–19
Crossover Distortion

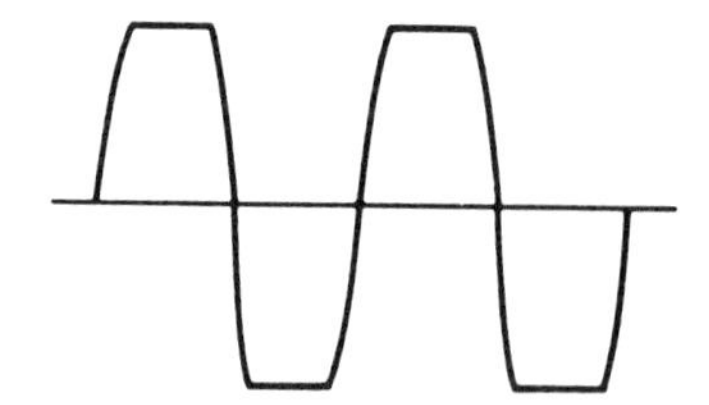

FIGURE 7–20
Clipping Distortion

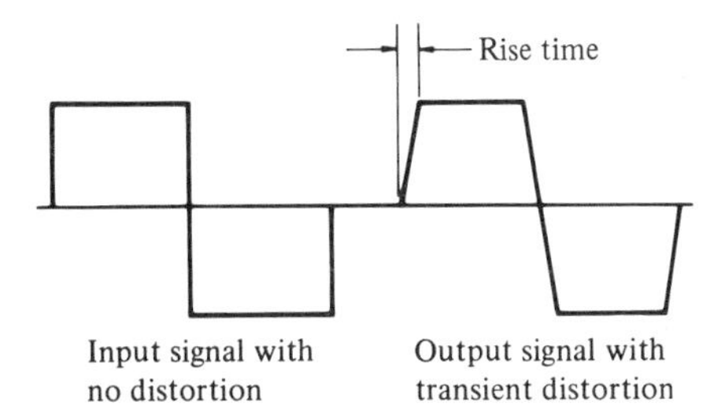

FIGURE 7–21
Transient Distortion

Transient Distortion

Transient distortion occurs in small-signal amplifiers when high frequencies appear. It often is the result of tones from bells, chimes, pianos, drums, or cymbals. Because these tones cause quick changes in input current, quick changes in output current must also occur. The audio amplifier may not be able to handle these quick changes in current. The difference in the response time of the input signal current as opposed to the response time of the output signal current is identified as a **rise time.** An input waveform and the resulting output waveform under transient distortion are given in Figure 7–21.

Transient distortion is generally the fault of poor amplifier performance or defective components within the amplifier. The amplifier with this type of distortion has poor linear operating characteristics. To test for transient distortion, a square wave is injected into the amplifier, and the output waveform observed on an oscilloscope. If the shape of the square wave changes between input and output, the amplifier has a good chance of developing transient distortion.

SUMMARY

Amplifiers are divided into small-signal and large-signal amplifiers. Small-signal amplifiers are generally found in preamplifiers of audio sections, IF amplifiers, and RF amplifiers.

It is often necessary to connect the output of one amplifier into another amplifier. This connection is called coupling. The purpose of coupling is to pass the maximum amount of signal from one amplifier stage to another.

In direct coupling, the output of one stage is fed directly into the next stage. Direct coupling has no effect on the amplitude of any frequency passed between amplifier stages. Directly coupled amplifiers are not very stable. That is, a change in one amplifier will affect the operation of the other amplifier.

The Darlington connection is one method of direct coupling

amplifier stages. Darlington amplifiers provide large signal gains. These amplifier configurations are generally found in a single casing.

Transformer coupling allows two amplifier stages to be isolated from each other. It also provides impedance matching between the two amplifier stages. Transformer coupling is not used with low frequencies, because transformers cannot pass those frequencies. Placing a capacitor across the primary or secondary causes the transformer to become tuned. The circuit can then be used to select frequencies. Generally, transformer coupling is found in the IF and RF sections of radios and televisions.

Capacitive coupling is another method of transferring the signal from one stage to another. The capacitor provides isolation between stages for dc bias voltages. Capacitive coupling creates a mismatch of impedances between the two amplifier stages. Capacitive coupling is the most common coupling method found in consumer products.

Feedback in amplifiers is either positive or negative. Positive feedback, also called regenerative feedback, is found where the gain of the amplifier must be increased, often in the RF amplifier section of radios and televisions. Negative feedback, also known as degenerative feedback, is developed in amplifiers to reduce the amount of gain. It is developed either through current feedback or through voltage feedback.

An emitter bypass capacitor is used to reduce the amount of negative feedback developed in amplifiers. This capacitor is placed across the emitter resistor to eliminate the current and voltage variations developed in the emitter lead.

Each amplifier is designed to amplify a certain set of frequencies. Some amplifiers are designed for low frequencies, while others are designed for higher frequencies. Because of dropoff of certain frequencies within the amplifier, compensating circuits are added to boost these frequencies during amplification.

Distortion in an amplifier occurs when the amplifier has a change in the shape of the waveform from the input to the output. Distortion comes in several different types: harmonic distortion, intermodulation distortion, crossover distortion, clipping distortion, and transient distortion.

KEY TERMS

capacitive coupling
cascading
clipping distortion
crossover distortion
current feedback
Darlington amplifier
direct coupling
distortion
double-tuned transformer

emitter bypass capacitor
frequency response
harmonic distortion
high-frequency cutoff
intermodulation distortion
low-frequency cutoff
negative feedback
positive feedback
RC compensating circuit
resonant frequency
rise time
single-tuned transformer
transformer coupling
transient distortion
voltage feedback

REVIEW EXERCISES

1. List several coupling methods found in consumer products.
2. What type of coupling should be used in a circuit where low frequencies are to be amplified?
3. What effect does a temperature change have in directly coupled amplifiers?
4. In a Darlington amplifier connection, each transistor develops a gain of 30. What is the total gain for the stage?
5. Which amplifier coupling should be used to deliver large current gains and high input impedance?
6. Why is a transformer an ideal component to use in amplifier coupling?
7. What turns ratio will be needed on a transformer to match a 30 kΩ amplifier output to an 8 Ω speaker?
8. List several disadvantages of transformer-coupled amplifiers.
9. What two things should be taken into consideration when selecting a capacitor for a coupling circuit?
10. What is the main disadvantage of capacitive coupling?
11. What phase relationship is developed in negative feedback?
12. Will an amplifier with negative feedback have an increased or decreased gain?
13. Will positive feedback increase or decrease amplifier gain?
14. What precaution should be taken with positive feedback?
15. Is the signal developed on the base and the emitter in phase or out of phase?
16. An emitter resistor will develop what type of feedback in an amplifier?
17. Negative feedback will reduce gain, but will it increase or decrease the frequency amplification range?
18. How can current feedback in amplifiers be removed?
19. Describe in your own words just how an emitter bypass capacitor operates. How is the size of the capacitor selected?
20. Which type of semiconductor material develops poor high-frequency response?
21. To make up for lost gain in the coupling stage, what type of circuit is used in the output circuit? Draw a schematic diagram to support your answer.
22. Name several different types of distortion developed in amplifiers. Draw a waveform of each.

8 LARGE-SIGNAL AMPLIFIERS

OBJECTIVES

Upon completing this chapter, you should be familiar with:

—Power amplifier operation classes
—Class-A power amplifier characteristics
—Class-B power amplifier characteristics
—Class-AB power amplifier characteristics
—Class-C power amplifier characteristics
—Troubleshooting power amplifiers

INTRODUCTION

Large-signal amplifiers are found in the output sections of radios, televisions, and a host of other electronic consumer products. Another name for the large-signal amplifier is the *power amplifier.*

An amplifier system consists of a signal pickup transducer, a small-signal amplifier, a large-signal amplifier, and an output transducer. The input transducer generally puts out a small-amplitude signal. The small-signal amplifier takes that signal and increases the amplitude. In these small-signal amplifiers, the primary function is to provide linear amplification of the input signal. Because the input signal from the transducer is small, the amount of signal power within small-signal amplifiers is generally low. Found between the small-signal amplifier and the large-signal amplifier is the voltage amplifier. This device will usually increase the voltage level to a distortion-free signal so that the large-signal amplifier can operate.

Power amplifiers must be efficient in operation. They must be able to transfer the maximum power drawn from the dc bias source, and develop it into maximum signal strength. The large-signal amplifiers described in this chapter range from amplifiers that handle only a few watts to those capable of handling several hundred watts. An important consideration of the power amplifiers is the power efficiency of the circuit, the maximum power capabilities of the circuit, and the impedance match between the amplifier and the load.

The first power amplifier discussed in this chapter is the class-A type. Single-ended, class-A and transformer-coupled, class-A amplifiers are considered. Other popular styles of power amplifiers are also introduced, such as the class-B, push-pull and the transformerless class-B power amplifier. Finally, the class-C power amplifier is discussed, including its uses in consumer products.

At the end of the chapter is a section devoted to troubleshooting power amplifiers. Because these large-signal amplifiers are found in many consumer electronic products, the service technician must be familiar with their operation.

AMPLIFIER CLASS

As noted in Chapter 6, amplifiers fall into different classes of operation. The amplifier can also be classified by the type of signal it amplifies. The large-signal or **power amplifier** is one type of amplifier based on signal amplification. The block diagram of a typical amplifier system is given in Figure 8–1. Notice that the input sound is converted into

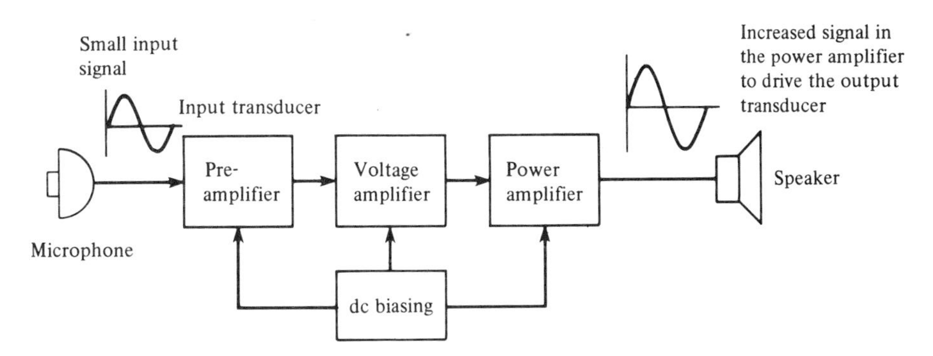

FIGURE 8–1
Block Diagram of a Typical Amplifier System

an electrical signal by the microphone, which is the **input transducer.** This signal developed at the input is only a few millivolts in amplitude. The two voltage amplifiers that follow are used only to increase the signal amplitude. The final stage is the power amplifier.

The power amplifier is designed to develop a large increase in the power amplitude of the input signal in order to drive the output transducer. Because the power developed at the output is made up of both voltage and current, the power amplifier must increase both voltage and current levels before driving the speaker, which is the output transducer.

It is important to understand where the power amplifier develops its power. The usable power for the power amplifier is developed from the available power seen in the dc power supply. Figure 8–2 contains a block diagram of a power amplifier and the dc bias supply needed for operation of the amplifier. Notice that the power supply is supplying 20 V at 1 A. Using Watt's law, it is shown in the figure that the power supply will deliver 20 W to the amplifier. These 20 W can be turned into signal power. However, note that only 10 W of signal power are being delivered to the load. Therefore, only half of the available power is being turned into signal power.

The ability of an amplifier to convert the dc power from the power supply into signal power is identified as the **amplifier efficiency.** A formula can be developed to determine the percent efficiency of a power amplifier. This formula is given as Equation 8–1.

$$\% \text{ efficiency} = \frac{\text{signal output power}}{\text{dc power input}} \times 100 \qquad \textbf{(8–1)}$$

For the power amplifier under study, the efficiency is calculated as follows:

$$\begin{aligned}\% \text{ efficiency} &= \frac{\text{signal output power}}{\text{dc power input}} \times 100 = \frac{10 \text{ W}}{20 \text{ W}} \times 100 \\ &= 0.5 \times 100 \\ &= 50\%\end{aligned}$$

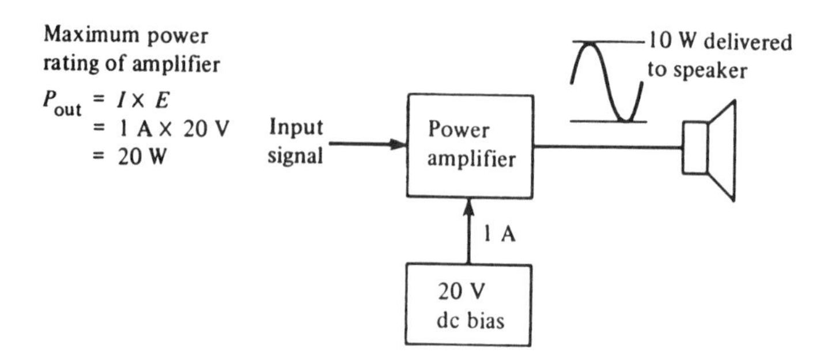

FIGURE 8–2
A dc Power Source Supplying Necessary Operating Bias

The maximum power rating of the amplifier is

$$P_{out} = I \times E = 1\text{ A} \times 20\text{ V}$$
$$= 20\text{ W}$$

Thus, in this case only 50% of the available power is being transferred into signal power. The remaining 50% of power is being wasted. This wasted power generally is given off as heat. It is for this reason that the power amplifier needs large heat sinks. The heat sinks are used to take the 50% wasted power away from the active device and keep that component operating within its temperature characteristics.

The type of power amplifier just discussed is a single-ended, **class-A power amplifier.** Remember that class-A amplifiers are biased so that current at the output flows for the entire input signal. Power amplifiers biased in the class-A mode waste a great deal of power, and therefore operate at low efficiency. However, in any amplifier, if the output signal is not a replica of the input signal, the amplifier distorts the signal. Since the class-A amplifier operates at the center of the load line, this type of amplifier will offer the lowest distortion to the signal than any other power amplifier configurations.

Figure 8–3 shows the point at which the class-A amplifier is biased. At this operating point distortion is low. This low-distortion characteristic is one of the greatest advantages of the class-A power amplifier.

Figure 8–4 shows another type of biasing point for the power amplifier. Here, the transistor is operated at cutoff. This is accomplished by placing zero bias between the base and the emitter of the transistor. The only way this transistor can conduct is if the input signal brings the transistor away from cutoff. Notice that as the input signal swings in the positive direction, the transistor is turned on, and the input signal is amplified. However, only one-half of the input signal is reproduced at the output. This type of amplifier is called a **class-B**

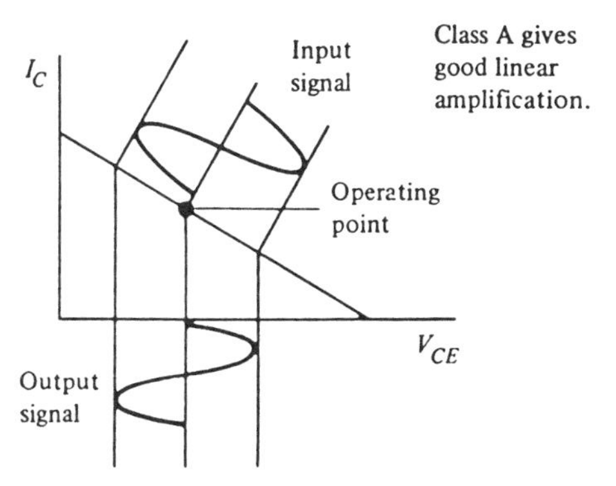

FIGURE 8–3
Biasing a Class-A Power Amplifier at Center of Load Line

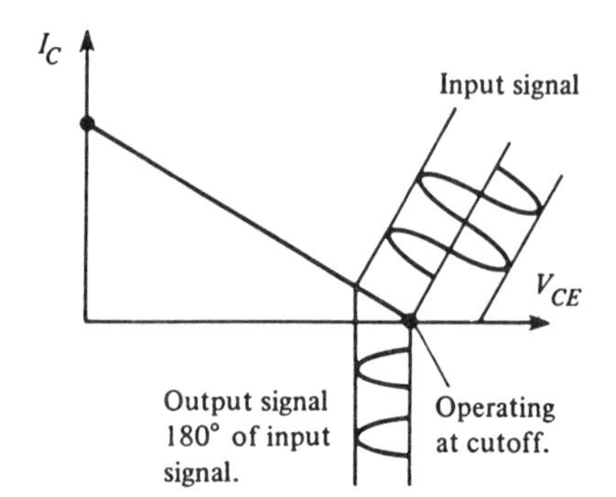

FIGURE 8–4
Operating an Amplifier in Class-B Area Near Cutoff

power amplifier. The class-A amplifier will conduct for the entire input cycle, whereas the class-B amplifier will conduct for only one-half of the input cycle.

Because the class-B amplifier amplifies only one-half of the input cycle, it develops a large amount of distortion. As stated in Chapter 7, distortion results when the output signal is not a reproduction of the input signal. In spite of this distortion, the class-B power amplifier will turn more dc power into signal power than the class-A. It is the biasing at cutoff that saves power. The reason is that the current flow through the power output stage is developed only half of the time. Since the class-B amplifier is biased at cutoff, there will be no current flow from the power supply until the signal appears at the input. Zero current flow in the transistor means zero wattage developed from the power source. Therefore, no current drain is drawn from the dc power supply until the signal appears at the input. Another characteristic of the class-B power amplifier is that the larger the input signal, the larger the current drain from the dc power supply. Larger current drain makes more dc power available to the amplifier to convert into signal power.

Because the class-B power amplifier draws more dc current when larger signals appear at the input, it is useful for large-signal amplification. Some of the distortion developed in the class-B amplifier can be removed by connecting two transistors in a *push-pull* arrangement. The class-B, push-pull amplifier is more efficient than the class-A, and develops better amplified signals at the output than a simple class-B amplifier. The push-pull connection is considered in more detail later in the chapter.

Additional classes of amplifiers can be developed by biasing the transistor at different points along the load line. Among these, the **class-AB amplifier** and the **class-C amplifier** can be used as power amplifiers. A summary of the different operation classes of power amplifiers is given in Table 8–1.

TABLE 8–1 Power Amplifier Class of Operation

Characteristic	Class A	Class AB	Class B	Class C
Distortion	low	moderate	high	extreme
Output signal vs. input signal	360° in 360° out	between class A and B	360° in 180° out	360° in about 90° out
Efficiency (average)	25%	between class A and B	60%	above 79%

TABLE 8–2 Power Amplifier Application

Class A	Class AB	Class B	Class C
Used in few audio power amplifiers. Generally found in small-signal amplifier stages.	High power stage found in audio and radio frequency application.	High-power audio output stages used in push-pull amplifier arrangement.	Needs tuned circuit in the output circuitry. Generally found in radio frequency application.

Many different types of amplifiers are used in consumer products. Each amplifier type performs a certain task. For example, **voltage amplifiers** are used for small-signal amplification, generally in **preamplifiers** in audio work, and in RF and IF sections of radios and televisions. Power amplifiers are usually found as the final stage in the consumer product. The common applications for the different classes of power amplifier are listed in Table 8–2.

CLASS-A POWER AMPLIFIERS

The simple fixed biased circuit found in Figure 8–5 is a class-A power amplifier. The only difference between the large-signal power amplifier and the small-signal amplifier is that the signal developed at the output in the power amplifier is larger in amplitude, and the output has larger amounts of power. As will be shown later, the class-A power amplifier is not the best type of amplifier to use in power applications.

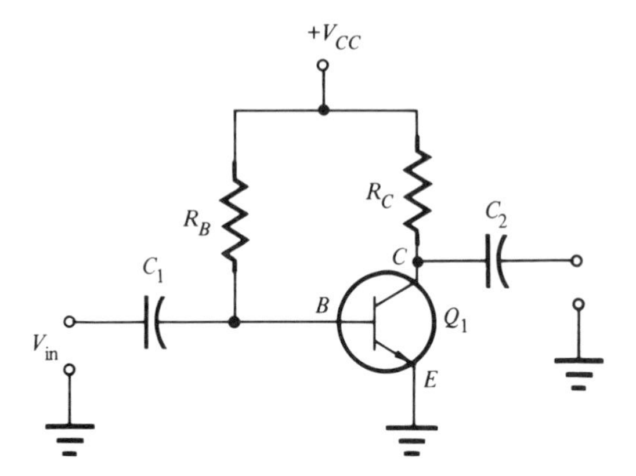

FIGURE 8–5
Biasing Established for a Class-A Power Amplifier

Series-Fed, Class-A Power Amplifier

Shown in Figure 8–6a is a **series-fed, class-A power amplifier** with the necessary biasing components. The characteristic curves for this transistor are shown in Figure 8–6b. Notice that the input signal to be amplified is only 0.5 V(p). The output signal is developed across a 20 Ω load resistance. The biasing of the base loop can be determined from the set of characteristic curves for this transistor. The voltage drop developed between the collector and the emitter is given as 10 V, and the collector current is given as 500 mA. Plotting these values on the curves gives a base current of 30 mA. This value of 30 mA will be the operating point of the transistor. The development of the load line for this transistor amplifier is the same as for the small-signal amplifier circuit.

An input signal amplitude is given for this amplifier. From this information, the change in base current can be calculated by the application of Ohm's law in the form of Equation 8–2.

$$I_t = \frac{V_1}{Z_t} \tag{8–2}$$

where:

I_t = total peak current developed at the base
V_1 = peak value of the input signal voltage
Z_t = total internal impedance of the base-emitter junction in forward bias. This impedance is about 50 Ω.

Thus, for the circuit under study,

$$I_t = \frac{V_1}{Z_t} = \frac{0.5\ \text{V}}{50\ \Omega}$$
$$= 10\ \text{mA(p)}$$

Therefore, a 10 mA change in current will develop in the base circuit.

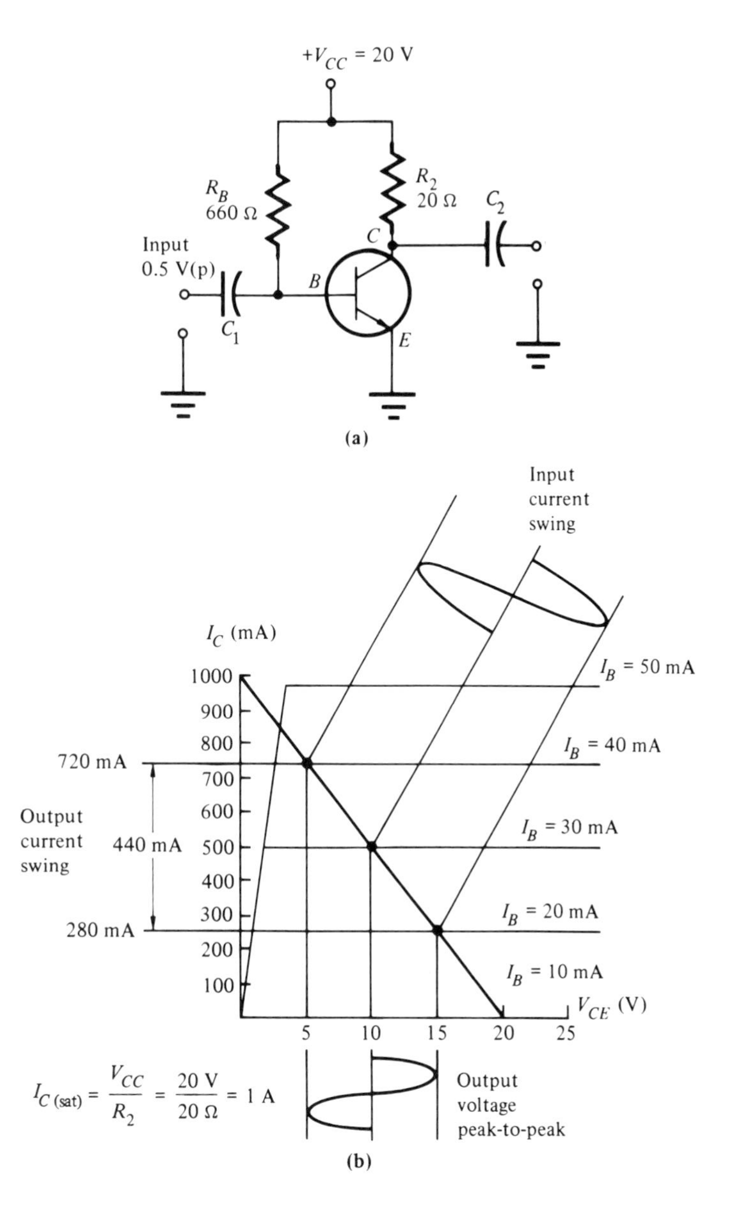

FIGURE 8–6
Operation of a Class-A Power Amplifier: (a) Schematic Diagram of Series-Fed, Class-A Amplifier, (b) Waveform Developed within Characteristic Curves

This value can be plotted on Figure 8–6b. This 10 mA value is doubled to find the peak-to-peak current change for the base circuit.

By placing this doubled value—20 mA—on the characteristic curve, the amount of collector current swing can be read. The change in collector current would be about 220 mA(p). Dropping down from the base current to the collector-emitter voltage line, it is seen that a change of 5 V(p) will be developed as the collector-emitter voltage drop. With this information, the gain of this transistor can be found by the following three-part equation:

$$A_v = \frac{\Delta V_o}{\Delta V_1} = 5 \div 0.5$$
$$= 10$$

$$A_i = \frac{\Delta I_o}{\Delta I_1} = 220 \div 10$$
$$= 22$$

$$A_p = A_v \times A_i = 10 \times 22$$
$$= 220$$

where:

A_v = voltage gain

ΔV_o = change in output voltage

ΔV_1 = change in input voltage

A_i = current gain

ΔI_o = change in output current

ΔI_1 = change in input current

A_p = power gain

Up until now, the power found for this amplifier is the signal power. As noted earlier, the efficiency of this amplifier depends upon the amount of dc power converted into signal power. This calculation for the efficiency of the amplifier was given in Equation 8–1. However, before this efficiency value can be found, the dc power delivered from the power supply must be calculated. This amount of power can be determined by the following equation:

$$P_{dc} = I \times E$$

Therefore, the dc power delivered in the amplifier under study is as follows:

$$P_{dc} = I \times E = 530 \text{ mA} \times 20 \text{ V}$$
$$= 10.6 \text{ W}$$

The signal wattage is found using a different procedure than that used for the dc signal. The signal power can be calculated using Equation 8–3.

$$P_{sig(out)} = I^2_{RMS} \times R_C \qquad \textbf{(8–3)}$$

where:

$P_{sig(out)}$ = power developed into signal power

I = RMS rating of the current

R_C = resistance found in the collector load

Checking the characteristic curves in Figure 8–6b, it is found that the output current is about 220 mA(p). This peak value must be converted into the RMS value. Recall that this change from peak to RMS is accomplished by multiplying the peak value times 0.707. Thus, in the circuit of Figure 8–6,

$$\text{signal current (RMS)} = 0.707 \times I_p = 0.707 \times 0.22 \text{ A} = 0.155 \text{ A}$$

The RMS value of current can now be placed into the signal power equation:

$$P_{\text{sig(out)}} = I^2 \times R_C = (0.155)^2 \times 20\ \Omega = 0.4805 \text{ W}$$

Because both the dc and the signal power values are known, the efficiency of this amplifier can be found. Values are substituted into Equation 8–1, as follows:

$$\%\ \text{efficiency} = \frac{P_{\text{sig}}}{P_{\text{dc}}} \times 100 = \frac{0.4805}{10.6} \times 100 = 0.0453 \times 100 = 4.53\%$$

Thus, the power efficiency of this class-A power amplifier is only 4.53%, which is extremely poor. The class-A amplifier shown in Figure 8–6 is not a good amplifier for handling large amounts of power. Of the total 10.6 W available for signal power, only 4.80%, or 0.453 W is turned into signal power. The other 95.52% is turned into heat. If a speaker were connected in place of the load, only 0.480 W of power would be developed across the speaker. In addition, the transistor used in this amplifier would need a very large heat sink to dissipate the heat from the transistor. The one advantage of this class-A amplifier is that since it operates at the center of the load line, there will be little distortion in the output waveform.

Transformer-Coupled, Class-A Amplifier

As we just learned, the series-fed, class-A amplifier has a low efficiency value. A more efficient circuit can be made with a class-A amplifier. By replacing the load resistance with a transformer, the efficiency of the amplifier can be improved. The **transformer-coupled, class-A power amplifier** uses a transformer to couple the signal from the output of the transistor to the load. Figure 8–7a shows an example of this type of connection.

The important characteristic of this type of connection is that the transformer will match the impedances between the primary and

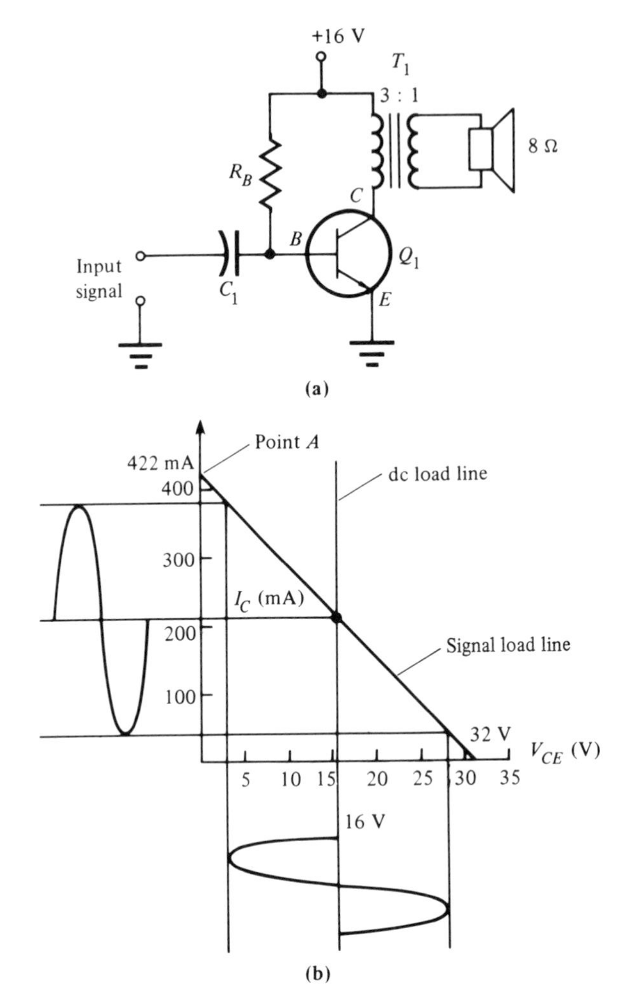

FIGURE 8–7
Transformer-Coupled, Class-A Power Amplifier: (a) Schematic Diagram of Class-A Amplifier, (b) Load Line Development to Determine Output Waveform

the secondary circuits. Typically, a speaker or output transducer is connected to the secondary of the transformer. When the collector of the transistor was connected directly to the load in the series-fed type, efficiency was poor. With the use of a transformer between the output of the transistor and the speaker, an impedance match is developed. Recall that when the impedances are matched between the output and the load, maximum power is transferred.

Transformer Matching The transformer in the transformer-coupled amplifier is used to match the impedance of the output circuit to the load. The impedance developed on the primary is related to the impedance developed on the secondary. This relationship is related to the number of turns found on the primary and secondary circuits. In

Chapter 1 some equations were presented that are used to determine this relationship between impedance and the number of turns. From these equations it was found that the input and output impedances of the transformer are directly related to the square of the turns ratio. This relationship can be expressed as follows:

$$\frac{Z_P}{Z_s} = \left(\frac{N_1}{N_2}\right)^2 = a^2$$

and impedance of the primary as

$$Z_p = a^2 \times R_L = \left(\frac{N_1}{N_2}\right)^2 \times R_L$$

where:

Z_p = impedance in the primary

Z_s = impedance in the secondary

a^2 = turns ratio of the transformer

R_L = impedance of the load

$a = \frac{N_1}{N_2}$ = turns ratio of the transformer windings. This ratio will allow the primary and secondary impedances to be equal.

The problems in Examples 1 and 2 will help illustrate the calculation of the effective impedance in the primary circuit, and the matching of the transformer impedance.

EXAMPLE 1

A step-down transformer has a 15:1 turns ratio, and an 8 Ω speaker is connected to the secondary. What is the effective resistance in the primary circuit?

Solution:

$$Z_p = \left(\frac{N_1}{N_2}\right)^2 \times R_L = \left(\frac{15}{1}\right)^2 \times 8 = 15^2 \times 8$$

$$= 1.8 \text{ k}\Omega$$

Thus, the effective impedance of the primary circuit is 1.8 kΩ. This means that the primary will develop a 1.8 kΩ load, which will determine current in the primary.

EXAMPLE 2

An amplifier has 20 kΩ of primary impedance, and the secondary is connected to an 8 Ω speaker. What is the turns ratio needed to match the primary and secondary impedance?

$$\frac{Z_p}{Z_s} = \left(\frac{N_1}{N_2}\right)^2 = \frac{20{,}000}{8}$$
$$= 2500$$

$$\frac{N_1}{N_2} = \sqrt{2500}$$
$$= 50$$

Therefore, for this circuit to match the impedance, the transformer must have a 50:1 turns ratio.

Transformer dc Load Line The power amplifier can best be described with the load line. As with small-signal amplifiers, the load line determines the operating characteristics of the amplifier. Load lines can also predict the operation of the power amplifier. To establish this load line, the transistor must be placed in cutoff, and also in saturation. Figure 8–7a shows an example of a transformer-coupled power amplifier.

Figure 8–7b shows the development of a dc and a signal load line used to predict the operation of the transformer-coupled power amplifier.

First, the transistor is placed in the cutoff bias. When this happens, all the supply voltage develops across the collector and emitter terminals. This point is marked on the V_{CE} axis. Next, the maximum current developed in this circuit must be found. The current flow can be calculated using Ohm's law as follows:

$$I_C = \frac{V_C}{R_C}$$

However, because the transformer is in the primary circuit, the dc load resistance is very low for the transistor. This near zero resistance makes the dc load line a vertical line from the voltage drop across the collector-emitter terminals. This line has no slope, and therefore makes it hard to predict the amount of amplification that this amplifier will develop. To find the amplification that will develop, an ac load line must be established.

Transformer Signal Load Line The development of the signal load line is also shown in Figure 8–7b. Again, note that the dc load line is vertical. The operating point for the base current in the signal load line is the same as that in the dc load line. The base current remains at the same point because the collector biasing does not affect the base current.

To develop the signal load line, the impedance offered by the primary of the transformer must be found. Because the transformer is a step-down transformer, the impedance is calculated as the square of the turns formula used earlier, and the impedance in the primary as follows:

$$Z_p = a^2 \times R_L$$

Using the circuit in Figure 8–7a, the amount of impedance offered in the primary is calculated as

$$Z_p = a^2 \times R_L = 3^2 \times 8 = 9 \times 8$$
$$= 72\ \Omega$$

Therefore, the impedance that the transformer primary will offer to the collector circuit is 72 Ω.

The following formula is used to locate the current point on the I_C axis:

$$I_C = \frac{V_{CE}}{Z_p}$$

This formula gives the maximum current swing that can be developed in the primary. Using the values from Figure 8–7, the following current swing in the primary is found:

$$I_C = \frac{V_{CE}}{Z_p} = \frac{16\ \text{V}}{72\ \Omega}$$
$$= 222\ \text{mA}$$

This 222 mA is the maximum swing that can be expected in the primary. This value must be added to the value found when the transistor was operating under the static dc bias condition to find the total signal and dc current in the circuit. The dc load line gives this as 200 mA. The total current value, then, is 422 mA. This value has been placed on the load line in Figure 8–7b, and has been designated as point *A*. From point *A*, a line has been drawn through the original operating point. This line is the load line for the signal condition in the amplifier. It shows that the maximum collector-emitter voltage is 32 V. The doubling of the collector-emitter voltage is the result of the magnetic field developed in the primary circuit. When this magnetic field collapsed, the voltage created by the collapsing field was added to the collector supply voltage. This is the basic inductor theory.

Because the swing on the primary voltage is double the supply voltage, the selection of the transistor is very important. The transistor must be able to handle large amounts of voltage between the collector and emitter terminals. Checking the transistor's characteristics will show whether the proper transistor has been selected.

Because the output voltage swing for this transistor amplifier circuit has doubled, the power must be quadrupled at the load. Remember that power varies directly as the voltage is squared. Therefore, if the voltage is doubled, the power is quadrupled. As stated earlier, the power in the primary must equal the power in the secondary. This means that class-A power amplifier circuits that are transformer coupled are about 50% efficient. This type of amplifier makes an excellent power amplifier in the 5 W and below range. This classifies the class-A as a medium power range power amplifier.

The largest drawback for the transformer-coupled, class-A amplifier is the transformer itself. The transformer is an expensive and bulky circuit component, often costing more than the rest of the amplifier circuit. For these reasons, this type of amplifier may not be the best choice for the final stage of the consumer product.

The preceding discussion has characterized the transformer as an efficient component. However, the transformers used in consumer products are only about 75% efficient, which is a very poor rating. This means that the secondary signal development will be lower than was originally stated.

Another problem with the class-A power amplifier is the fixed drain placed on the power supply. Even when there is no signal at the input of the amplifier, the transistor is still conducting current. Because the collector is still conducting, there is a fixed drain, and this fixed drain causes power development in the transistor. Finally, the class-A power amplifier exhibits poor amplification between the input and the output when the signal at the input is low in amplitude. Because of the overall poor showing of the class-A power amplifier, other types of power amplifier circuits should be used.

CLASS-B POWER AMPLIFIERS

The class-B power amplifier is biased at cutoff. No current flows in this amplifier until the input signal brings the transistor away from the cutoff bias. The signal must be of sufficient amplitude for conduction to occur. Because the transistor conducts only when the signal is applied, the fixed drain of current is removed from the power supply. The efficiency of the amplifier is therefore improved.

The major drawback of the class-B power amplifier is the amount of distortion created. Because only one-half of the input signal is amplified, one class-B transistor by itself is useless. To overcome this problem, two transistors are connected in a **push-pull arrangement.**

In the push-pull arrangement, one of the transistors amplifies the positive half of the waveform, and the other transistor amplifies the negative half of the waveform. The two halves of the amplified waveform appear at the output. Here, the parts are combined in their amplified states.

Note in Figure 8–8a that two transformers are used for the push-pull connection. Transformer T_1 is the **driver transformer.** A center-tapped transformer is used for T_1 so that the signal is split into two phases. Figure 8–8b shows that from center tap to terminal A, the

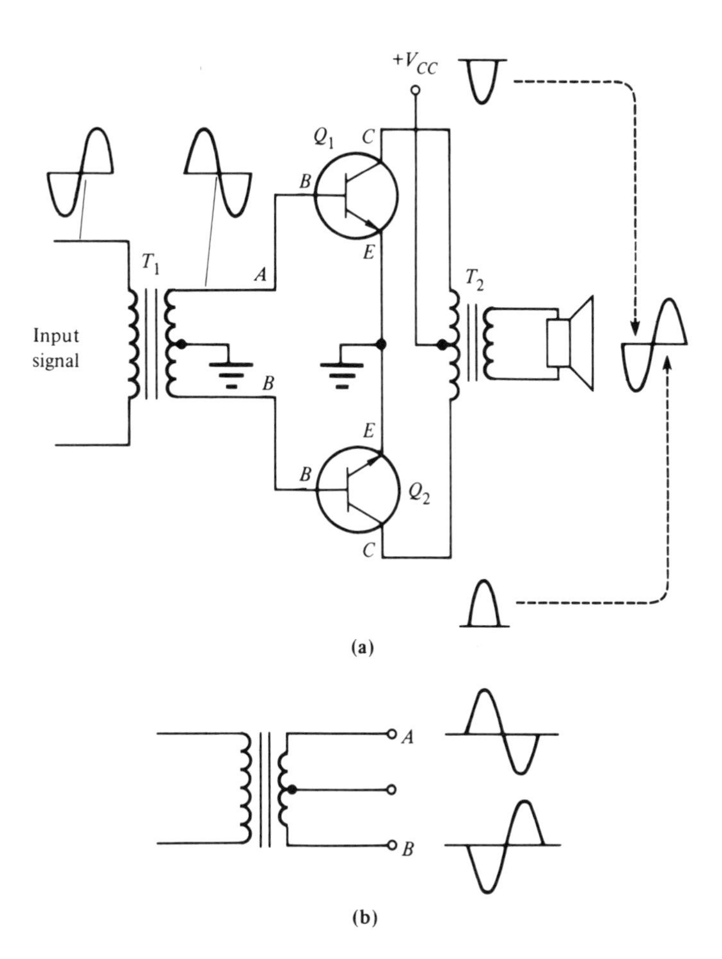

FIGURE 8–8
Class-B Push-Pull Power Amplifier: (a) Schematic Diagram of Push-Pull Arrangement, (b) Phase Shift in Center-Tapped Transformer

signal is developed in the positive polarity. From the center tap to terminal *B*, the signal is developed in the negative polarity. Terminal *A* waveform is fed to transistor Q_1. Since this polarity is positive, it turns on transistor Q_1. The negative polarity signal developed from terminal *B* on the transformer is fed to Q_2, which is an NPN transistor. This negative waveform will keep Q_2 in cutoff. Transformer T_2 is the **output transformer.** Notice that this transformer is also center-tapped.

Figure 8–9 shows what happens when a signal is applied at the input of the amplifier in Figure 8–8a. With no signal applied at the input, Q_1 and Q_2 are biased at cutoff. Therefore, no current drain is placed on the power supply. When the input signal appears at T_1, secondary transistor Q_1 is turned on by the positive polarity. Current is drawn through the top half of T_2. Because current flows through the top part of T_2, the positive alternation is developed at the secondary of T_2. Since current is drawn through the top half, Q_1 is biased on, and Q_2 is still biased off.

As the polarity of the input signal changes, the transistors reverse their biasing states. Transistor Q_2 is now turned on, and transistor Q_1 is turned off. Figure 8–10 shows this action for the circuit in Figure 8–8a. Note that the current flows up through the secondary. This action develops the negative polarity at the secondary of T_2. Now, both the positive and negative alternations are developed at the transformer secondary. Since both the negative and positive alternations have been reproduced, the distortion of this amplifier has been greatly reduced.

To explain the operation of this amplifier, a load line for the circuit can be developed. Shown in Figure 8–11 is a load line developed

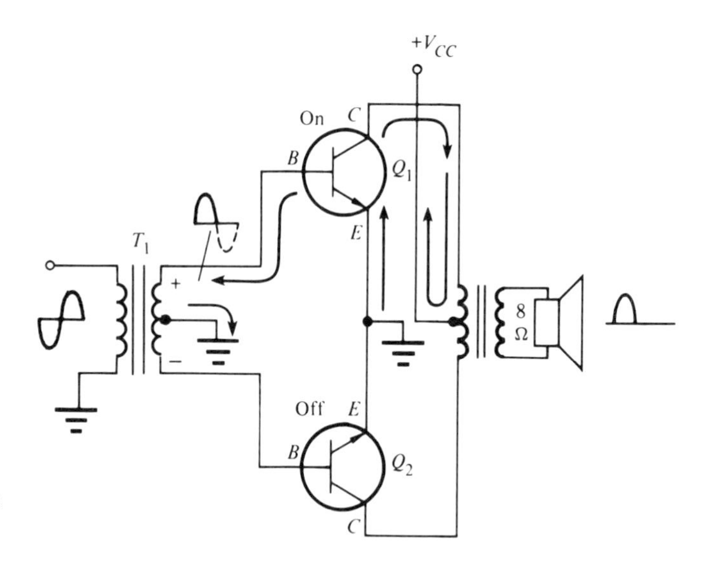

FIGURE 8–9
Current Path Established during the Positive Input Cycle on the Class-B Power Amplifier in Figure 8–8

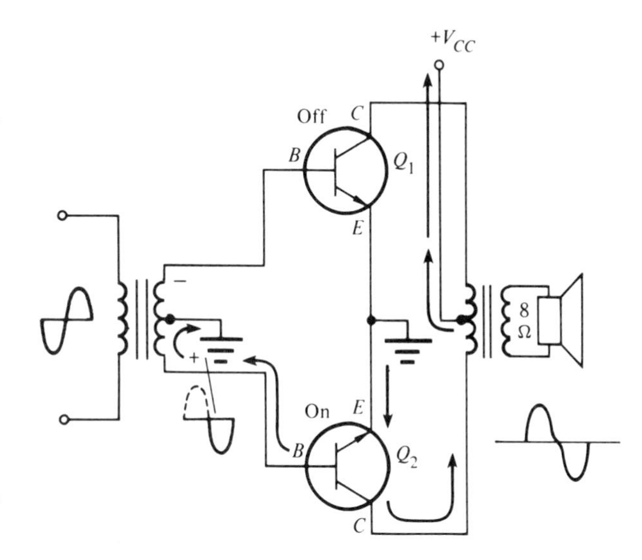

FIGURE 8–10
Current Path Established during the Negative Input Cycle on the Class-B Power Amplifier in Figure 8–8

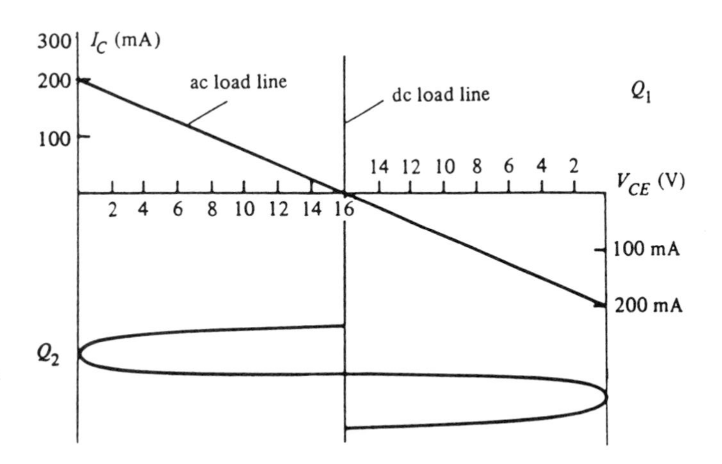

FIGURE 8–11
Load Line Analysis of the Class-B, Push-Pull Power Amplifier in Figure 8–8

for the class-B, push-pull power amplifier in Figure 8–8a. As shown earlier the dc load line for the class-A, transformer-coupled amplifier is completely vertical because of the near zero resistance factor of the primary windings.

To develop the load line for the class-B amplifier, a signal load must be established. First, the primary impedance must be found. The procedure is the same as that used to find the transformer-coupled, class-A signal load line. Assume the transformer in Figure 8–8a has a 6:1 turns ratio. This ratio establishes the amount of impedance that will be seen in the primary circuit. Since the primary is center-tapped, the turns ratio must be divided in half. The turns ratio can now be shown as

$$\frac{6}{2} = 3{:}1$$

Now the load for the primary circuit can be calculated. Using the square of the turns times the load, this impedance for the circuit under study is as follows:

$$Z_p = a^2 \times Z_s = 3^2 \times 8 = 9 \times 8$$
$$= 72\ \Omega$$

Thus, the primary circuit will develop a 72 Ω impedance.

Collector current can now be calculated for one of the two transistors:

$$I_C = \frac{V_{CE}}{Z_p} = \frac{16\text{ V}}{72\ \Omega}$$
$$= 222\text{ mA}$$

A line is drawn from the I_C point to the V_{CE} point in Figure 8–11. It must be remembered that this line represents conduction of only one transistor. Since both transistors in the push-pull conduct, the line must be extended into the negative part of the graph. Note that in this region of the graph, the collector-emitter voltage goes from 16 V toward zero, while the collector current goes from zero to 200 mA. This allows a full swing in output voltage of 32 V when both transistors are conducting.

Once the voltage swing has been identified, the power developed from this circuit can be found. Since the 32 V reading is a peak-to-peak value, it must be converted into an RMS rating. This is done as follows:

$$V_{\text{p}} = \frac{V_{\text{p-p}}}{2} = \frac{32\text{ V}}{2}$$
$$= 16\text{ V}$$

$$V_{\text{RMS}} = V_{\text{p}} \times 0.707 = 16\text{ V} \times 0.707$$
$$= 11.31\text{ V}$$

Next, current can be found in this circuit. Again, the current value from the graph is rated peak-to-peak. To calculate the power, the current must be converted into an RMS reading. The conversion is done in the following manner:

$$I_{\text{RMS}} = \frac{I_{\text{p-p}}}{2} \times 0.707 = \frac{444\text{ mA}}{2} \times 0.707 = 222\text{ mA} \times 0.707$$
$$= 156.9\text{ mA}$$

Finally, the power developed in this amplifier can be calculated as follows:

$$P_{sig} = V \times I = 11.31 \text{ V} \times 156.9 \text{ mA}$$
$$= 1.7 \text{ W}$$

This power value can be used to find the efficiency of this amplifier. First, however, the power drawn from the dc power source must be determined. The two necessary factors can be obtained from Figure 8–11. The collector-emitter voltage is 16 V, and the collector current is 222 mA. Again, the 222 mA is the current draw when only one transistor is conducting. Therefore, to identify the total current draw for both transistors, the average value of this current flow must be calculated. The average value can be found by multiplying the peak rating of the current flow times 0.636:

$$I_{av} = I_p \times 0.636 = 222 \text{ mA} \times 0.636$$
$$= 141.1 \text{ mA}$$

Now, the power draw from the dc supply can be calculated. This can be done as follows by using Watt's law:

$$P_{dc} = I \times V = 141.1 \text{ mA} \times 16 \text{ V}$$
$$= 2.25 \text{ W}$$

Note in this calculation that the average current flow value was used to find wattage.

The efficiency of this class-B amplifier can now be found by the following calculation:

$$\% \text{ efficiency} = \frac{P_{sig}}{P_{dc}} \times 100 = \frac{1.7 \text{ W}}{2.25 \text{ W}} \times 100$$
$$= 75.5\%$$

Therefore the class-B power amplifier delivers about 75% of the dc power to the transistors to turn into signal power. As noted previously, the class-A power amplifier has an efficiency of only 5%. The class-B has improved that efficiency rating by over 70%. This greater efficiency enables the push-pull amplifier to amplify the lower-level input signals. In addition, only about 25% of the power is being turned into heat. Therefore the class-B power amplifier requires a smaller heat sink than the class-A power amplifier. This characteristic makes the push-pull power amplifier very useful in high-power applications.

The excellent efficiency rating of the class-B, push-pull power amplifier means that only a small amount of power is being wasted. The wattage rating for the class-B amplifier is low—only about one-fifth that developed in the class-A amplifier. For example, to build a 100 W class-B power amplifier, the transistors needed would only have

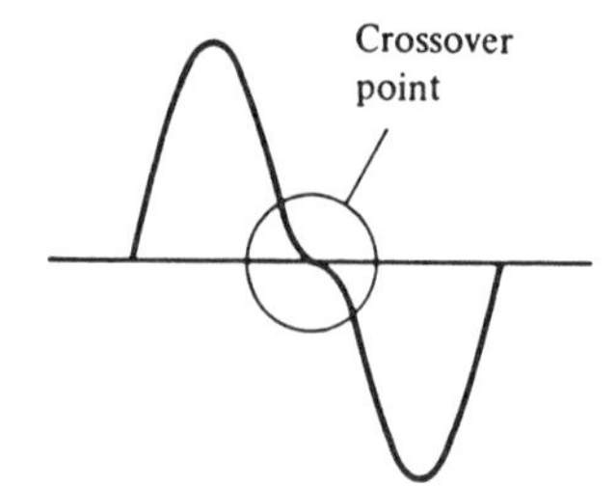

FIGURE 8–12
Class-B, Push-Pull Power Amplifier Developing Crossover Distortion

to be rated at 10 W each. First of all, the class-B amplifier develops only about one-fifth of the power at the transistors. Therefore in the case at hand,

$$100\text{ W} \times \frac{1}{5} = 20\text{ W}$$

This figure of 20 W is a total. Since there are two transistors in the push-pull arrangement, the 20 W must be divided in half. Therefore each transistor will develop 10 W. If a class-A power amplifier were used instead of this push-pull configuration, the wattage rating of the transistor would have to be larger because of the inefficiency of the class-A type.

There is only one major drawback to class-B power amplifier operation. This disadvantage is called **crossover distortion.** Figure 8–12 shows crossover distortion developed in the push-pull waveform. The term *crossover* refers to the point at which one transistor turns off and the other turns on. Crossover distortion is very noticeable at low amplitudes, and less noticeable when the signal is of larger amplitude. When troubleshooting the class-B power amplifier, it is useful to know that crossover distortion occurs at low amplitudes.

The high efficiency rating of the class-B amplifier makes this type of amplifier very attractive for audio output stages of many consumer products. However, transformers are involved again, and as was noted for class-A power amplifiers, transformers are large and expensive. The two transformers in the push-pull amplifier may be more expensive than all the other components taken together. The lower efficiency ratings of the transformers will develop lower output signals.

CLASS-AB POWER AMPLIFIERS

Because the class-B power amplifier is biased at cutoff, it develops crossover distortion. The way to reduce this distortion is to place a small forward bias to the output transistors. This small forward bias

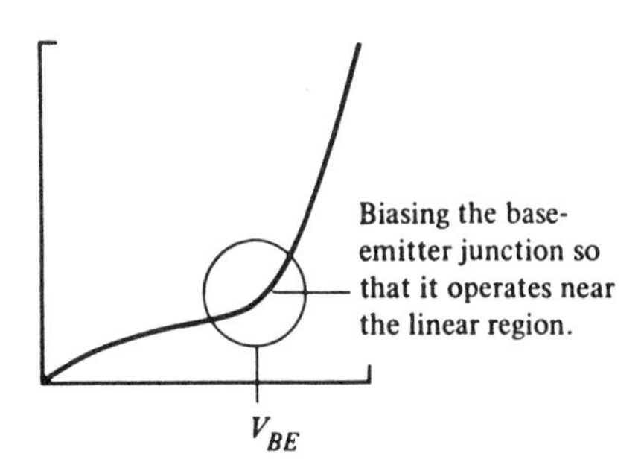

FIGURE 8–13
Forward Bias Curve for Class-AB Power Amplifier

applied to the base-emitter junctions causes the transistors to operate closer to the center of the linear region. Figure 8–13 shows the forward bias characteristic curve of a diode. Because the base-emitter junction operates like a diode, by applying a small amount of forward bias, the junction is made to operate closer to its linear region. This means that in the transistors, the small forward bias keeps the transistors on for more than 180° but less than 360° of the input cycle. The small forward bias moves the transistors into the class-AB power amplifier class just above the cutoff point. Class-AB is between class-A and class-B operation.

Figure 8–14 shows two transistors connected in the class-AB power amplifier arrangement. Notice in the drawing that R_1 and R_2 comprise a voltage-divider network. This network is used to develop the forward bias on Q_1 and Q_2. Capacitor C_1 is used to ground the

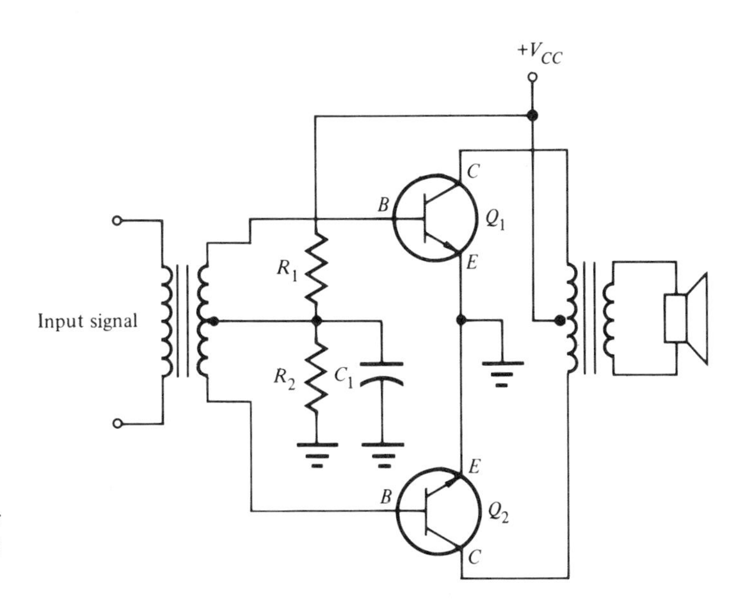

FIGURE 8–14
Biasing a Push-Pull Power Amplifier in the Class-AB Range

center tap of the driver transformer. This capacitor allows the signal to pass to ground without being dropped across R_2, thus raising the gain of the amplifier.

Class-AB power amplifiers are very popular for high-power audio work. These amplifiers will deliver slightly less efficiency than class-B power amplifiers. In fact, the specifications for the class-AB amplifier are the same as those for the class-B arrangement. The only real difference is that the class-AB amplifier removes much of the crossover distortion developed in the class-B amplifier. The class-AB power amplifier is found in the output sections of portable radios, tape recorders, and other consumer products.

Again, the major drawback of this amplifier type is the presence of transformers. Transformers are expensive, and are not the most efficient electronic components.

Complementary Symmetry Amplifier

The driver transformer and the output transformer of the class-AB power amplifier can be eliminated by replacing Q_1 and Q_2 with opposite-polarity transistors. Figure 8–15 shows an example of this amplifier connection, called the **complementary symmetry power amplifier.** The action of this circuit is very similar to that of the push-pull amplifier. Positive dc voltage is developed at the base-emitter junctions to bring these transistors into class-AB operation. A positive-going signal applied to the base of the NPN transistor, Q_1, turns the transistor on. The same positive-going signal applied to the PNP transistor, Q_2,

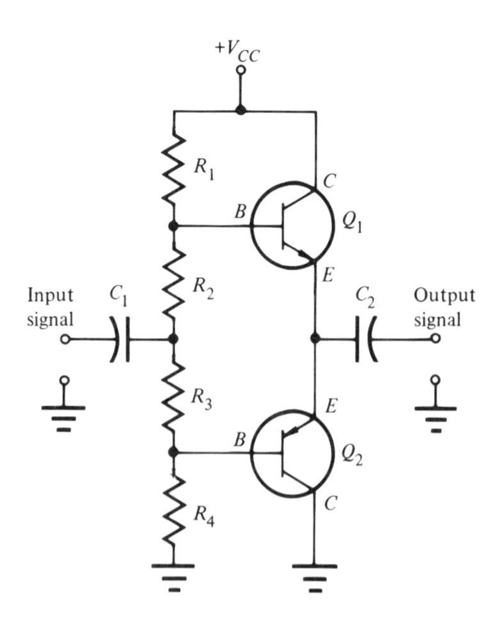

FIGURE 8–15
Complementary Symmetry Power Amplifier

turns this transistor off. The polarity of the input signal is therefore responsible for turning the two transistors on and off. Remember that this is the responsibility of the driver transformer in the push-pull amplifier connection.

The output transformer of the class-AB amplifier is eliminated by the complementary symmetry amplifier. The output from the amplifier is taken from the emitters. Because the output is taken from the emitters, the transistors are connected in a common collector configuration. Since the common collector amplifier develops low output impedance, the output of this amplifier can be connected to a speaker.

The NPN and the PNP transistors are *complements* of one another, hence the name of the amplifier. For these transistors to operate at maximum efficiency, they must have the same characteristics. The matching of these two transistors is important to the operation of this amplifier. For this reason, when a problem develops in the amplifier, both transistors should be replaced. Generally these two transistors will come as a pair, and should be replaced as a pair.

Figure 8–16 shows the biasing action for the complementary symmetry amplifier. This amplifier is biased just above cutoff, and therefore is biased at the class-AB region. Both transistors are biased on, to establish a rest current. *Rest current* describes the condition of the amplifier when no signal is applied. The base biasing is established by three resistors and a driver transistor. These components are connected in series, and are connected to a 32 V supply. The driver transistor Q_6, is biased to operate as a class-A amplifier. This type of biasing on Q_6 allows a linear signal to be applied to the output transistors. The conduction of current through Q_6 develops voltage drops across R_{44} and R_{42}. These voltage drops, in conjuction with Q_{10}, set the rest current state for this power amplifier.

Resistor R_{46} develops a voltage drop of about 0.1 V. This, then, results in an emitter voltage on Q_{10} of about 15.9 V. This small voltage drop has forward biased Q_{10}, the PNP transistor. The emitter voltage of Q_8 is developed by the center bus of this amplifier. Resistor R_{44} develops a base voltage of 16.1 V. This higher base voltage provides a small forward bias for Q_8. These small forward bias voltages have raised the amplifier into class-AB operation. The small forward bias eliminates the crossover distortion.

The center-bus voltage of the complementary symmetry amplifier reveals an important operating characteristic. When the amplifier is not driven by an input signal, the bus voltage is about one-half the supply voltage.

The *predriver stage,* Q_4, is an inverted amplifier configuration used as a buffer for the driver stage. Notice that the emitter is operating at about one-half the supply voltage, and the collector is grounded

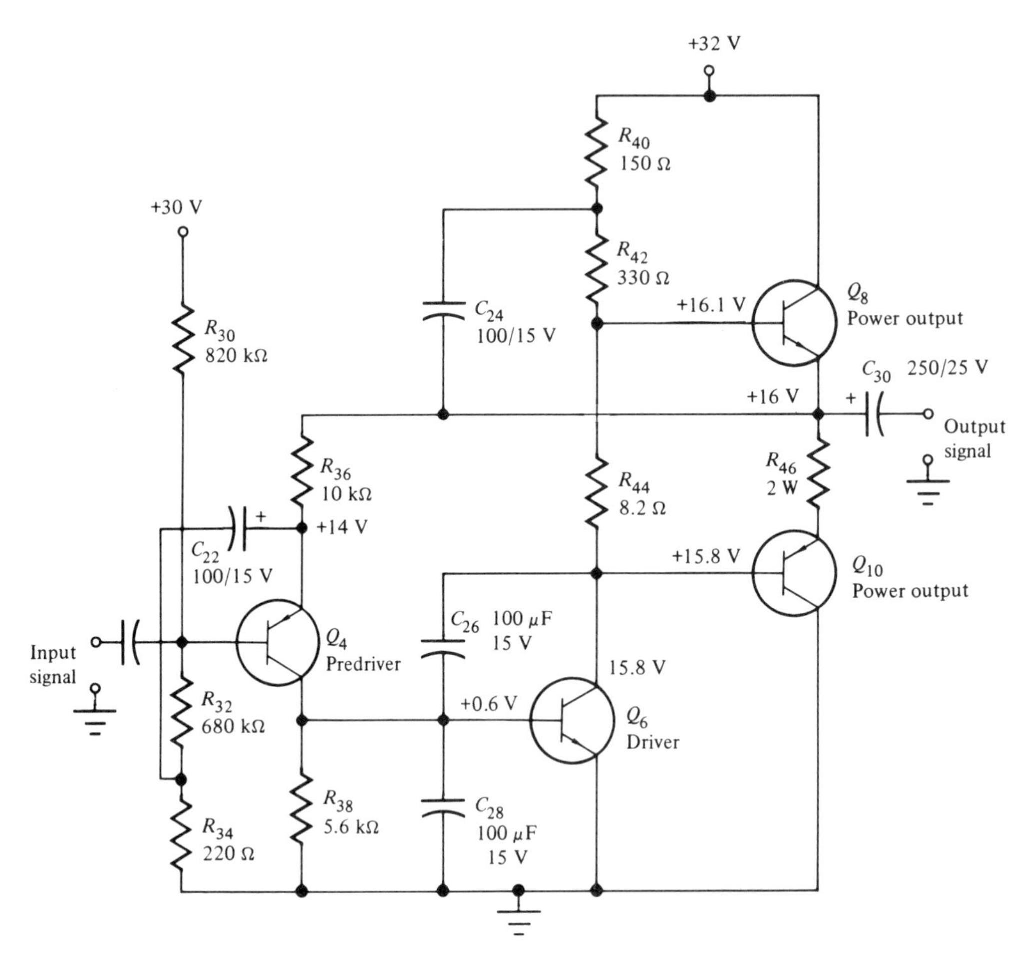

FIGURE 8–16
Biasing a Complementary Symmetry Amplifier

through a 5.6 kΩ resistor. As the input signal swings negative, it increases conduction through this transistor. This, then, increases the voltage drop across R_{38}. Then Q_6, the driver stage, increases conduction. As Q_6 starts to conduct harder, the voltage at the collector lowers. Because the collector of Q_6 is directly coupled to the base of Q_{10}, the voltage at the base of the output transistor causes Q_{10} to increase its forward bias. At the same time, the voltage at the base of Q_8 is being lowered. Lowering the base voltage on an NPN transistor will cause the transistor to turn off. Thus, by the application of the input signal, Q_{10} has turned on, and Q_8 has turned off.

Capacitor C_{24} acts as a filter in the circuit. It will supply the collector (dc) supply voltage. The capacitor will charge up to the applied voltage, depending on how much conduction is seen through Q_{10}.

Quasi-Complementary Symmetry Amplifier

Another type of amplifier configuration is shown in Figure 8–17. This type of class-AB power amplifier is called the **quasi-complementary symmetry power amplifier.** Notice that the output transistors, Q_3 and Q_4, are not complementary. They are both NPN types. The complementary types of transistors have been moved back to the driver stage. This reduces the cost of the amplifier network.

Because Q_3 and Q_4 are both NPN types, they should have the same characteristics. Transistors Q_1 and Q_2 are complementary, and drive the output.

Getting proper bias for Q_1 and Q_2 might seem difficult since both receive their bias from Q_5. Because the center-bus voltage is the reference for the emitters of Q_1 and Q_2, the potential difference between them will tend to remain constant. Resistors R_1 and R_2 set the

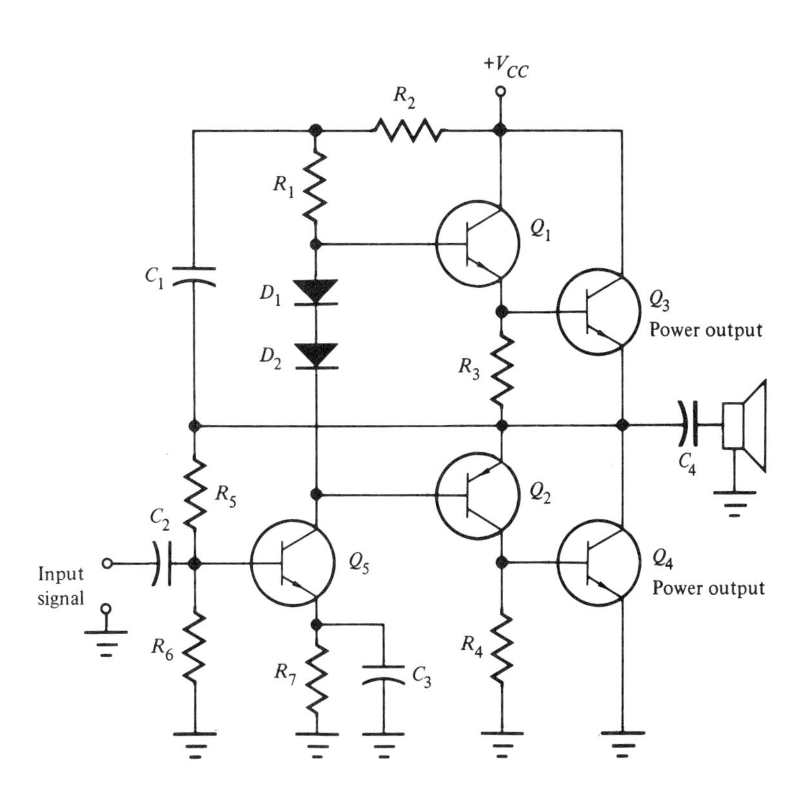

FIGURE 8–17
Quasi-Complementary Symmetry Power Amplifier

proper bias on Q_1. Diodes D_1 and D_2 are forward biased. This forward bias action keeps Q_2 at the proper biasing point.

When the audio signal appears at the input, it appears at the collector of Q_5. The signal is then passed to the bases of Q_2 and Q_1. The signal is allowed to appear at Q_1 because the diodes are forward biased and have low resistance. Because the forward biasing has been applied to Q_1 and Q_2, crossover distortion is removed from the output signal.

Capacitor C_1 prevents collector-base feedback problems in Q_1. If the power supply is highly regulated, C_1 is not needed. However, because the regulation in the majority of consumer products is not good, C_1 must be included. This capacitor will prevent hum and signal variations of the power supply from being amplified.

The diodes in Figure 8–17—D_1 and D_2—are used as biasing points for the two transistors. These biasing diodes are responsible for keeping the transistors forward biased, and keeping them operating in the class-AB region.

The diodes also function as **temperature-compensating diodes.** The transistors tend to heat up when conducting, and when a semiconductor heats up, the operating point of the amplifier changes. This shift in operating point will cause distortion in the amplifier. The diodes in the configuration tend to conduct more as the temperature in the circuit starts to rise. As the temperature increases in the power amplifier, the voltage drop across the diodes also decreases. This decrease in voltage drop decreases the forward bias at the transistor, thus reducing the current conduction. A reduction of current flow brings the operating point of the circuit back to a stable condition. Thus, with the diodes in the output section, the amplifier is more stable.

The class-AB, push-pull power amplifier is very popular in home entertainment products because of its good gain qualities and low distortion.

CLASS-C POWER AMPLIFIERS

The final type of power amplifier found in consumer products is the class-C power amplifier. A schematic drawing of the typical class-C amplifier is shown in Figure 8–18. Notice the bias voltage applied to the base. This voltage is negative, which places the NPN transistor far into cutoff. The only way that this amplifier will conduct is if a very positive signal is applied to the base. Generally this will occur when the input signal reaches the most positive part of the input waveform. Therefore it can be stated that the class-C amplifier will reproduce less than 90° of the input signal at the output.

Because the input signal allows the output only a short conduc-

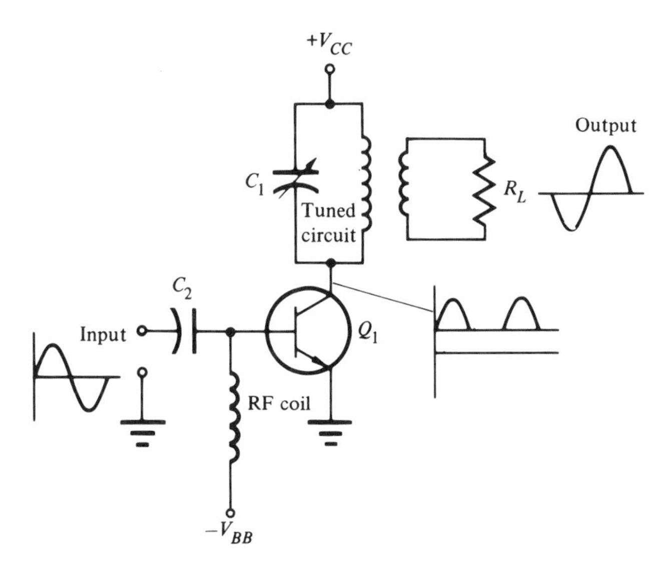

FIGURE 8–18
Schematic Diagram of a Class-C Power Amplifier

tion period, the output waveform is reduced. As shown in Figure 8–18, the output waveform is not a replica of the input waveform. There is a great deal of distortion in the output signal. The great amount of distortion makes this amplifier unsuitable for audio application. The main application of this amplifier is in RF communication products.

To become an effective amplifier, the class-C power amplifier must restore the input waveform. The reconstruction of the sine wave is accomplished by a tuned circuit. Refer to Figure 8–18. As the collector voltage is turned on by the input waveform, C_1 is charged. After the pulse has charged C_1, the transistor turns off, thus removing the pulse at the collector. The capacitor begins to discharge. The path of the discharging capacitor is through the inductor. A magnetic field is maintained around the windings of the inductor.

Once the capacitor has discharged completely, the magnetic field around the inductor collapses. The collapsing magnetic field maintains the current flow in the circuit. This current flow recharges C_1, but this time in the opposite polarity. After the magnetic field has collapsed completely, C_1 begins to discharge, developing a current in the opposite direction through the inductor's winding. This discharging capacitor creates another magnetic field around the inductor in the opposite polarity. The action of the discharging capacitor and magnetic building up and collapsing is the action of this tuned circuit. This back-and-forth motion establishes the missing sine wave at the output of the circuit. In real circuit operation, this does not happen because of the circuit impedance. The waveform that will actually appear at the output of this tuned circuit is shown in Figure 8–19. This type of waveform is called a *damped waveform.*

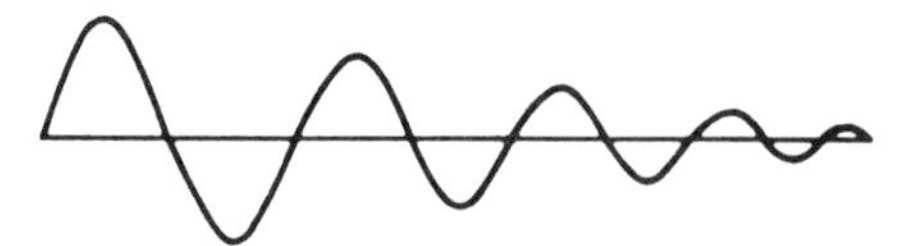

FIGURE 8–19
Damped Waveform Developed at the Output of a Class-C Tuned Amplifier

The input signal on the class-C amplifier will bring the amplifier from the deep cutoff region into the operating region for a short period of time. During this short period of time, a pulse is generated at the output of the amplifier. This pulse is then fed to the tuned circuit, which is then excited by the pulse. The tuned circuit will resonate, producing a sine wave at the output. Before the tuned circuit causes the waveform to be damped, another pulse is delivered by the class-C output that excites the tuned circuit into resonating and producing another sine wave output. The sine wave developed by the tuned circuit is then fed to the load.

The values of the capacitor and inductor are important for the establishment of the output frequency. When the input frequency is known, it is easy to select the size inductor and capacitor necessary to re-create the incoming frequency of the incoming waveform. With knowledge of the size of inductor and capacitor, the **resonant frequency,** the frequency at which the circuit will become resonant, can be calculated. The formula for this operation frequency is given in Equation 8–5.

$$f_r = \frac{1}{6.28\sqrt{LC}} \qquad \textbf{(8–5)}$$

Example 3 illustrates the use of this formula.

EXAMPLE 3

A 10 μH coil and a 0.01 μF capacitor are found in the tuned circuit at the output of a class-C amplifier. What will be the operating frequency for this circuit?

Solution:

Using Equation 8–5, the resonant frequency of this circuit can be calculated as follows:

$$f_r = \frac{1}{6.28\sqrt{LC}} = \frac{1}{6.28\sqrt{(10\ \mu\text{H})(0.01\ \mu\text{F})}}$$
$$= \frac{1}{6.28\sqrt{0.1 \times 10^{-12}}} = \frac{1}{(6.28)(.00000031)} = \frac{1}{.0000019}$$
$$= 503547\ \text{Hz}$$

In some cases tank circuits (parallel *LC* network) can be tuned to operate at two or three times the incoming frequency. These types of circuits are called **frequency doublers** and **frequency triplers,** repectively. They are commonly used when higher frequencies are required. For example, suppose a 120 MHz transmitter is in need of repair. First, the input (fundamental) frequency is checked. Then, as the frequency is passed through the different doubling and tripling stages, frequency checks are made to indicate which section is not operating. Figure 8–20 shows the basic class-C tuned amplifier system for reproducing the necessary output frequency.

The power efficiency for the class-C amplifier is much greater than for other power amplifier circuits. During the class-C amplification period, only a small signal is developed at the output. All of the signal is developed in the tuned circuit, and then in the load. Therefore the class-C power amplifier is turning the great majority of dc signal into signal power. However, in practical operation the class-C power amplifier operates at only about 85% efficiency. This is still a higher efficiency rating than found in the other types of power amplifiers.

A practical biasing circuit for the class-C power amplifier is illustrated in Figure 8–21. As the input signal goes positive, it forward biases the transistor. The current flow in the base charges C_1. Resistor R_1 is used to discharge C_1 between the positive peaks of the input waveform. Resistor R_1 cannot fully discharge C_1. This charge buildup on C_1 will keep the transistor turned off until the next positive peak is present, and turns on the transistor again.

Also shown in Figure 8–21 is a different type of tuned circuit that might be found in the output circuit. The circuit shown is an *L network*. This circuit enables the amplifier to develop the characteristic low output impedance, 50 Ω, in order to match the output of the amplifier to the RF circuits.

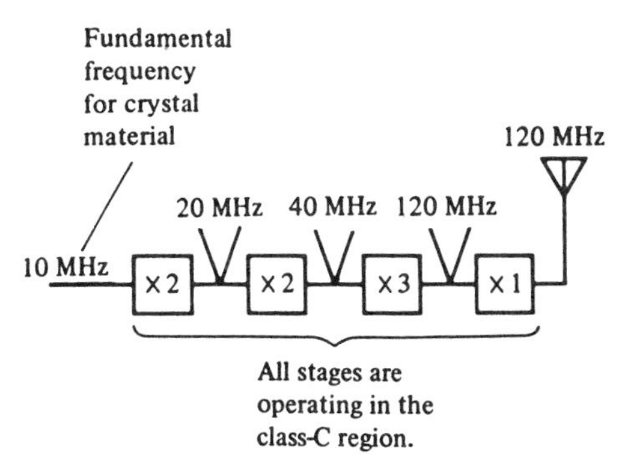

FIGURE 8–20
Doubling and Tripling Circuits Used to Increase Frequency

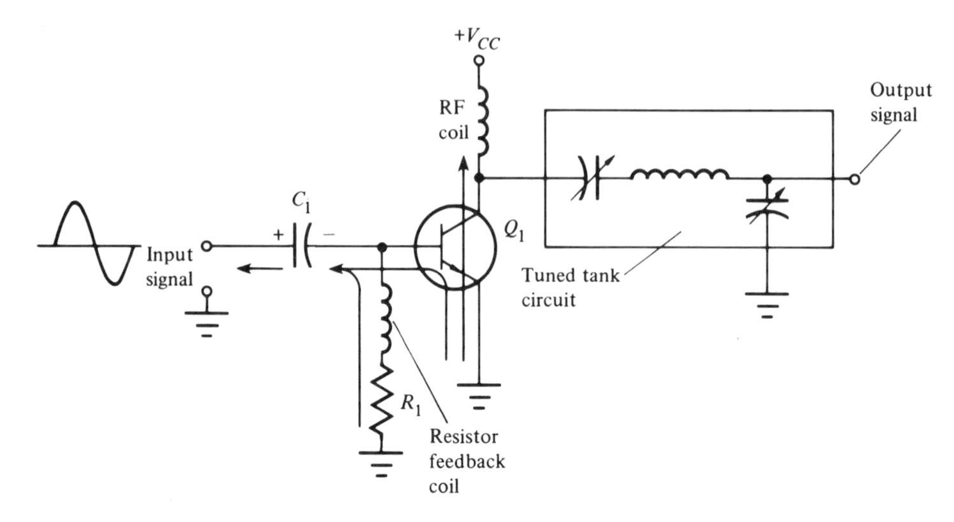

FIGURE 8–21
Class-C Amplifier Using Single-Source Biasing

TROUBLESHOOTING

POWER AMPLIFIERS

More and more solid-state power amplifiers are used in consumer products. Many problems arise in the power amplifier section. A few of them are covered in this discussion.

Hum in Output

A loud hum heard from the output section of an AM-FM stereo receiver generally indicates that one of the transistors has been driven into saturation, or one is shorted. The hum is created by a steady drain of current from the dc power supply. This large drain from the power supply means that the filter capacitors cannot charge and discharge properly, and the ripple voltage has increased. Referring to Figure 8–22, the components that might go bad are Q_3 and Q_4. These two transistors are placed under the greatest strain when the amplifier is in operation. When an ohmmeter check or voltage reading at the transistor terminals shows a defective component, the transistors should be replaced. Since these two transistors operate together, they must both be replaced even if one reads good.

A voltage check is the best test to use for the power amplifier circuit. Again, refer to Figure 8–22. If the output section is suspected

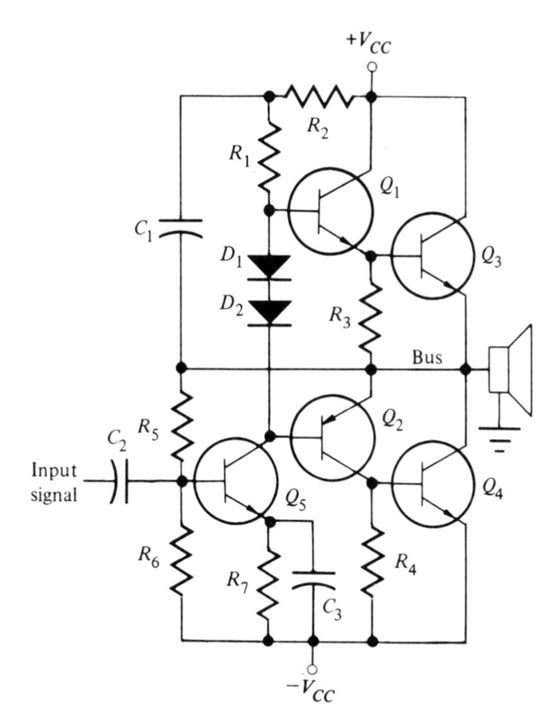

FIGURE 8–22 Troubleshooting Power Amplifiers

of being bad, voltage readings are taken at the center bus. If these readings indicate abnormally high or low voltage, there is an open or a short in one of the components. For example, if Q_5 develops a short from the collector to the base, the voltage drop across R_1 and R_2 will be higher than normal. That will make the readings on the base and emitter lower than normal. Since Q_1 has been placed in cutoff, Q_2 cannot drop an equal amount of voltage, and the bus voltage will develop a negative voltage. The normal reading on the bus should be about zero. The heavy conduction through Q_2 causes a greater than normal voltage drop across R_4, which will cause Q_4 to go into saturation.

Intermittents

Intermittents, sometimes operational and sometimes nonoperational, are another problem found in the power amplifier section. Refer again to Figure 8–22. First, the power amplifier section should be isolated from the rest of the receiver. This is accomplished by disconnecting C_2. The amplifier section should be turned on, and static or noise generated within the power amplified listened for. If a noise is heard, a check of Q_1 through Q_5 should be started. When checking these transistors with an ohmmeter, look for lower than normal reverse resistance readings between the collector and base. If all the transistors read good, then C_1 and C_2 should be checked. Then the diodes should be checked, although normally they will be good.

The intermittent problem also arises when a soldering connection is cracked. In addition, the circuit board should be checked to insure that proper trails have been established. Cold solder joints, where poor contact is made between the circuit board and component terminal, or cracked circuit boards should be suspected if other tests fail to locate the problem.

Distortion

Distortion appears in any amplifier whose operating point has shifted. Again, look at Figure 8–22. The first components that should be checked are the temperature-compensating diodes. If these diodes are open or shorted, a great deal of distortion will develop at the output of the amplifier. If these diodes check good, then a check of C_1 and C_3 should be made. In addition, if the forward bias voltage drop across a diode is in the area of 0.5 V, there will be crossover distortion in the amplifier.

A good check of power amplifiers characterized by excessive distortion is the center-bus voltage. If the bus voltage is several volts away from being normal, say 10%–20% from normal, resistors are the likely defective components. However, if the bus voltage is 50%–75% off, the problem lies in the transistors.

SUMMARY

Large-signal amplifiers are found in the output section of audio devices, televisions, and communications equipment. Large-signal amplifiers are also called power amplifiers.

Power amplifiers are used to drive output transducers. These amplifiers are designed to increase voltage and current at the output.

The efficiency of a power amplifier is based on the amount of dc power turned into signal power. Some classes of amplifiers have higher efficiency ratings than others.

Class-A power amplifiers operate with the lowest distortion level. This amplifier type has the lowest distortion because it operates at the center of the load line, that is, in the linear region.

Class-B power amplifiers are characterized by amounts of distortion because they develop only one-half of the waveform at the output. When the class-B amplifier is made to operate in the push-pull arrangement, distortion is greatly reduced.

Class-C power amplifiers are not suited for audio equipment. This type of amplifier generates the largest amounts of distortion.

Class-A power amplifiers are medium-range power amplifiers, generally developing 5 W or less to the load. The series-fed, class-A power amplifier requires large heat sinks to dissipate heat away from the transistor.

Class-A power amplifiers can couple the signal from the amplifier to the load through a transformer. The efficiency of this amplifier with the transformer is increased over class-A amplifiers without transformers. Transformers are used to match the impedance of the load to the output of the amplifier.

Class-B power amplifiers are biased at cutoff. No current will flow through the circuit until a signal is present at the input.

Class-B, push-pull amplifiers require two transformers in the circuit: an input transformer called the driver, and an output transformer. This type of amplifier develops crossover distortion in the output signal. The efficiency of the class-B, push-pull is higher than that for the class-A power amplifier.

Class-AB power amplifiers are used to reduce crossover distortion. Another advantage is that they do not require the use of expensive transformers.

Power amplifiers that do not need transformers and are biased in the class-AB region are called complementary symmetry amplifiers. This type of amplifier uses an NPN- and a PNP-type transistor in the output section.

Quasi-complementary symmetry amplifiers are characterized by a complementary pair of transistors in the driver stage.

Class-AB power amplifiers are found in the output section of many consumer products.

Class-C power amplifiers are generally found in high-frequency applications. The class-C amplifier is biased far into the cutoff region. The only way to bring this circuit out of cutoff is to have a large positive input signal.

Frequency doubler and tripler circuits consist of a class-C power amplifier plus a tuned circuit placed at the output of the circuit.

The efficiency of the class-C power amplifier is highest of all the power amplifiers, about 85%.

Troubleshooting power amplifier circuits is the same as for any transistor circuit. The best approach is to measure the center-bus voltage found on some classes of amplifier.

KEY TERMS

amplifier efficiency
class-A power amplifier
class-AB power amplifier
class-B power amplifier
class-C power amplifier
complementary symmetry amplifier
crossover distortion
driver transformer
frequency doubler
frequency tripler
input transducer
output transformer
power amplifier
preamplifier
push-pull arrangement
quasi-complementary symmetry amplifier
resonant frequency
series-fed, class-A power amplifier
temperature-compensating diode
transformer-coupled, class-A power amplifier
voltage amplifier

REVIEW EXERCISES

1. Are power amplifiers generally used to drive input or output transducers?

2. A dc power supply delivers 10 V at 1 A to a power output stage. The power amplifier delivers 7.5 W to the load. What is the efficiency of this amplifier?

3. Which class of power amplifier delivers the lowest amount of distortion to the load?

4. Which class of power amplifier is biased at cutoff? Draw a waveform of the ouput of this amplifier.

5. At which area on the load line is the class-A power amplifier biased?

6. A dc power supply develops 20 V at 1 A. How much dc power is available to be converted into signal power?

7. Why is a transformer placed between the output of the power amplifier and the transducer?

8. What is the typical efficiency rating of

- **a.** a transformer-coupled, class-A power amplifier?
- **b.** a class-B, push-pull power amplifier?
- **c.** a class-AB power amplifier?
- **d.** a class-C power amplifier? Why is it so high?

9. What are some disadvantages of using transformer-coupled, class-A amplifiers?

10. What brings the class-B power amplifier out of cutoff?

11. What is the major disadvantage of the class-B power amplifier?

12. Name the type of power amplifier that requires two transistors, one to amplify the positive half of the input waveform and one to amplify the negative half.

13. In a class-B, push-pull power amplifier, what kind of transformer is used to connect the output to the load?

14. What type of distortion is developed in a class-B, push-pull power amplifier?

15. Which class of power amplifier delivers a small amount of forward bias to the output transistor?

16. What is the major disadvantage of using the class-AB power amplifier?

17. Why does the complementary symmetry amplifier have a matched pair of PNP and NPN transistors?
18. What is the center-bus voltage of a complementary symmetry amplifier when no signal is applied to the amplifier?
19. Class-C power amplifiers are used primarily in which type of application?
20. Which components in the class-C power amplifier are used to re-create the waveform?
21. List some of the common problems that occur in power amplifiers.
22. In Figure 8–22, which voltage reading should be checked first?
23. Referring to Figure 8–22, which components should be checked if distortion develops?
24. If the center-bus voltage in Figure 8–22 is 50% away from the correct value, which components should be suspected of being defective?

9 FIELD-EFFECT TRANSISTORS

OBJECTIVES

Upon completing this chapter, you should be familiar with:

—JFET characteristics
—MOSFET characteristics
—FET amplifier characteristics
—FET dc biasing characteristics
—General JFET and MOSFET circuit construction
—JFET and MOSFET packaging

INTRODUCTION

The **field-effect transistor (FET)** is a special type of semiconductor device. Although the FET is similar to the bipolar transistor, it differs in many ways. The most important of these differences is the high **input impedance** found on the FET, compared with the low input impedance of the bipolar transistor. This means that the input signal for the FET can be close to zero power, yet still result in an output signal.

The FET, like the bipolar transistor, operates on low dc supply voltages. It also operates by sending current carriers through a solid material. Its light weight, small size, mechanical ruggedness, and cool operation make it very useful in modern consumer products.

The transistor is called a bipolar device because both electrons and holes are required for conduction. The FET is called a

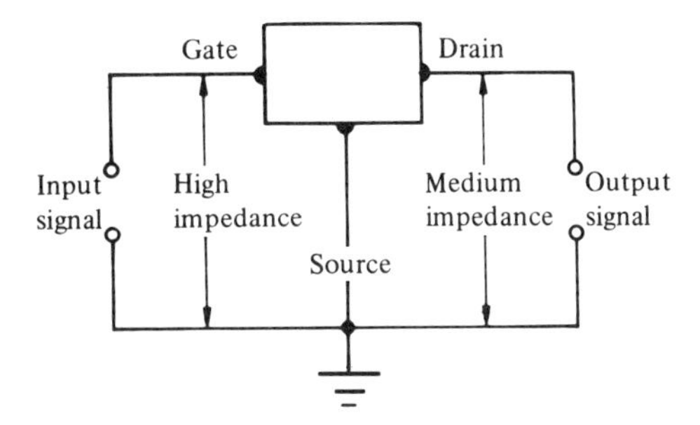

FIGURE 9–1
Block Diagram of a Single-Gate FET

unipolar device because it requires only one type of current carrier—holes for the P-type material and electrons for N-type material.

The FET has three terminals: the **gate (*G*),** the **source (*S*),** and the **drain (*D*).** These terminals correspond to the base, emitter, and collector of the bipolar transistor. In practical terms, the drain and source could be called the anode and cathode, and the gate the current-controlling device. Figure 9–1 shows a block diagram representing a single-gate FET.

An FET can be described as being symmetrical or asymmetrical. **Symmetrical** refers to the fact that the source and drain can be interchanged without affecting the operation of the circuit. **Asymmetrical** means that drain and source are so labeled, and that they cannot be interchanged without greatly affecting circuit operation. Most FETs used in consumer products are asymmetrical.

A given FET is one of two major types: a junction FET (JFET) or a metal oxide semiconductor FET (MOSFET). Both types will be discussed in detail.

JFETs

The **junction FET (JFET)** can be classified according to the type of material that comprises its channel. The channel can be made from either P-type or N-type material.

JFET Construction

The basic construction of the N-channel JFET is shown in Figure 9–2. The N-channel JFET consists essentially of a small, thin bar of silicon (in some cases, germanium may be used). Terminals are attached to the ends of this silicon bar. The source terminal is attached

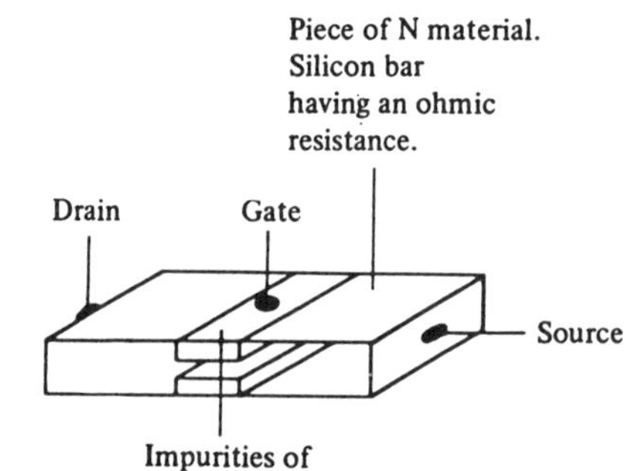

FIGURE 9–2
Basic Structure of an N-Channel JFET

to one end, and the drain terminal is attached to the other. These connections give an ohmic value to the silicon bar. The faces of the silicon bar are doped with P-type material. These two strips of P material are connected internally. Their external connection is the gate.

A P-channel JFET is of the same basic construction. The bar is made of P-type material, however, and the impurities are N-type material.

Several examples of JFET current conduction are given in Figure 9–3. First, a bias voltage is applied to the three terminals. In the figure, two supplies are used: one for biasing the drain-source, and the other for biasing the gate-source. The power supply used to bias the drain-source is identified as V_{DS}. Typically, the amount of voltage used to bias the drain-source is about one-half the value of V_{DD}. The second half of biasing is from a power source identified as V_{GG}. This voltage is used to supply the necessary reverse bias to the gate and source of the JFET. In the example of Figure 9–3, V_{DD} shall remain constant.

In Figure 9–3a the gate-source voltage, V_{GS}, is 0 V. The drain current, I_D, will be relatively high. The conduction path between the source and drain of the silicon bar is called the **channel.** The drain current in this zero gate voltage state is determined by the resistance of the channel. As stated, V_{DD} remains constant.

In Figure 9–3b a small amount of voltage is applied to the gate and source. This small amount of voltage is enough to place the gate-source in reverse bias. (The zero voltage state of the gate in Figure 9–3a means that the gate-source was not in the reverse bias state.) When this reverse bias is established, it sets up a depletion region in the channel. Recall that a depletion region is free of current carriers and acts as a resistance to current flow. The depletion region penetrates into the channel, and narrows the width of the channel. This action reduces the current flow through the channel, and reduces drain cur-

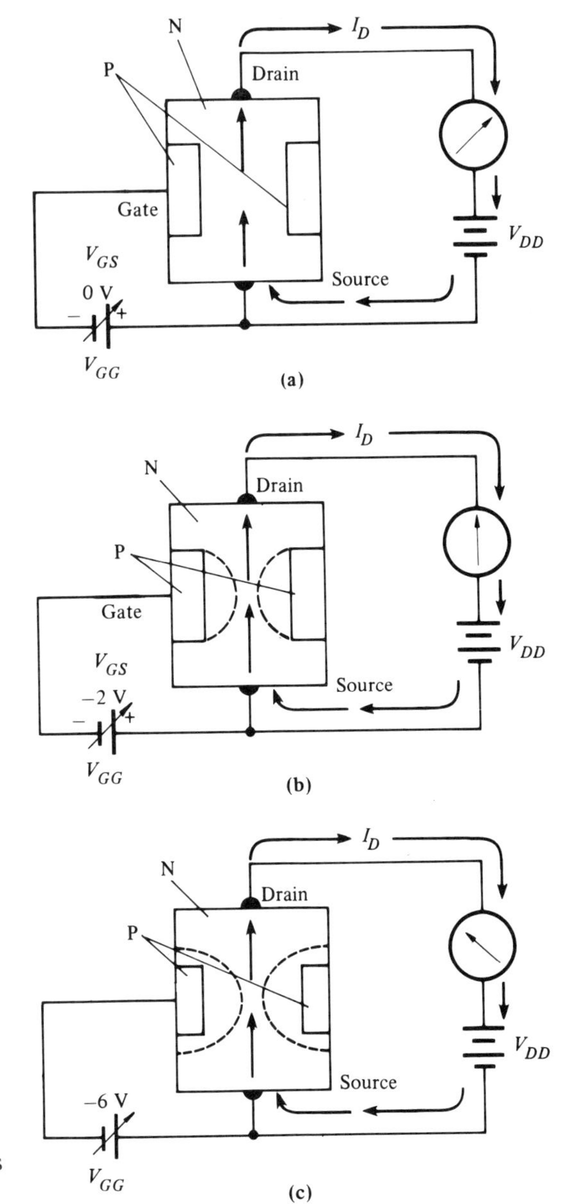

FIGURE 9–3
FJET Conduction Path: (a) 0 V Applied to Gate Allows Conduction, (b) Increasing Gate-Source Voltage Reduces Drain Current, (c) Increasing Gate-Source Voltage Cuts Off Current Drain

rent. Because the gate-source is reversed biased, there is no current flow. This causes the input resistance to be very high. Input resistances for the JFET are typically around 100 MΩ.

When the gate-source voltage is increased, the depletion region penetrates deeper into the channel, causing the channel to be reduced in size. This reduction in size decreases the drain current for the JFET.

Figure 9–3c shows the deep penetration. Increasing the gate-source voltage eventually narrows the channel until the two depletion regions almost touch, cutting off current flow to the drain. Therefore, the controlling device for the JFET is the applied voltage between the gate and source. This voltage controls the width of the depletion area, which in turn controls output current.

JFET Characteristic Curves

Just as the bipolar transistor has a set of characteristic curves, so also does the JFET. These curves describe the action of the JFET when different gate supply and drain supply voltages are applied to the JFET. Figure 9–4 shows a sample of a set of characteristic curves for the JFET. Note that the top line represents 0 V applied to the gate-source. As the gate-source voltage is made more negative, the drain current is reduced.

As can be seen from Figure 9–4a, all the curves have a common characteristic. The drain current increases very fast, and then levels

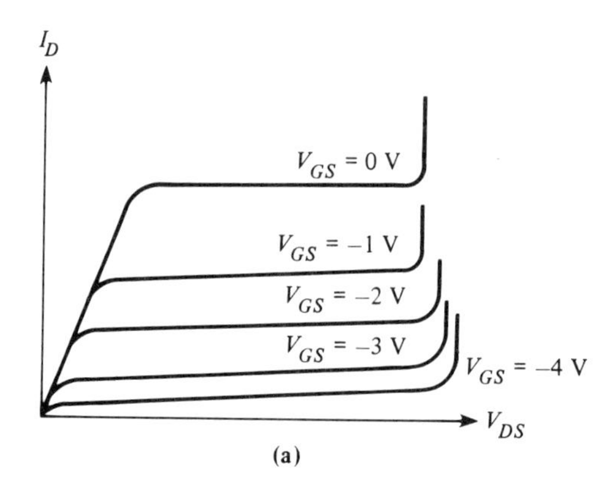

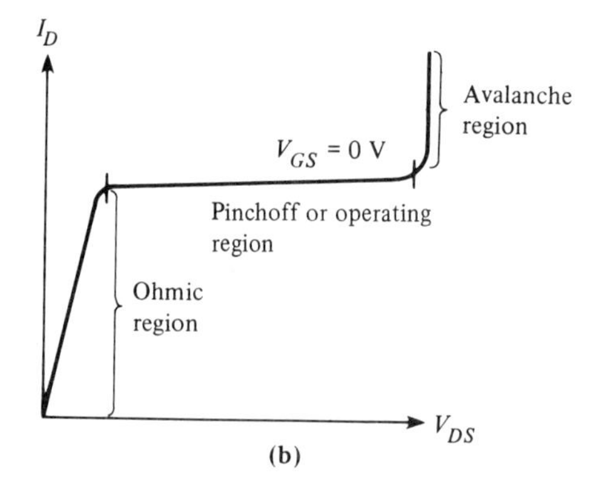

FIGURE 9–4
Characteristic Curves for the JFET: (a) Set of Curves for an N-Channel JFET, (b) Ohmic, Pinch-off, and Avalanche Regions

off. Now refer to Figure 9–4b. The first area where the drain current rises very rapidly and then rises again is called the **ohmic region** of the JFET. In this region, the drain current is controlled by the drain-source voltage and the resistance of the channel. The flattened part of the graph is called the **pinchoff region,** or the **operating region.** In this area of operation, the drain current is controlled primarily by the width of the channel. The second region where the drain current takes a sharp increase is called the **avalanche,** or **breakdown, region.** At this point large amounts of current are allowed to pass through the channel. If this current is not controlled by an external series resistor, the JFET can be damaged. Operation in this area will result in a burned-out JFET.

As shown in Figure 9–4a, in operation the JFET must have either 0 V applied to the gate-source or a negative voltage. If the voltage applied to the gate-source is made positive, certain conditions will result. The drain current will increase to very high levels. Also, making the gate-source positive will forward bias the gate-source. This forward bias action helps current flow through the channel. The forward bias action also reduces the resistance of the input. This reduction gives the input lower resistance, and the JFET will lose one of its attractive qualities.

JFET Schematic Symbol

Figure 9–5a shows the schematic symbol for the JFET. As stated earlier, the source and drain terminals are connected to the channel.

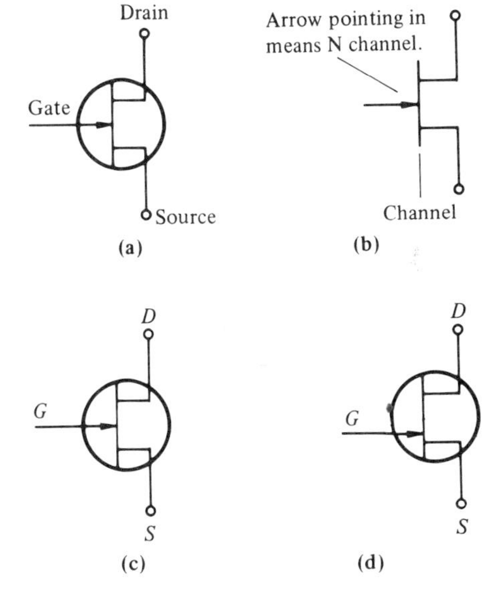

FIGURE 9–5
Schematic Symbols for the JFET: (a) Schematic Diagram of a Symmetrical JFET, (b) Polarity of the Channel, (c) Abbreviations of Terminal Leads, (d) Schematic Diagram of an Asymmetrical JFET with the Gate Lead Close to the Source

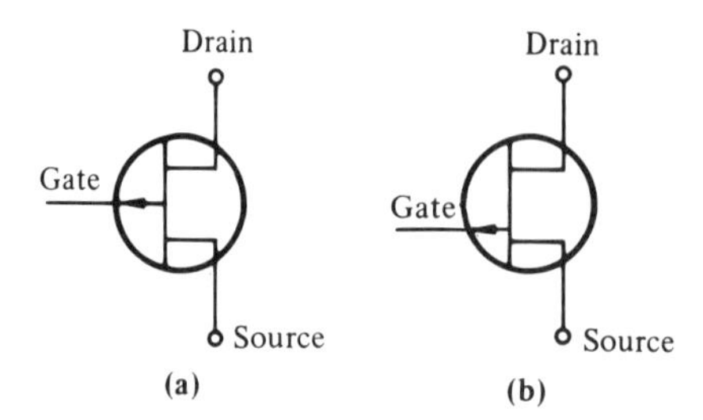

FIGURE 9–6
Schematic Diagrams of the P-Channel JFET: (a) Symmetrical P-Channel JFET, (b) Asymmetrical P-Channel JFET

The arrow is always found on the gate. As with the bipolar transistor, the arrow points to the N material. As shown in Figure 9–5b, the arrow on the gate points in, which means that the channel is N-type material. Often the terminals will be labeled with abbreviations, as illustrated in Figure 9–5c. As mentioned in the introduction to this chapter, an FET is symmetrical or asymmetrical. A symmetrical JFET is shown in Figure 9–5a, and an asymmetrical JFET is illustrated in Figure 9–5d. Note that the gate is drawn closer to the source in Figure 9–5d.

To represent the interchangeability of the N channels and P channels of JFETs, certain schematic symbols are assigned. The P-type JFET is shown in schematic diagram form in Figure 9–6. Note that the arrow on the gate is pointing away from the channel. Therefore, the channel is made of P-type material.

As with the N-channel JFET, the P-channel JFET can be symmetrical or asymmetrical. Figure 9–6b contains the schematic symbol for the asymmetrical JFET.

MOSFETs

As noted previously, another type of FET is the **metal oxide semiconductor FET,** or **MOSFET.** The MOSFET also is sometimes referred to as the **insulated-gate FET (IGFET).** The MOSFET contains the same three terminals as the JFET: drain, source, and gate. The gate voltage still controls the drain current, and the input impedance of the MOSFET is of larger ohmic value than the JFET. The main difference between the JFET and the MOSFET is that the MOSFET can have positive voltage applied to the gate and still develop zero gate current.

MOSFETs come in two basic types, classified according to mode of operation. A MOSFET will operate in the depletion mode or in the enhancement mode.

MOSFET Construction

Figure 9–7 illustrates the different parts of the MOSFET. In Figure 9–7a the drain and source are connected to a block of N material. As before, a positive voltage applied to the drain-source terminals causes a current flow between the source and the drain. In Figure 9–7b impurities have been added to the N material. However, unlike the JFET, the MOSFET has only one block of P-type material. A terminal is connected to this P-material block, and is called the **substrate.** The substrate creates a narrow channel between the source and the drain. Electrons flowing from the source to the drain must pass through this narrow channel.

In Figure 9–7c a metal oxide has been added to the outside of the N-material block. This metal oxide is an insulator. As shown in Figure 9–7d a metallic gate is added on top of the insulator. The gate is thus insulated from the channel. This explains why the MOSFET is sometimes called an insulated-gate FET, or IGFET.

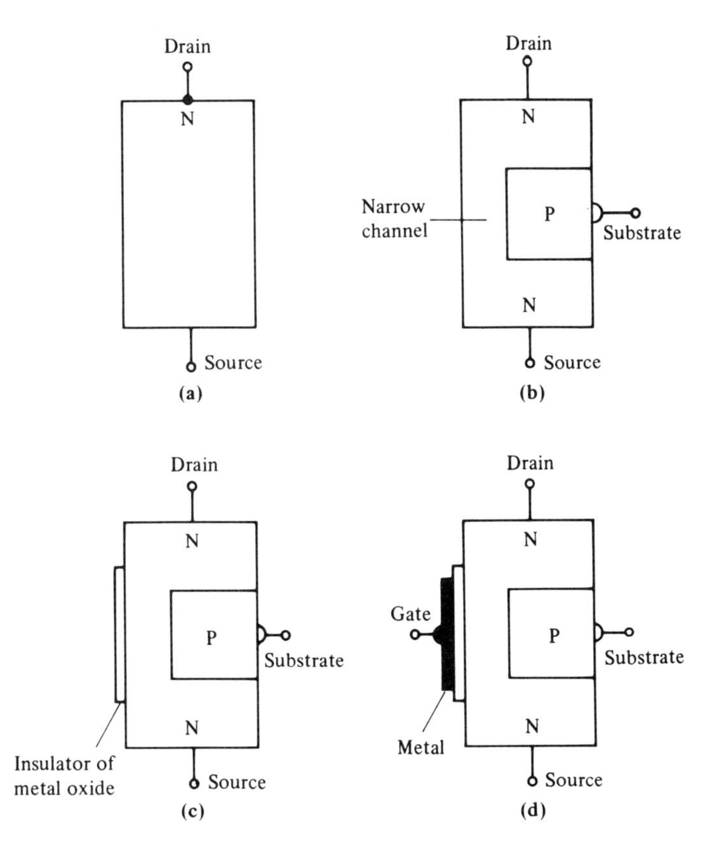

FIGURE 9–7
Construction of the MOSFET: (a) Block of N-Type Material, (b) Formation of Substrate by Addition of P Impurities, (c) Addition of Metal Oxide Insulation, (d) Attachment of Gate to Metal Oxide Insulator

MOSFET Depletion Mode

The MOSFET operates in generally the same manner as the JFET. The drain is connected to a positive voltage, V_{DD}, and the source is connected to the negative voltage, V_{DD}. The substrate is connected to the source voltage supply. In some applications the substrate can be connected to an external power source, but in this description the substrate will be connected to the source internally. This biasing has set up an electron flow between the source and the drain. Current flows through the narrow channel created by the substrate.

As for the JFET, the gate in the MOSFET is used to control current through the channel. However, because the MOSFET gate is insulated from the channel, a positive or negative voltage can be applied to the gate. This is the major difference in biasing methods between the JFET and the MOSFET. Figure 9–8 shows an example of how biasing is established on the MOSFET.

To show how this operates, the gate and the N-type material can be thought of as two plates of a capacitor. The metal oxide is the dielectric material between the plates. As the negative voltage is applied to the gate, a negative charge is developed on the gate. Since this assembly acts as a capacitor, the other plate develops a positive charge. Figure 9–8b illustrates this action. The positive charge creates a depletion area, and restricts current through the narrow channel.

The more negative the gate voltage, the wider the depletion region becomes in the N channel. With enough negative voltage applied to the gate, current between the source and the drain can be cut off.

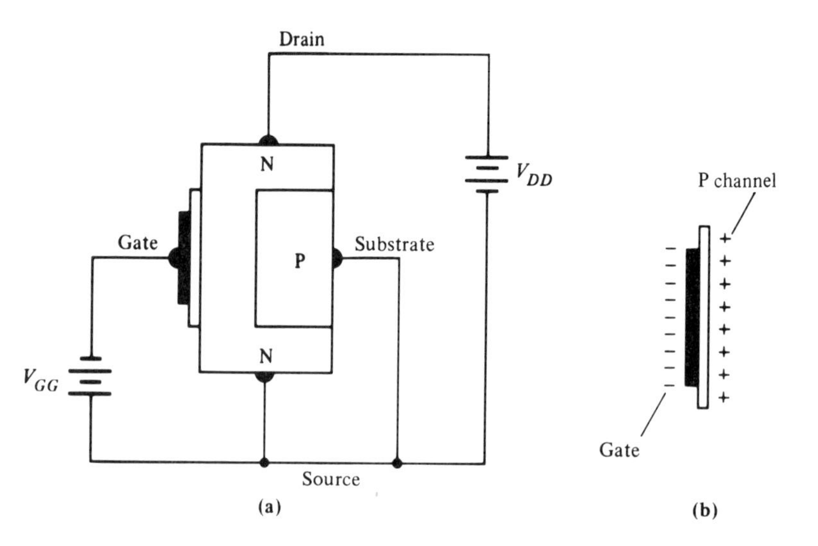

FIGURE 9–8
Biasing a Depletion-Mode MOSFET: (a) Biasing Diagram, (b) Charge Developed on Gate

This action of developing a positive charge in the N channel *depleted* the current-carrying electrons. This is why the operation of the MOSFET is called **depletion mode.**

MOSFET Enhancement Mode

The gate of the MOSFET is insulated from the channel. Thus a negative or positive voltage can be applied to the gate. In the **enhancement mode** of operation, a positive voltage is applied to the gate. As in the depletion mode, the gate, insulator, and channel act like a capacitor. However, in this case the gate has developed a positive charge. This means that the channel has a negative charge. These negative charges developed in the channel are current carriers in the N material. Therefore, they aid in the number of electrons that reach the drain. In this way the current has increased, that is, the current flow in the channel has been *enhanced.* The more positive the gate voltage becomes, the more current flow is seen in the drain. Figure 9–9 illustrates the enhancement mode of operation.

Because of the insulator between the gate and channel, relatively small amounts of current flow in either mode of MOSFET operation. Therefore, the input resistance of the MOSFET is extremely high. This resistance value ranges from about 10,000 MΩ to over 10,000,000 MΩ.

The devices shown in Figures 9–8 and 9–9 are N-channel MOSFETs. The P-channel MOSFET operates in the same manner as the N-channel MOSFET (except that a reverse of bias is necessary for the operation of the P-channel type), so a detailed explanation of the P-

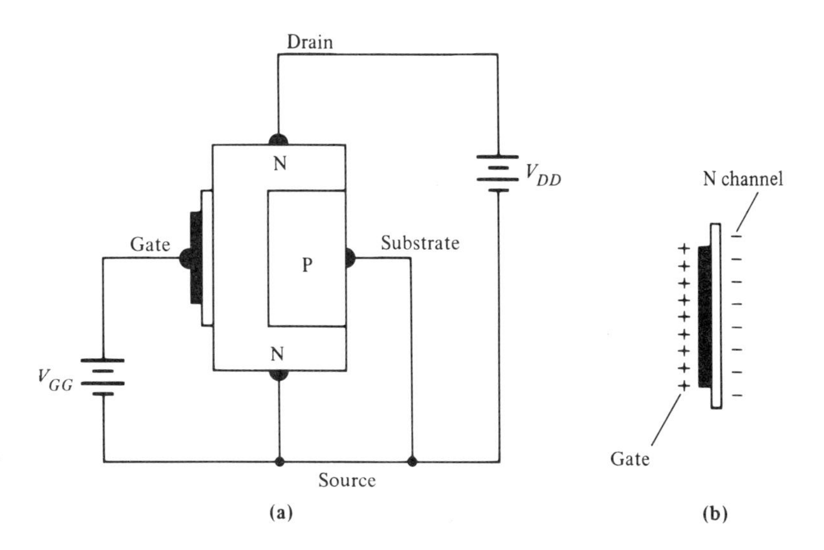

FIGURE 9–9
Biasing an Enhancement Mode MOSFET: (a) Biasing Diagram, (b) Charge Developed on Gate

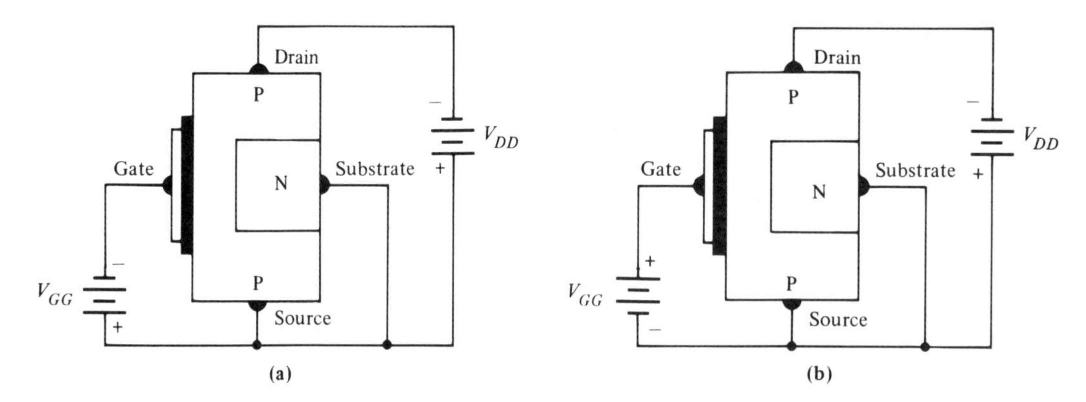

FIGURE 9–10
Biasing a P-Channel MOSFET: (a) in Enhancement Mode, (b) in Depletion Mode

channel MOSFET will not be included here. Figure 9–10 provides an illustration of P-channel bias operation.

A tiny capacitor is formed by the gate. The metal oxide is one plate, and the oxide film is the other plate. This capacitance is small, on the order of 1–10 pF. Because the insulating material is thin, it can be damaged by static voltages picked up by the MOSFET. This static charge can do irreparable damage to the MOSFET. For protection of this unit, manufacturers short the leads of the device together before shipment. This shorting device should not be removed until the MOSFET is placed in the circuit. MOSFETs are now manufactured with zener diodes between the gate and the source to help protect them from static discharge.

MOSFET Characteristic Curves

Figure 9–11 shows a typical set of characteristic curves for the MOSFET. The starting point for viewing this graph is at $V_{GS} = 0$ V. When the voltage falls below 0 V, into the negative region, the MOSFET operates in the depletion mode. On the other hand, when the voltage increases above zero, into the positive region, the MOSFET operates in the enhancement mode. Because a MOSFET can operate in either the depletion or the enhancement mode, it is sometimes called a **depletion-enhancement MOSFET.** In addition, since the MOSFET conducts drain current at $V_{GS} = 0$ V, it is also known as a **normally-on MOSFET.**

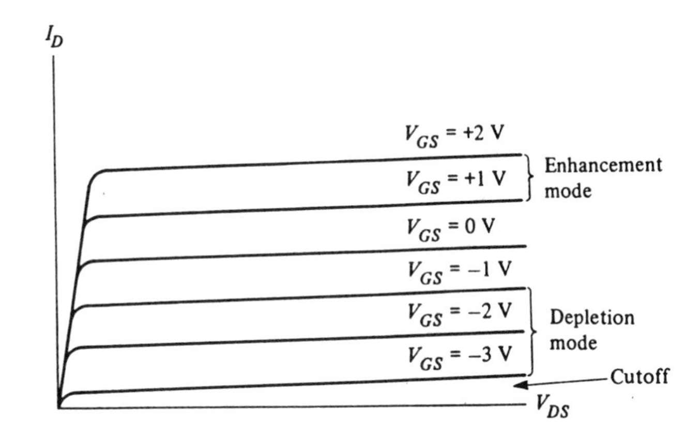

FIGURE 9–11
Set of Characteristic Curves for an N-Channel MOSFET

MOSFET Schematic Symbol

The schematic symbol for the normally-on N-channel MOSFET is given in Figure 9–12a. The gate is separated from the thin line where the source and drain are connected. This separation represents the insulation found between the gate and the channel. The clue to whether this MOSFET is P-type or N-type is given on the substrate lead, as shown in Figure 9–12b. The arrow on the substrate points toward the N material. In Figure 9–12c note that the substrate is not connected to the source.

Figure 9–13 shows the schematic symbol for the P-channel MOSFET. As before, the way to determine the polarity is to look at the arrow. This time the arrow is pointing away from the channel. Thus, the channel is made of P-type material.

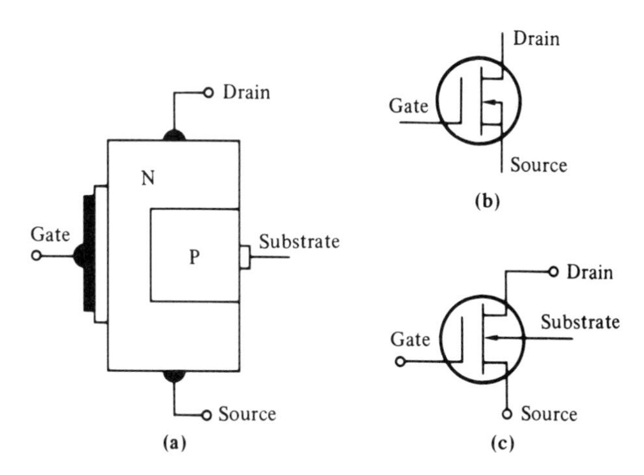

FIGURE 9–12
Schematic Diagrams of an N-Channel MOSFET: (a) Block Diagram of the MOSFET, (b) Three-Terminal MOSFET, (c) Four-Terminal MOSFET

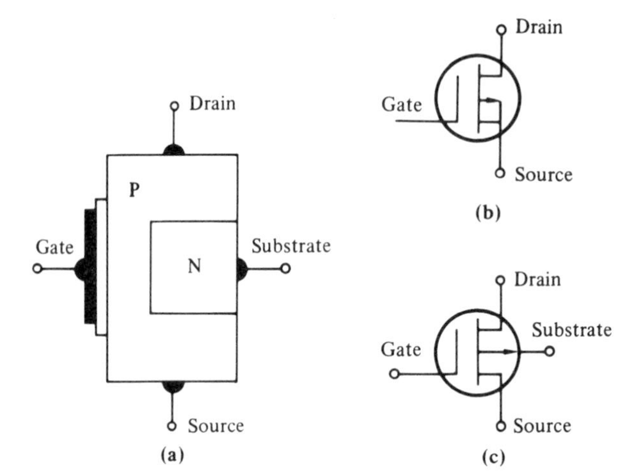

FIGURE 9–13
Schematic Diagrams of a P-Channel MOSFET: (a) Block Diagram of the MOSFET, (b) Three-Terminal MOSFET, (c) Four-Terminal MOSFET

As shown in Figures 9–12 and 9–13, the substrate can either be connected to the source, or it can be separate. If the MOSFET is a three-terminal device, as shown in Figure 9–13b, the manufacturer has already connected the substrate during assembly. If the MOSFET is a four-terminal device, the substrate is separate, and not internally connected, as shown in Figure 9–13c.

Enhancement-Only MOSFETs

Another type of MOSFET is called the **enhancement-only MOSFET.** As its name implies, this MOSFET operates in the enhancement mode only. This type of MOSFET is frequently found in digital circuits and switching circuits.

Figure 9–14 shows a block diagram of the enhancement-only

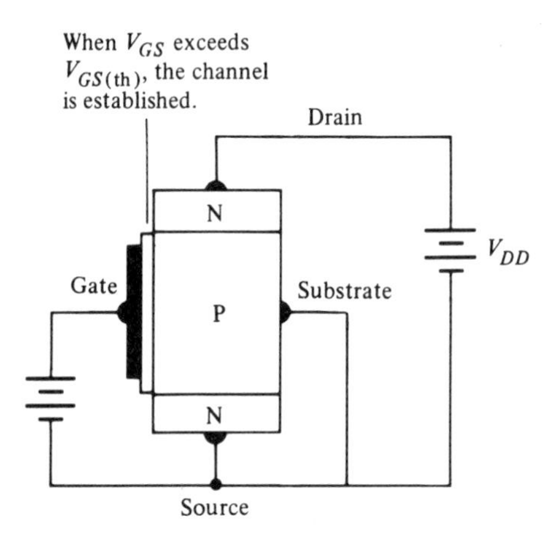

FIGURE 9–14
Block Diagram of an Enhancement-Only MOSFET

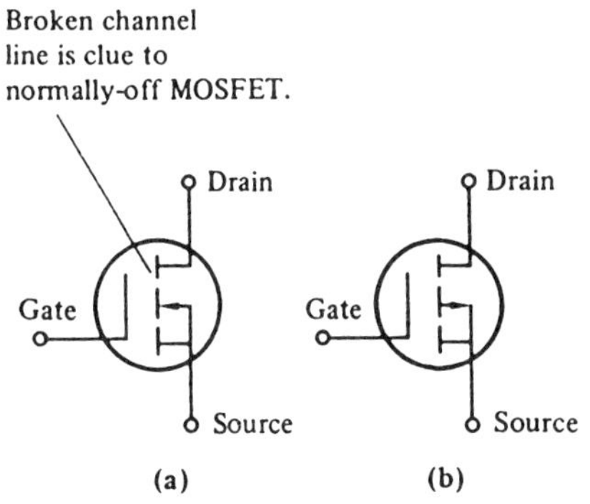

FIGURE 9–15
Schematic Diagrams of a Normally-Off MOSFET: (a) N-Channel Diagram, (b) P-Channel Diagram

MOSFET. Notice that the substrate extends all the way to the metal oxide. Structurally, there is no channel in the MOSFET.

In operation this MOSFET is different from the other MOSFETs. When normal biasing is set up on the source and the drain, and $V_{GS} = 0$ V, current tries to conduct between the source and the drain. However, the substrate blocks any current flow. For this reason, the enhancement-only MOSFET is also called the **normally-off MOSFET.**

To get a significant current flow in the enhancement-only MOSFET, a high positive voltage must be applied to the gate. As the positive gate voltage is increased, the gate and substrate material become the plates of a capacitor. As this voltage on the gate becomes more positive, the substrate develops a negative charge. When this negative charge becomes large enough, it creates its own channel, and conduction is established between the source and the drain. The level of voltage necessary to create this channel is called the **threshold voltage ($V_{GS(th)}$).** When V_{GS} is less than $V_{GS(th)}$, no current flows between the source and the drain. When V_{GS} exceeds $V_{GS(th)}$, a significant current flow is established in the MOSFET. Threshold voltages in enhancement only MOSFETS vary widely, from as little as 1 V to as much as 5 V.

As stated, when $V_{GS} = 0$ V, there is no channel established between the source and the drain. In the schematic symbol, the channel is indicated by a broken line between these terminals. This broken line signifies that this type of MOSFET is a normally-off MOSFET. Figure 9–15a shows the schematic symbol for the N-channel, normally-off MOSFET. The schematic symbol for the P-channel, normally-off MOSFET is given in Figure 9–15b.

Dual-Gate MOSFETs

MOSFET devices are also constructed with two gates. Figure 9–16 shows a layer diagram of this **dual-gate MOSFET** arrangement. Current

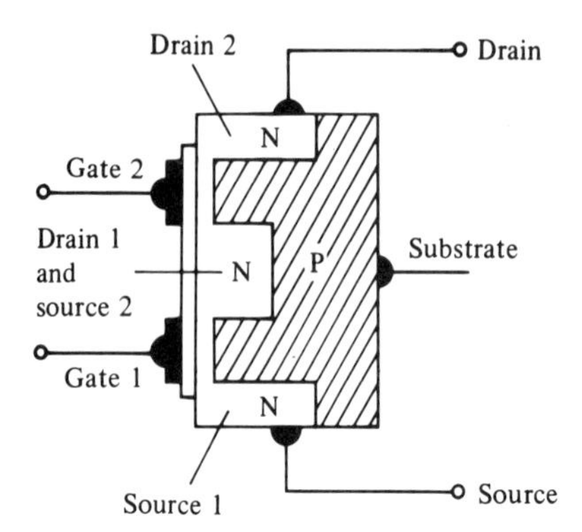

FIGURE 9–16
Layer Diagram of the Dual-Gate MOSFET

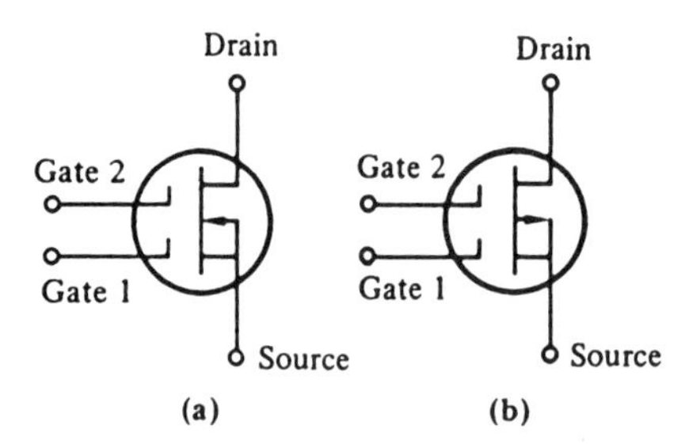

FIGURE 9–17
Schematic Diagrams of the Dual-Gate MOSFET: (a) N-Channel Diagram, (b) P-Channel Diagram

through this device can be cut off by either gate. The device operates in the same manner as the MOSFETs studied earlier.

Because of the dual-gate arrangement, this MOSFET can be used as an RF amplifier, a gain control amplifier, mixer, and demodulator circuit. For example, in a gain control circuit, the signal is applied to gate 1, and the gain control voltage is applied to gate 2. The gain control voltage will be used to control the output from the drain. A schematic symbol for the dual-gate MOSFET is given in Figure 9–17.

FET AMPLIFIERS

All active circuit devices increase the signal between the input and the output. The FET amplifier is one such active device. Because the voltage applied to the gate controls current through the drain, the FET amplifier has several advantages over the bipolar transistor. Some of the advantages are as follows:

1. A voltage-controlled amplifier means high input impedance.
2. Linearity qualities are good, which makes the FET amplifier useful for low-distortion amplifiers.

3. The interelectrode capacitance is low, which makes the FET good for RF amplifier circuits.
4. The noise output is low, which makes the FET attractive for preamplifier circuits.
5. FET amplifiers can be manufactured with two gates so that they can act as a gain-controlling device, or can be used for mixing two signals.

The bipolar transistor is still the workhorse in consumer electronic products, but the FET serves its purposes well.

FET BIASING

To understand better how the FET operates as an amplifier, the dc biasing states should be explored. As with the bipolar transistor, the FET can be connected in several different configurations for amplification. These different circuit configurations can be divided into three basic types: the common source, the common gate, and the common drain.

Biasing JFETs

Figure 9–18 shows an example of **fixed biasing**—biasing on a common source amplifier. Notice that the source is common to the input of the amplifier and to the output of the amplifier. This circuit is similar to the common emitter amplifier circuit. The supply voltage, V_{DD}, is positive with respect to ground. A current path is established between the common ground of the source through the N channel to the drain. This arrangement is the dc bias circuit for the drain-source circuit.

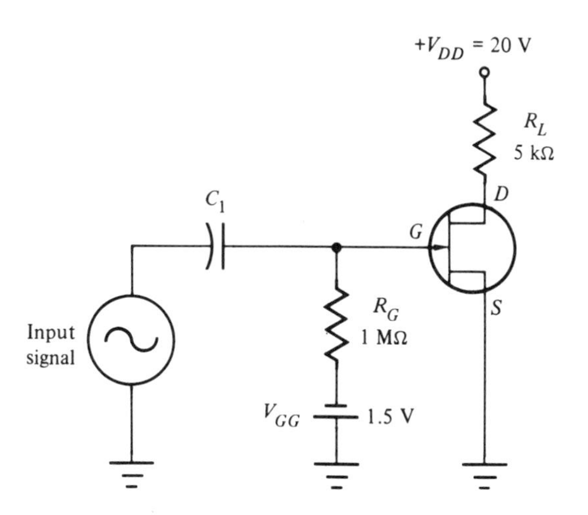

FIGURE 9–18
JFET Biasing Using Fixed Bias

To bias the gate, a separate power supply, V_{GG}, is used. This power supply delivers the necessary reverse bias to operate the gate of the JFET. Because of the reverse-biasing action, no gate current will flow.

Also shown in the gate circuit is the large gate biasing resistor. This resistor has a value of about 1 MΩ. There will be no current flowing in the gate. Therefore no voltage will drop across the resistor in the gate. This resistor is used primarily to keep the input impedance of the JFET constant. If the gate resistance is equal to 1 MΩ, then the signal source will see that large input impedance.

Because the active device is used to increase signal between the input and the output, the load resistor plays a very large part. The load resistor allows the circuit to produce voltage gain. A load line will help explain the operation of this device in the linear region. To establish the load line, one end of the line is placed at the supply voltage. For the JFET in Figure 9–18, $V_{DD} = 20$ V. Next, the current in the drain is calculated. This is accomplished by using Ohm's law as follows:

$$I_{\text{sat}} = \frac{V_{DD}}{R_L} = \frac{20 \text{ V}}{5000\ \Omega}$$
$$= 4 \text{ mA}$$

Thus the saturation level for this JFET is at 4 mA. This point is plotted on the I_D axis. The two points are connected. The resulting load line is shown in Figure 9–19.

Because the amplifier is to operate in the linear region, a point in the center of the load line is chosen for the gate voltage. This point is picked at -1.5 V. Dropping a line down to the V_{DS} axis gives a value of about 10 V. This voltage is equal to one-half of the supply voltage, V_{DD}, a relationship mentioned earlier in the chapter. This value means that the JFET is operating in its linear region.

Next, a 1 $V_{\text{(p-p)}}$ signal is injected into the gate of the JFET. Starting from the operating point of -1.5 V, it is seen that the signal causes the operating point to move in the positive direction, to -1 V, then swing in the negative direction, to -2 V. This swing in voltage on the gate controls current through the channel. This change in channel current can be verified from the graph. Drain current will swing between 1.00 mA and 2.5 mA. Because of this drain current swing, a change in voltage across the load resistor will be seen. Projecting the lines down to the V_{DS} axis indicates a change of 8 V on V_{DS}. Thus the voltage across the output will change from 7 V to 13 V.

The amplifier has accomplished its task. It has changed a 1 $V_{\text{(p-p)}}$ signal on the input to an 8 $V_{\text{(p-p)}}$ signal in the output. This establishes a voltage gain between input and output. This voltage gain can be calculated as follows:

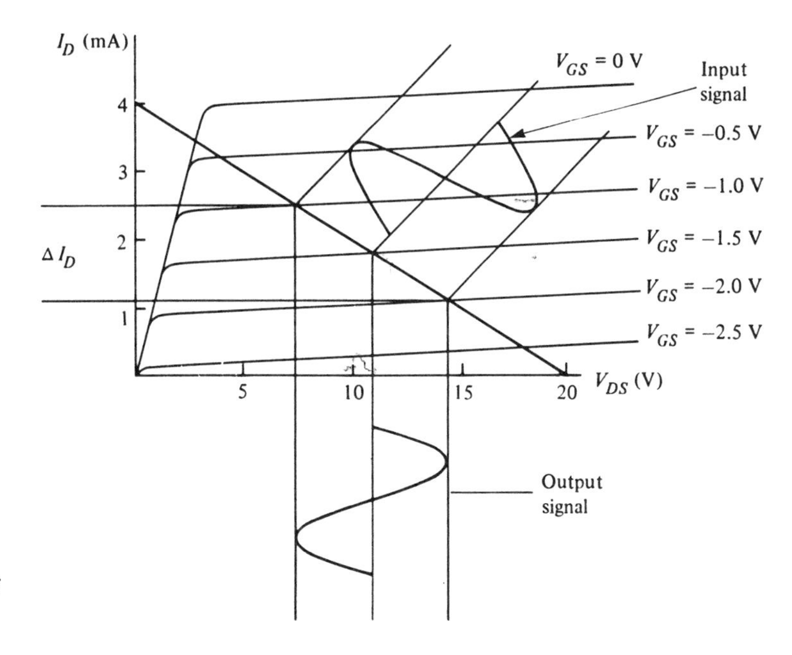

FIGURE 9–19
Set of Characteristic Curves and Load Line for the JFET in Figure 9–18

$$A_v = \frac{\text{signal out}}{\text{signal in}} = \frac{8}{1}$$
$$= 8$$

This shows that the JFET will produce a gain of 8 under these biasing conditions.

Also of importance is the phase of the input signal to that of the output signal. In the common emitter, a 180° phase shift is seen between input and output. The common source JFET amplifier also provides a 180° phase shift between input and output. A look at Figure 9–19 will explain this action. As the input signal changes from −1.5 V to −1.0 V, the output signal changes from about 11 V to 7 V. The drain voltage has gone negative. The other circuit configurations, the common gate and the common drain, do not provide a phase shift between input and output.

As was mentioned earlier, the load resistor for the JFET is important in determining the gain of the amplifier. A general statement can be made: If the load resistor is increased, the gain of the amplifier will be increased. As can be seen from Figure 9–20, if the load resistor is increased, the load line will be lowered. Line *A* shows what happens when the load resistance is increased to 8.2 kΩ. Line *B* shows the load resistance decreased to 1 kΩ. Note that Line *B* is steeper and will not be able to develop as large an output as higher load resistances.

Other biasing configurations can be used to keep the JFET

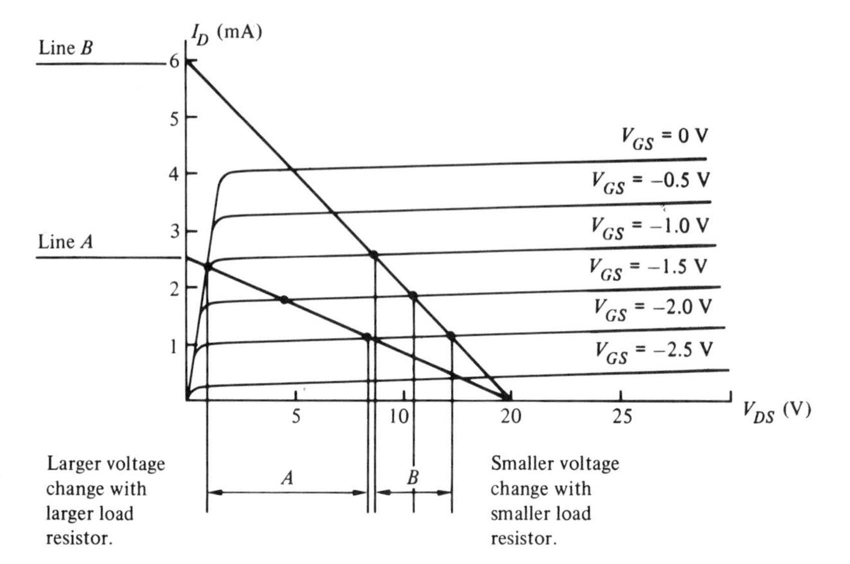

FIGURE 9–20
Gain of a JFET Amplifier versus Size of Load Resistor

stable in its operation as an amplifier. Figure 9–21 shows an example of **source biasing.** Notice that the power source V_{GG} has been eliminated, and that a resistor has been added into the source circuit. As current flows from the source to the drain, a current is seen through the source resistor, R_S. Since a current flow is developed, a voltage drop is seen across this resistor. The voltage drop serves to develop the gate voltage.

To establish the proper biasing in this circuit, several factors must be known. With these values, it becomes a simple process to find

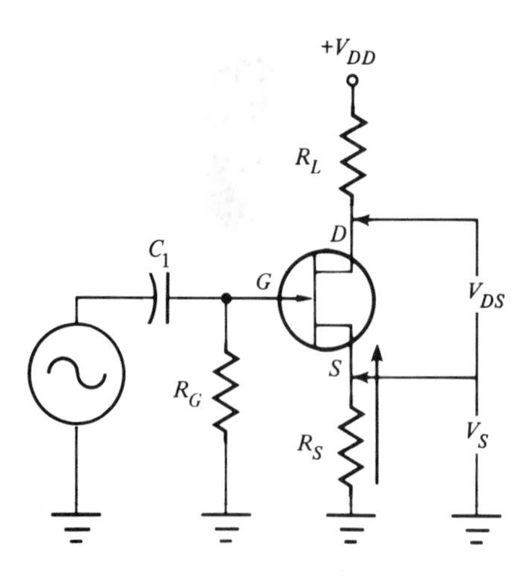

FIGURE 9–21
Source Biasing the JFET

the needed source resistor. Equation 9–1 shows the formula for this calculation.

$$R_S = \frac{V_{GS}}{I_D} \tag{9–1}$$

A sample problem is given in Example 1.

EXAMPLE 1

A JFET amplifier develops 1.5 V between the gate and the source. This develops 1.75 mA of drain current. What size source resistor should be used?

Solution:

The known values are placed into Equation 9–1:

$$R_S = \frac{V_{GS}}{I_D} = \frac{1.5\ \text{V}}{1.75\ \text{mA}}$$
$$= 857\ \Omega$$

The source bias is developed by the current flow through the source resistor. This current flow develops a voltage drop, V_S. This voltage drop at the source is positive with respect to ground. Since the gate has no current flow, the gate is at ground potential. The reverse bias for the gate is developed from the difference of potential between gate and source. The voltage on the source is made positive, which develops a difference in potential between source and gate. Since the gate is negative, this accomplishes the reverse-biasing action of the gate without a power source.

Source biasing is a simple method of biasing the JFET. It eliminates a power supply, and adds only an inexpensive resistor. The one disadvantage of using the source-biasing method is that the voltage gain of the amplifier is reduced. To study this action of decreased voltage gain, refer to Figure 9–22. As the signal goes positive at the gate, it causes the reverse bias on the gate to be reduced. The reduction of the reverse bias results in an increase in drain current. More drain current leads to an increase in source current. This means more voltage drop across the source resistor. Since Kirchhoff's voltage law states that the voltage will drop across each load, some of the drop will be absorbed by the source resistor, thus reducing the voltage gain of the JFET.

Two types of biasing the JFET have been discussed: fixed biasing and source biasing. The technician should be aware that when the

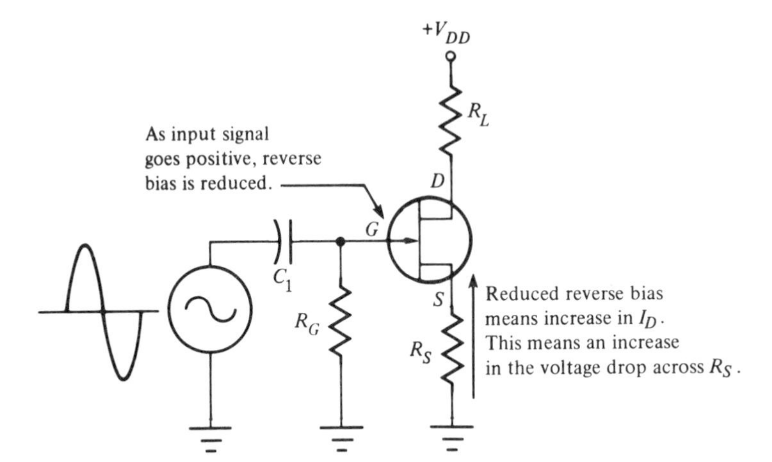

FIGURE 9–22
Reduction in Gain through Source Biasing

fixed-biasing method is used, the JFET circuit will be expected to operate close to the set standards of the JFET. In this case, simply replacing the JFET could cause problems with the amplification. The fixed-biasing method is not generally used in consumer products. Source biasing allows removal of the V_{GG} supply. However, the addition of a source resistor reduces the overall gain of the stage. Because of the source resistor, the operation of the JFET becomes more stable. If the JFET starts to conduct more current, the source resistor will keep the device operating within its normal range.

The selection of the source resistor is very important. Generally, the larger the resistor, the more stable the JFET will be in operation. However, when a large-value resistor is chosen, the JFET will operate near the cutoff region, not in the linear region. The way to combat this operation near cutoff is to use a **combinational biasing** network. Figure 9–23 shows an example of this type of biasing action. Notice that fixed-biasing methods are applied to the source and the gate.

This fixed-biasing network develops a positive voltage on the gate, and a positive voltage on the source. The positive voltage is developed by the resistor network of R_{G_1} and R_{G_2}. These two resistors will generally have high values. These high values of resistance will keep the impedance at the input high for this biasing method.

To illustrate the biasing action of this circuit, a sample problem will be used to identify the terminal voltages. The drain current for this circuit will be 2.5 mA. With this knowledge, the voltage drop across R_S can be found. Using Figure 9–23, the size of resistance can be plugged into Ohm's law equation, as follows:

$$V_S = I_D \times R_S = 2.5\text{ mA} \times 3.3\text{ k}\Omega$$
$$= 8.25\text{ V}$$

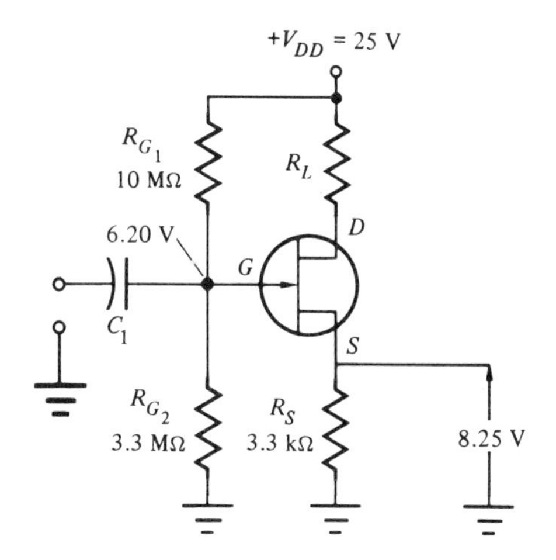

FIGURE 9–23
Combinational Biasing for the N-Channel JFET

The voltage at the gate can now be calculated. Using the voltage divider equation, the voltage drop across R_{G_2} is as follows:

$$V_{RG2} = \frac{R_{G2}}{R_{G_1} + R_{G_2}} \times V_{DD} = \frac{3.3\ \text{M}\Omega}{10\ \text{M}\Omega + 3.3\ \text{M}\Omega} \times 25\ \text{V}$$

$$= \frac{3.3\ \text{M}\Omega}{13.3\ \text{M}\Omega} \times 25\ \text{V}$$

$$= 6.20\ \text{V}$$

Thus the voltage at the gate will be 6.20 V. Both the gate and the source voltages will be positive with respect to ground. If this is so, the JFET cannot operate. Recall that the gate must have negative voltage in order to operate. Therefore, when comparing the two voltages, the gate voltage must be lower than the source voltage in order for the JFET to be in the proper bias. The lower voltage on the gate means that the gate is reversed biased with respect to the source.

Biasing the P-channel JFET is just the opposite of biasing the N-channel JFET. Figure 9–24 shows the combinational biasing used for the P-channel JFET. The only difference in the circuit diagram is that the V_{DD} power supply has a different polarity.

Biasing MOSFETs

In general, the biasing methods for MOSFETs are the same as those used in JFET biasing. However, since the gate in the MOSFET is insulated from the channel, the gate current will not control the current flow between the source and drain as in the JFET.

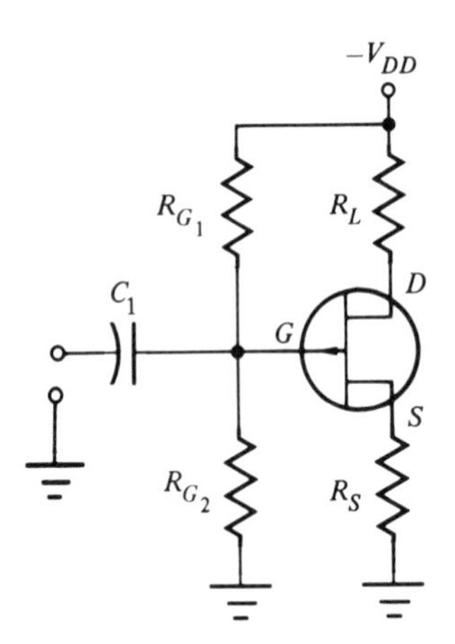

FIGURE 9–24
Biasing the P-Channel JFET

When a signal is applied to the gate, it develops the same polarity between input signal and output current. As the input signal increases, the drain current increases. As the signal goes negative, the drain current is reduced.

Figure 9–25 shows an example of using a dual-gate MOSFET amplifier. This amplifier is used as the RF amplifier in a receiver. The RF signal is coupled from L_1 through C_2 to G_2 of the MOSFET. In this MOSFET, G_1 is fed by the automatic gain control (AGC), which is used in television receivers to control gain of RF and IF sections. This is generally the connection when one of the gates is used for AGC. With the AGC fed to this gate, the gain of the MOSFET can be controlled. The other gate of this MOSFET is responsible for the amplification of the incoming signal. The dc biasing voltage for this MOSFET is developed through R_5 and R_7.

Because of its dual gates, the MOSFET makes an excellent component for the RF section of radios and television receivers. One signal can be applied to G_1, and a controlling signal can be applied to G_2. This is a very attractive component characteristic.

GENERAL FET CIRCUIT CONFIGURATION

In consumer products there are three general types of circuit configurations for the MOSFET and JFET: the common source, the common drain, and the common gate. Each of these different circuits has advantages and disadvantages.

Common Source

Shown in Figure 9–26 is the schematic drawing for the **common source** connection. This circuit connection is the most common type found in

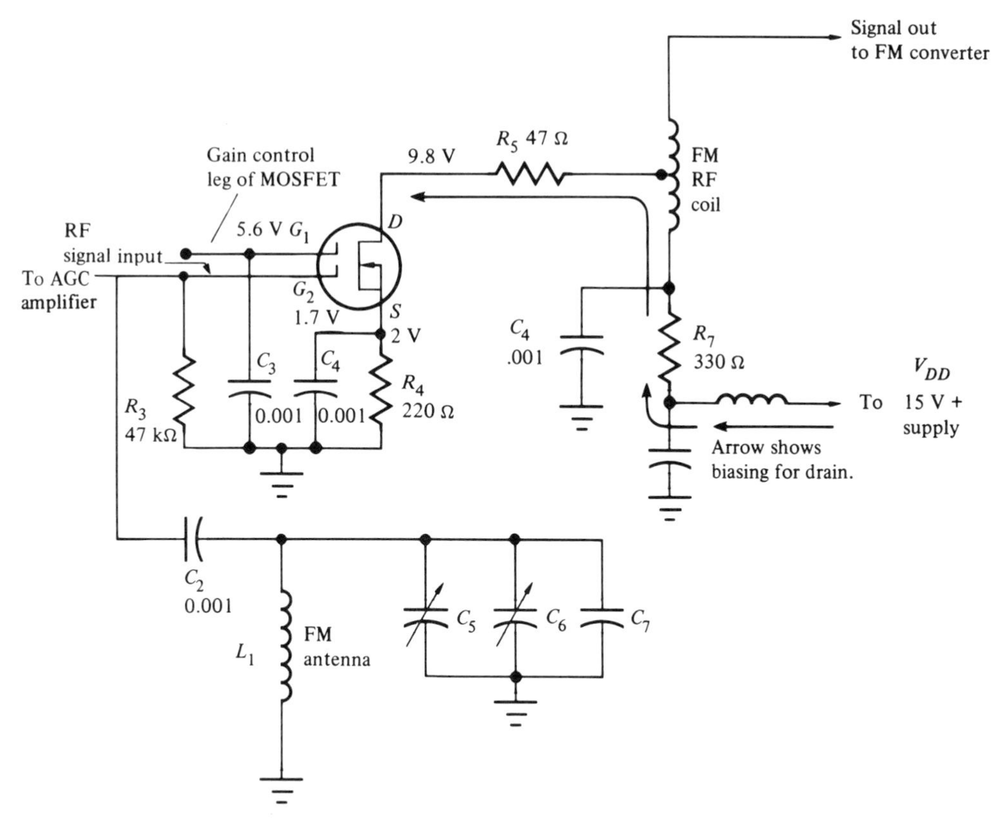

FIGURE 9–25
Dual-Gate MOSFET as an RF Amplifier

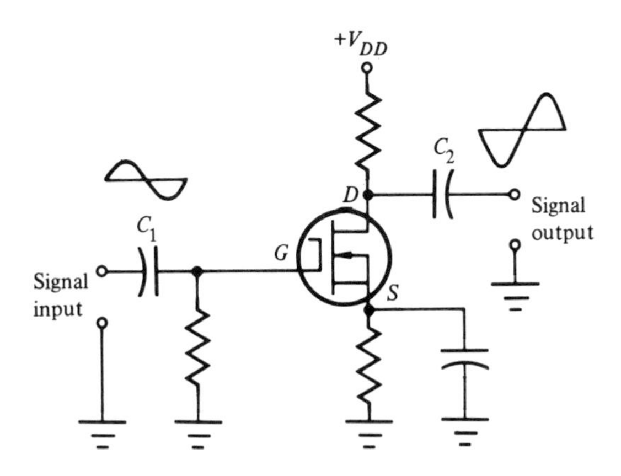

FIGURE 9–26
MOSFET Connected in the Common Source Mode

consumer products. It provides a high input impedance, a medium to high output impedance, and a voltage gain greater than one. The signal is placed between the gate and the source. This is the input of the amplifier. The output of this amplifier is found on the drain and the source. The fact that the source is common to the input and the output signal gives this amplifer its name.

Common Gate

Figure 9–27 shows the **common gate** circuit. The input signal is developed between the source and the gate. This amplifier develops low input impedance. The output signal is developed between the gate and the drain. This means a high output impedance for this amplifier. One important characteristic of this amplifier is that it can operate at high frequencies. Because of its high-frequency operation and low input impedances, this amplifier will offer low gain to the signal. Because of this, the amplifier does not require large neutralization by feedback capacitors, which reduce positive feedback and, thus, reduce oscillation at high frequencies.

Common Drain

Figure 9–28 illustrates the third type of configuration, the **common drain** circuit. This connection is also called the *source follower*. The input impedance is higher in this circuit than in the common source circuit, while the output impedance is very low. As seen with other circuits, there will be no signal shift between the input and the output. Because of this action, the voltage gain for this type of MOSFET amplifier is less than one.

FET Ratings

The service technician will continually deal with the rating of components. Each bipolar transistor has a certain rating, and every unipolar

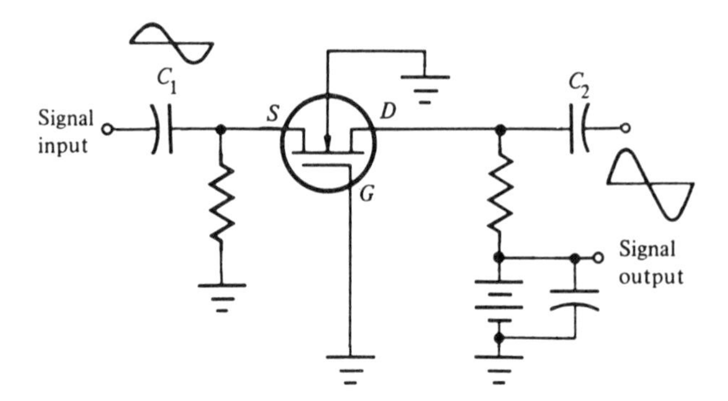

FIGURE 9–27
MOSFET Connected in the Common Gate Mode

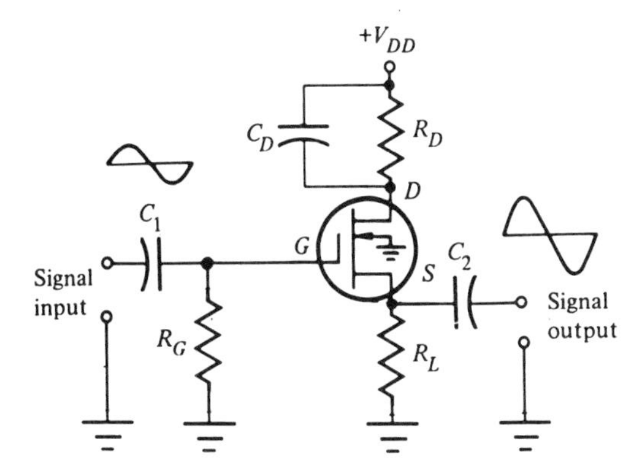

FIGURE 9–28
MOSFET Connected in the Common Drain Mode

transistor also has its own rating. The main rating used for FETs is called **transconductance (*gfs*).** Transconductance shows the relationship between a change in gate voltage and drain current. This rating is calculated by Equation 9–2.

$$gfs = \frac{\Delta I_D}{\Delta V_{GS}} \tag{9-2}$$

where:

gfs = transconductance (micromhos, μmhos)
ΔI_D = change in drain current (mA (dc))
ΔV_{GS} = change in gate-source voltage (V (dc))

For example, assume that V_D is held constant at 10 V, and that drain current changes from 15 mA to 10 mA. This drop in gate current is caused by a gate voltage change of 1 V. Therefore, the change in gate voltage has changed the drain current by 5 mA. The transconductance of this circuit is calculated as follows:

$$gfs = \frac{I_D}{V_{GS}} = \frac{5 \text{ mA}}{1 \text{ V}} = 0.005$$
$$= 5000\ \mu\text{mhos}$$

Transconductance is not the only rating applied to FETs, however. The following list summarizes the ratings found on FETs:

—*Transconductance*: Ratio of I_D to V_{GS}. Typical range of transconductance is between 35 μmhos and 50,000 μmhos.
—*Drain current at zero gate voltage* (I_{DS}): Current in the drain-source circuit. Typical range of current is between 0.1 mA and 250 mA. This rating is taken when V_{DS} = 35 V.
—*Drain current cutoff* ($I_{D(\text{off})}$): Leakage current developed in the channel when V_{GS} is set at cutoff.

—*Gate-drain voltage* (V_{GD}): Maximum voltage that will be developed between the gate and the drain.
—*Gate-source breakdown voltage* (BV_{GS}): Voltage range in which the JFET will enter avalanche. Typical voltage is between 25 V and 50 V.
—*Gate-source pinchoff voltage* (V_p): Gate-source voltage rating at which the electrostatic field closes the channel to conduction.
—*Total device dissipation* (P): Maximum power rating. Typical ratings are found between 200 mW and 0.8 W in free air.

FET Packaging

Like the bipolar transistor, the FET comes in many different packaging styles. Because of the different styles, the technician should consult the data sheets to find the ratings and lead identification of a given JFET device.

Figure 9–29 shows several different cases for the JFET and the MOSFET. When working with these components in the circuit, the technician should take care when measuring the pin voltages and when handling the FETs.

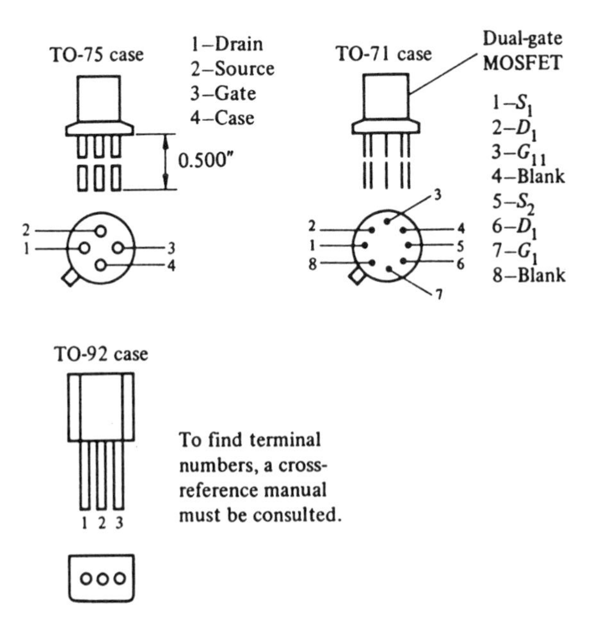

FIGURE 9–29
Example Case Styles and Lead Arrangements

SUMMARY

The FET is a unipolar device. Its three terminals are the gate, the source, and the drain. Most FETs are constructed from silicon material. Part of this material is doped to form a channel between the source and the drain. The channel may be made of P or N material. The channel is the current carrier path in this component. The current through this channel is called the drain current.

The FET conducts when a reverse bias or zero gate voltage is applied. A small change in gate voltage will develop large current changes through the channel. Because the input voltage controls the output current, the FET is said to have high input impedance.

FETs are classified as JFETs and MOSFETs. JFETs can either be symmetrical or asymmetrical. The source and drain of symmetrical JFETs are interchangeable. In asymmetrical JFETs, the source and drain cannot be reversed. In MOSFETs the gate is insulated from the channel. MOSFETs operate on the same principle as JFETs. MOSFETs operate in one of two modes: the depletion mode or the enhancement mode.

JFETs and MOSFETs are excellent active devices for use in electronic products. Their large input impedance makes them suitable for RF amplifiers in radio receivers. During amplification they offer low noise to the signal. This means a reduction in distortion.

MOSFETs have one or two gates. If the MOSFET has two gates, one gate is used for input, and the other is used for signal control.

Biasing of FETs is accomplished through fixed biasing, source biasing, or combinational biasing. Each of these biasing modes operates the FET at the operating region. FETs can be biased to operate in the linear region. This allows the FET to be an excellent amplifier. The FET offers good gain at low distortion levels to the signal.

Gain of the FET can be controlled by selecting different values of load resistance. The larger the load resistance, the greater the gain from the FET.

Like the bipolar transistor, the FET has three common amplifier connections. In the FET these connections are the common drain, the common source, and the common gate circuits. Each configuration has its own characteristics.

Each FET and MOSFET has a specific rating under which it will operate. When replacing these components, the characteristics of the new component should be matched to the faulty component.

JFET and MOSFET packaging comes in a wide variety of casings. Data sheets should be consulted to identify the terminals and ratings of these active devices.

KEY TERMS

asymmetrical
avalanche region
breakdown region
channel
combinational biasing
common drain
common gate
common source
depletion-enhancement MOSFET
depletion mode
drain
dual-gate MOSFET
enhancement mode
enhancement-only MOSFET
FET
fixed biasing
gate
input impedance
insulated-gate FET (IGFET)
junction FET (JFET)
metal oxide semiconductor FET (MOSFET)
normally-off MOSFET
normally-on MOSFET
ohmic region
operating region
pinchoff region
source
source biasing
substrate
symmetrical
threshold voltage
transconductance
unipolar

REVIEW EXERCISES

1. Why is the FET called a unipolar device?
2. Draw a schematic diagram for a P-channel and an N-channel FET. Label each of the terminals.
3. What is a symmetrical FET?
4. Name two modes of operation for the MOSFET.
5. What is the name of the conduction path established between the source and the drain of an FET?
6. Which terminal of the MOSFET controls drain current?
7. Draw a schematic diagram of an enhancement mode MOSFET, and apply the proper voltages to each terminal.
8. What is the typical input resistance found on a MOSFET?
9. What value of capacitance is developed by the gate of the MOSFET?
10. What protection is included between the gates of the MOSFET?
11. In what application is the enhancement mode MOSFET frequently found?
12. Explain why the enhancement mode MOSFET is generally considered an "off" device.
13. What are some applications of the dual-gate MOSFET?
14. List five characteristics of FET amplifiers.
15. What bias should be applied to the gate of a JFET?
16. Name the two types of JFET amplifier connections that do not provide a phase shift between the input and the output signal.
17. How can the gain of a JFET amplifier be increased?
18. What is one major disadvantage of using the source-biasing method?

19. Draw schematic diagrams for the following circuits, and list advantages and disadvantages of each circuit:

  a. Common source
  b. Common gate
  c. Common drain

20. What is the main rating used for JFETs? In what unit is it measured?

21. Refer to load line *A* in Figure 9–20. At what V_{GS} point would the JFET be cut off? At what V_{GS} point would the JFET be saturated?

10 SPECIAL SEMICONDUCTOR DEVICES— THYRISTORS

OBJECTIVES

Upon completing this chapter, you should be familiar with:

—Silicon-controlled rectifier characteristics
—Triac characteristics
—Diac characteristics
—Unijunction transistor characteristics
—Troubleshooting thyristors

INTRODUCTION

The **thyristor** has become an important component in many new consumer products. The term *thyristor* describes a broad range of electronic components used as electronic switches. Each of these components has the ability to switch between two states very rapidly. These two states are the conducting (on) state, and the nonconducting (off) state. Some of the thyristor devices can control current in both directions, whereas other thyristors conduct current in only one direction.

Thyristors are often used in applications where dc and ac power must be controlled. These components are used to control the amount of power delivered to a load. For example, the speed of a motor can be controlled by a thyristor.

Thyristors operate differently than diodes and transistors. In this chapter the different thyristors, including the SCR, the diac, and the triac, are studied. Their construction and application will be considered, along with their operation in consumer products. Thyristor troubleshooting is also covered.

SILICON-CONTROLLED RECTIFIER

The most popular of all thyristors is the **silicon-controlled rectifier (SCR).** The SCR is a three-junction device that functions as a switch. As its name implies, the SCR is a basic rectifier. In other words, it will conduct current in only one direction. The important part of SCR operation is that the device can be turned on or off. This on/off action makes the SCR an excellent device to control current.

SCR Construction

The construction of a component will provide information about component operation. This maxim holds true for the SCR. All solid-state devices are made by joining P and N material into a junction or junctions. The bipolar transistor, the diode, and the FET are all constructed in this manner. The SCR is made by stacking four alternating layers of P and N silicon material.

A block diagram of the SCR is shown in Figure 10–1. The four layers of the SCR are sandwiched together to form the three junctions. Leads for external connection are attached to these layers. These three connections are called the **anode (*A*),** the **cathode (*K*),** and the **gate (*G*).**

Figure 10–2 gives the schematic symbol for the SCR. This schematic symbol is basically the same as that for the rectifier diode. The main difference is the additional lead, the gate. Note the circle around the schematic symbol. This circle may or may not be included. Also, the leads may not be identified on the schematic drawing. When

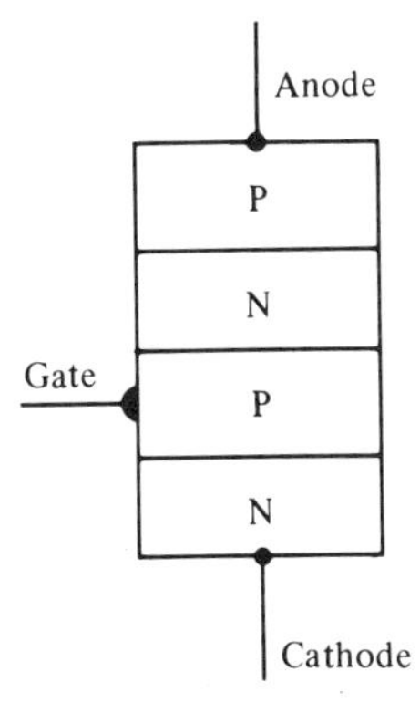

FIGURE 10–1
Block Construction of the SCR

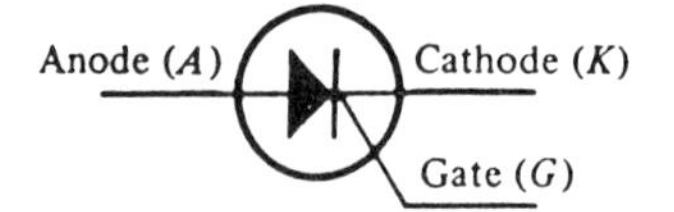

FIGURE 10–2
Schematic Diagram of the SCR

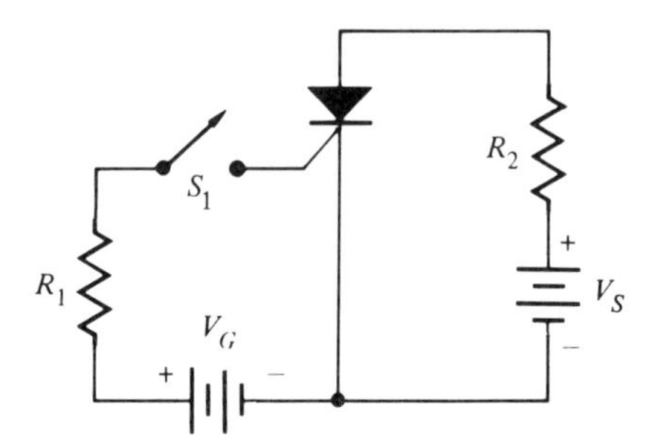

FIGURE 10–3
Biasing the SCR

the leads are identified, they will be marked with the letters *A, K,* and *G,* as shown.

As with any other component, the SCR must have proper biasing to operate. Figure 10–3 shows the proper biasing applied to the SCR. A switch is used in the gate circuit to apply voltage to the gate. Resistor R_1 is used to limit the current flow in the gate circuit. A second voltage source supplies the necessary forward bias to the anode and cathode of the SCR. A resistor (R_2) is placed in series with the anode-cathode circuit. This resistor is also used as a current-limiting resistor. It will prevent high currents from causing damage to the SCR. Without this resistor, the SCR would conduct hard in forward bias, and burn out after a short operating period.

SCR Operation

A volt-ampere graph will be useful in explaining SCR operation. A sample graph is shown in Figure 10–4.

The SCR is turned on when two conditions are met. The first is that the anode-cathode terminals be placed in forward bias. The second is that a positive voltage be applied to the gate.

The graph in Figure 10–4 shows that when voltage is first applied to the SCR, only a small amount of current flows. This small current flow is due to leakage between the three junctions. As the voltage on the anode-cathode of the SCR is increased, the current continues to rise, but by only a small amount. As soon as point *A* is reached, the forward current through the SCR increases rapidly. This means that SCR resistance has been reduced. If the resistance is reduced, the voltage drop across the diode will also be reduced. Again, this is shown on the graph. The point where the current increases rapidly is called

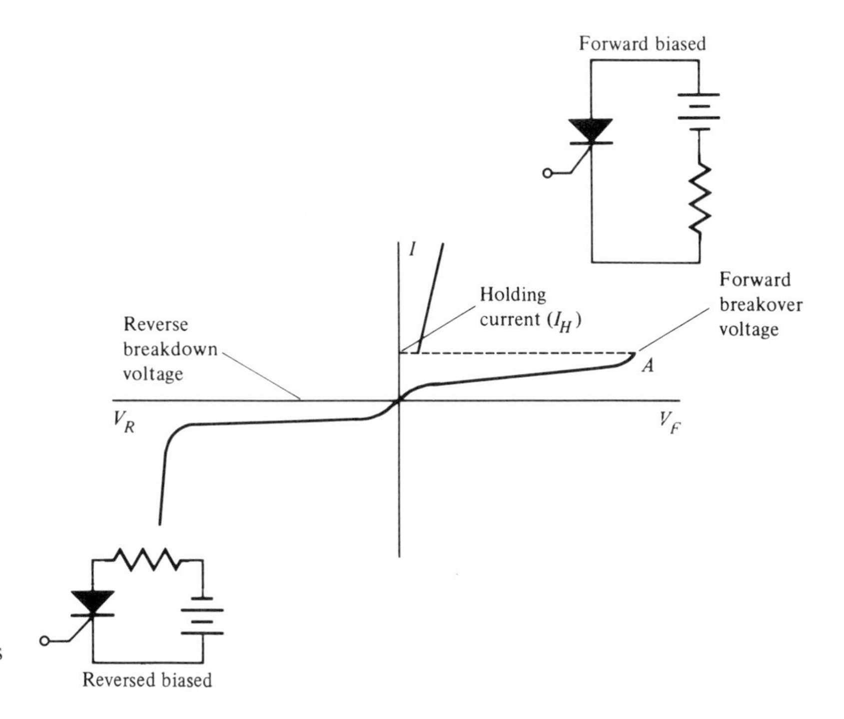

FIGURE 10–4
Forward and Reverse Bias to the SCR

the **forward breakover voltage.** When this voltage is increased, the SCR becomes a conductor and starts to conduct large amounts of current. The only protection the SCR has is the series resistor.

When the SCR is in the breakover voltage state, a small change in voltage will cause a large change in current flow. The SCR will remain in this state as long as the large foward current is maintained. Only when the forward current is reduced below the holding current will the SCR stop conducting. The minimum current that holds the SCR in the breakover voltage state is an important characteristic of this component. This current is called the **holding current,** and is generally abbreviated I_H on specification sheets.

Since the SCR has a forward operating point, it must also have a reverse operating point. Figure 10–4 also shows the reverse bias characteristic of the SCR. The SCR placed in this bias will act like a PN junction diode. As the reverse voltage increases across the SCR, a small leakage current is developed. This current remains at about the same level until the voltage has increased enough to break down the SCR. At this point the reverse current will increase rapidly with a small change in voltage. The voltage required to break down the SCR in the reverse direction is called the **reverse breakdown voltage.** If the reverse voltage is allowed to exist for long periods of time, the SCR will be damaged. The breakdown rating for the SCR is rarely reached in consumer products because of the low levels of voltage.

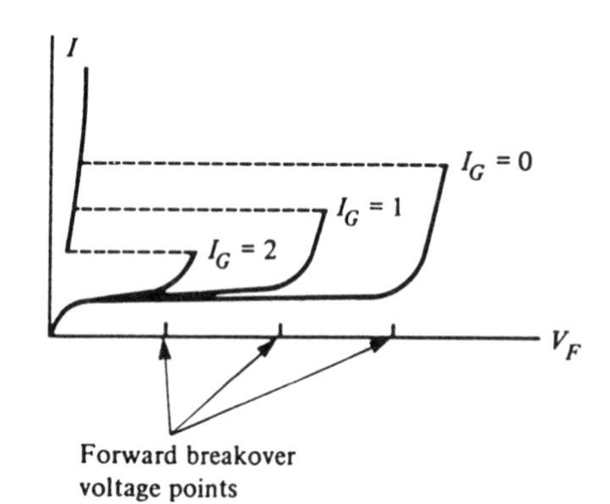

FIGURE 10–5
Different Breakover Points Controlled by Gate Voltage

The SCR operation presented thus far has not included the gate control. All of the discussion has been with the open gate. For the gate to have control over the SCR, the gate must develop a positive voltage with respect to the cathode. When this voltage is established, a gate current is developed, and the forward breakover point is changed. Figure 10–5 shows a volt-ampere graph with different breakover points. Notice that when the gate voltage is changed, the breakover points are moved. The graph also shows that when no current is developed in the gate, the voltage at which breakover occurs is large. As the gate voltage is increased, the breakover voltage is reduced. The graphed curves have about the same shape. As soon as breakover occurs, the resistance of the SCR reduces, and the current increases. The only major difference in the curves is in the location of the breakover voltages.

The graph in Figure 10–5 shows that the forward breakover voltage generally decreases as the gate current increases. If the gate current is made large enough, the SCR will behave like a PN junction diode. The ability of the gate current to control the current between the anode and the cathode makes the SCR a useful component in electronic circuits.

The different breakover voltage points in Figure 10–5 illustrate an important SCR characteristic: A specific gate current must be reached before an SCR will become a conductor. Each SCR has its own breakover voltage. Therefore, each SCR must have the proper forward bias applied, and the proper gate current, in order to function effectively as a switch in the circuit.

SCR Circuit Connection

The SCR has many circuit applications. The forward bias applied to the SCR is just low enough to keep the SCR from the breakover voltage point. Therefore, the only method that will turn on the SCR is to develop a suitable gate current. This gate voltage insures that the SCR will turn on at the proper time. Also, as has been shown, the SCR conducts as soon as the gate current is applied. This conduction continues until the forward bias is lowered below the holding current point. Even if the gate voltage is removed, the SCR will still be a conductor. The only way to shut off the SCR is to reduce the forward bias voltage to below the holding current.

SCR dc Operation

The SCR is generally used to control ac and dc power to loads in the circuit. It can be used as a switch to turn power on and off at the load, or it can be used to control the amount of power delivered to the load.

One of the most basic SCR circuits is shown in Figure 10–3. When power is applied to this circuit, the SCR does not conduct until the gate voltage is developed on the gate. This simple circuit allows the SCR to be turned on, but has no means to turn it off. If a switch were added in series with the load R_2, it could be opened to break the circuit operation. This would be a way to remove power from the load, and therefore turn off this device.

Because this method of turning off the SCR is not used in circuit application, another method must be used. This method is shown in Figure 10–6. Notice that when a positive voltage is applied to SCR_1, power is applied to the load. This action causes the plate of C_1 to develop ground potential. Capacitor C_1 then starts to charge, causing the right-hand plate of capacitor C_1 to develop a positive charge. When a small positive pulse is applied to SCR_2, it turns on this SCR_2. This action places C_1 in parallel with SCR_1. The voltage developed across the capacitor now places SCR_1 in reverse bias. This action reduces the

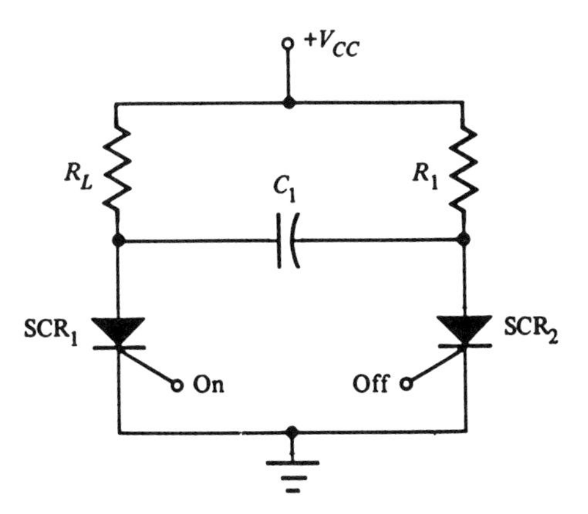

FIGURE 10–6
Connection of Two SCRs to Act as an On/Off Switch

forward bias and the holding current, and turns off the device. The current flow through the load is stopped. To start the circuit conducting, a small pulse is applied to the gate of SCR_1, and the process begins all over again. To start this circuit, a pulse is applied to SCR_1; to turn it off, a pulse is applied to SCR_2.

In several applications other biasing components will be used to turn on and off the SCR. The method just described is only one of many methods utilized.

SCR ac Operation

The SCR is also used to control ac power. Like the dc power, the ac power can also be turned on and off by the SCR.

Because the SCR is a rectifier, it operates on only one ac input alternation. Figure 10–7 shows a simple SCR connected as a switch circuit. In this circuit the SCR will conduct only when the input cycle makes the anode positive and the cathode negative. Also, S_1 must be closed to develop positive gate voltage, which will turn on the SCR. As in dc operation of the SCR, a series resistor is placed in the gate circuit. This resistor is a current-limiting resistor. Diode D_1 is placed in the circuit to protect the anode and cathode during the reverse voltage operation.

When S_1 in Figure 10–7 is closed, the SCR will conduct when the proper polarity appears at the anode. If the gate switch is opened, the SCR will continue to conduct until the voltage between the anode and the cathode falls below the holding current. Once the voltage falls below this level, the SCR will remain off until S_1 is closed.

Figure 10–7 illustrates the basic operation for the SCR in an ac circuit. Unlike in the dc circuit, the SCR in the ac circuit requires no additional parts to turn on or off. The positive and negative alternations of the ac waveform will turn on and off the SCR.

In ac operation the largest concern is that the SCR conduct at the proper time. To insure proper timing, additional circuits may have to be used.

The power developed at the load is also important. In ac op-

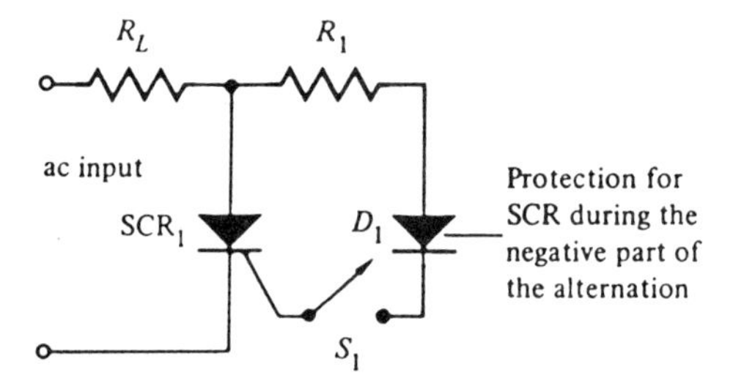

FIGURE 10–7
SCR Used to Control ac Voltage to the Load

eration the SCR will develop only one-half of the power at the load, because it conducts for only one-half cycle. However, additional circuits can be added to the basic SCR circuit to insure that the SCR will conduct for the entire input cycle, and thus deliver the full amount of power to the load.

The circuits discussed so far are simple control circuits. The SCR is turned on and off by the application of power to the circuit. This on/off state relates to the use of the SCR as a switch. The SCR has many advantages over the mechanical switch. The SCR will not wear out, it will not develop contact arcing, nor will it stick in one position. Overall, the SCR is a more reliable component than the mechanical switch, especially in high-current application. The SCR may be controlled by an electrical switch, or it may be controlled by an electrical pulse. Another important characteristic is that a small amount of power applied to the gate will control large amounts of power to the load.

SCR Circuit Application

The SCR is used widely in consumer products. It is found in low-voltage power supplies, and in scanning circuits of television receivers. A description of the scanning circuit for the horizontal sweep circuit will give a good idea of the SCR in use.

The horizontal scanning of a television receiver can be divided into two basic parts: the trace across the screen, and the retrace to return the beam to develop a new trace. To handle the necessary currents and voltages, two SCRs are used. One is used to control the trace voltage and current, and the other is used to control the retrace voltage and current. This operation is shown in Figure 10–8. The trace

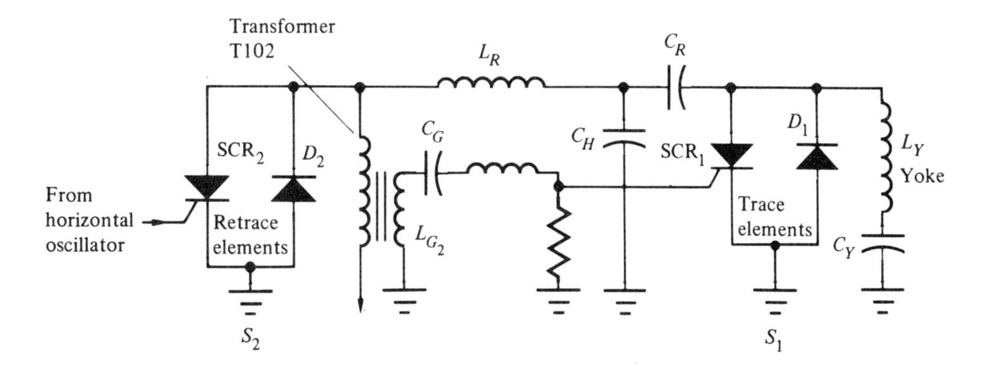

FIGURE 10–8
Two SCRs Used to Control the Trace and Retrace Cycle in the Horizontal Sweep Circuit

element, SCR_1, together with D_1, develops a *single-pole, single-throw switch,* which is labeled S_1. The retrace element, SCR_2, together with D_2, comprises another single-pole, single-throw switch, which is labeled S_2. The components L_R, C_R, C_H, and C_Y are the necessary energy storage and timing elements to develop the trace and the retrace action.

During the first half of the trace action, a magnetic field is collapsing around the yoke, labeled L_Y. This causes the D_1 to conduct and recharge C_Y.

Also during the first half of the trace interval, SCR_1 is made ready to conduct by the application of a gate pulse. However, this SCR will not conduct until the second half of the trace waveform. An example of this action is given in Figure 10–9. This second part of the trace cycle will forward bias the anode and cathode of SCR_1.

When the current through the yoke has reached zero due to the depletion of the yoke inductive energy, C_Y begins to discharge into the yoke. Figure 10–10 shows an example of this reverse action. The current is now reversed, which in turn reverse biases D_1, and simultaneously forward biases SCR_1. This discharging action of C_Y results in the necessary energy to complete the trace on the screen.

Once the scanning beam has completed its trace, the beam must be retraced to the starting point on the screen. This process is called **retrace.**

Figure 10–11 shows the waveform present from the horizontal oscillator at the gate of SCR_2. When T_3 is reached on the waveform, SCR_2 turns on, releasing a charge already built up on C_R. This charge is released into the circuit comprised of L_R and C_R. This release allows SCR_1 to continue to conduct. For a short time, SCR_1 and SCR_2 are conducting. This current flow is only for a short time, however, and

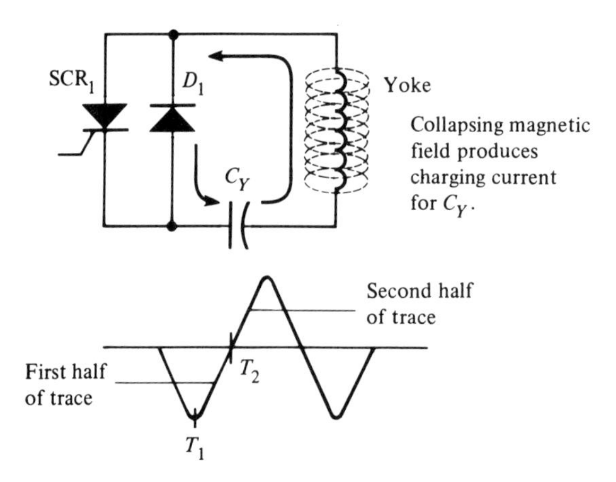

FIGURE 10–9
Charging Path for C_Y

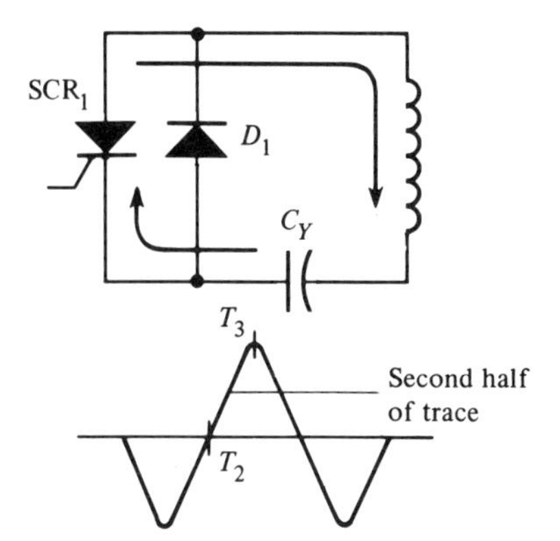

FIGURE 10–10
Discharging Path That Forward Biases SCR_1

soon the gate pulse is removed from SCR_1, causing it to stop conducting.

With SCR_1 and D_1 biased off, the retrace cycle can begin. A reverse current is developed in the yoke, causing the beam to be retraced back to the starting point. Figure 10–11 shows an example of this reverse current.

This application is only one of many in which the SCR is used as a directional switch. Directional switches are used to keep the proper current flow through the yoke, and to keep the beam scanning properly across the face of the picture tube.

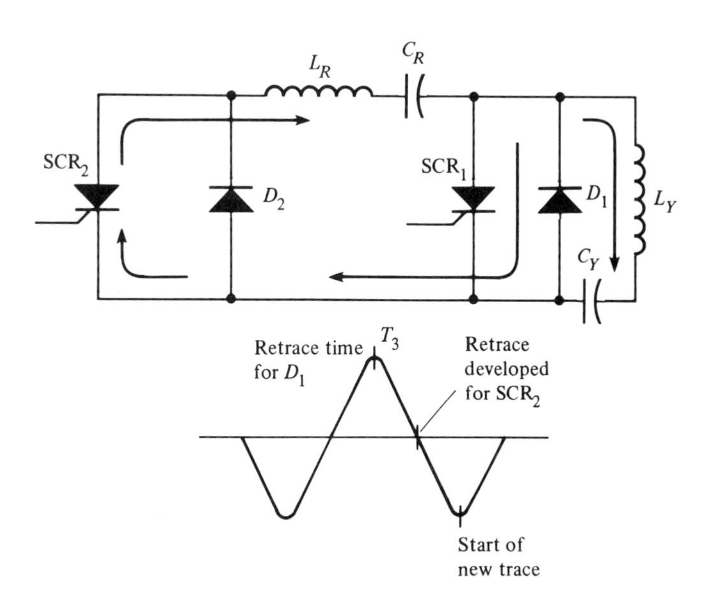

FIGURE 10–11
Retrace Path Developed through SCR_2

TRIAC

Also found in the thyristor family is the component called the **triac.** Because the SCR can control current in only one direction, the SCR is limited in its use. The triac can conduct in both directions. The triac has the same characteristics as the SCR. Therefore, the triac can be thought of as two SCRs placed in parallel but connected in the opposite direction.

Triac Construction

A simplified block drawing of the triac is shown in Figure 10–12. The triac has three terminals: **main terminal 1 (MT_1), main terminal 2 (MT_2),** and the **gate (G).**

If the entire PN structure of the triac is examined, it can be seen that current can pass through a PNPN layer, or it can pass through an NPNP layer. The device can be described as having an NPNP layer in parallel with a PNPN layer. This arrangement of four-layer material gives the triac a connection of two SCRs in parallel. This connection is illustrated in Figure 10–13. It must be pointed out that the connection in Figure 10–13 is not actually how the triac operates, because the triac gate voltage will respond differently than the SCR gate voltage.

Figure 10–14 shows the schematic diagram of the triac. Because the triac can conduct current in both directions, the schematic diagram contains two diodes facing in opposite directions.

Triac Volt-Ampere Relationship

As in any semiconductor, the volt-ampere characteristic of the triac is important to understand. Figure 10–15 shows a typical volt-ampere graph for the triac. This graph represents normal triac operation with no gate current applied. Notice that the graph is almost identical to

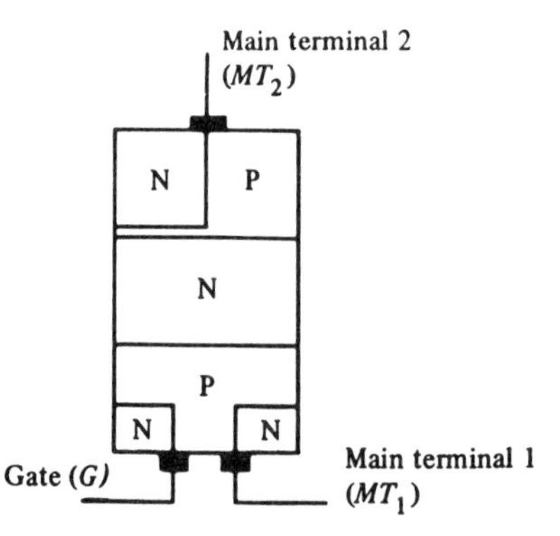

FIGURE 10–12
PN Block Structure for the Triac

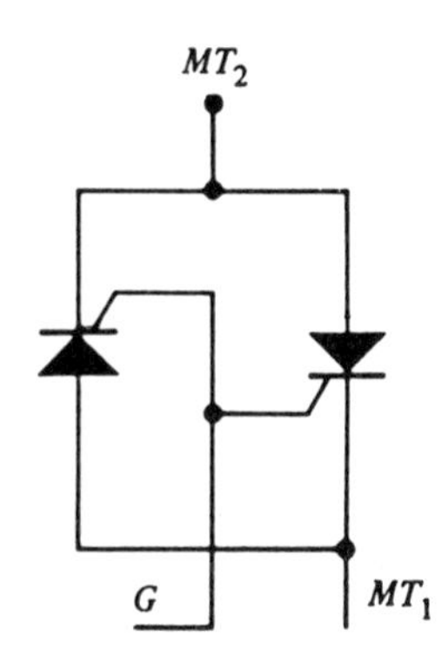

FIGURE 10–13
Connection of Two SCRs in Parallel to Form the Basic Triac

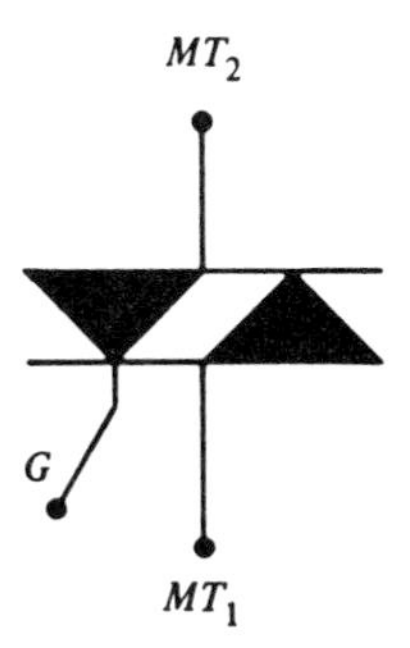

FIGURE 10–14
Schematic Diagram of the Triac

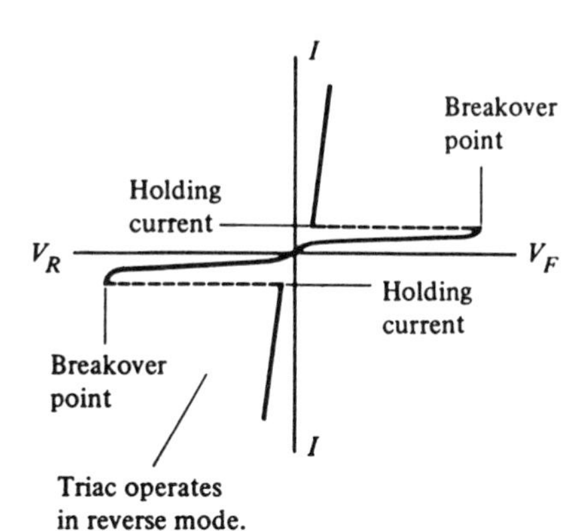

FIGURE 10–15
Volt-Ampere Characteristic Graph for the Triac

that for the SCR (see Figure 10–4). The same terminology is used for the triac as for the SCR. In the graph, MT_2 is positive with respect to MT_1. As the forward voltage is increased, a small current flows in the triac. This small current flow is leakage current. Once the forward voltage reaches the breakover point, the triac begins to conduct. Notice that the voltage has dropped across the triac. This voltage is applied to the load.

When the polarities of MT_1 and MT_2 are switched, the characteristics are different from those found in the SCR. Here, the triac will conduct in the reverse bias voltage mode. Notice that as the voltage is increased, the triac develops a small leakage current. Again, once the breakover voltage is reached, triac resistance drops, and the device begins to conduct. The triac current is limited only by the load resistance.

The graph in Figure 10–15 shows that the triac will conduct in either direction. However, this graph does not take into account the ability of controlling triac operation by gate current. As with the SCR, the gate current controls the breakover point. When the gate current is increased, the breakover point is lowered.

Another difference between SCR and triac operation is that the SCR requires a positive gate voltage. Because the triac conducts in both directions, the triac gate can be triggered by a positive or a negative voltage. To determine if the gate is positive or negative, the gate voltage is referenced against MT_1.

In circuit operation the triac is always subjected to voltage levels below the breakover point. Therefore, it is turned on by current flows in the gate. To turn off the triac, the holding current is reduced. The triac can be turned on again by another gate pulse.

Triac Circuit Application

Because the triac can conduct in either direction, it is best suited to the control of ac power. This device can be used as shown in Figure 10–16. In this circuit the full power will be applied to the load when the gate is triggered on. When S_1 is open, the triac cannot conduct, because the voltage applied to the triac is below the breakover point. When S_1 is closed, the triac is triggered on, and both halves of the ac power are applied to the load. This differs greatly from SCR operation.

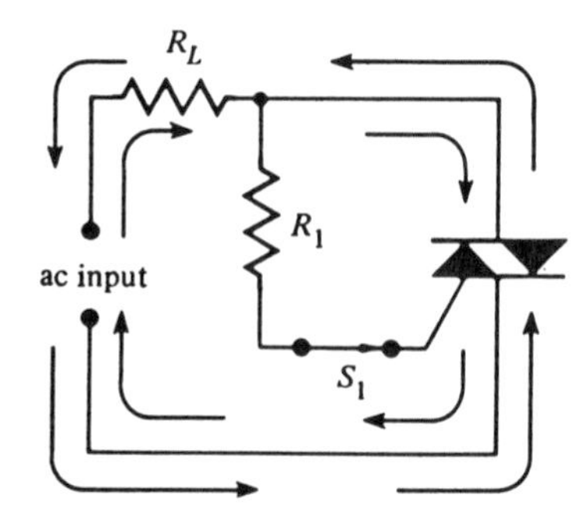

FIGURE 10–16
Small Gate Current Controlling Larger Current Flow between the Terminals in the Triac

The SCR can apply only half of the power to the load because it conducts in only one direction. Again, the advantage of all thyristors is that small gate currents can control large load currents.

Although the triac can conduct in both directions, and requires a small current to operate, it has some disadvantages when compared to the SCR. In general the SCR has higher current ratings than the triac. The triac can handle currents up to 25 A, whereas the SCR can safely handle currents of around 800 A. Therefore, when large currents are required, the SCR is the better component.

The SCR and triac also have different frequency-handling capabilities. The triac is often slower in turning on when applied to inductive-type loads, such as when it is used to control a motor. The triac is designed to operate mainly in the low-frequency range, 30–400 Hz, whereas the SCR can safely handle frequencies up to 30 kHz.

DIAC

The **diac** is a bidirectional device used to control the gates of triacs. This bidirectional device operates as described in the following sections.

Diac Construction

The diac is constructed in basically the same manner as a bipolar transistor. This device has three doped materials, as shown in Figure 10–17. However, the doping arrangement differs from that for the bipolar transistor. The doping in the diac is concentrated around the junction areas, and only two leads are attached to the outer layers. Since the diac has just two leads, in physical shape it resembles a diode. Sometimes the diac is packed in a plastic or metal case with axial leads. The diac could also be packaged in a housing that resembles a transistor with only two leads. Figure 10–18 shows the physical drawing of the diac with the axial leads.

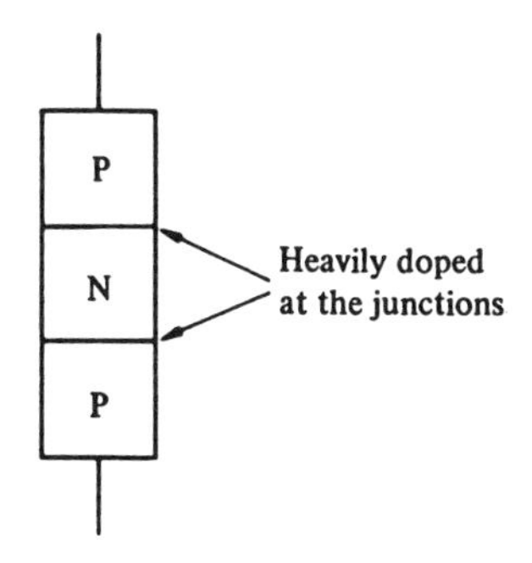

FIGURE 10–17
PN Block Structure of the Diac

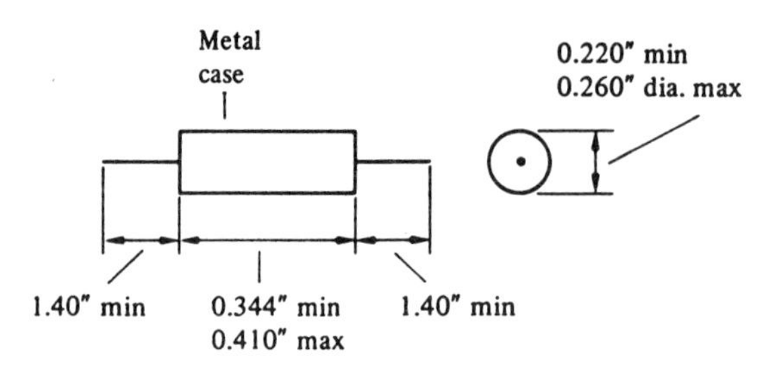

FIGURE 10–18
Physical Structure of the Diac

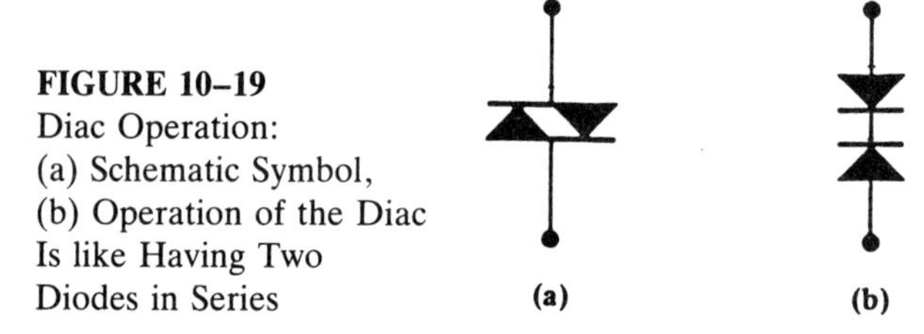

FIGURE 10–19
Diac Operation:
(a) Schematic Symbol,
(b) Operation of the Diac Is like Having Two Diodes in Series

The diac allows current to flow in both directions. Because of this action, it can be thought of as two diodes placed in parallel. The schematic symbol for the diac represents this two-way current flow, as shown in Figure 10–19a. In operation the diac can be considered as two diodes connected in series. This connection is illustrated in Figure 10–19b.

Diac Operation

Thinking of the diac as two diodes connected in series, but in opposite directions, clarifies the operation of the diac. For the diac to conduct, one of the diodes must be forward biased, and the other must be reversed biased. Figure 10–20 shows this configuration. When voltage is applied to this circuit, only a small amount of current flows. This current is leakage current. It is not enough current to cause D_1 to conduct. Once the voltage has risen sufficiently, D_1 will conduct. When this happens, D_2, because it is forward biased, also begins to conduct. Conduction will also occur if the polarity on the diac is reversed. Therefore, the conduction path is in both directions. The point at which the diac starts to conduct is called the **breakover point.** Generally this voltage is 26–38 V, depending upon the diac.

As with all thyristors, the diac has a volt-ampere characteristic curve. Figure 10–21 shows an example of this graph. Notice that once the breakover voltage is reached, only a slight voltage drop occurs, and current begins to conduct.

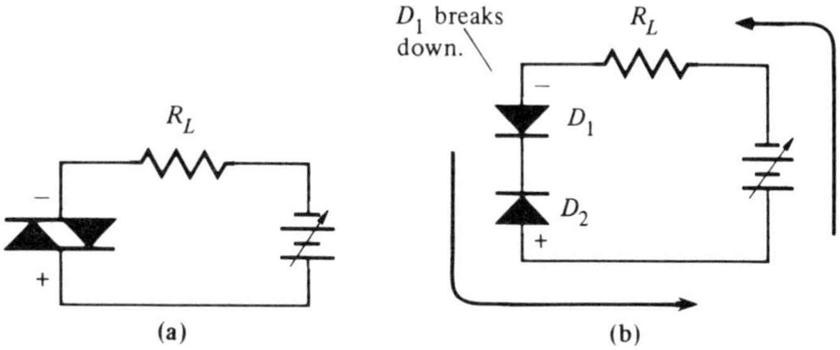

FIGURE 10–20
Biasing of the Diac:
(a) Schematic Diagram,
(b) Diac Diagram as Two Diodes

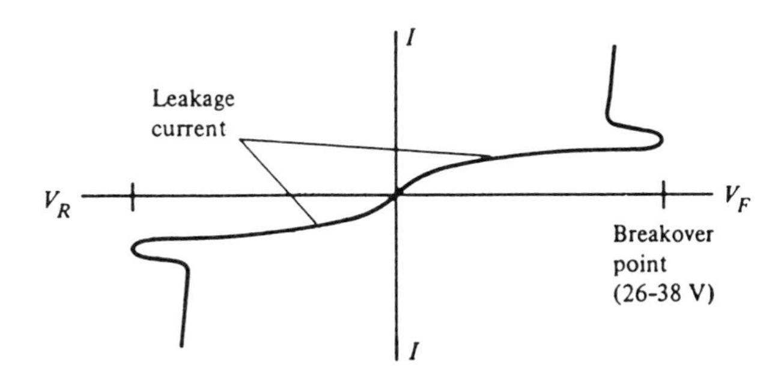

FIGURE 10–21
Characteristic Curve of the Diac

The diac is said to be in the off state until the breakover point is reached and the diac begins to conduct. The diac is able to conduct very large amounts of current. To turn off the diac, the current in the device is reduced to below the holding current.

Diac Circuit Application

Figure 10–22 shows a typical use of the diac. Here the diac is applied primarily as a gating device for the triac. When the voltage is applied, the C_1 voltage rises. In this case the voltage rises above the turn-on level of the diac. This rise in voltage allows the diac to be turned on, and then triggers the triac into conduction. This short interval of conduction by the triac is allowed because of the diac. Generally the

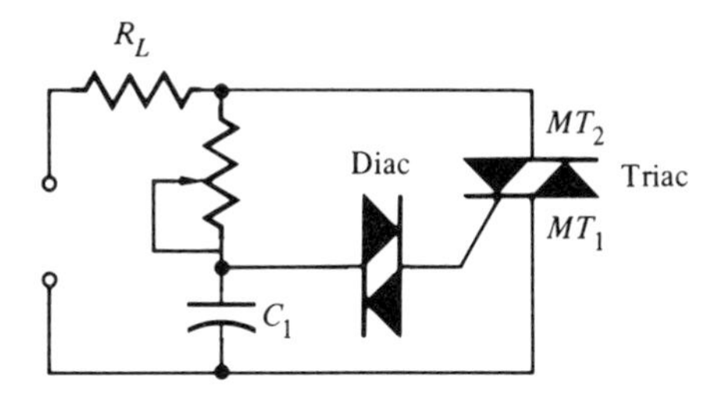

FIGURE 10–22
Diac Used to Control the On/Off State of the Triac

diac is used in phase controls for motors. The diac is also found in the controls for motor speed and the direction of motor rotation, and in the control of temperature in motor circuits.

UNIJUNCTION TRANSISTOR

The final member of the thyristor family is the **unijunction transistor (UJT).** The UJT has different characteristics than the diac, triac, and SCR. The UJT is a special type of transistor that is used as a switch, not as an amplifying device. Because of its switching action, the UJT is a very useful component in oscillator circuits.

Related to the UJT is the **programmed unijunction transistor (PUT).** The PUT operates basically as an electronically controlled switch.

UJT Construction

As the name implies, the UJT has only one junction. This component consists basically of a block or bar of N-type semiconductor material with a small pellet of P-type material attached. Leads are connected to the ends of the N-type bar. The leads are designated as **base 1 (B_1)** and **base 2 (B_2).** Another lead is attached to the pellet of P material. This lead is the **emitter (E).** Figure 10–23a shows the basic construction of this device. Figure 10–23b shows the schematic diagram of the UJT.

The N-type bar is lightly doped, and therefore will have generally moderate resistance. Typical resistances for this bar are between 5 kΩ and 9 kΩ. This resistance is found between B_1 and B_2.

To better understand the operation of the UJT, an equivalent circuit can be drawn. Figure 10–24 shows this circuit diagram. Notice that the N-type material is shown by two resistances, R_{B_1} and R_{B_2}. In addition, since the P-type pellet is connected to the N-type bar, this junction can be thought of as a diode. Also shown in the drawing is a

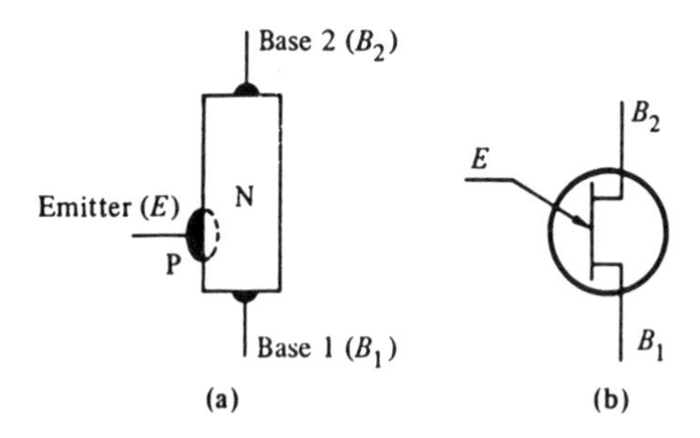

FIGURE 10–23
UJT: (a) Block Diagram of the PN Structure, (b) Schematic Diagram

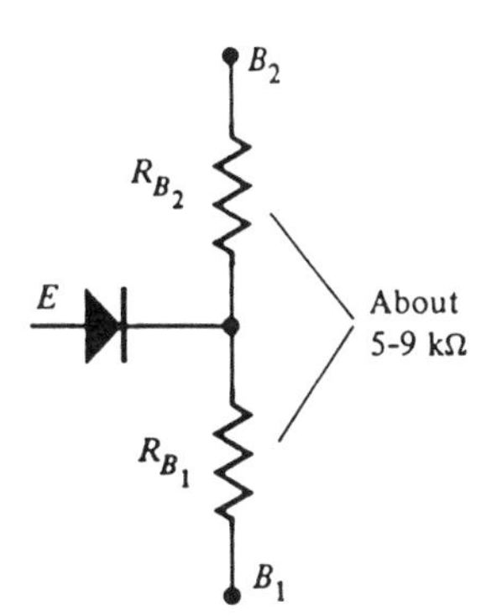

FIGURE 10–24
Equivalent UJT Circuit

diode connected to the emitter lead. This diode represents the connection of the P-type pellet to the N-type bar.

UJT Operation

As with any other component found in the thyristor family, a proper biasing voltage must be applied to the UJT. Without the proper biasing, the UJT will conduct only small amounts of leakage current. These leakage currents will not perform any work in the circuit.

Figure 10–25 shows the equivalent UJT circuit with the proper voltage applied to the terminals. The positive lead of the dc power source is applied to B_2. The negative lead of the power source is connected to B_1. This connection properly biases B_1 and B_2. For the emitter lead of the UJT to be biased properly, a positive voltage is applied to the emitter lead of the UJT. To limit current flow through the emitter lead, a series resistor is placed between the emitter and the positive power supply lead. The voltage placed on the emitter causes the emitter to be positive with respect to B_1 of the UJT. All voltage polarities in the UJT are referenced against the B_1 lead.

Once the power is applied, only a small amount of leakage current flows through the UJT. For conduction of the UJT to increase,

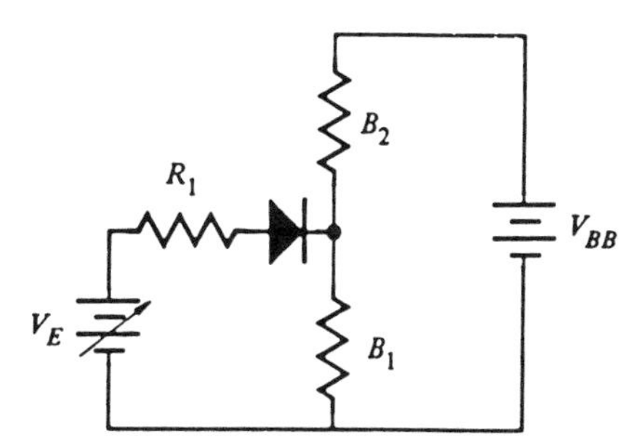

FIGURE 10–25
Biasing the Equivalent UJT Circuit

the voltage on the emitter lead must be increased. Since the emitter is acting like a diode, it must have the same voltage drops as a rectifier diode. Therefore, the voltage drop at the diode must be 0.3–0.7 V, plus the voltage drop across the resistance of B_1 caused by the V_{BB} supply. Once the proper voltage level is reached, the internal resistance between B_1 and B_2 decreases, and the UJT starts to conduct. The UJT will continue to conduct until the emitter current is increased.

UJT Circuit Application

Because the UJT has the ability to control current flow as well as resistance in the base region, it is an excellent component to use in oscillators, trigger circuits, sawtooth generators, and the like. Figure 10–26 shows a typical UJT oscillator application.

The oscillator in Figure 10–26 is a **relaxation oscillator.** This oscillator is capable of producing sawtooth waveforms at the output.

Once S_1 in Figure 10–26 is closed, C_1 charges through R_1. The charge in the capacitor brings the emitter voltage above the point at which the UJT starts to conduct. Once the UJT conducts, the resistance between B_1 and B_2 drops. This causes the capacitor to discharge through R_3 and the UJT. After C_1 is discharged, the emitter current is reduced, turning off the UJT. The waveform developed at the output of the UJT is a sawtooth waveform. This output waveform has positive spikes of voltage that can be used to trigger triacs and SCRs. The frequency of this waveform can be adjusted by R_1, which controls the charging time of C_1.

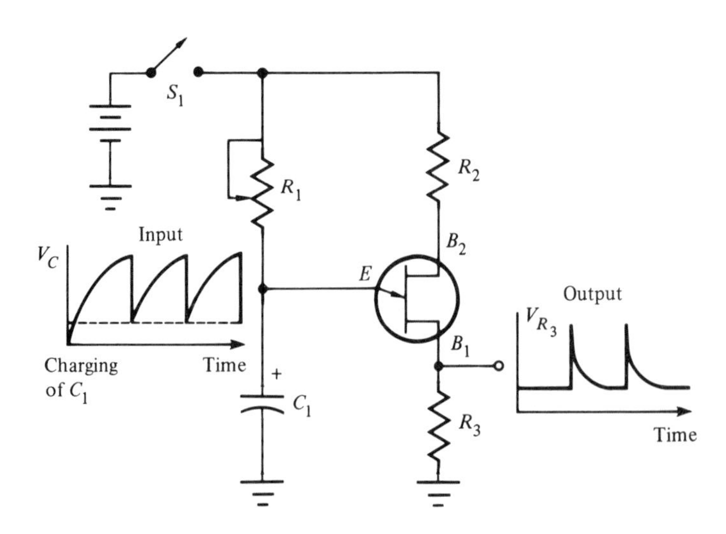

FIGURE 10–26
UJT as a Relaxation Oscillator

TROUBLESHOOTING

THYRISTORS

More and more thyristors are being found in modern consumer products. Still one of the best troubleshooting tools a technician can use is the ohmmeter. An ohmmeter can help find most defective thyristors in circuit operation.

Testing the SCRs

The resistance of an SCR can be tested in or out of circuit. Common troubles are anode-cathode shorts; opens; and, less frequently, failure to trigger or failure to hold once triggered. When testing an SCR, the $R \times 1$ range of the ohmmeter should be used.

First the polarity of the ohmmeter leads should be determined. Usually the red lead is positive, and the black lead is negative.

A small, low-wattage SCR generally exhibits diode characteristics between the anode and cathode when the gate is triggered on. When checking the SCR, the positive ohmmeter lead should be connected to the anode, and the negative ohmmeter lead to the cathode. This connection is illustrated in Figure 10–27. Then a jumper lead should be placed momentarily between the gate and the anode. This allows the positive voltage to trigger the SCR into forward conduction. The SCR resistance should be low. This low resistance state should remain even after the jumper lead has been removed. To stop conduction, ohmmeter leads should be removed from the SCR. The test should be repeated. If the SCR triggers on when the short is applied between the anode and the gate, it might not hold the SCR in conduction once the short is removed. This

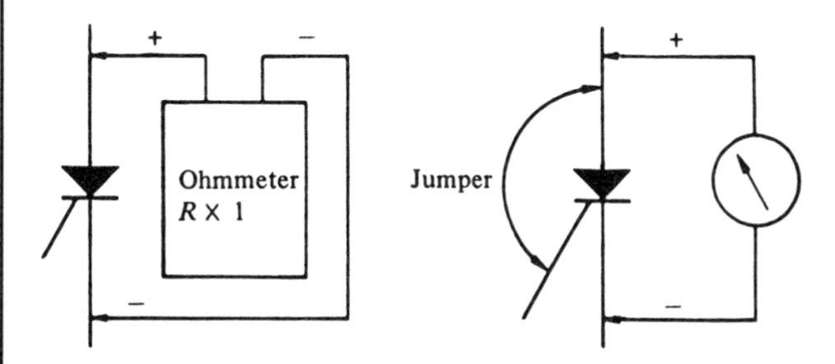

FIGURE 10–27 Testing the SCR Ohmic Value between the Anode and the Cathode

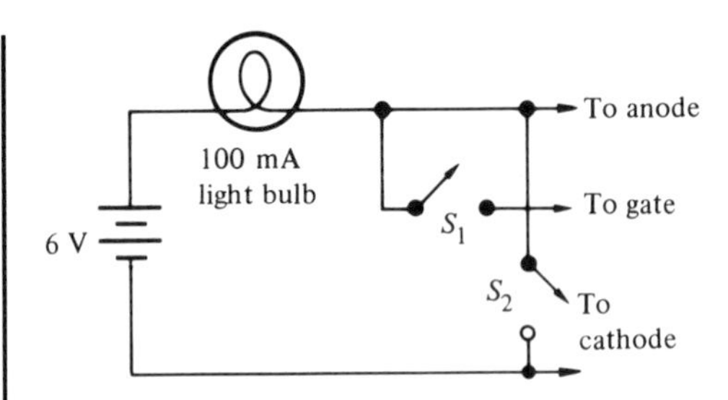

FIGURE 10–28 SCR Go–No-Go Tester

might mean that the SCR is bad. However, in some cases the ohmmeter cannot supply enough current to hold the SCR in the conducting state. Some of the larger SCRs require 50 mA or more of holding current, whereas the smaller ones require 25 mA or less of holding current.

Figure 10–28 shows a simple *go–no-go tester* for an SCR. This tester would be good for a larger SCR. A conventional 6 V battery supplies the necessary current. The lamp is an indication of conduction in the SCR. When S_1 is pushed in, the gate triggers conduction, and the lamp should glow to almost full brightness. When S_2 is pressed, the SCR should be unlatched, and the light should go off. This is a quick little circuit to build to check whether the SCR is good or bad.

Testing the Triac

As stated earlier, the triac is basically two SCRs connected in parallel. To test this component, the ohmmeter is used as in SCR testing. Between MT_1 and the gate of the triac, there should be a low resistance measurement (Figure 10–29a). Between MT_2 and the gate, the resistance should be high (Figure 10–29b). When the resistance is measured between MT_1 and MT_2, the ohmmeter again should measure high (Figure 10–29c).

The next measurement of resistance is taken by connecting the ohmmeter between MT_1 and MT_2, as shown in Figure 10–30a. Notice that with the triac, the polarities of MT_1 and MT_2 do not matter. Next, in Figure 10–30b the gate lead is shorted to either MT_1 or MT_2. This short should cause the triac to develop low resistance. When the short is removed, the low resistance state should remain. When the ohmmeter is removed, the low resistance will be removed. Again, with a larger triac the current from the ohmmeter might not be large enough to hold the triac on. If the test in Figure 10–30c shows the triac to be good, the leads should be reversed, and the test repeated.

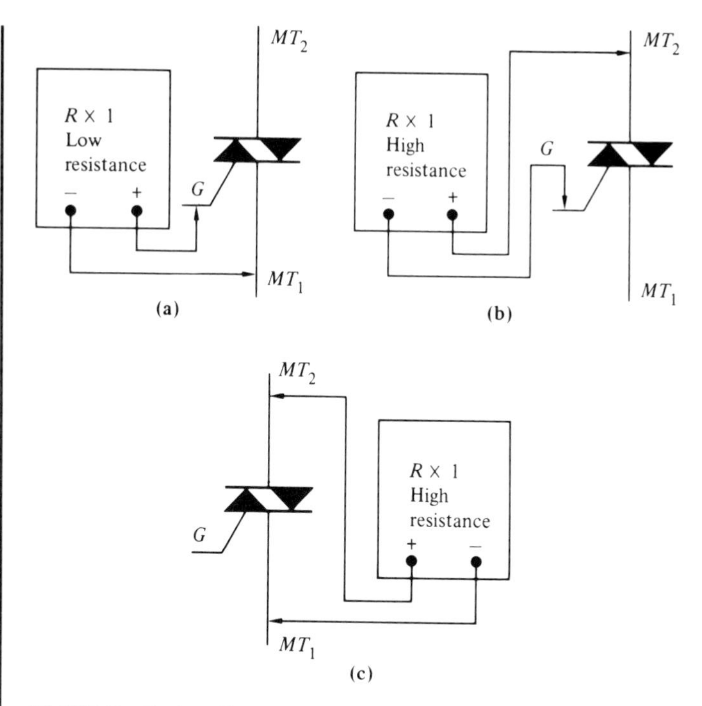

FIGURE 10–29 Terminal Resistance Measurements: (a) Resistance between the Gate and MT_1, (b) Resistance between the Gate and MT_2, (c) Resistance between MT_1 and MT_2

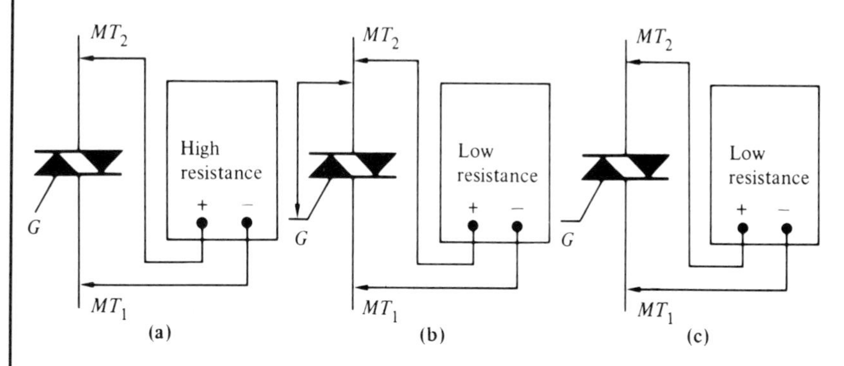

FIGURE 10–30 Testing Resistance of the Triac for Holding Resistance and On/Off State

Testing the Diac

The simple ohmmeter test can also be performed on the diac. Because the diac is essentially two diodes connected in parallel, it

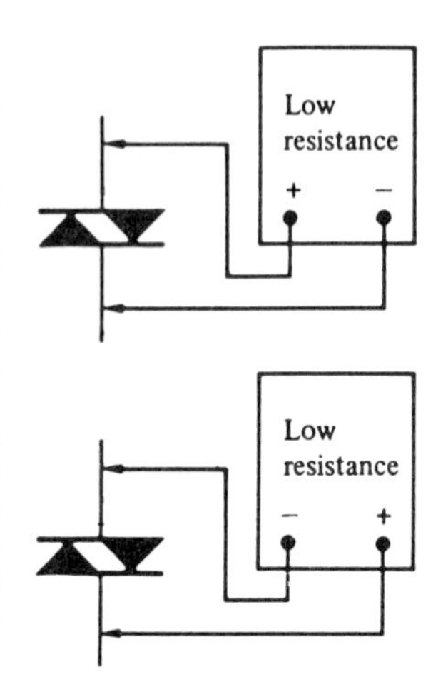

FIGURE 10–31 Testing the Resistance of the Diac

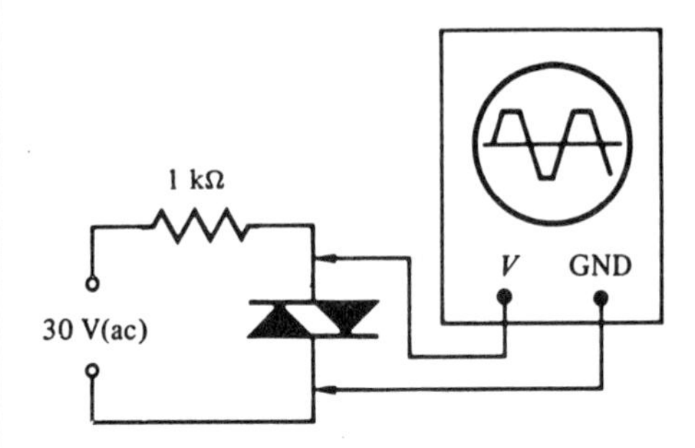

FIGURE 10–32 Waveform Developed at the Output of the Diac

should show low resistance in either direction. However, testing the diac in this fashion will show only that this component is not open. It will not demonstrate if the component is shorted. Figure 10–31 shows an example of testing the diac in this manner.

Further tests of the diac must be performed by using an external power source, a series-limiting resistor, and an oscilloscope. An example of this test, and the waveform developed at the output of the diac, are shown in Figure 10–32.

SUMMARY

Many new components are being used in consumer products. Among these are the SCR, the triac, the diac, and the UJT, all of which are thyristors.

The SCR is the most popular of the thyristor family. The component has three leads: the cathode, the anode, and the gate.

The SCR acts like an electronic switch that can be turned on by a positive gate voltage, and the cathode and anode are placed in forward bias. To turn off the SCR, the voltage across the SCR is lowered, which causes the current through the SCR to drop below the holding current.

Many new circuits found in consumer products use the SCR as a regulating device. Scan-derived, low-voltage power supplies and trace circuits use the action of the on and off states of the SCR.

The triac is another component found in the thyristor family. It behaves like the SCR, except that it conducts in both directions. The triac operates in this fashion because it acts like two SCRs connected in parallel.

The terminals on the triac are called main terminal 1, main terminal 2, and the gate. Because the triac will conduct in both directions, it is generally found in motor control circuits.

Another type of thyristor is the diac. The diac is a bidirectional switch. This switch turns on in either direction once the proper breakover voltage is reached. The main use of the diac is to turn on the triac.

The UJT is primarily a bar of N-type material that varies its resistance. When the emitter voltage is raised, the UJT acts like a diode, and will conduct large amounts of current.

The terminals on the UJT are base 1, base 2, and the emitter. This component is found primarily in oscillator circuits, and as a controlling device for the SCR.

To check if the thyristor is operating properly, resistance checks can be made on the terminals. In the SCR, resistance should be high between the anode and the cathode. In the triac, resistance should read high between MT_1 and MT_2. Once the gates develop positive voltage, the resistances between these terminals will drop.

The diac test with an ohmmeter will only prove that the diac is not open. To test for a short, a voltage check should be made on the terminals.

KEY TERMS

anode
base 1
base 2
breakover point
cathode
diac
emitter
forward breakover voltage
gate
holding current (I_H)
main terminal 1
main terminal 2
programmed unijunction transistor (PUT)
relaxation oscillator

retrace
reverse breakdown voltage
silicon-controlled rectifier (SCR)
thyristor
triac
unijunction transistor (UJT)

REVIEW EXERCISES

1. In what two states will thyristors operate?
2. Name three different thyristors, and draw the schematic symbol of each.
3. Name the terminals on the SCR.
4. Draw a schematic of the SCR, and properly bias each of the terminals.
5. What is the name given to the current that keeps the SCR operating?
6. List some advantages of the SCR over a mechanical switch.
7. Using Figure 10–8, describe the process used to develop the trace and retrace cycle.
8. What two advantages does the triac have over the SCR?
9. What is the safe current rating for a triac?
10. What is the frequency range of the triac? What is the main application of this device?
11. In how many directions will current flow in a diac?
12. Within what voltage range does the diac generally begin to conduct?
13. What is the primary function of the diac?
14. Name four typical applications for a UJT.
15. Describe the method used to test the SCR.
16. Draw a schematic of a simple tester that can be constructed to test the SCR.
17. What resistance should be measured between MT_1 and MT_2 of a triac?
18. Draw the waveforms that would be developed at the output of a UJT relaxation oscillator.
19. Draw the volt-amp characteristic curve for a forward bias SCR circuit. Identify the breakover voltage point on the graph, and the holding current point.
20. Draw the schematic diagram of the UJT and label all terminals.
21. Draw the waveform output developed across a diac when an ac voltage is developed at the input.

11 OSCILLATORS

OBJECTIVES

Upon completing this chapter, you should be familiar with:

—Basic oscillator operation
—Oscillator biasing points
—Phase-shift oscillator characteristics
—*LC* sine wave oscillator characteristics
—Nonsinusoidal oscillator characteristics
—Basic oscillator frequency networks
—Positive feedback characteristics
—Wien bridge oscillator characteristics

INTRODUCTION

The basic signal-generating source in electronic circuits is the **oscillator.** An oscillator will convert a dc signal into an ac signal. Oscillators come in various electronic circuit configurations, and can generate almost any value of frequency required for the circuit. In most cases the oscillator can be classified by the frequency range it delivers. These ranges are the *low-frequency range* and the *high-frequency range.*

In this chapter the basic oscillator circuits found in consumer products are explored. Operation and circuit configuration are stressed. The starting point of the oscillator will be the amplifier. Almost any amplifier can be turned into an oscillator if certain conditions are met.

BASIC OSCILLATORS

Oscillators are devices that convert dc voltage into ac voltage. Because of the different demands placed on oscillators in consumer products, oscillators are divided into two basic categories: **sinusoidal oscillators,** and **nonsinusoidal oscillators.** The terms *sinusoidal* and *nonsinusoidal* describe the types of output waveform. Figure 11–1a shows the sinusoidal oscillator and its sine wave, and Figure 11–1b shows the nonsinusoidal oscillator and its square wave. Because digital circuits are being used more and more in consumer products, and because these circuits require timing, the nonsinusoidal oscillator is becoming a very important device.

The basic oscillator is made up of three parts: the amplifier, the feedback, and the timing circuit. Figure 11–2 shows an example of the oscillator in block diagram form.

All transistor oscillator circuits have the same basic requirement: The **amplifier** section must have a gain large enough to overcome circuit loss. The gain of the amplifier between input and output must be greater than one. There must be a **timing circuit,** such as a tank circuit, an *RC* circuit, or a crystal control network, to set the frequency of the oscillator. Finally, the oscillator must have **feedback.** In this type of feedback, a sample of the output signal is fed back in-phase to the input to sustain oscillator operation. This feedback signal is called **regeneration,** or **positive feedback.**

In most cases an amplifier can be made to oscillate. One of the best examples is shown in Figure 11–3. When the controls of a public address system have been set too high, squeals and howls are developed from the output of the speaker system. These howls and squeals mean

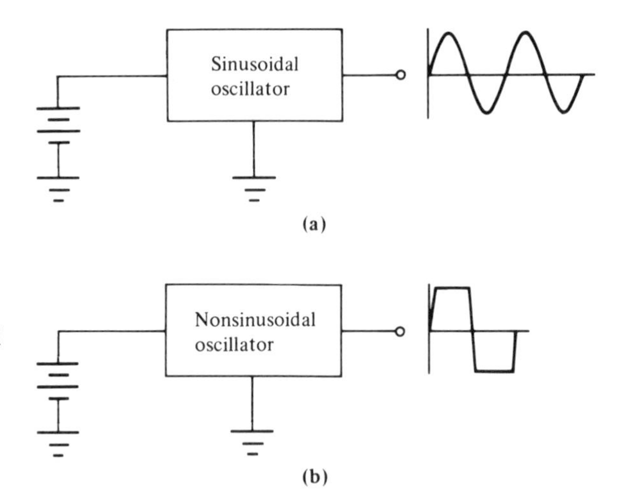

FIGURE 11–1
Basic Oscillator Types and Output Waveforms (a) Sinusoidal Oscillator and Sine Wave, (b) Nonsinusoidal Oscillator and Square Wave

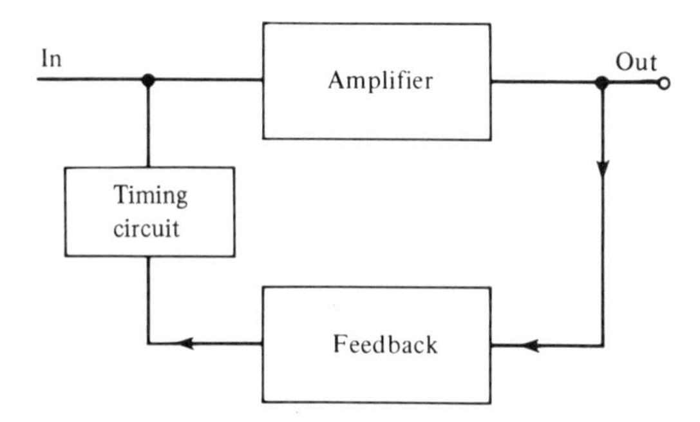

FIGURE 11–2
Basic Block Diagram of an Oscillator

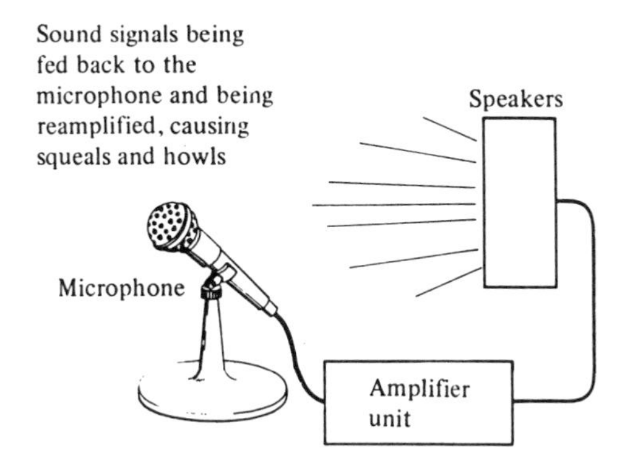

FIGURE 11–3
Public Address Amplifier System Being Turned into an Oscillator

that the amplifiers have broken into oscillation. This is generally the result of a part of the output being fed back to the microphone, and then reamplified. The input signals have distorted the amplifiers' input signals, and have made them useless.

The feedback found between output and input in most oscillator circuits is different than that just described. The feedback path is usually coupled through capacitors, inductors, or resistors. This coupling allows the amplifier to develop its own input signal to regenerate itself for another oscillation.

The feedback signal and the amplifier will not guarantee oscillations. As the block diagram in Figure 11–2 shows, the oscillator requires a frequency-controlling section. Because these oscillators are required to operate over a large frequency range, the timing or frequency-controlling networks are very important. Generally, oscillators found in consumer products are divided into two categories: those oscillators that generate audio frequencies, and those that generate radio frequencies. In the audio range, a frequency network of an *RC* (resistor-capacitor) network will be used. In radio frequency oscillators, an *LC* (inductor-capacitor) or crystal network will be used.

Different oscillators have different requirements. They do, however, have one common requirement—stability. To be stable, an oscillator must be able to generate a constant amplitude signal and constant frequency. A requirement for some oscillators is that they have the ability to be tuned. The local oscillator section of an AM radio must be able to be tuned through a variety of frequencies. This type of oscillator is called a **variable-frequency oscillator (VFO).**

FREQUENCY NETWORKS

The major function of an oscillator is to generate a set frequency, or to generate a range of frequencies. This important task is accomplished in one of several ways. The frequency is developed by an *LC* tank circuit, by an *RC* circuit, or by crystal control.

LC Circuit

Because many consumer product oscillators generate frequencies of 100 kHz and above, the ***LC* circuit** design is used. The inductor-capacitor is connected to form a **tank circuit.** This connection is then connected to the input of an active device. Figure 11–4 shows an example of this type of connection. The *LC* network will carry the oscillator output well into the megahertz range, and will generate perfect sine waves for the oscillator output. Also, because the *LC* combination is very frequency selective, this type of oscillator is very frequency stable.

The *LC* network operates by the capacitor storing an electrical charge, and the stored energy that surrounds the inductor. Figure 11–5 shows an example of the development of the sinusoidal output in a tank circuit. In Figure 11–5a it is assumed that the capacitor is already charged. As the capacitor begins to discharge through the inductor, a magnetic field is built up around the inductor. After the capacitor has completely discharged, the magnetic field around the inductor begins to collapse. This collapsing keeps the current flowing in the same direction and recharges the capacitor, but in the opposite polarity.

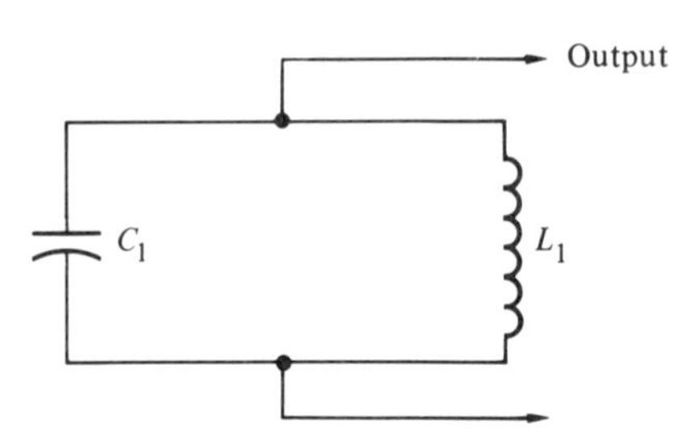

FIGURE 11–4
Parallel *LC* Network

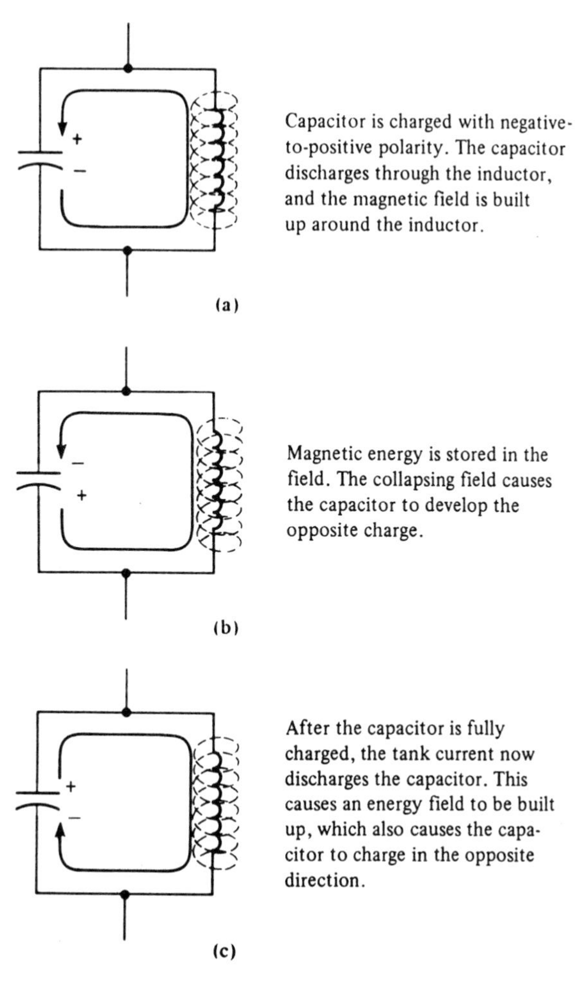

FIGURE 11–5
Charging and Discharging Process of the Capacitor in a Tank Circuit: (a) Discharging of the Capacitor Establishes a Magnetic Field around the Inductor, (b) Collapsing Magnetic Field Recharges the Capacitor, (c) Discharging Capacitor Develops a Magnetic Field around the Inductor to Repeat the Cycle

Figure 11–5b shows this action. The capacitor is now recharged, and is acting as the power source for the circuit. The capacitor then discharges again, but in the opposite direction from the first discharge. The magnetic field is built up around the inductor. When the capacitor discharges, the inductor field collapses and continues to charge the capacitor in the opposite direction. Figure 11–5c shows this action. The *LC* circuit is now charged with the original polarity. The process repeats itself. This back-and-forth charging and discharging of the capacitor establishes a rate of oscillation. This rate of oscillation can be translated into the frequency of the oscillator.

Because the capacitor and the inductor both store energy in the *LC* circuit, they are the frequency-setting components. A formula can be used to find the approximate frequency developed by this network. This formula is shown in Equation 11–1.

$$f_r = \frac{1}{6.28\sqrt{LC}} \tag{11–1}$$

where:

f_r = resonant frequency of the LC network (Hz)
L = size of the inductor (H)
C = size of the capacitor (F)
$6.28 = 2\pi$

This formula is generally used to find the resonant point of the tank circuit.

The technician must be aware of the different values of inductance and capacitance in the oscillator circuit, and in what general frequency range the oscillator will operate. An inductive and capacitive combination of relatively large value means lower frequencies being generated by the oscillator, while small values of inductance and capacitance relate to larger frequencies. Take, for example, a 4 H inductor and 10 μF capacitor combination. By placing these values into Equation 11–1, it is found that this oscillator will operate at about 25 Hz. By reducing the values to a 2 μH inductor and a 3 pF capacitor, the operating frequency of the oscillator increases tremendously. The operating frequency of this oscillator can be calculated in Example 1.

EXAMPLE 1

At what frequency will an oscillator with a 2 μH inductor and a 6 pF capacitor operate?

Solution:

$$f_r = \frac{1}{6.28\sqrt{LC}} = \frac{1}{6.28\sqrt{2 \times 10^{-6}\,(6 \times 10^{-12})}}$$

$$= \frac{1}{6.28\sqrt{12 \times 10^{-18}}} = \frac{1}{(6.28)(3.5 \times 10^{-9})}$$

$$= 45 \text{ MHz}$$

Thus this oscillator will operate at about 45 MHz.

Because the LC circuit has resistance, it is unable to sustain the oscillation process. The resistance seen from the alternating charging and discharging causes the waveform to decay in amplitude. Figure 11–6 shows an example of this action. This decaying of the waveform

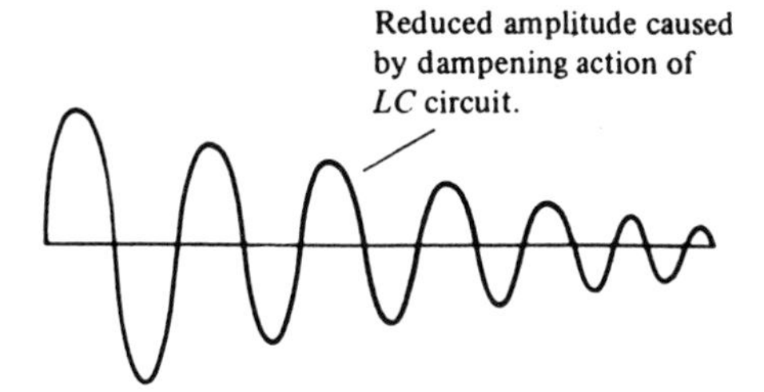

FIGURE 11–6
Amplitude of LC Action Being Dampened

is called **dampening.** Dampening action is the reason that an amplifier is required to sustain the oscillating action.

RC Circuit

Another method of frequency control is to use an ***RC* circuit.** Figure 11–7 shows the resistor-capacitor combination used most often with these types of oscillators.

The RC network operates on the principle of the capacitor charging and discharging. The rate at which the capacitor charges and discharges is controlled by the size of the resistor. Again, a mathematical equation can be used to express this relationship. The frequency at which this oscillator will operate is given by Equation 11–2.

$$f_r = \frac{1}{6.28RC} \tag{11–2}$$

where:

f_r = resonant frequency of the RC network (Hz)
R = size of the resistor (Ω)
C = size of the capacitor (F)
$6.28 = 2\pi$

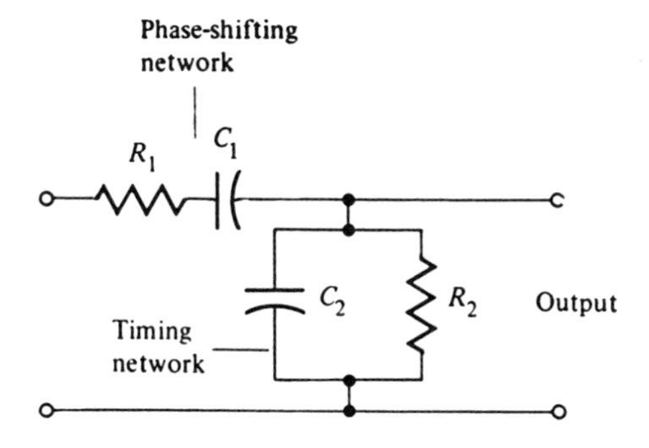

FIGURE 11–7
Circuit Configuration for an RC Oscillator

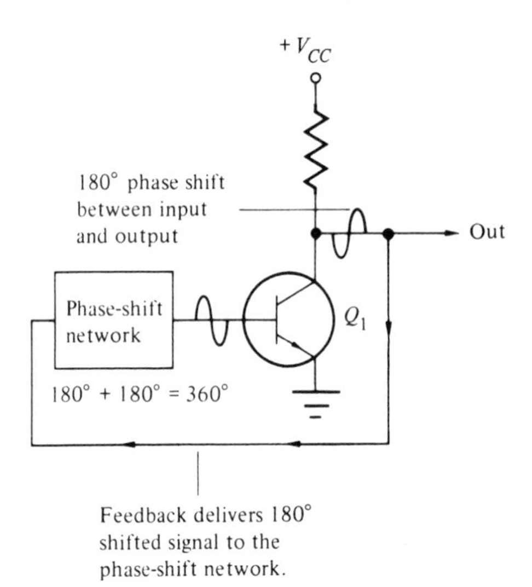

FIGURE 11–8
Phase-Shift Network of the Oscillator

Generally, *RC* types of circuit combinations are used in low-frequency audio oscillators.

Also found in some *RC* circuits is a series connection of a resistor and a capacitor. In Figure 11–7 this is shown as a **phase-shift network.** This network is commonly used in the feedback path. Remember that feedback is required to sustain oscillation. Often the output of the amplifier gives a phase shift of 180°. If this signal were fed back to the input, it would subtract from the input signal. This **negative feedback** would remove the feedback signal, thus stopping the action of the oscillator.

Using a phase-shift network will result in shifting the output signal phase between 0° and 90°. Using several of these networks will cause positive feedback to be developed for the oscillator. Figure 11–8 shows an example of this type of network in the feedback path. Transistor Q_1 is the amplifier of this circuit. It is known that at the collector a 180° phase shift is developed between input and output. To have positive feedback, a phase-shift network is used to develop the proper phase into the input of the oscillator.

Phase-shifting networks will be considered in more detail later in the chapter.

Crystal Devices

LC circuits are used in a wide variety of frequency-determining networks. However, these circuits are unsuitable in applications where oscillators are required to operate at exact frequencies. Oscillators needed in transmitters are a good example. If the *LC* network were used, the oscillator would tend to drift from the desired frequency when high frequencies are needed. **Drifting** in frequency means that frequency developed by the frequency network will go above and below the desired output frequency of the oscillator. To overcome this problem, crystals are used as the frequency-determining component in certain types of oscillators.

The crystal is generally made from quartz. Quartz is a **piezoelectric material.** This type of material changes energy forms. For example, the quartz crystal can change mechanical energy into electrical energy, and it can change electrical energy into mechanical energy. The quartz crystal will vibrate at a certain frequency. This resonant frequency is determined when the crystal is manufactured.

When an ac voltage is applied to the quartz crystal, the crystal vibrates mechanically. Due to the internal resistance of the quartz, the mechanical vibrations will dampen out if the ac voltage is removed. However, if enough electrical energy is fed back to the crystal's input, the vibrations will continue, thus overcoming the dampening action of the crystal's internal resistance. The crystal will continue to vibrate, and will develop the same action and waveform as the *LC* circuit.

The amplitude of the quartz crystal operation will depend upon the natural mechanical frequency of the crystal itself. Several factors are involved. Some of these factors are the type of material, how it is cut to develop resonant frequency, and its physical dimensions. The one factor that plays heavily in the frequency determination is the **crystal thickness.** In general, the thinner the crystal material, the higher the resonant frequency. Figure 11–9 shows the thickness of the crystal as it relates to the resonant frequency. Shown in Figure 11–10 is the schematic symbol for the crystal.

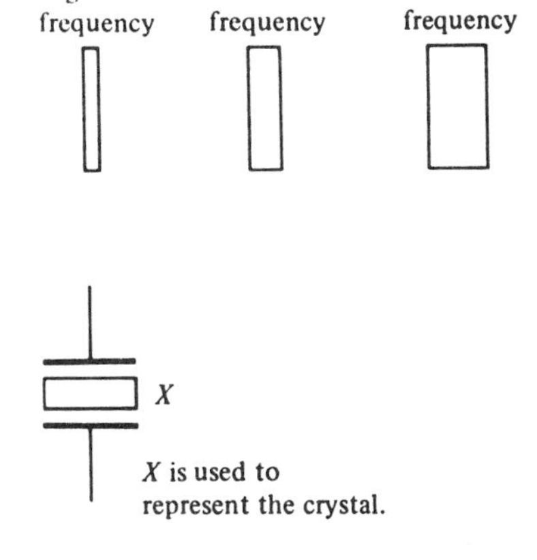

FIGURE 11–9
Thickness of Crystal Material Related to Resonant Frequency

FIGURE 11–10
Schematic Symbol for a Crystal

Most crystal materials used in circuits are oven baked. These crystal types are temperature stable. In other words, as the temperature changes around the crystal, the shape of the crystal does not change. If the shape were to change, there would be a shift in frequency. This frequency shift would cause the oscillator to drift in frequency.

Crystal-controlled oscillator circuits outperform the other types of frequency-determining networks. Changes in the frequency of the *RC* and *LC* oscillator circuits are greatly reduced by the crystal circuits. Some of the causes of change in oscillator frequency are:

—Supply voltage to the oscillator
—Mechanical vibration near the oscillator
—Temperature change near the oscillator
—Resistors, capacitors, inductors, or transistors changing in value
—Metal parts near the oscillator

Crystal-controlled oscillators, which will be discussed later in the chapter, can greatly reduce these effects.

AMPLIFIER BIASING

Because the amplifier is an important part of the total operation of the oscillator, the biasing of the amplifier section will be discussed. Transistors or any other active devices used in the oscillator can be biased at one of several points along the characteristic curve. The amplifiers can be biased to operate at class A, class B, or class C. Generally, one might think that the class-A amplifier would be the best amplifier to use for the oscillator. The class-A amplifier requires a small feedback signal to keep within its active region. This small feedback signal may not be of sufficient amplitude to sustain oscillation, however. Therefore other classes of operation are used for the oscillator's amplifier section. Typically, the class-C biased amplifier is used in the oscillator's amplifier section.

Much of the biasing for the oscillator is done in the class-C region of amplifier operation. This biasing allows the amplifier to operate between cutoff and saturation. Also, any distortion generated in the amplifier can be removed by filtering circuits in the oscillator or feedback line.

One important consideration in the oscillator is the amount of feedback signal. Too little feedback signal will make the oscillator operate improperly. With ample feedback signal, the oscillator will continue to operate, even when external factors such as supply voltage or temperature are changed.

POSITIVE FEEDBACK

As was mentioned earlier, an important consideration in oscillator operation is achieving the proper feedback signal. The oscillator requires positive or regenerative feedback to operate. In a single-stage common emitter amplifier circuit, the output is taken from the collector. This signal is 180° out of phase from the input signal. Before this output signal can be fed back to the input, the phase must be shifted. Then the signal will arrive at the base, developing regeneration. If the signal is not shifted in phase, the phase will cause cancellation of the input signal, or degeneration. A method used to invert or phase-shift the signal is to pass the feedback through resistor-capacitor networks. These networks are called **lag lines.** Figure 11–11 shows a lag line connection. Each of the *RC* networks will shift the signal 60° in phase. By passing the signal through three networks, a 180° phase shift can be accomplished.

PHASE-SHIFT OSCILLATOR

The necessary phase-shifting action is accomplished by the **phase-shift oscillator,** also called the **lag line oscillator.** Figure 11–11 shows the schematic diagram for this circuit.

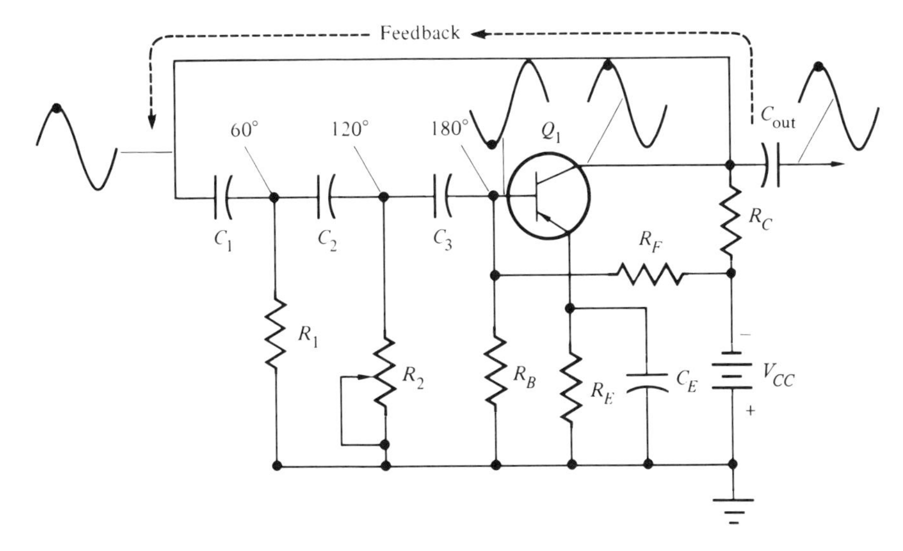

FIGURE 11–11
Schematic Diagram of a Phase-Shift Oscillator

The phase-shift oscillator is a sine wave generator that uses an *RC* network as its frequency-determining device. Besides determining the frequency, the *RC* network provides the proper feedback to sustain oscillation.

Notice in Figure 11–11 that a common emitter amplifier is being used. There will be a 180° phase shift between the base and the collector of this amplifier. To take the feedback from the collector to the base, the signal must be shifted in phase. An *RC* network consisting of three sections provides the proper phase inversion and feedback for the oscillator. Each section will supply a phase shift of 60°.

In ac theory it is shown that phase angles develop between voltage and current in an ac circuit. By the addition of different components, this phase angle across a capacitor develops a phase shift of 90°. If a resistance is added to the capacitor, the phase-shift angle can be reduced. Therefore small values of resistance have been added to develop a 60° phase shift.

Figure 11–11 also shows the biasing required for circuit operation. Resistors R_B, R_C, and R_F provide the necessary base and collector bias. Capacitor C_E is used as the emitter bypass capacitor to swing any ac variations around the emitter resistor to ground. The phase-shifting network is made up of capacitors C_1, C_2 and C_3, and resistors R_1, R_2, and R_B. Resistor R_2 is a variable resistor used for fine-tuning to compensate for any small changes in the values of the other components that might cause a shift in frequency or a shift in phase development.

When power is applied to the circuit in Figure 11–11, oscillations are started by random noise within the circuit. Random noise is generated from charging capacitors, electrons traveling through the transistor, or any other noise within the circuit. This change in base current causes a change in collector current, which, it will be remembered, is shifted 180°. The signal is returned to the base. Figure 11–11 shows the different phase-shift points within the circuit. The signal is passed to C_1 and R_1, where a 60° shift is developed. Resistor R_2 and capacitor C_2 cause another 60° phase shift. This signal is finally fed to C_3 and R_B, where the final 60° phase shift is accomplished. With the correct amount of capacitance and resistance in the phase-shift network, regenerative feedback occurs. Also, the *RC* networks oscillate at only one frequency. At any other frequency, they develop the wrong phase shift, and the circuit becomes degenerative. Thus the oscillator will operate at only one frequency.

The frequency of the phase-shift network is very difficult to calculate, mainly because of the loading effects. To achieve a rough approximation of the output frequency, Equation 11–3 can be used.

$$f_r = \frac{0.092}{RC} \tag{11–3}$$

where:

f_r = output frequency of the oscillator (Hz)
C = size of the capacitor in the phase-shift network (F)
R = size of the resistor in the phase-shift network (Ω)

This formula is based on the fact that in the phase-shift network, all the resistors are of equal value, as are all the capacitors. A sample frequency calculation for this circuit is given in Example 2.

EXAMPLE 2

An RC phase-shift network contains three 0.01 μF capacitors and three 100 kΩ resistors. What is the approximate resonant frequency of this network?

Solution:

$$f_r = \frac{0.092}{RC} = \frac{0.092}{(0.01 \times 10^{-6})(100 \times 10^{3})} = \frac{0.092}{1 \times 10^{-3}}$$
$$= 92 \text{ Hz}$$

Thus the resonant frequency of this oscillator is 92 Hz. If the size of the resistor were increased, the output frequency would decrease, and if the resistance were decreased, the output frequency would increase.

The amplifier circuit is very important for the operation of this type of oscillator. A high-gain transistor must be used because of the high losses of power within the RC networks. If additional RC networks are used, the phase shift of each section is reduced. Also, the power of the signal to drive this type of oscillator must be increased.

Looking at Equation 11–3, it can be seen that high frequencies require low values of capacitance and resistance. As a result, the phase-shift networks load the output of the amplifier. This loading, if increased, will stop the operation of the oscillator. To sustain oscillations at high frequencies, an emitter-follower amplifier needs to be inserted. This amplifier is positioned between the output of the oscillator and the feedback line. Figure 11–12 shows an example of this type of circuit operation. The emitter follower will provide no gain to the signal, but will give the signal enough driving power to overcome the losses in the RC phase-shifting network.

WIEN BRIDGE OSCILLATOR

Another oscillator that uses resistors and capacitors as the frequency-determining network is called the **Wien bridge oscillator.** This oscillator

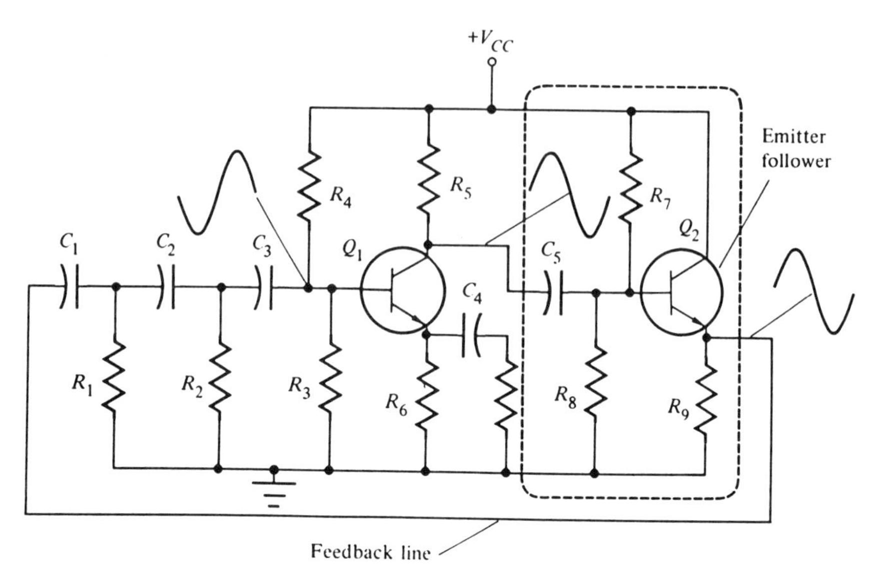

FIGURE 11–12
Emitter Follower Used to Reduce Loading Effects

provides a sine wave at the output. Like its counterpart, the phase-shift oscillator, this oscillator requires positive feedback, an amplifier, and a timing network.

The Wien bridge oscillator has three general characteristics:

—Frequency stability
—Low output distortion
—Amplitude stability

These three characteristics make this oscillator very useful as a master timing oscillator. The Wien bridge oscillator is commonly used in audio generator circuits.

Figure 11–13 shows an example of the Wien bridge oscillator. As seen in this schematic, the oscillator contains two transistors. The first transistor, Q_1, is used as an amplifier, and the second transistor, Q_2, is used to shift the signal phase to develop positive feedback to sustain oscillations. The schematic diagram in Figure 11–13a shows the development of the feedback signal path. The feedback signal path is developed through C_3. Figure 11–13b shows a schematic diagram of the typical operation of this circuit.

Both transistors Q_1 and Q_2 use the same type of biasing and stabilization. Notice that R_1 and R_2 in the base circuit of Q_1 develop the base bias voltage, while R_3 and R_4 develop the base bias for transistor Q_2. The collector biasing for these transistors is found through R_5 and R_6. The stability for operation of this oscillator is produced

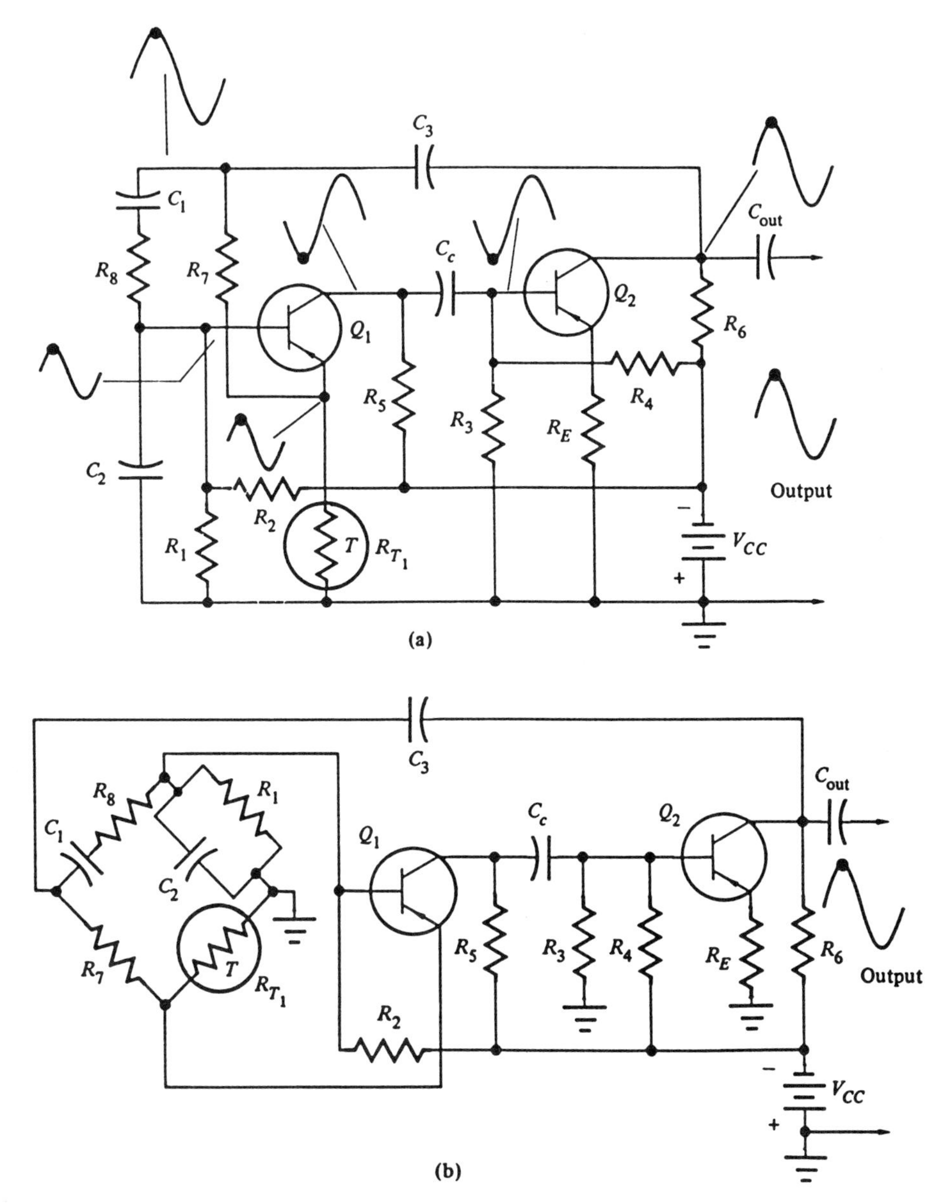

FIGURE 11–13
Wien Bridge Oscillator: (a) Development of Feedback Paths, (b) Typical Circuit Schematic Diagram

through negative feedback in the emitter circuit of Q_1. Notice that the emitter circuit R_{T_1} and R_7 are connected to the emitter of Q_1. As Q_1 starts to generate more current, the thermistor develops more resistance. This action will keep Q_1 very stable in operation. Stability in

Q_2 is established through the unbypassed resistor R_E. This causes degeneration and decreases the output waveform in Q_2. The signal from Q_1 is then coupled to the base of Q_2 through coupling capacitor C_c. The feedback signal is developed from the collector of Q_2 to the resonant frequency network by C_3. The output sine wave is coupled to the next stage via C_{out}.

The frequency of the Wien bridge oscillator is determined by the resonant frequency circuit comprised of capacitors C_1 and C_2, and resistors R_8 and R_1. Oscillations start when power is applied. A change in base current of Q_1 results in an increase in collector current, and a 180° phase shift is seen between the input and output. The signal is then applied to Q_2. The amplified output is also phase-shifted 180°. In other words, this phase shift has developed the proper phase for regenerative feedback. This output is then fed back to sustain oscillations.

The frequency selector circuit in the feedback path prevents the output frequency from changing. The frequency network becomes selective to only one frequency. If the frequency being developed from the output of Q_2 is not the proper frequency, the network will not respond, and there will be no output frequency. When the feedback signal is in phase, the Wien bridge oscillator is a very stable timing oscillator. This stability makes this oscillator a useful device in many audio frequency generators.

Because of the need for stability within the amplifier circuit, a thermistor is used in the emitter leg of Q_1. This thermistor has a positive temperature coefficient. In some circuits a lamp may be used as the thermistor. In all cases the thermistor is used to stabilize the output amplitude of the amplifier. When the amplitude of the output signal increases above its predetermined range, an increased feedback voltage results. As a result of the increased feedback voltage, the current through the thermistor will increase. This increased current causes a generation of heat at the thermistor, which results in more resistance in the emitter resistance. This increase in resistance causes more negative feedback in the Q_1 amplifier circuit. The gain of the amplifier is then reduced. Figure 11–13b shows a schematic diagram of the audio Wien bridge oscillator that is capable of producing frequencies from 20 Hz to 20 kHz.

LC SINE WAVE OSCILLATORS

As was mentioned earlier, the *RC* sine wave oscillators have a very limited frequency range. To achieve higher frequency considerations, an inductor-capacitor combination is used. As noted previously, such *LC* circuits are called tank circuits. There are several types of these oscillators. Some that the technician may come across are the Hartley oscillator, the Colpitts oscillator, the Clapp oscillator, and the crystal-controlled oscillator.

Hartley Oscillator

The **Hartley oscillator** is shown in schematic diagram form in Figure 11–14. This type of oscillator provides a wide range of frequencies, and is very easy to tune. However, its stability leaves something to be desired. This oscillator, like all other oscillators, has an amplifier, a tuned circuit, and positive feedback.

The amplifier circuit in Figure 11–14 is being taken care of by Q_1. The tuned circuit is being handled by the split inductor L_1-L_2. Variable capacitor C_2 serves as the other part of the tuned circuit. The feedback circuit is developed from the collector of Q_1 through C_3 and back to the tank circuit. To achieve regenerative feedback in the circuit, the inductor L_1-L_2 causes a 180° phase shift. This phase shift develops regenerative feedback for the base action. Because the dc bias line and the ac feedback line are the same line, this type of oscillator is called a **series-fed Hartley oscillator.**

The series-fed Hartley oscillator consists of the tank circuit of L_1, L_2, and C_2. The feedback circuit is developed through C_1 to the

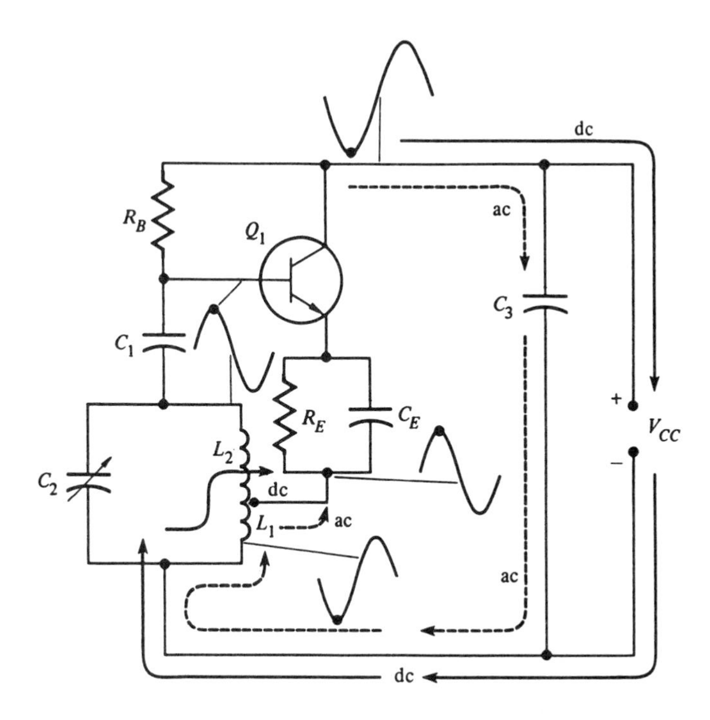

FIGURE 11–14
Series-Fed Hartley Oscillator

base of Q_1. Components C_E and R_E are used to stabilize the transistor amplifier circuit, and to keep the transistor from going into thermal runaway. This parallel circuit helps prevent degeneration within the amplifier. The biasing network for the base is developed by R_B. Capacitor C_1 also serves as a dc *blocking capacitor*, preventing dc current from shorting to ground through the low resistance path of L_2.

When voltage is applied to the circuit, current flows from the power source through L_1 and to R_E. Current through the emitter then flows through the collector back to the power source. This large surge of current through L_1 induces a voltage in L_2 to start oscillations within the tank circuit.

When current first starts to flow, the top of L_1 becomes positive, and the bottom of L_1 becomes negative. This induced voltage in L_1 makes the top of L_2 positive. This positive potential on the base of Q_1 causes an increased forward bias, and more current flow through the transistor. This action continues until the coil and transistor have reached saturation and C_2 has a positive charge on the top plate. Then the current flow starts to decrease. C_2 will then discharge through L_2. This building up and decreasing of voltage around L_1 develops the oscillator action.

This action is continued by the magnetic field built up around the inductor. The next half-cycle of waveform is generated by the tank circuit action. The internal resistance of the tank circuit prevents the tank circuit from continuing its oscillations. Therefore the regenerative feedback signal returns from the collector to develop another cycle. This regenerative action causes the cycle to repeat itself until the bias is removed.

Figure 11–15 shows another type of Hartley oscillator, called the **shunt-fed Hartley oscillator.** This circuit operates in the same basic fashion as the series-fed Hartley oscillator. The main difference is that the dc bias voltage does not flow through the tank circuit, as shown by the solid dc line. When voltage is applied to the circuit, Q_1 starts to conduct. As the collector current of Q_1 increases, the change is coupled through C_3 to the tank circuit, starting the oscillator action. Capacitor C_3 also acts as an *isolation capacitor* to prevent dc from flowing through the feedback coil. The feedback path is shown by the dotted line through C_3. Phase-shifting of the signal is again accomplished through the tapped inductor L_1-L_2.

Colpitts Oscillator

Another sine wave oscillator is called the **Colpitts oscillator.** Compared to the Hartley oscillator, the Colpitts oscillator is a fairly stable device. It also has the ability to be tuned over a wide range of frequencies,

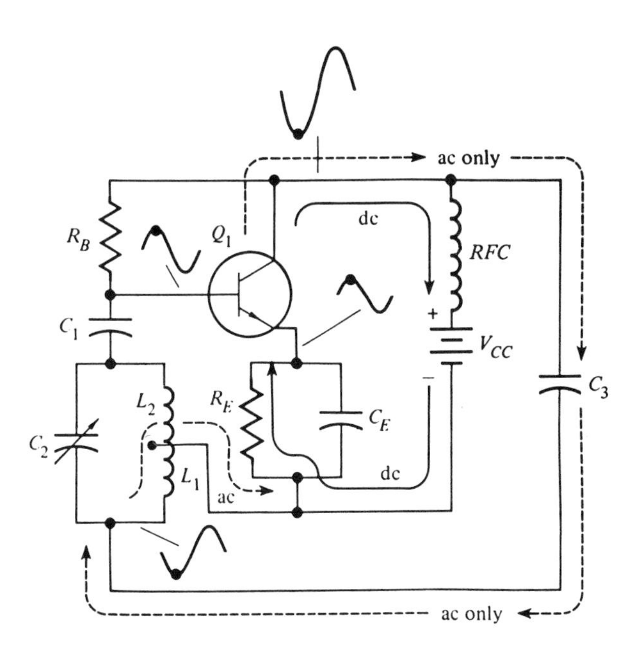

FIGURE 11–15
Shunt-Fed Hartley Oscillator

and can produce a stable output frequency. Figure 11–16 shows a schematic diagram for the Colpitts oscillator.

The Colpitts oscillator is very similar to the shunt-fed Hartley oscillator, except that it uses two capacitors instead of the tapped

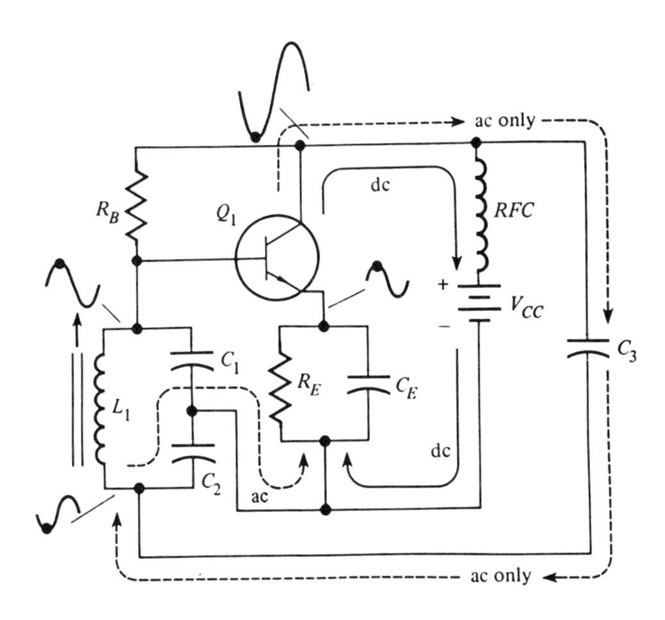

FIGURE 11–16
Schematic Diagram of the Colpitts Oscillator

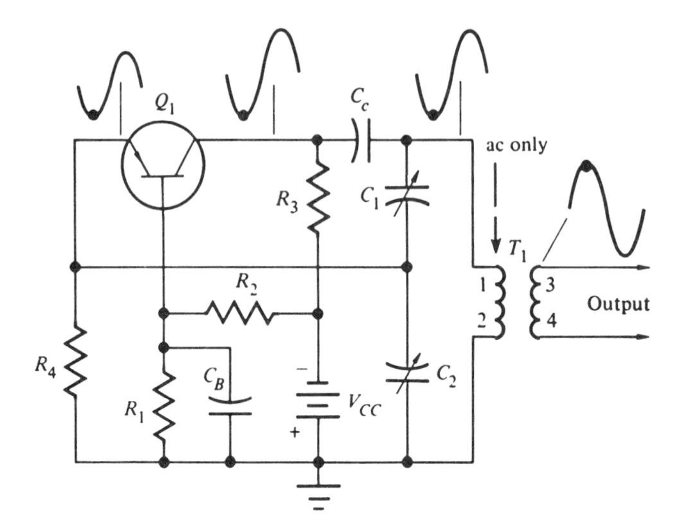

FIGURE 11–17
Common Base Amplifier Connected as a Colpitts Oscillator

inductor. The Hartley oscillator has a tap between two coils, whereas the Colpitts oscillator has a tap between the two capacitors. To change the frequency of this oscillator, the capacitors can be adjusted, or the inductor can be made to vary. Notice in Figure 11–16 that no coupling capacitor is needed between the output of the tank circuit and the base of Q_1. The reason for this is that C_1 and C_2 serve as isolating capacitors.

Figure 11–17 shows an example of a common base connection of the Colpitts oscillator. Base bias is provided by R_1 and R_2. Resistor R_3 serves as bias for the collector circuit. Resistor R_4 serves as the emitter bias resistor to stabilize amplifier operation, and couples the feedback signal from the tank circuit.

The tank circuit is comprised of C_1 and C_2, and the primary winding of T_1. The feedback signal need not develop a phase shift because of the phase relationship between the emitter and collector of a common base amplifier. When the emitter swings negative, the collector also swings negative. This causes C_2 to develop a negative charge. This negative signal is then fed back to the emitter. The regenerative process allows the oscillator to continue its operation.

Clapp Oscillator

Through a simple modification, the Colpitts oscillator circuit can be transformed into a **Clapp oscillator.** The only difference between the two is the addition in the Clapp oscillator of a variable capacitor in series with the inductor in the tank circuit. The addition of this capacitor causes increased frequency stability. Figure 11–18 shows an example of the Clapp oscillator.

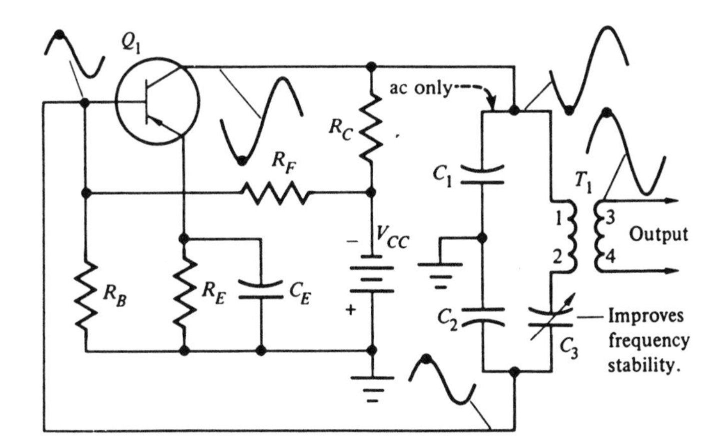

FIGURE 11–18
Clapp Oscillator

Crystal-Controlled Oscillator

As was mentioned earlier, the crystal is a device that converts mechanical strain into an electrical signal. Because of its exact frequency consideration, the crystal makes an excellent material for the tuned circuit in an oscillator. Figure 11–19 shows an example of a **crystal-controlled oscillator.** The amplifier used for this circuit is the standard common emitter configuration. This means that in order to achieve regeneration within the circuit, a phase shift must be accomplished. This phase shift is effected by the crystal, X_1. Here the crystal operates in the parallel mode, and produces a 180° phase shift. Capacitor C_2 is

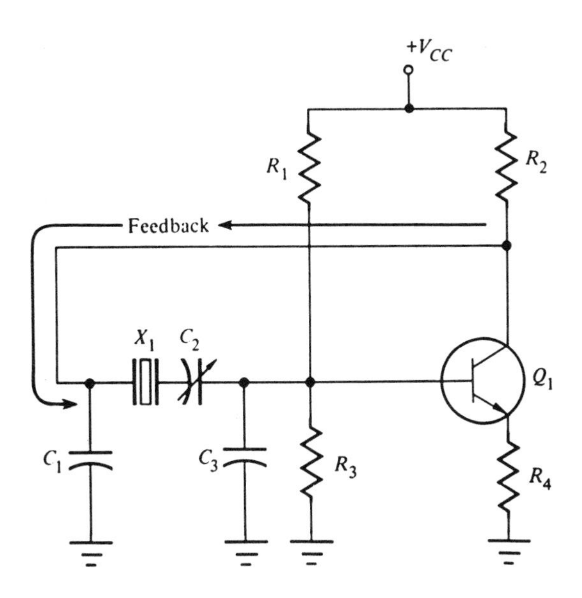

FIGURE 11–19
Schematic Diagram of a Crystal-Controlled Oscillator

a *trimmer capacitor*. It is used as a frequency adjustment for the oscillator. Capacitors C_1 and C_3 are used to form a voltage divider that controls the amount of feedback in the circuit. If large amounts of feedback voltage are developed, the oscillator will produce distortion, and if too little feedback is developed, the oscillations will not be sustained. The remaining components are used to bias the common emitter amplifier.

As shown in Figure 11–9, the thickness of the crystal material is directly related to the output frequency of the oscillator. When crystal oscillators are used to generate very high frequencies, the crystal material presents a problem. As the oscillator approaches 15 MHz, the crystal material becomes very thin, and very fragile. To overcome this problem, a technique using **overtone crystals** is employed. This technique allows the frequency of the oscillator to extend to 150 MHz.

Figure 11–20 shows a schematic diagram of the overtone crystal oscillator circuit. Notice that an inductor and a split-capacitor tuned circuit are used. These components are part of the selection process of the proper frequency. Capacitors C_3, C_4, and C_5 form the tank circuit. Capacitor C_2 is a trimmer capacitor used to set the crystal frequency.

The base of Q_1 is grounded for ac signals through C_1. This grounding action makes the amplifier circuit a common base amplifier. Because of the common base connection, no feedback signal inversion is required. Therefore, the signal can be fed directly from collector to emitter to develop regeneration in the oscillator. Capacitors C_3 and C_4 also serve as a voltage control for the feedback signal. Crystal X_1 is in the feedback path, and is operated in the series mode.

Crystal-controlled oscillators are found in many circuits used in

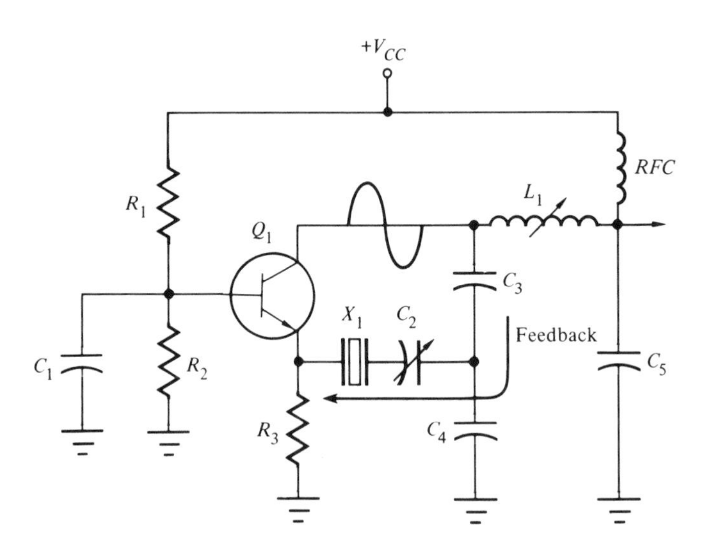

FIGURE 11–20
Schematic Diagram of an Overtone Crystal Oscillator

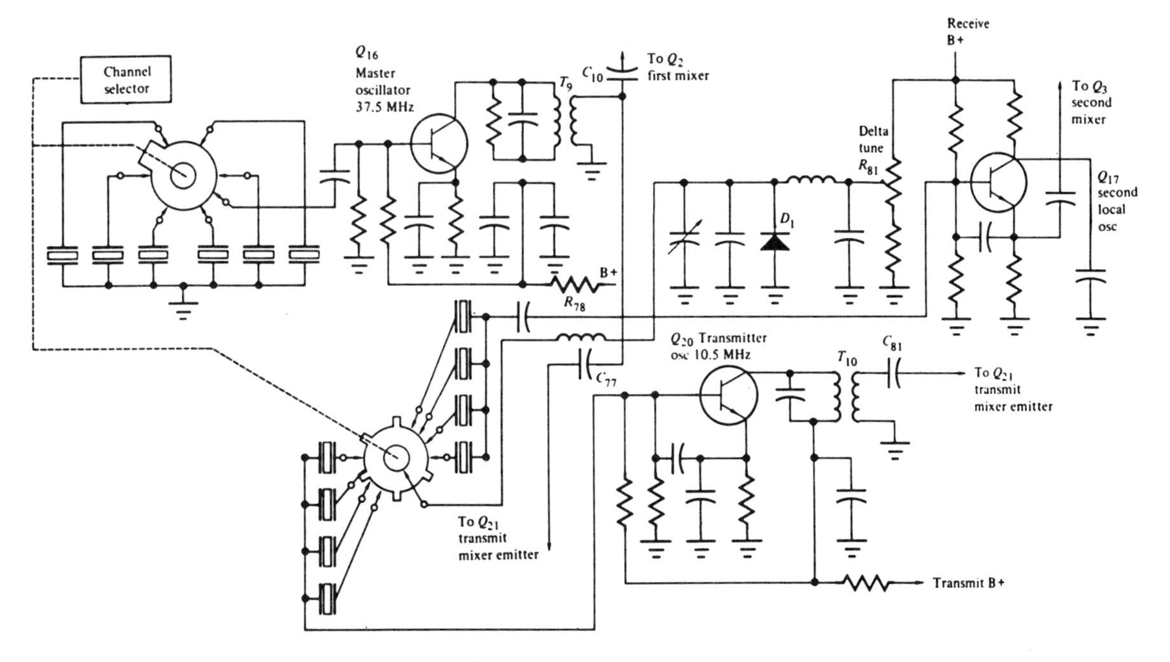

FIGURE 11–21
Mobile CB Frequency Synthesizer for 23-Channel Transmit and Receive

consumer products. They are found in the chroma circuit of color receivers to regenerate the color subcarrier. These crystal circuits are also found within CB transceivers. Here the circuits are used to develop the transmitting frequencies.

Figure 11–21 illustrates the process of *frequency synthesis* used in some older brands of CB transmitters. The crystal values are combined to generate the 27 MHz needed for transmission. This oscillator circuit is comprised of two circuits: a master oscillator and a local oscillator. The master oscillator circuit consists of crystals in the 37.5 MHz range, and the local oscillator contains crystals in the 10.5 MHz range. As these frequencies are combined in the mixer circuit, they generate frequencies within the CB 27 MHz band. Newer *phase-lock loop (PLL) circuits* have taken over the generation of frequencies in CB units.

RELAXATION OSCILLATOR—NONSINUSOIDAL

Another important oscillator in consumer products is called the **relaxation oscillator.** This type of oscillator is a nonsinusoidal oscillator. It contains a resistor-capacitor timing network to provide switching action. The charging and discharging of the capacitors develop sawtooth, square, or other pulse-shaped waveforms.

Relaxation oscillators can be divided into **multivibrators** and **blocking oscillators.** They are further classified as free-running or triggered. A **free-running oscillator** is an oscillator that starts to develop waveforms as soon as power is applied. The **triggered multivibrator** is controlled by synchronizing or by externally triggered signals. This latter type operates only once, and will not produce output waveforms until another triggered pulse is received. The triggered multivibrator is sometimes referred to as a **one-shot.**

Free-Running Multivibrator

The free-running or **astable multivibrator** is mainly a nonsinusoidal, two-stage oscillator. *Astable* means that the oscillator has two temporary states. In this oscillator one stage will conduct while the other is off until it reaches a point, at which time the multivibrator switches stage condition. The stage that was off starts to conduct, and the other stage shuts off. This on/off process of stages results in a square wave output, or a pulse-shaped output. The shape of the output waveform depends upon the type of circuitry. Figure 11–22 shows a typical astable multivibrator.

The key to astable multivibrator operation is the charging and discharging of C_1 and C_2. With the use of resistors, this process can be controlled and adjusted to develop a frequency.

When supply voltage is applied, both transistors begin to draw current. Assume that the voltage on the collector of Q_2 rises more quickly than the voltage on the collector of Q_1. This positive-going rise

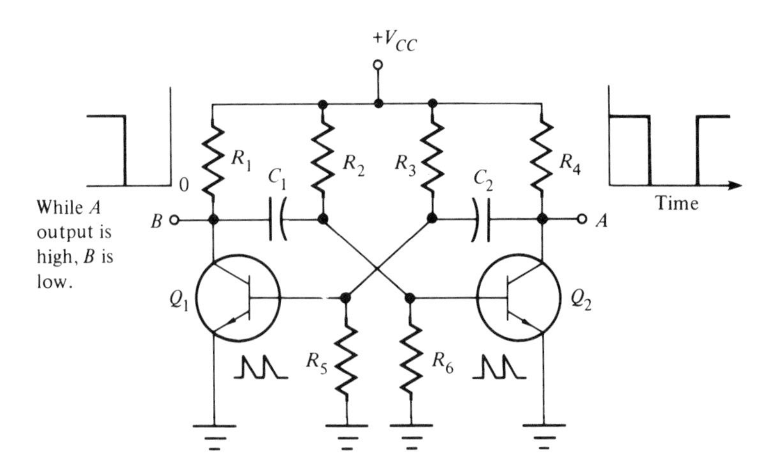

FIGURE 11–22
Schematic Diagram of an Astable Multivibrator

on Q_2 is then coupled via C_2 to the base of Q_1. With this positive rise of Q_1 base voltage, the transistor starts to conduct harder. Since Q_1 is a common emitter amplifier, this increase is seen also at the collector. The signal on the collector is now going negative. The large negative signal is then coupled through C_1 back to the base of Q_2. This negative signal voltage at the base of Q_2 turns off Q_2, thus allowing C_2 to charge. The charging path for C_2 is from ground, through the base-emitter junction of Q_1, through R_4, and back to V_{CC}.

After a period of time, any charge that has been built up on C_1 is discharged to ground through R_2. Soon the base current is allowed to flow again in Q_2. The collector voltage of Q_2 then begins to fall. This negative-going collector voltage is transferred to the base of Q_1, which is then driven to cutoff. Output B goes high. Capacitor C_1 begins to charge through the base-emitter junction of Q_2, through R_1, and back to V_{CC}. As soon as the charge that was developed on C_2 discharges via R_3, the process starts over again.

The time during which Q_1 and Q_2 are held at cutoff is developed through capacitors C_1 and C_2. The discharge paths through the resistors are used to develop the RC time constant. This time can be related to a frequency.

Astable multivibrator circuits are found in consumer products as well as in computer circuits. In computer circuits they are called *clocks*.

Monostable Multivibrator

A **monostable multivibrator** is considered stuck in one state until its state is reversed. For there to be a change in state, the monostable multivibrator must be triggered by an external pulse into the next state. Figure 11–23 shows a schematic diagram of this type of circuit.

When power is applied to the circuit in Figure 11–23, Q_1 conducts because of the steady current flow to the base from R_2. With this on condition, the transistor is close to saturation, and the voltage at the collector is relatively low. Since Q_1 acts as a low-resistance path for current, no current flow is developed through the base-emitter of Q_2. This condition makes Q_2 turn off, and thus the collector voltage is near supply. Therefore output A is zero, and output B is high. Capacitor C_1 begins to charge through the base-emitter junction of Q_1. Transistor Q_3 is also at cutoff, because it receives no base current.

When a triggered signal is applied to Q_3, a pulse of base current is developed. This base current causes conduction in Q_3. As soon as Q_3 conducts, it develops a low-resistance path to ground between the base and the emitter. This action discharges C_1 to ground through this low-resistance path. This action then develops negative signal voltage at the base of Q_1, causing this transistor to go into the cutoff range.

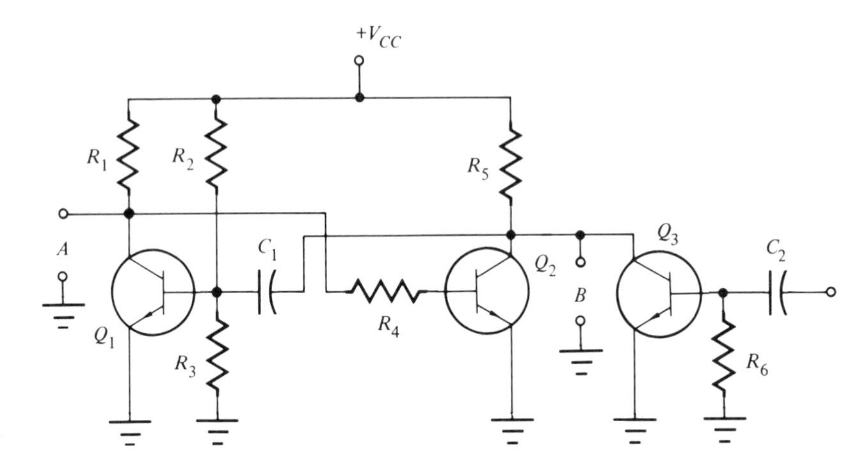

FIGURE 11–23
Schematic Diagram of a Monostable Multivibrator

The collector voltage of Q_1 rises sharply in the positive direction, causing output A to go high. The positive voltage at the collector of Q_1 is then applied through R_4 to the base of Q_2. This positive voltage starts Q_2 conducting, and the output B drops to a low value of voltage. This state is maintained until the charge developed across C_1 is bled off. Until the capacitor has been bled of all charges, the monostable multivibrator will not change state. Once the charge is bled off C_1, the multivibrator is ready to change condition. This delay in time is determined by the RC discharge time constant of C_1 and R_2 and R_3. This action gives the circuit its name—the one-shot, or **single-shot multivibrator.**

Bistable Multivibrator

Another type of relaxation oscillator is called the **bistable multivibrator.** As its name implies, the bistable multivibrator has two states in which it operates. Figure 11–24 shows a schematic diagram of this type of multivibrator. Because of its actions, it is used as a memory circuit in computers, and in many home entertainment products.

As shown in Figure 11–24, Q_1 and Q_2 depend upon each other for the supply of base current. Assume that Q_1 conducts more than Q_2 when power is applied. This negative voltage at the collector of Q_1 is transferred to the base of Q_2 via R_4. This negative voltage reduces conduction through Q_2. The reduction causes the collector voltage of Q_2 to rise. This action continues until Q_1 is fully on, and Q_2 is at cutoff.

To change states, an external pulse is fed to the base of Q_2. This positive pulse developed at Q_2 starts Q_2 conducting. This action turns off Q_1. The multivibrator has changed states. To change states again, a positive pulse must be fed into the base of Q_1.

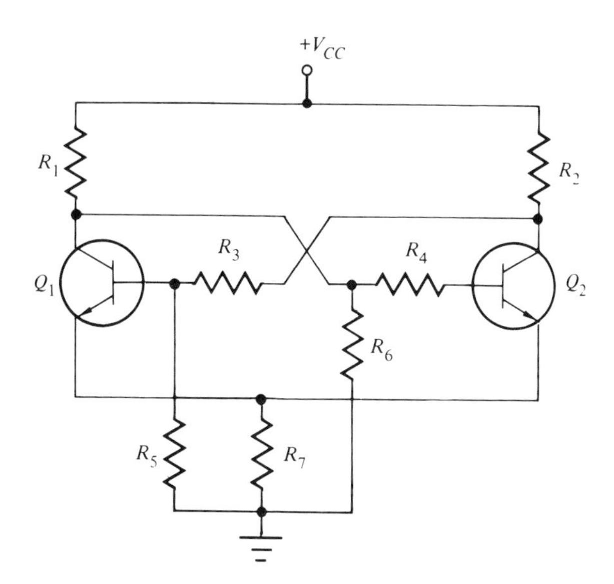

FIGURE 11–24
Schematic Diagram of a Bistable Multivibrator

Blocking Oscillators

Blocking oscillators are used in many circuits as master oscillators. They are capable of generating pulses of very short duration that can be used to trigger and synchronize other circuits. Blocking oscillators come in two forms: the free-running blocking oscillator and the triggered blocking oscillator. The free-running blocking oscillator is used in many television receivers, and will be discussed here.

The schematic of the free-running blocking oscillator is shown in Figure 11–25. When voltage is applied, Q_1 starts to conduct. Collector current flows through the primary windings of T_1 terminals 3 and 4. The magnetic field that builds up across the primary windings induces a voltage into the other two windings. This results in a positive

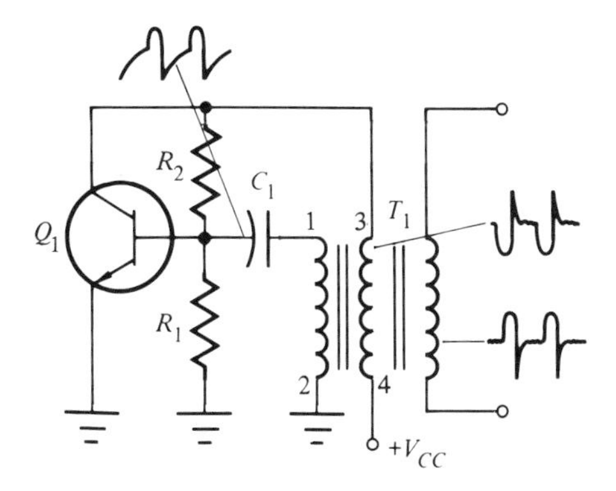

FIGURE 11–25
Schematic Diagram of a Free-Running Blocking Oscillator

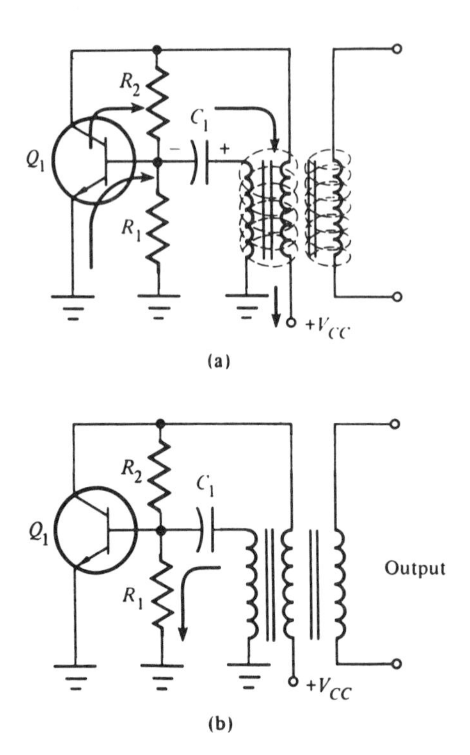

FIGURE 11–26
Blocking Oscillator Current Paths: (a) Charging Path, (b) Discharging Path

voltage on terminal 1 of T_1, which is coupled through C_1. This positive voltage causes Q_1 to go into saturation. Capacitor C_1 can then charge quickly through the low base-emitter resistance. The charge developed across the capacitor is shown in Figure 11–26a. When Q_1 reaches saturation, collector current stops increasing and remains at the same value. Since the current through the transformer remains the same, the magnetic field around the transformer begins to collapse. Once the magnetic field reduces, C_1 begins to discharge through R_1. This discharging action sends the base into the negative area, thus cutting off the transistor as illustrated in Figure 11–26b. The time required to discharge C_1 keeps Q_1 cut off. This cutoff action can be controlled by the size of R_1. Once the voltage on the capacitor is nearly discharged, the base voltage turns positive, and the action repeats itself. This process occurs over and over, causing a frequency to be developed.

SUMMARY

Oscillators are devices that convert dc into ac voltage. They are classified into two categories based on waveform shape: sinusoidal oscillators and nonsinusoidal oscillators.

A basic oscillator has an amplifier, a timing circuit, and regenerative or positive feedback. All amplifiers within the oscillator must have a gain greater than one. The common emitter amplifier is used most often.

The timing circuit further classifies the oscillators into categories—those oscillators that generate frequencies in the audio range, and those that operate in the RF range.

The resonant frequency of an oscillator can be developed from three types of networks: an *RC* network for lower frequencies, *LC* circuits for higher frequency, and the crystal-controlled oscillators used for high-frequency application.

Crystals convert mechanical vibrations into electrical current. Generally, the crystal is cut to resonate at a specific frequency. The thickness of the crystal material determines at what frequency the material will operate. The thinner the material, the higher the frequency. Typically, the crystals will operate up to about 15 MHz. After that, they are generally sent into overtone operation.

Amplifier bias must maintain the oscillator within the operating range of the active device. Usually, the amplifier is operating as a common emitter amplifier, and is in the class-C range.

Positive or regenerative feedback is needed to sustain oscillations. Without regeneration, oscillations would cease because of the internal resistance of the circuit.

The phase-shift oscillator contains networks of resistors and capacitors to shift phase, and also to establish the frequency of the oscillator. This oscillator is an *RC* sine wave oscillator.

The Wien bridge oscillator is another *RC* sine wave oscillator. It has two stages to develop the necessary amplification and regenerative feedback. This oscillator is characterized by frequency stability, low output distortion, and amplitude stability.

The tuned circuit of the Hartley oscillator is a tapped inductor and parallel capacitor. This oscillator is an *LC* sine wave oscillator.

The Colpitts oscillator uses a split capacitance to develop its resonant circuit. As with the Hartley oscillator, the Colpitts oscillator can be tuned over a wide range of frequencies.

The Clapp oscillator has a variable capacitor in series with an inductor as its resonant circuit. The Clapp oscillator is like the Colpitts oscillator in both structure and performance.

Crystal-controlled oscillators generate a wide range of frequencies. If oscillator frequencies that exceed the size of the crystal are needed, a crystal overtone or a crystal synthesizer circuit is used. Crystal oscillators are used in transmitters, color receivers, and other consumer products as frequency generation systems.

Relaxation oscillators generate a variety of waveshapes. These oscillators are found in many of the newer receivers and computer circuits. They are also called multivibrators.

The free-running and astable multivibrators have two states. They are either on or off. Other multivibrators—the one-shot and the bistable multivibrator—and the blocking oscillator are widely used in home entertainment products. As the consumer market shifts toward digital circuits, the relaxation oscillator circuit will become more and more important.

KEY TERMS

amplifier
astable multivibrator
bistable multivibrator
blocking oscillator
Clapp oscillator
Colpitts oscillator
crystal-controlled oscillator
crystal thickness
dampening
drifting
feedback
free-running oscillator
Hartley oscillator
lag line oscillator
lag lines
LC circuit
monostable multivibrator
multivibrator
negative feedback
nonsinusoidal oscillator
one-shot
oscillator
overtone crystals
phase-shift network
phase-shift oscillator
piezoelectric material
positive feedback
RC circuit
regeneration
relaxation oscillator
series-fed Hartley oscillator
shunt-fed Hartley oscillator
single-shot multivibrator
sinusoidal oscillator
tank circuit
timing circuit
triggered multivibrator
variable-frequency oscillator
Wien bridge oscillator

REVIEW EXERCISES

1. Name the two classifications of oscillators found in consumer products. Draw their waveforms.
2. Draw a block diagram of the basic oscillator, and label each part.
3. In oscillators that operate in the audio range, which components control the frequency?
4. Describe three methods by which the frequency of an oscillator can be controlled.
5. A 1 μF capacitor is placed in parallel with a 10 μH coil. What will be the resonant frequency for this circuit?
6. Draw an example of a network that could be used to develop positive feedback in an amplifier. Include capacitors and resistors.
7. Why are crystals used to operate oscillators at the higher frequencies?
8. List several factors that will change the frequency of an oscillator.
9. Explain why oscillators are commonly biased in the class-C region.
10. Explain the terms *regeneration* and *degeneration.* What type of feedback is necessary for sustaining oscillation?

11. Draw a schematic diagram of a lag line oscillator, and label the feedback line, the frequency network, and the amplifier.

12. Why is it so difficult to predict the output frequency in a lag line oscillator network?

13. Why is an emitter-follower amplifier inserted at the output of a lag line oscillator?

14. Draw a schematic diagram of a Wien bridge oscillator, and list several of its characteristics.

15. Draw a schematic diagram of the series-fed Hartley oscillator, and label the amplifier section, the timing circuit, and the feedback circuit.

16. Draw a schematic diagram of the timing circuit section for the Colpitts oscillator.

17. Why is it a good idea to place a variable capacitor in a crystal-controlled oscillator?

18. Name some circuits that may contain crystal-controlled oscillators.

19. Name several waveforms that can be generated at the output of a relaxation oscillator.

20. What develops the timing network for the relaxation oscillator?

21. How many states does a monostable multivibrator have? What changes the state of this oscillator?

22. Describe the operation of a blocking oscillator, and list some typical uses of this device.

12 LINEAR INTEGRATED CIRCUITS

OBJECTIVES

Upon completing this chapter, you should be familiar with:

—Operational amplifier characteristics
—Basic op amp operation
—Basic op amp construction
—Basic op amp dc biasing methods
—Op amp circuit gain
—Op amp gain-versus-frequency relationship
—Op amp applications in consumer products
—Characteristics of linear circuit application
—Troubleshooting linear integrated circuits

INTRODUCTION

Integrated circuits have had a large impact on the consumer product market. They lead the way in circuit design, offering the lowest cost and the highest performance. For this reason the consumer repair technician must be aware of their operation within the circuit.

Integrated circuits can be divided into two major categories: digital circuits and linear integrated circuits. In this chapter linear circuits are explored. The operational amplifier, linear systems, and linear circuit troubleshooting are considered.

OPERATIONAL AMPLIFIER CHARACTERISTICS

The most common of the integrated circuit (IC) amplifiers is the **operational amplifier,** or **op amp.** This circuit will give good linear amplification to input waveforms. The gain of this amplifier ranges from a low of one to several thousand. This wide range of operation makes the op amp an attractive component for use in consumer products. External feedback is the means by which this wide range in gain is accomplished.

The op amp without feedback is described as being in an *open-loop mode.* The characteristics of the ideal op amp are as follows:

—*Open-loop gain*: The op amp can deliver high amounts of gain. Using feedback will control the gain of the op amp.
—*High input impedance*: Op amps will not load the input signal source.
—*Low output impedance*: The op amp can deliver an amplified signal to low impedance sources without signal distortion.
—*Common-mode rejection*: The op amp has the ability to reject noise and hum present in the signal.

These qualities describe the op amp as an ideal amplifier.

BASIC OP AMP OPERATION

The fundamental operation of the op amp can be illustrated by a circuit called the **differential amplifier.** The schematic symbol for this amplifier is shown in Figure 12–1.

The differential amplifier is designed to respond to the difference between two input signals. An amplifier that has two input signals and gives an output by comparing the difference between the two signals is a differential amplifier. Notice in Figure 12–1 that the differential amplifier requires two power supplies for operation, one output of the power supply being positive, and the other output being negative. Notice that labels A and B identify the output terminals for the op amp. Figure 12–2 shows the schematic diagram for this dual low-voltage power supply.

Although the differential amplifier has two inputs, it can be driven with only one input. With only one input waveform, an output will be developed for Q_1 and Q_2.

In operation an input signal appears at the base of Q_1. This signal is going in the positive direction. Because this signal is increasing,

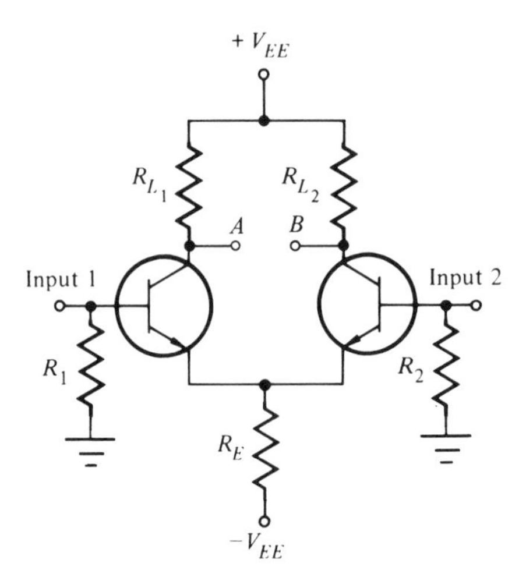

FIGURE 12–1
Schematic Diagram of a Differential Amplifier

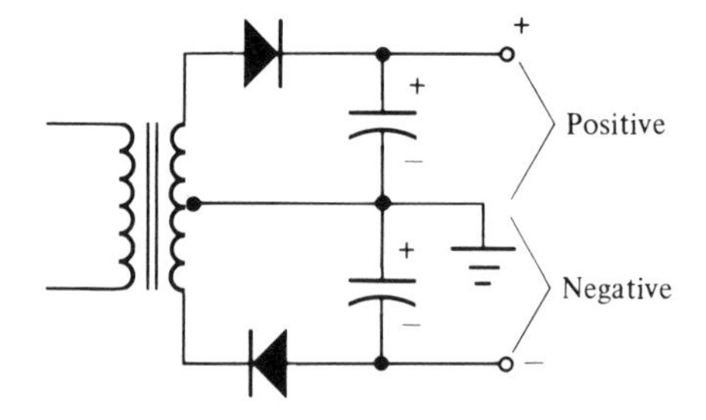

FIGURE 12–2
Schematic Diagram of a Dual-Polarity Power Supply

forward bias on Q_1 conduction through the transistor is also increasing. With an increase in conduction, the resistance of Q_1 must be decreasing between the collector and emitter terminals. Therefore, more current will develop through the transistor, and a larger voltage drop will occur at the load resistor. Because the resistance across the collector-emitter junction is decreasing, the collector's voltage is also decreasing. The voltage is said to be going in the negative direction. This negative-going voltage causes an inverted signal at the output. Figure 12–3 shows an example of the output developed at Q_1.

The conduction through Q_1 has an effect on the conduction through Q_2. As Q_1 is turned on by the positive-going input signal, the current through the emitter resistor is also on the rise. This increase in current flow is causing an increased voltage drop across the emitter resistor R_E. Thus the voltages at the emitters of Q_1 and Q_2 will be going in the positive direction. This increase in emitter voltage has the same effect as lowering the base voltage to the cutoff region. This results in less conduction through Q_2. A lower conduction means less

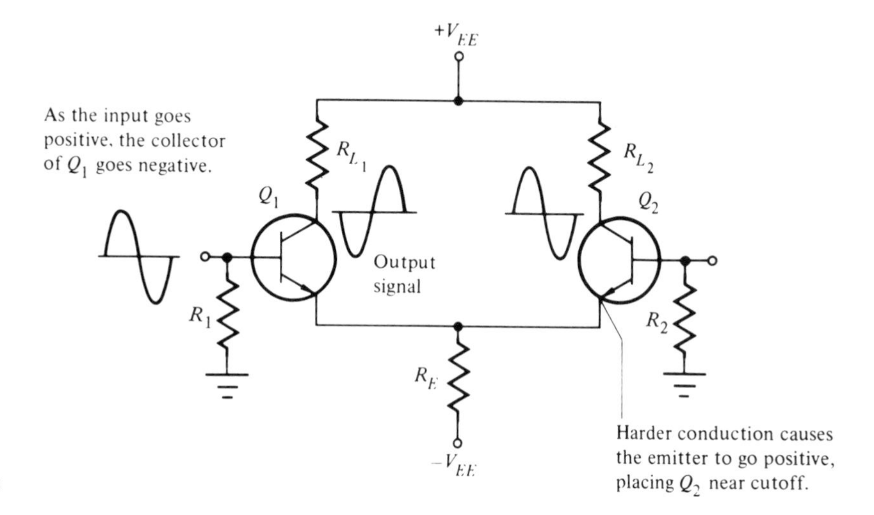

FIGURE 12–3
Waveforms for Q_1 and Q_2

current through the load, and therefore a smaller voltage at the load. The collector voltage of Q_2 then is being made more positive. Figure 12–3 shows this action.

With the input on the base of Q_1, an inverted signal (180° phase-shifted signal) and a noninverted signal (signal in phase with the input) are available at the differential amplifier output. In addition, this amplifier circuit is very selective in the signals it will amplify. Unwanted signals from the filtering circuits in the power supply, or a line hum developed in the circuit, are rejected. This ability to reject unwanted signals is characteristic of the differential amplifier, and is called the **common rejection mode.** It is one of the qualities of the general op amp.

The gain of the differential amplifier, $A_{v(\text{diff})}$, can be found from the formula for basic gain of an amplifier. This formula is given in Equation 12–1.

$$A_{v(\text{diff})} = \frac{\text{signal output}}{\text{signal input}} \tag{12–1}$$

where:

$A_{v(\text{diff})}$ = gain of the op amp

A sample calculation follows in Example 1.

EXAMPLE 1

The input signal developed at the op amp is 0.1 V, and the output signal is 10 V. What is the gain of this op amp?

Solution:

The known factors are placed into Equation 12–1 as follows:

$$A_{v(\text{diff})} = \frac{\text{signal output}}{\text{signal input}} = \frac{10\text{ V}}{0.1\text{ V}}$$
$$= 100$$

Figure 12–4 shows the general circuit diagram of the op amp. Notice that the differential amplifier is the first block in the op amp circuit. The differential amplifier is placed first because it develops high input impedance and good gain. Also note that there are two inputs to the differential amplifier: the inverting input and the noninverting input.

The next block in the op amp circuit is a high-gain amplifier. This high-gain amplifier develops additional gain to the input signal.

The final block in the op amp is a single-ended amplifier. Generally, a common collector amplifier is placed in this section. Because of the low output impedance of the common collector ampli-

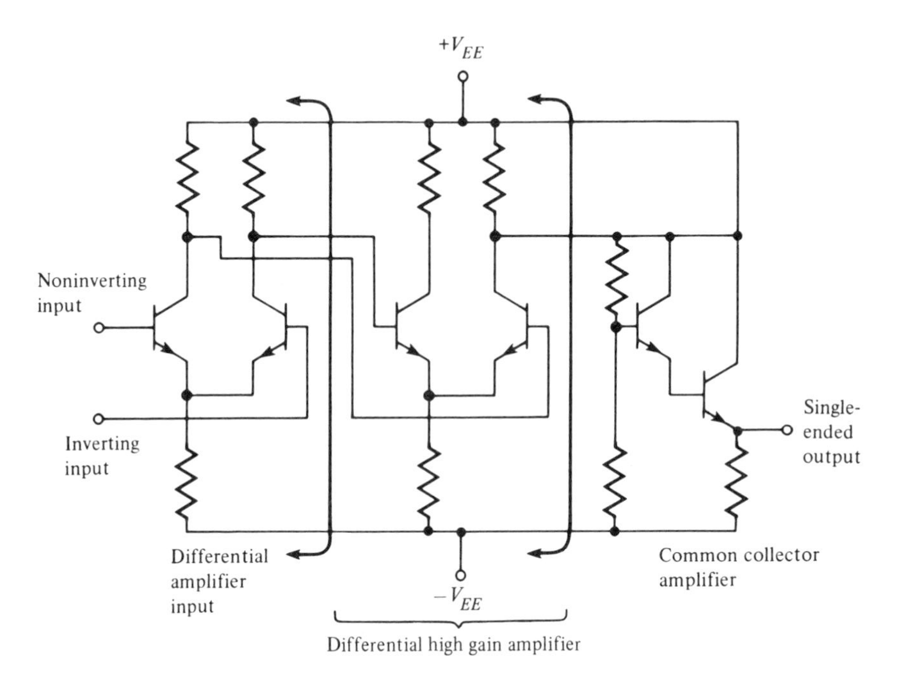

FIGURE 12–4
Three Sections of the Op Amp

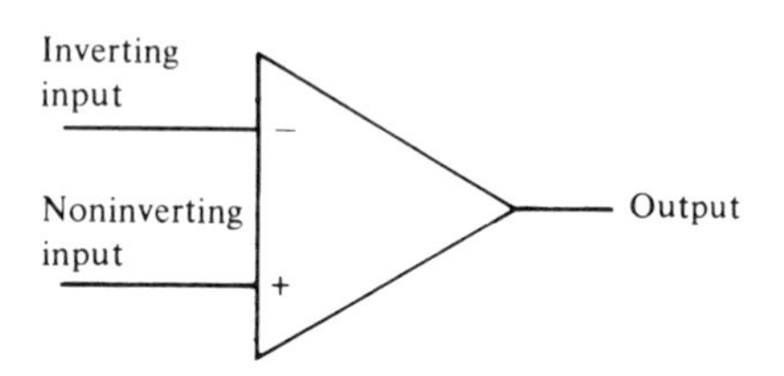

FIGURE 12–5
Schematic Symbol of the Op Amp

fier, the op amp can be a good impedance match to input of any amplifier. Note that the output to the load is only one terminal. This is why this circuit is described as single ended.

A single-ended output will develop only one output phase in the signal as it relates to ground. This is the reason there are two terminals at the input of the op amp. The op amp can develop either an inverted or a noninverted signal, depending upon the terminal connection at the input.

Figure 12–5 shows the standard schematic symbol for the op amp. One terminal is marked positive, and the other is marked negative. When the input signal is connected to the positive terminal, the signal is not inverted. When the input signal is connected to the negative terminal, the signal is inverted between the input and the output. This schematic symbol is the standard symbol used in linear circuit application.

OP AMP CONSTRUCTION

Several different transistors and resistors can be used to build an op amp. See Figure 12–4 for an example of building this circuit with discrete components. Constructing the circuit in this fashion is time consuming, and the circuit itself is relatively large.

One way to reduce both the size and the cost of the op amp is to use **integrated circuits.** The process of placing the three or more different amplifier circuits of the op amp into one package is called **integration.** The circuits are placed on a thin layer of silicon, and then housed in a plastic package.

The integrated circuit (IC) chip is made by the **monolithic** process. In the monolithic process the transistors, resistors, diodes, and other circuit components are placed on a single thin layer of silicon material by a photographic process. Integrated circuits are not only smaller and cheaper than discretely built circuits, they are also more efficient. The technology of integration has led to the lowering cost of electronic products.

Once the small silicon chip is manufactured, it must be placed

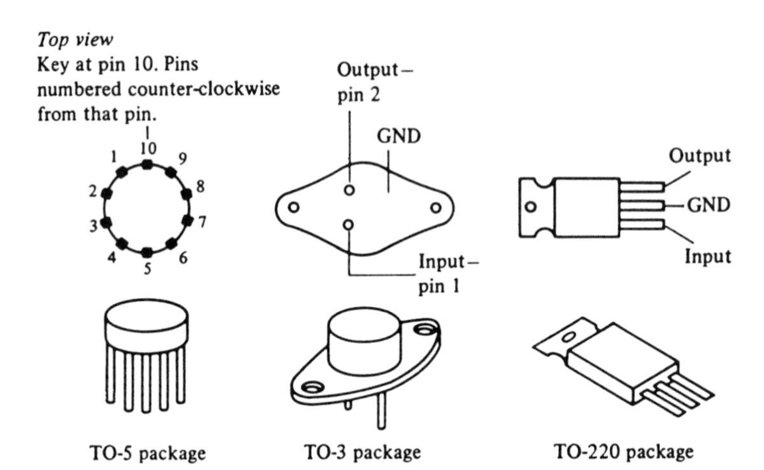

FIGURE 12–6
TO Packaging Styles of Integrated Circuits

into a housing. Integrated circuits come in a wide variety of packaging styles. Some of these packages look like transistor housings. Figure 12–6 shows examples of the TO style of packaging IC chips.

Another type of packaging used with integrated circuits is shown in Figure 12–7. This style is called the **dual in-line package,** or **DIP.** The package in Figure 12–7a has 16 pins connected externally. These 16 pins are connected to the internal circuit on the silicon wafer. Generally, a DIP will have 24, 16, or 14 pins connected to the external shell. Shown in Figure 12–7b is a *mini DIP.* This package style has only 8 pins connected to the external shell.

Also of importance to the technician is the pin numbering system on the op amp. In the DIP series of packaging, the numbering of the pins is always the same. There is a notch at the top of the chip. To

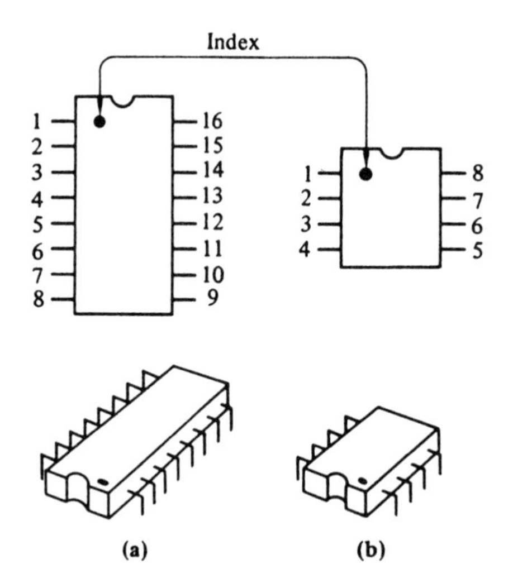

FIGURE 12–7
Dual In-Line Packaging (DIP) Styles: (a) 16-Pin DIP, (b) 8-Pin DIP (mini DIP)

the left of this notch is a circle. This circle is in line with pin 1. The pins are numbered counterclockwise around the chip. Note the pin arrangements for the two chips in Figure 12–7.

As more and more integrated circuits are used in consumer products, pin identification becomes a more important part of the service technician's knowledge. If a pin identification is not clear, an IC manual should be consulted.

OP AMP BIASING

The technician must be aware of the different components used to bias the op amp. Figure 12–8 shows the schematic symbol for the op amp,

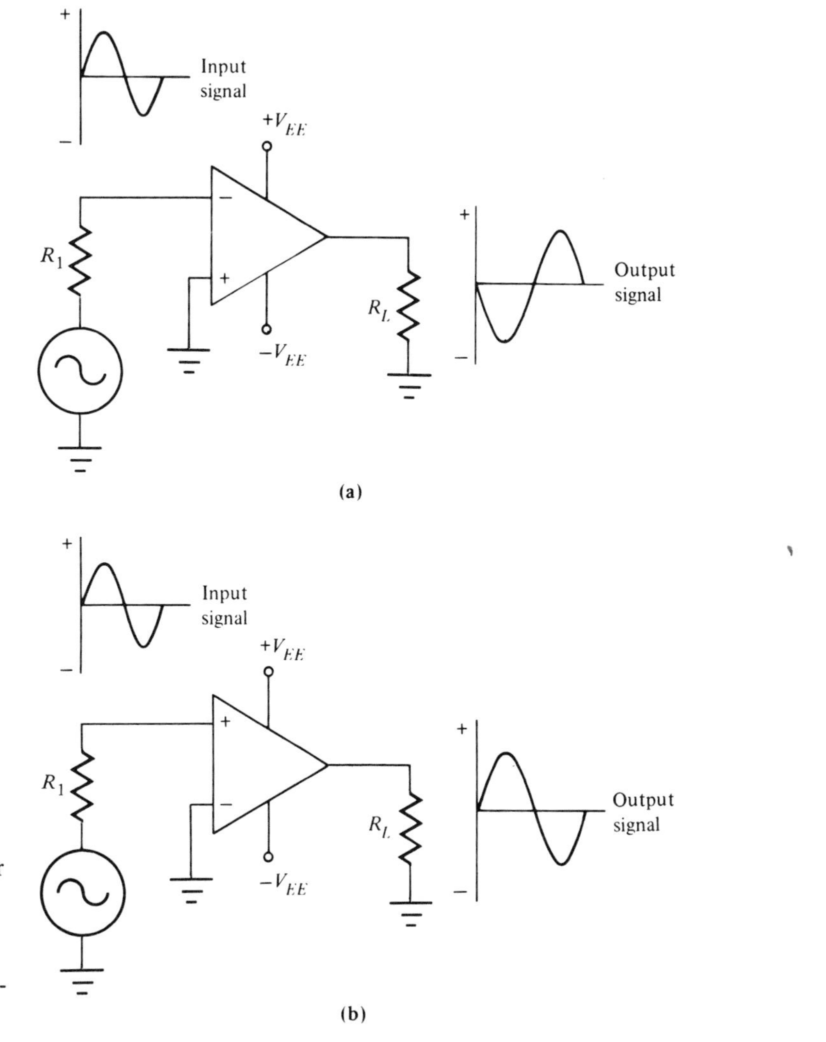

FIGURE 12–8
Typical Biasing Circuit for an Op Amp: (a) Signal Connected to Inverting Input, (b) Signal Connected to Noninverting Input

and the different output waveforms that are developed. In Figure 12–8a the signal source is shown connected to the negative or inverting input. The output waveform developed across the load resistor is amplified and inverted. In Figure 12–8b the inverted input is grounded, and the waveform is connected to the positive input. The output waveform developed across the load resistor is not inverted.

Also notice that connected to the op amp are the positive and negative voltage sources. Without these voltage sources, the op amp would not function.

As noted earlier, the bases of the transistors in the differential amplifier are grounded through resistors. When these resistors are used in an op amp, they cause a current flow. This current flow produces a voltage drop at the input of the op amp. Figure 12–9 shows what happens in this circuit. The signal is applied to the positive input terminal, and the negative input terminal is tied to ground. With no signal applied to the input terminals, a current flow is established through the resistors. The no-signal voltage developed at the terminals is called the **offset voltage.** This dc voltage developed at the terminals causes an imbalance between the input dc voltage and the dc voltage on the output terminal.

To overcome this offset voltage, a resistor is commonly connected between the input terminal and ground. Figure 12–10 shows the connection of this resistor. The value of this resistor is equal to the parallel connection of R_1 and R_2. To find the value of R_3, the simple parallel resistor formula can be used:

$$R_3 = \frac{R_1 \times R_2}{R_1 + R_2}$$

The voltage developed by R_3 will cancel the voltage developed by R_1 and R_2. Another method of cancelling the offset voltage is to connect a variable resistor to the terminal marked on the op amp as the offset **null terminal.** Figure 12–11 shows an example of this connection. If the offset voltage remains low on the op amp, there is no problem. However, when the offset voltage reaches above the 8 mV

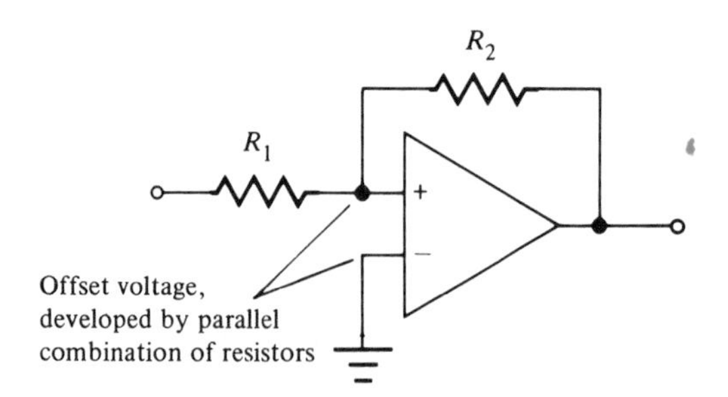

FIGURE 12–9
Development of Offset Voltage between Input Terminals

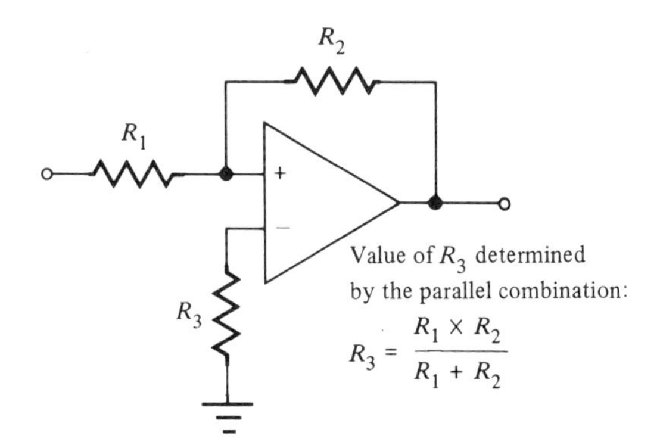

FIGURE 12–10
Method to Reduce Offset Voltage

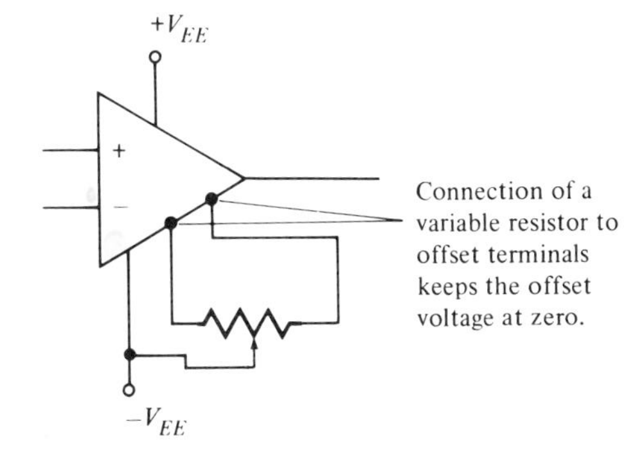

FIGURE 12–11
Null Terminal Connection Used to Reduce Offset Voltage

range, additional circuits must be added. Most op amps contain such extra circuitry.

As with other amplifier circuits, the relationship between gain and signal frequency in the op amp is important. In the op amp circuit this relationship can be expressed by the **slew rate.** The slew rate is the maximum output swing from one voltage to another in a given period of time. Figure 12–12 shows an example of the slew rate measurement.

The slew rate is an important consideration for circuit operations, especially in high-frequency applications. Assume, for example, that a 500 kHz sine wave is applied to the input of an op amp. The time required for this frequency to swing voltage polarities is as follows:

$$t = \frac{1}{f} = \frac{1}{500 \text{ kHz}}$$
$$= 2\ \mu\text{s}$$

Thus, the period for one cycle is 2 μs. For the circuit in Figure 12–12, the voltage swing was developed during only one-half of the time

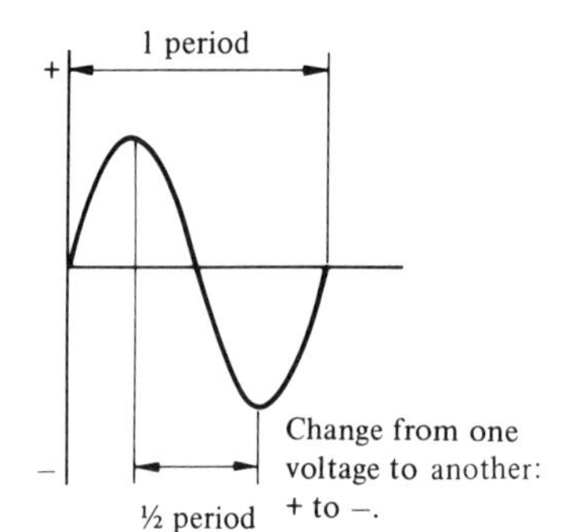

FIGURE 12–12
Measuring the Slew Rate of the Op Amp

period. Therefore the time for the voltage swing would be 1 μs. Such a fast change of voltage in the op amp circuit might develop distortion in the amplified output waveform. Figure 12–13a illustrates a high-frequency square wave input. Notice that the op amp has distorted the output because of the slew rate of the amplifier. Figure 12–13b illustrates a high-frequency sine wave applied to the input, causing a triangular wave at the output. Again, the waveshape change is due to the slew rate of the op amp. Therefore, a general-purpose op amp is not suited for high-frequency operation. A special high-frequency op amp must be used to handle the high-frequency application. The general op amp is an excellent component for amplification in the audio frequency range.

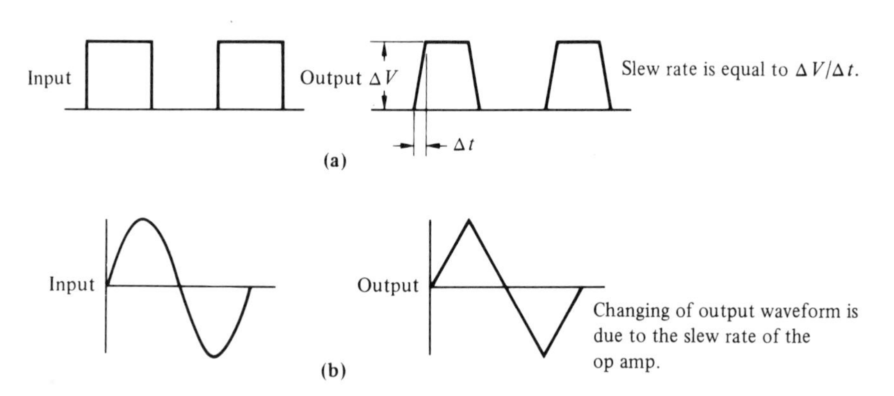

FIGURE 12–13
Distortion Caused by Slew Rate Developed in the Op Amp: (a) Square Wave Slew Rate, (b) Sine Wave Slew Rate

OP AMP GAIN

As mentioned earlier in the chapter, op amps provide excellent gain. With some op amps the gain could reach into the millions. This much gain is not normally needed in circuit application.

Again, an important relationship in the op amp is gain versus frequency. As was noted, the slew rate can greatly affect the amplification factor of the op amp. Therefore the op amp should be operated at the lower frequencies in order to achieve maximum gain. At these frequencies, the op amp will deliver good linear gain to the signal.

To achieve a controlled gain on the op amp, a feedback path is developed. This feedback developed in the op amp is negative feedback. Recall that negative feedback was also used to control gain in discrete transistor amplifiers.

The gain of the op amp can be found by looking at the op amp specification sheet. There, the gain is given as a range, and is identified as the **open-loop gain.** Open-loop gain means that there is no feedback path developed in the circuit. Open-loop gain seldom occurs in actual circuit application. All circuits used in circuit application will have some type of feedback.

Feedback is developed when part of the output signal is fed back to the input. In the op amp feedback operation, a signal is fed back to the inverting input terminal. This loop will develop negative feedback. If positive feedback is required, the signal is fed back to the positive terminal.

Figure 12–14 shows a diagram of an op amp with negative feedback established. This loop will reduce the gain of the op amp, but will increase the band of frequencies that the op amp will be able to amplify.

The gain of the op amp is generally established by two resistors. These resistors are designated R_1 and R_2 in Figure 12–14. The resistance values of these components will set the gain for the op amp. To

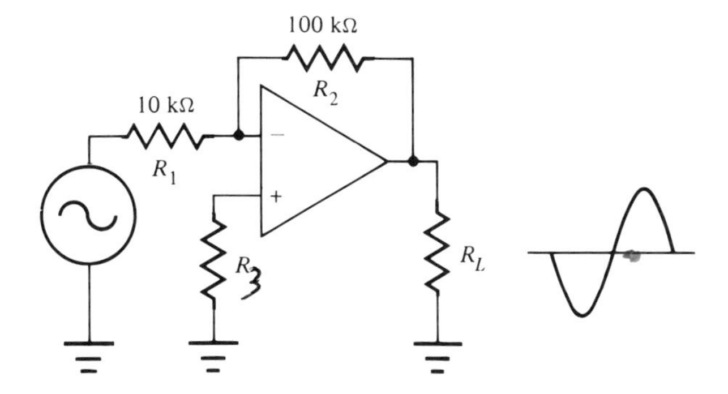

FIGURE 12–14
Gain Established with Negative Feedback

calculate the gain developed by the op amp, Equation 12–2 can be used.

$$A_v = \frac{R_2}{R_1} \tag{12–2}$$

Consider the following example.

EXAMPLE 2

Find the gain for the op amp circuit in Figure 12–14.

Solution:

Values are placed in Equation 12–2, as follows:

$$A_v = \frac{R_2}{R_1} = \frac{100\ \text{k}\Omega}{10\ \text{k}\Omega}$$
$$= 10$$

Therefore the gain for this amplifier is set at 10. The gain here would be increased if the value of resistor R_1 were reduced. For example, if R_1 were reduced to 1 kΩ, the gain of the circuit would be increased to 100. Conversely, if the value of R_2 were reduced, the gain of the op amp would also be reduced. A **unity gain,** or a gain of 1, can be established on this op amp if both resistors are of equal value.

The gain of the op amp can also be found when the input is on the positive terminal. Figure 12–15 shows this circuit construction. Notice that the feedback path is developed through R_2. However, now the feedback is placed on the negative terminal. Because of this, the

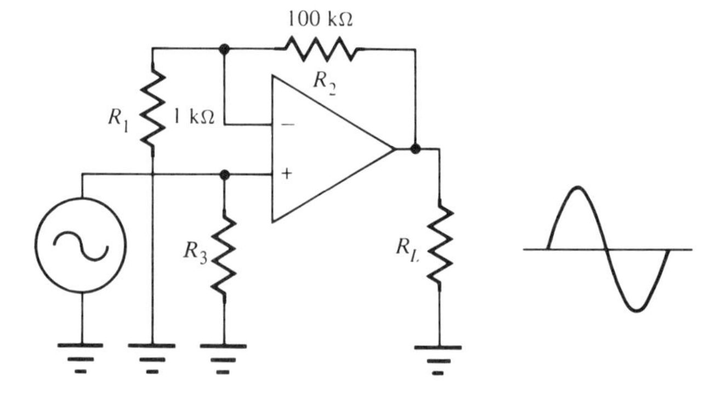

FIGURE 12–15
Gain Established with Positive Feedback

method of calculating gain is different than that used for the negative feedback path. The appropriate formula is shown in Equation 12–3.

$$A_v = 1 + \frac{R_2}{R_1} \tag{12–3}$$

EXAMPLE 3

Find the gain for the circuit in Figure 12–15.

Solution:

$$A_v = 1 + \frac{R_2}{R_1} = 1 + \frac{100\ \text{k}\Omega}{1\ \text{k}\Omega}$$
$$= 101$$

As can be seen from Equation 12–3, the gain in this mode will always be greater than one. No matter what the size of the resistors, the gain of this op amp will never be unity.

From these formulas, the gain of the op amp can be made variable. For example, the circuit in Figure 12–16 shows how the gain of an op amp can be adjusted by varying the resistor value in the feedback loop.

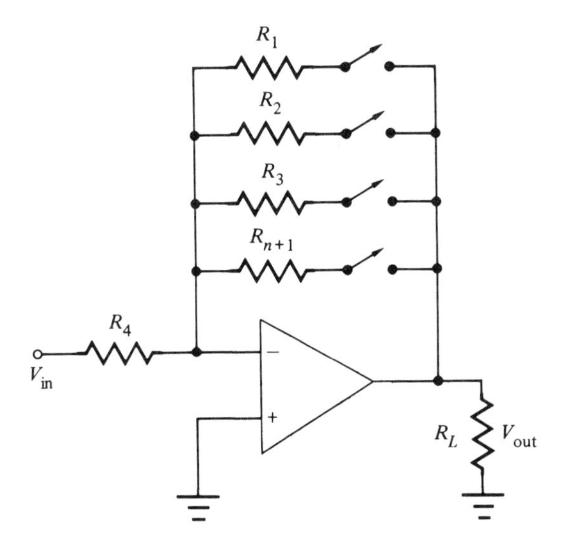

FIGURE 12–16
Making the Gain of an Op Amp Variable

GAIN VERSUS FREQUENCY

Unlike the ideal op amp, the practical op amp has limitations. The graph in Figure 12–17 shows gain of the op amp plotted against frequency. On the horizontal axis is the frequency response of the op amp. This frequency response ranges from the low level of dc, or 0 Hz, to 1 MHz. The vertical axis represents the gain of the op amp. The graph is for the op amp in the open-loop mode. This mode of operation is the ideal, with unlimited gain.

Notice that the gain of this op amp starts at 100 dB. It continues along at this value until a frequency of 10 Hz is reached. At 10 Hz, the gain of the op amp starts decreasing, and continues to drop until the 1 MHz frequency is reached. The point where the gain starts to decrease is called the *rolloff* point. As can be seen from the graph, the gain of the op amp is reduced by 20 dB for every 10-fold increase in frequency.

To improve this relationship between gain and frequency, negative feedback is used. For example, at the 40 dB point on the graph in Figure 12–17, the op amp will deliver frequency amplification up to 10 kHz. Notice in the graph that if the gain is dropped to 20 dB, the frequency range of the op amp is increased to 100 kHz. Thus when negative feedback is used, the frequency operation is increased.

The gain of the op amp drops as frequency increases because of the design of the op amp itself. As the frequency increases, the phase shift between the input signal and the output signal increases. Table 12–1 shows the amount of phase shift. Notice in the table that as the frequency approaches 10 MHz, the phase shift becomes closer to 180°. When the phase shift reaches 180°, the op amp will become unstable. The feedback will become positive, and the op amp will become an oscillator.

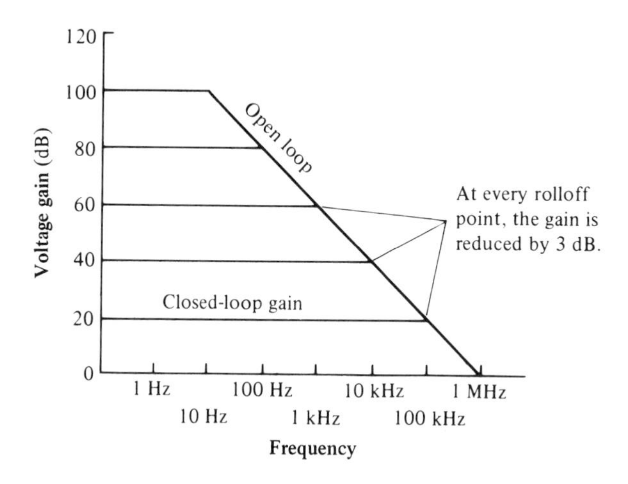

FIGURE 12–17
Graph of Gain versus Frequency

TABLE 12–1 Frequency versus Phase Shift in the Op Amp

Frequency	Phase Shift
0–1 kHz	0°
10 kHz	45°
100 kHz	90°
1 MHz	135°
10 MHz	180°

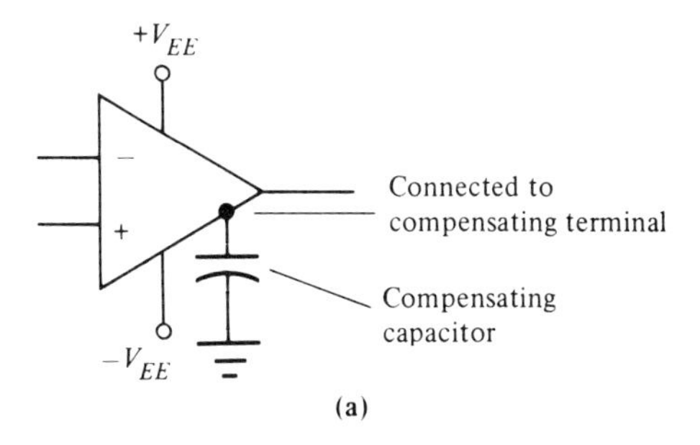

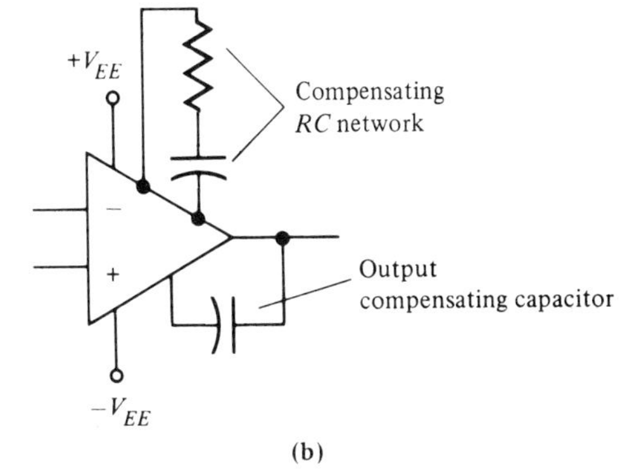

FIGURE 12–18 Compensating Networks for Op Amp Operation: (a) Capacitor Used to Correct Phase Shift, (b) *RC* Network Used to Correct Phase Shift

To correct for this phase shift within the op amp, an external **compensating network** is used. It generally consists of a capacitor connected from the output terminal back to the compensating terminal of the op amp, as shown in Figure 12–18a. Another type of circuit used is the *RC* network, illustrated in Figure 12–18b.

OP AMP CIRCUIT APPLICATION

Op amps are used in many circuit applications. Because of their low cost, small physical size, and electrical characteristics, they are very desirable for circuit use. First used in computer circuits, op amps have spread into the consumer market.

The op amp has excellent characteristics when operating at low frequencies. The output of the op amp can develop an in-phase or out-of-phase signal. The gain of the op amp can be established between unity and infinity, as in the open-loop mode. The consumer technician should be aware of these different operating modes and the op amp's general use in circuits.

The op amp in Figure 12–19 is used as a summing amplifier. Two signals are applied to the inverting terminal. The summed signals are taken from the output terminal, but the signal sum will be inverted. This type of amplifier can be used to add dc or ac voltages. A formula can be used to determine the output voltage. This formula is shown in Equation 12–4.

$$V_{out} = R_F\left(\frac{V_1}{R_1} + \frac{V_2}{R_2}\right) \quad \textbf{(12–4)}$$

where:

V_{out} = output voltage

R_F = resistor feedback

V_1, V_2 = input signals

R_1, R_2 = resistor inputs

EXAMPLE 4

All the resistors in an op amp circuit are equal to 20 kΩ. The input voltages are V_1 = 4 V and V_2 = 2 V. What is the output voltage?

Solution:

The output is calculated as follows:

$$\begin{aligned} V_{out} &= R_F\left(\frac{V_1}{R_1} + \frac{V_2}{R_2}\right) = 20\text{ k}\Omega\left(\frac{4\text{ V}}{20\text{ k}\Omega} + \frac{2\text{ V}}{20\text{ k}\Omega}\right) \\ &= \frac{4\text{ V} \times 20\text{ k}\Omega}{20\text{ k}\Omega} + \frac{2\text{ V} \times 20\text{ k}\Omega}{20\text{ k}\Omega} = 4\text{ V} + 2\text{ V} \\ &= 6\text{ V (negative)} \end{aligned}$$

Thus, the resulting output is 6 V, but, as noted, the output of the summing op amp is always phase shifted 180°.

Another example of the op amp as a summing amplifier is shown in Figure 12–20. Here the inputs have different values of resistance. This then establishes a different gain for each of the input waveforms.

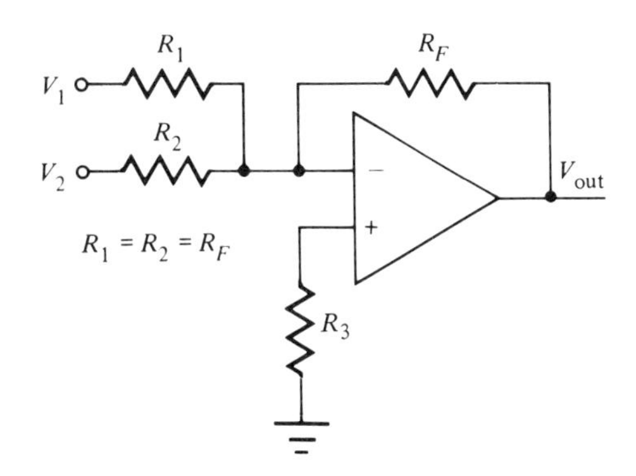

FIGURE 12–19
Op Amp Used as Summing Amplifier

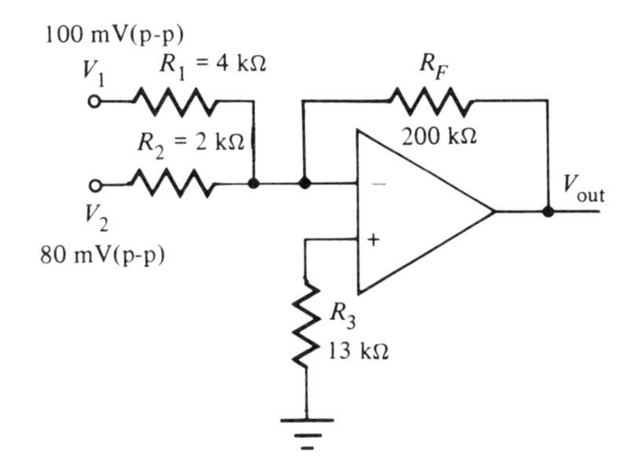

FIGURE 12–20
Summing Amp Used to Supply Different Gains

Based on the values given in the figure, the gain of the amplifier is as follows:

$$\begin{aligned}
V_{out} &= R_F\left(\frac{V_1}{R_1} + \frac{V_2}{R_2}\right) = 200\ k\Omega\left(\frac{100\ mV}{4\ k\Omega} + \frac{80\ mV}{2\ k\Omega}\right) \\
&= \frac{100\ mV \times 200\ k\Omega}{4\ k\Omega} + \frac{80\ mV \times 200\ k\Omega}{2\ k\Omega} \\
&= (100\ mV \times 50) + (80\ mV \times 100) \\
&= 5\ V(p\text{-}p) + 8\ V(p\text{-}p) \\
&= 13\ V(p\text{-}p)
\end{aligned}$$

Thus, the output of this summing amplifier would be 13 V(p-p).

Op Amps Used as Filters

Op amps are also used as filters. A filter allows certain frequencies to pass, and rejects all other frequencies. The op amp has this ability to select and reject built into its operation.

Filters can be divided into two categories: *passive* and *active*. In passive filters, capacitors, resistors, and inductors are used to pass and reject frequencies. In active filters, a transistor or an op amp is used to improve the circuit performance. The op amp, because of its low cost and frequency characteristics, makes an excellent device for active filters.

Figure 12–21a shows an op amp being used as a **low-pass filter.** The response of this circuit is shown in Figure 12–21b. As the frequency across this op amp starts to increase, the gain drops. Then the op amp will reject the higher frequencies.

Figure 12–22a shows an example of a high-pass circuit. Note that by rearranging the capacitors and the resistors, the op amp can be changed into a **high-pass filter.** As with the low-pass filter, this circuit will allow only a certain set of frequencies to pass. Figure 12–22b illustrates the band of frequencies developed in this filter.

Often in circuit design, a group of frequencies will be passed. This arrangement is called a **bandpass filter.** This filter will permit a range, or *band*, of frequencies to pass, while rejecting any frequency above or below that range. Figure 12–23a shows an example of the bandpass filter. Notice that the frequency passed through the filter will have an upper band and a lower band. Anything outside of the band-pass will be rejected. As seen in the graph in Figure 12–23b, the bandpass is established by multiplying the peak of the waveform by 0.707. Dropping two points on the graph will establish the point where the filter will attenuate or reduce any frequencies outside of the band-pass. Also, dropping the points down to the frequency line will clearly show the upper and lower limits that will pass through the filter, while those frequencies that are outside of the bandpass are rejected.

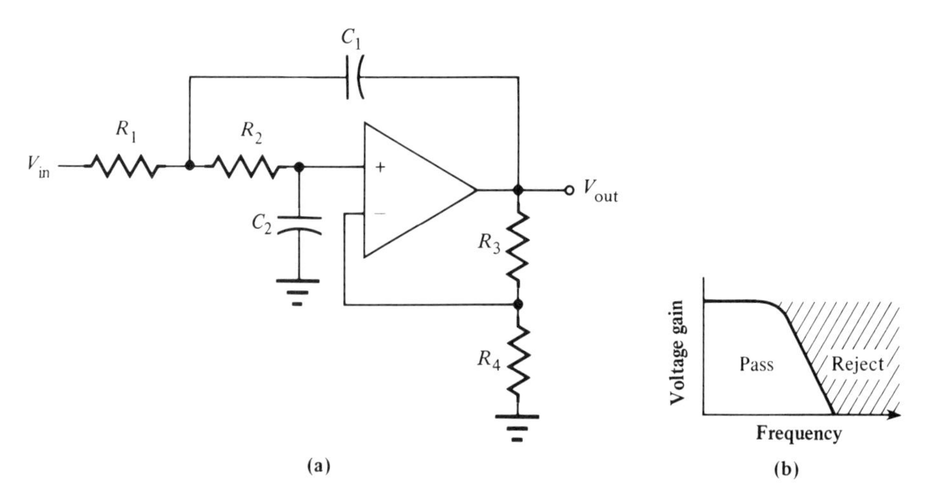

FIGURE 12–21
Op Amp Connected as a Low-Pass Filter

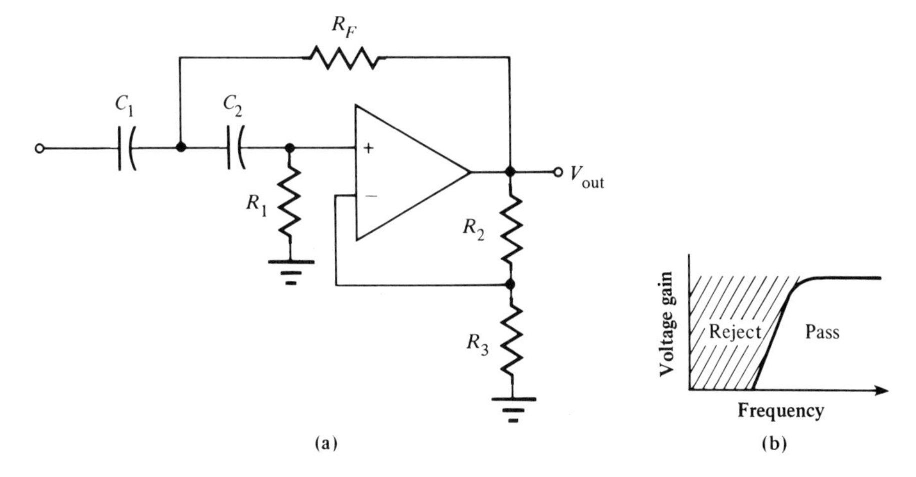

FIGURE 12–22
Op Amp Connected as a High-Pass Filter

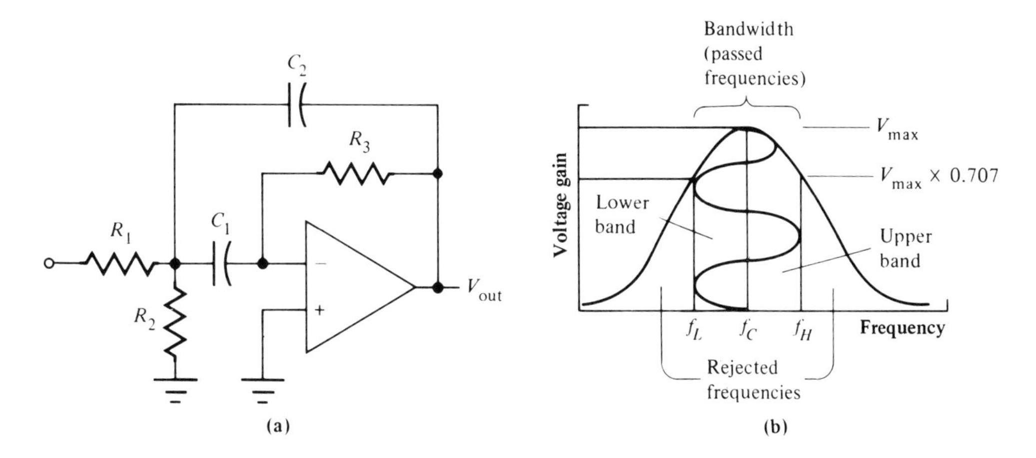

FIGURE 12–23
Op Amp Connection for Bandpass Filter: (a) Schematic Diagram, (b) Frequency Curve

This bandpass filter also can have a **Q factor** built into its operation. The Q factor relates to the sharpness of the curve. Figure 12–24a illustrates a low Q factor, and Figure 12–24b illustrates a high Q factor. By changing the sizes of resistors and capacitors in the circuit, different Q levels can be obtained.

In many communication circuits, it becomes necessary that only a very narrow band of frequencies be processed. To accomplish this,

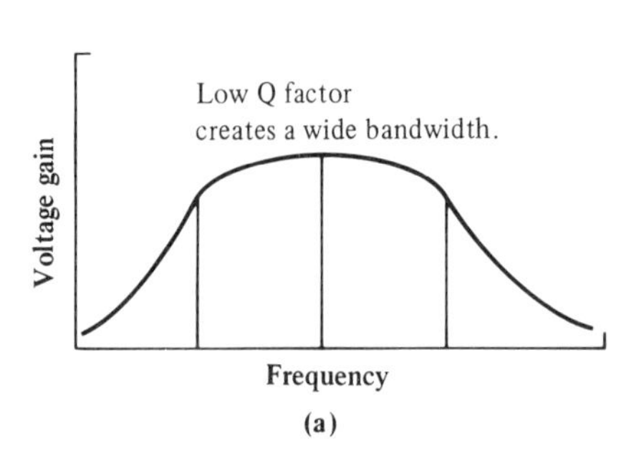

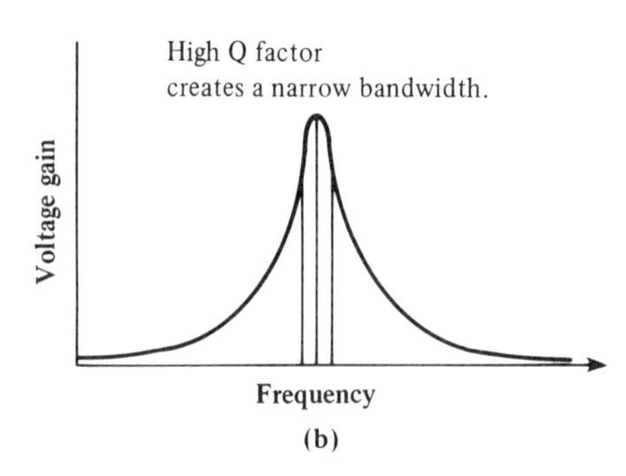

FIGURE 12–24
Selecting the Q Factor for a Filter: (a) Low Q Factor Results in Wide Bandwidth, (b) High Q Factor Results in Narrow Bandwidth

a **notch filter** is used. Figure 12–25a shows an example of the notch filter, and its gain-versus-frequency graph is given in Figure 12–25b. As shown in the graph in Figure 12–25b, the filter will get rid of a particular frequency. The gain of this frequency is greatly reduced. Filters of this type are useful when one frequency is causing a problem within the receiver.

Another use of the op amp is illustrated in Figure 12–26. The op amp here is used as a voltage regulator. The purpose of voltage regulators is to hold the output voltage constant under varying input voltages and varying load currents. Many outputs of power supplies use some type of voltage regulation to keep the voltage constant.

Op amps are excellent voltage regulators. Their positive and negative inputs make them ideal for comparing two voltages. If one of the input voltages changes, the output of the op amp will change.

In Figure 12–26 the op amp's noninverting input is connected to a zener diode. Therefore the voltage will remain constant. The inverting input of the op amp is connected to R_2 off the emitter of Q_1. Transistor Q_1 is called a *pass transistor* because all load current passes through its emitter-collector circuit. The output of the op amp is connected to the base of Q_1.

The voltage regulator in Figure 12–26 operates in the following fashion. If load current increases through the pass transistor, the output voltage begins to drop. This causes a smaller voltage drop to develop across series resistors R_2 and R_3. Then the voltage at the inverting input of the op amp drops. Because the input goes negative, a difference is developed at the op amp's two input terminals. This causes the output to go positive. The base of Q_1 also becomes positive. This makes the transistor conduct harder, and develop more voltage at the load.

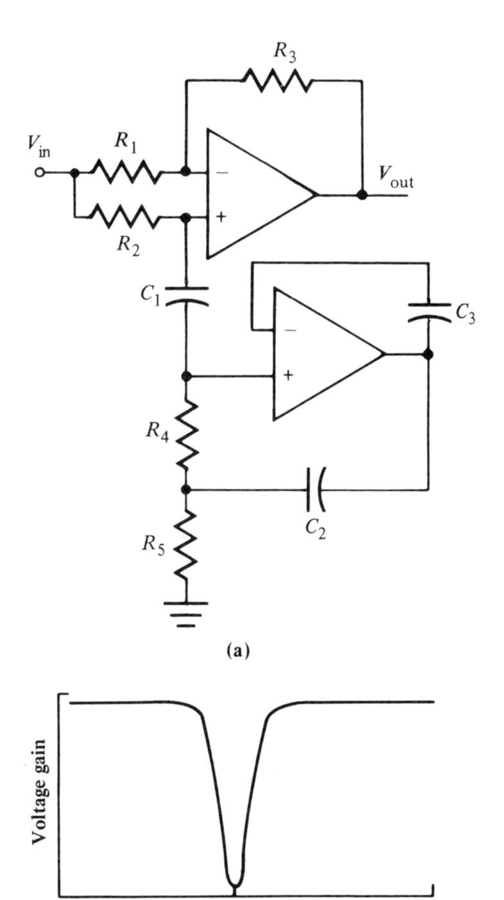

FIGURE 12–25
Filtering Circuit Found in Communication Circuits: (a) Notch Filter, (b) Voltage Gain versus Frequency for the Notch Filter

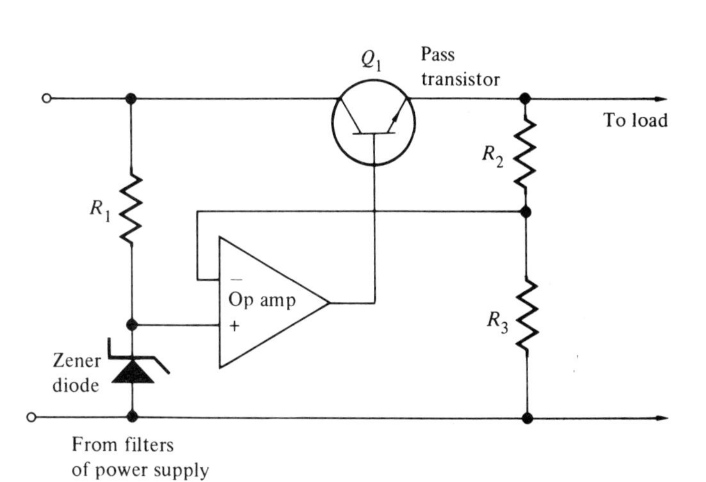

FIGURE 12–26
Op Amp Used as Voltage Regulator in a dc Power Supply

If voltage at the load increases, the op amp will sense the shift. At the same time, it will correct the pass transistor voltage at the source. Any change will allow the op amp to control Q_1.

The applications of the op amp described in the preceding paragraphs are just a few of the many applications of this device. The ones that were covered are those found most often in consumer products, and therefore the operations the technician should understand. Because of the op amp's low cost and high performance, its use is on the rise.

LINEAR ICs

Up to this point, the discussion has centered around the op amp and the accessory components needed to make it function. Operation was described using discrete components—resistors, capacitors, transistors, and op amps assembled together to perform a circuit operation.

Another process of operation involves the integration of all these components onto a single IC chip. Like the op amp, the linear integrated circuit is formed on a silicon wafer. This monolithic IC chip will be used to perform many different circuit operations. These integrated circuits are called linear circuits, and have various applications in consumer products.

Linear Circuit Operation

Linear ICs have many different uses in consumer products. They can perform almost all functions associated with linear electronics. Remember that linear electronics deals with amplified waveform reproduction. Many of the newer consumer products are composed entirely of linear ICs and digital circuits. Digital circuitry is discussed in Chapter 13.

The linear IC can be used to replace entire transistor stages within the consumer product. A linear circuit used in the IF section of an AM-FM stereo receiver is shown in Figure 12–27. Found within this linear IC is the complete IF amplifier section for the AM-FM radio. This one IC provides the amplification required for this receiver. No single transistor could approach the gain qualities of this stage. In some of the newer materials on the market, these ICs are referred to as **gain blocks.**

Gain blocks are ICs that contain several **linear amplifier** stages. For example, the gain block in the IF section of a radio receiver might contain linear amplifiers, but it might also contain a voltage regulator for the chip, and automatic gain control (AGC) and automatic frequency control (AFC) sections. Because of these gain blocks, the cost

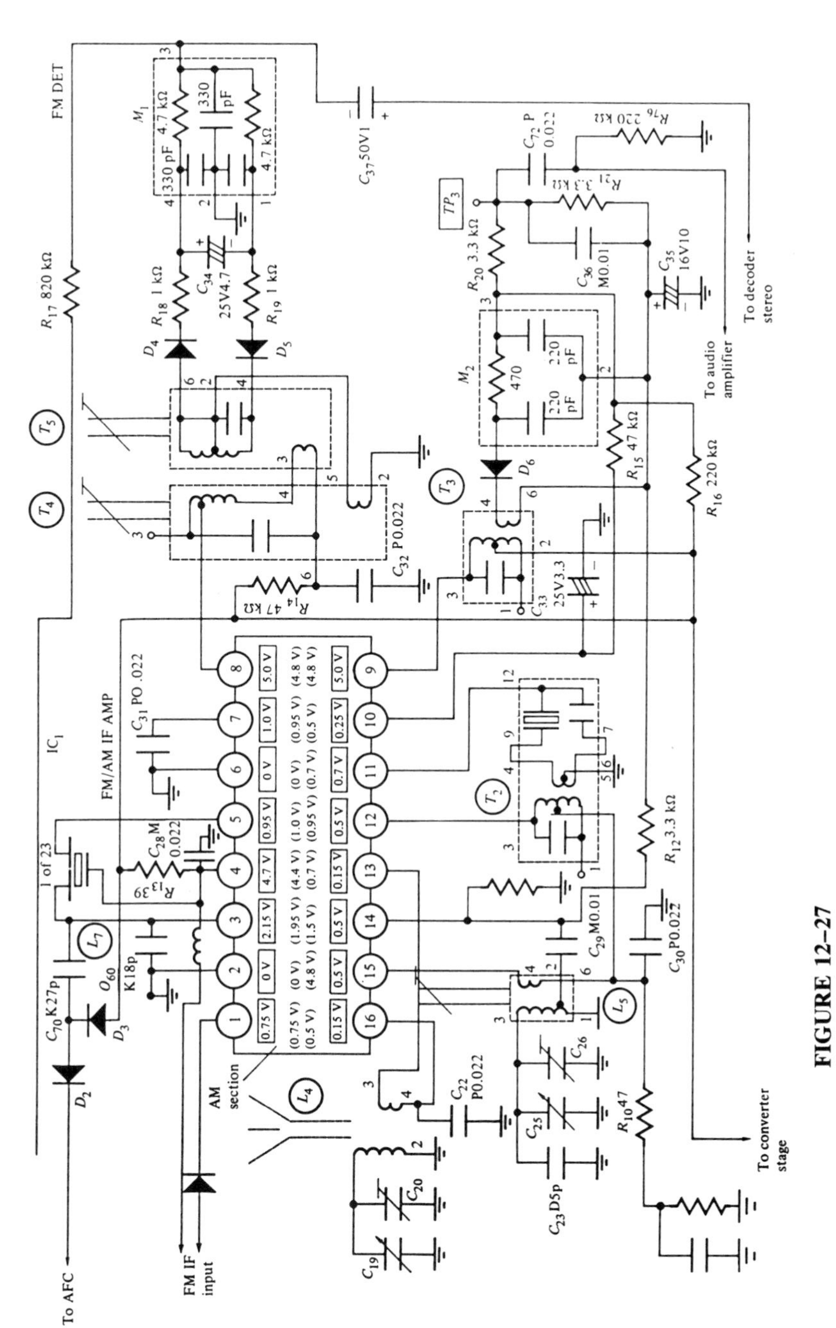

FIGURE 12–27
Linear Circuit Used as an IF Amplifier Section in an AM-FM Receiver

of the receiver has been reduced, and product reliability has been improved.

The linear amplifier has proven popular even in the sound output section. In many radios on the market today, the final audio output section is contained on one IC. Refer to the typical audio output section shown in Figure 12–28. This amplifier section is able to deliver 5 W to the load. Of course, this integrated circuit must have proper heat sinks applied.

Another popular use of linear ICs is as voltage regulators. In op amp circuits a pass transistor and an op amp are needed to perform the voltage regulation task. The linear circuit voltage regulator is built into one package. These three-terminal devices are fixed to deliver one voltage to the source. Their regulation qualities range from small amounts of voltage to around 50 V. They are also available in either positive or negative output voltage polarities. Finally, these devices can deliver up to 3 A to the load. Figure 12–29 shows the typical diagram of an IC voltage regulator.

A special set of decoders is used in FM and CB receivers to detect the intelligence from the carrier. These devices are called **phase-lock loop (PLL) circuits.** The basic operation of this system can be seen from the block diagram in Figure 12–30.

The phase-lock circuitry is made up of a phase detector, a low-pass filter, an error amplifier, and a voltage-controlled oscillator. The main function of the phase detector is to compare the phase of the

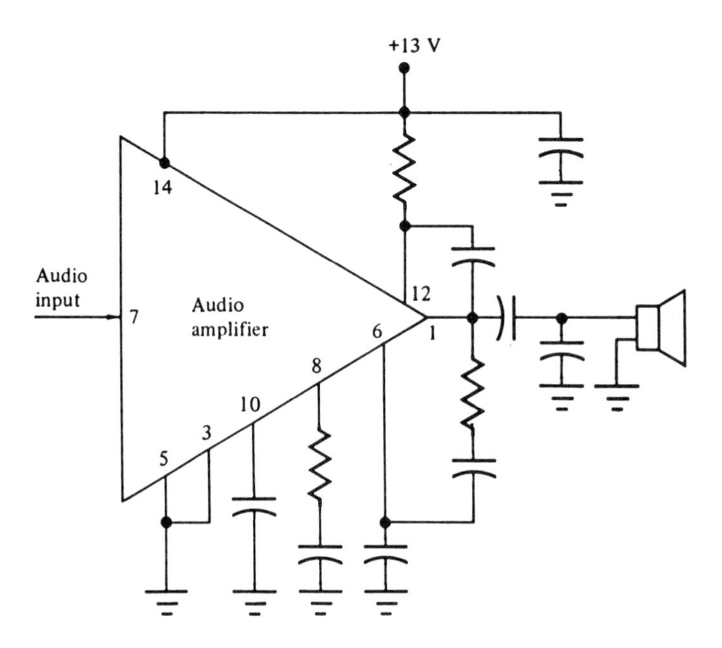

FIGURE 12–28
Linear Circuit Used as a Final Audio Output in Receivers

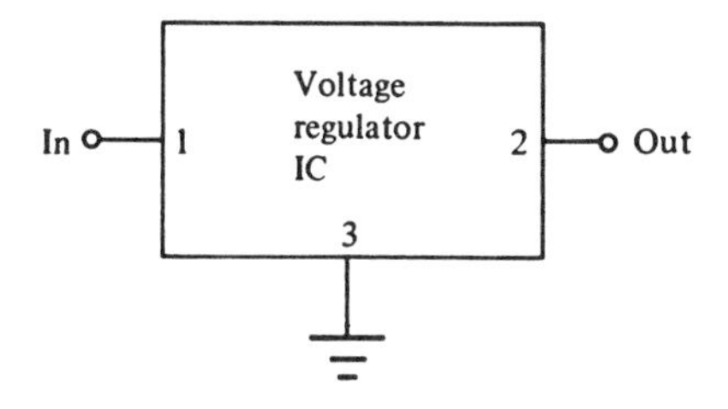

FIGURE 12–29
Linear Circuit Used as a Voltage Regulator

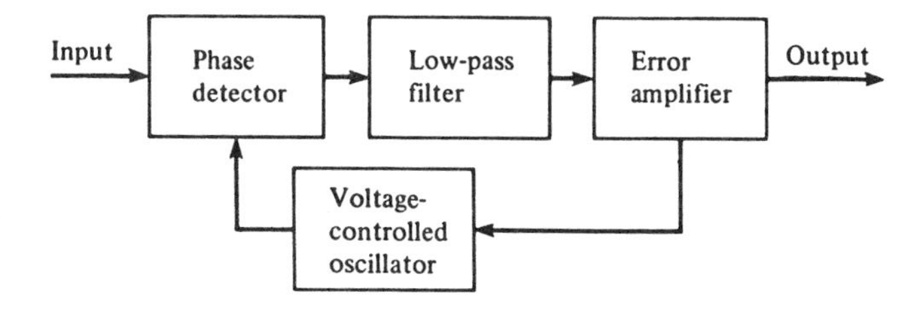

FIGURE 12–30
Basic Block Diagram of a Phase-Lock Loop (PLL) System

input signal to the phase of the voltage-controlled oscillator signal. Any difference between the phases, or frequencies, of these two signals is sent to the error amplifier. The error amplifier sends a correction voltage back to the voltage-controlled oscillator. When the incoming and voltage-controlled oscillator signals are on track with each other, the system "locks in" to that frequency. Now there is no error developed between the two signals, and no correction voltage is sent to the voltage-controlled oscillator.

PLL circuits are often used in consumer products as demodulators. Figure 12–31 shows a diagram of an FM demodulator using a PLL circuit. The input of the unmodulated signal is found on pin 12, and output is taken on pin 9. A variable capacitor is connected between pins 2 and 3, and is used to adjust the voltage-controlled oscillator to the center IF frequency.

PLL circuits are also used in CB radios. In the CB radio the PLL sets the receive frequency and the transmit frequency. An example of this PLL operation is given in Figure 12–32. In this circuit a monolithic IC is used for the PLL device. The heart of this system is a dc-controlled, voltage-controlled oscillator. The main job of this oscillator is to provide the transmit and receive frequencies for the receiver.

During the transmit mode, the overtone crystal oscillator generates a 35.42 MHz frequency. The voltage-controlled oscillator can generally operate at any frequency; here it is set to 37.66 MHz. These two frequencies mix in the loop mixer. This action produces a difference in frequency of 2.24 MHz. This 2.24 MHz frequency is fed through a buffer amplifier to a programmable counter. The programmable divider

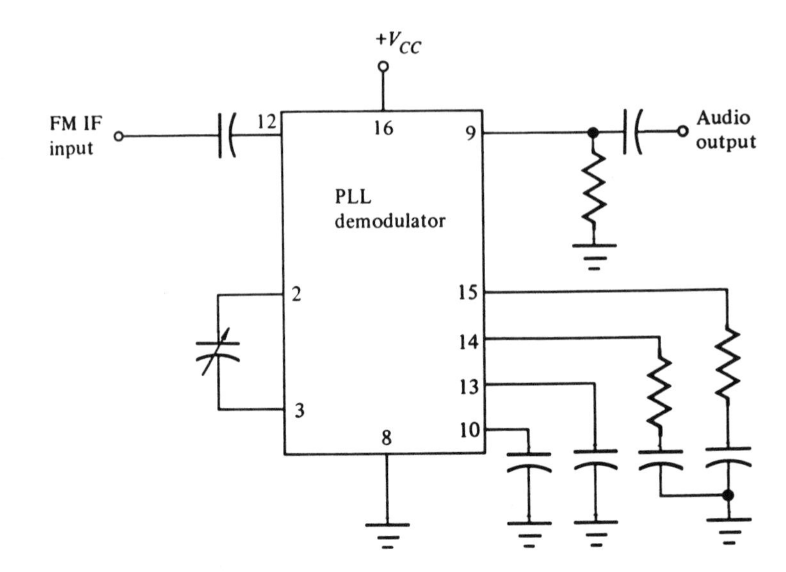

FIGURE 12–31
PLL Circuit Used as an FM Demodulator

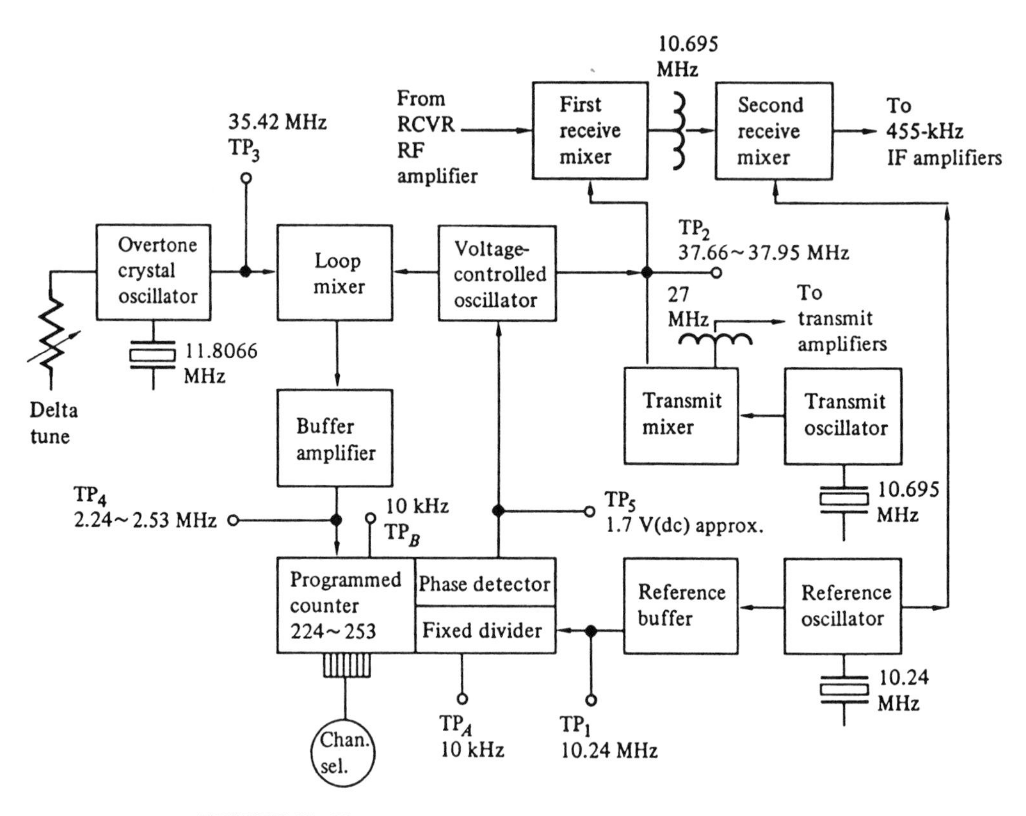

FIGURE 12–32
PLL Circuits Used in CB Radio Receivers

is used to divide the 2.24 MHz signal down to a 10 kHz signal. Note in the figure that the programmable divider allows the frequency from the loop mixer to be divided by numbers from 224 to 253. These numbers are established by placing ones and zeros on the programmable divider.

In the example in Figure 12–32, channel 1 is selected. The programmable divider is set to divide the 2.24 MHz signal by 224. This results in a 10 kHz signal. This signal is then fed to the phase detector.

The next part of the signal to be compared comes from the reference oscillator. This reference oscillator is another crystal-controlled oscillator. This oscillator operates at 10.24 MHz. This signal is fed to a fixed divider circuit, where it is divided by 1024. The divider circuit puts out a constant 10 kHz signal.

The two 10 kHz signals are compared in the phase detector. The phase detector checks to see that both signals have the proper phase and frequency. If their phase and frequency match, as shown in the example, a stabilized voltage is established to keep the voltage-controlled oscillator operating at the correct frequency.

When the channel selector is changed to channel 9, the transmit frequency must adjust. The channel selector now sets the programmable divider to divide by another number. Channel 9 is now dividing the frequency from the loop mixer by 234. The signal created and sent to the phase detector is not a 10 kHz signal. Thus an error is established between the reference signal and the signal from the programmable divider. A correction voltage is sent to the voltage-controlled oscillator. The voltage-controlled oscillator adjusts the frequency. Again, the process of frequency comparison begins. Now the signal coming from the loop is 2.34 MHz. The divider reduces the voltage to 10 kHz. In the phase detector the two signals again match in phase and frequency, and the voltage-controlled oscillator is kept on frequency.

Only three crystals are used to control the oscillators in this PLL system. Other types of frequency synthesis systems require a great many crystals to perform the receive and transmit function, sometimes more than 75. Therefore manufacturers have been forced, because of cost and availability of crystals, to go to the PLL system.

TROUBLESHOOTING

LINEAR CIRCUITS

Monolithic linear integrated circuits are used widely in modern consumer electronics products. The radio's entire IF section has been replaced by the monolithic IC. Transistor amplifier circuits have been totally replaced by the linear IC. This use of the IC has lead the technician to a new method of troubleshooting. The increased use of oscilloscopes, voltmeters, and signal generators are being used in the troubleshooting procedure.

Again, the key to effective troubleshooting is knowledge of the block diagram of the unit under repair. Next is the ability to quickly identify the faulty block or blocks. Finally, the technician must understand how each of the ICs serves to produce output within the unit.

Figure 12–33 shows a monolithic IC used in a television receiver. The entire sound system is contained within this circuit. Notice that the input for this circuit comes in on pin 1, and leaves to the speaker output on pin 9.

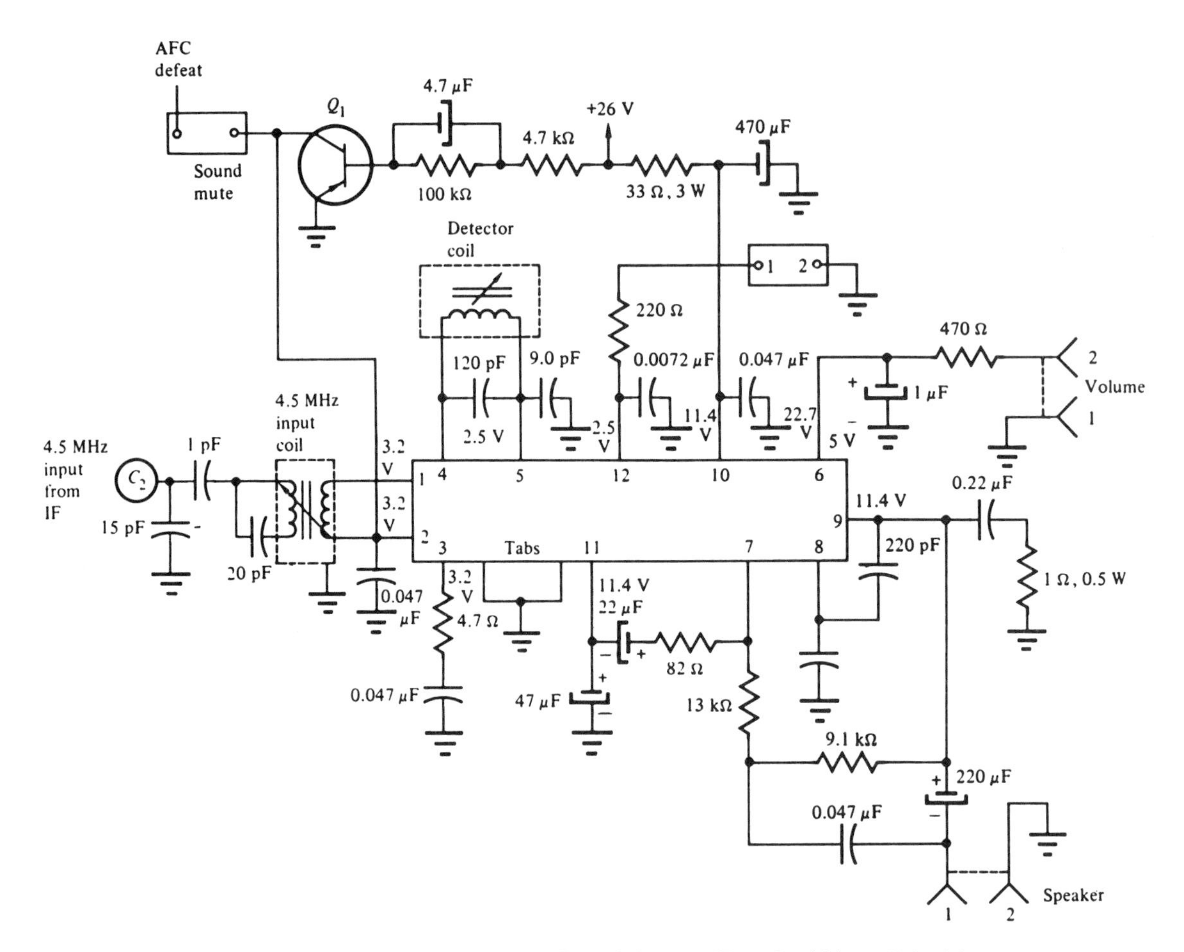

FIGURE 12–33 Sound System Found within a Television Receiver—All on One Chip

To troubleshoot a receiver with no audio output, the technician must be able to troubleshoot this IC. The first step is to make a visual inspection to insure that none of the supporting components are broken. Then the dc voltage used to operate the integrated circuit should be measured. Notice that a +26 V source is being used to supply dc bias voltage. A quick measurement of pin 10 will show whether the supply voltage is present. Another voltage measurement can be taken at the ground terminals, which are designated as tabs on the schematic diagram. Because these tabs are connected to ground, they should read 0 V. If they do not, a broken solder connection should be checked for. Next, a check of pin voltages around the circuit should be made.

When measuring voltages around the IC, the technician should take care not to touch two pins at once. For example, if when measuring supply voltage at pin 10, the probe slips to another pin with a lower potential, higher voltages would be placed on the lower potential pin. This might cause irreparable damage to the circuit.

Another point to remember when measuring voltages around the IC is that the read voltages may differ from the values given on the schematic. Generally, measured voltages will vary only about 5% from the rated values.

On some schematic diagrams, voltages are given as ranges. For example, pin 3 of the IC might be marked 3.2 V/3.6 V. Any voltage that falls in this range is acceptable. Any reading outside the range indicates trouble. Because the voltage readings given on schematics are rounded to the tenths or hundredths place, many service technicians use digital voltmeters.

Another important troubleshooting step is the measurement of waveforms around the input and output of the IC. Figure 12–34 shows a typical schematic diagram of a chroma IC found in a television receiver. The schematic gives the service technician valuable information about the operation of the circuit. Shapes and amplitudes of waveforms are shown. Measuring pins 9, 10, and 11 shows the waveforms necessary to produce the chroma information on the CRT. If these waveforms are missing, distorted, or low in amplitude, the IC should be suspected. Also, notice the waveform at pin 1. This is an input waveform. If this waveform is missing, distorted, or low in amplitude, components outside the IC should be checked. If it is missing, the coupling capacitor should be checked.

If the IC is shown to be at fault, it must be removed and replaced. In many circuit applications, the IC is connected directly to the board. To remove and replace this circuit requires unsoldering all the pins. Other times the IC is mounted in a socket. Removal of this circuit assembly is easy. In either case, when the IC is being removed and replaced, the power in the circuit must be disconnected, or damage to the chip may occur.

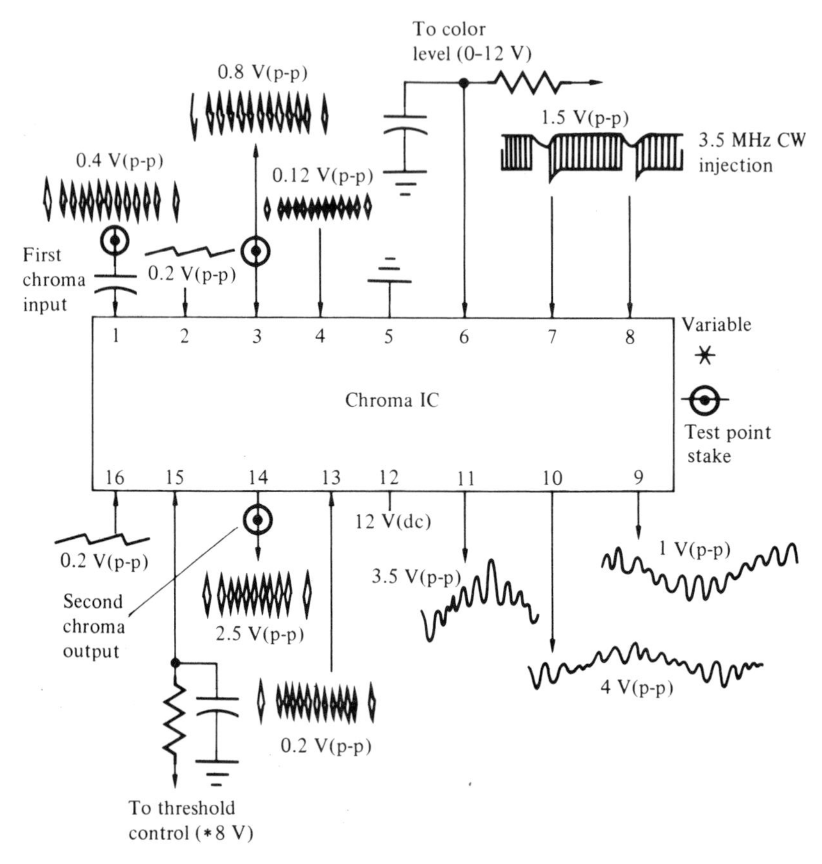

FIGURE 12–34 Linear IC Used in the Chroma Section of a Television Receiver

SUMMARY

Integrated circuits are classified as linear or digital. The most common linear circuit is the operational amplifier, or op amp. The op amp has several characteristics that make it ideal for consumer products. These qualities are as follows:

—High open-loop gain
—High input impedance

—Low output impedance
—Rejection of unwanted signals

The foundation of the op amp is the differential amplifier. This amplifier will provide an inverting input and a noninverting input.

The op amp is built on a single silicon chip. This type of construction is called monolithic construction. The chip is housed in one of many plastic packages, the most common being the DIP style.

Biasing of the op amp is established to provide the proper gain for the amplifier.

To insure proper amplification, the input terminals of the op amp must be balanced. To develop this balance, an offset resistor is used to connect the unused terminal to ground.

Amplification is affected by the frequency applied to the amplifier. The rate at which the output voltage changes is called the slew rate.

The gain of the op amp can be calculated as the ratio of the feedback resistor to the input resistor. To control the amount of gain, the op amp uses negative feedback.

In the closed-loop operation, the gain of the op amp is limited by the frequency. Generally, op amps are designed to operate within the audio range.

Op amps have a wide variety of applications useful in consumer products. Op amps are most commonly found in amplifier circuits, but are also used in filtering networks. One of the most popular uses of the op amp is as a voltage regulator.

Integrated circuits are also assembled by the monolithic process. An IC is the combination of various circuits into a single package.

Sometimes it may become necessary to combine digital and linear functions into one circuit. An example is the PLL system. This system is used to reduce the number of components, along with providing the necessary circuit application.

Troubleshooting monolithic linear circuits is basically measuring pin voltage, frequency, and signal injection. When the schematic readings differ from the actual circuit readings, the IC should be checked.

KEY TERMS

$A_{v(\text{diff})}$
bandpass filter
common rejection mode
compensating network
differential amplifier
dual in-line package (DIP)
gain blocks
high-pass filter
integrated circuit (IC)
integration
linear amplifier
low-pass filter

monolithic
notch filter
null terminal
offset voltage
open-loop gain
operational amplifier (op amp)
phase-lock loop (PLL) circuit
Q factor
slew rate

REVIEW EXERCISES

1. Draw a schematic diagram of a differential amplifier. Give a brief description of differential amplifier operation.
2. Draw the basic block diagram of an op amp, and give a brief description of the function of each block.
3. Why must the input of an op amp have an inverted and a noninverted input?
4. Describe the monolithic building of the op amp.
5. Draw examples of the TO and DIP packaging styles of the op amp.
6. How many power supplies are required to operate an op amp? What are their polarities?
7. Define the term *offset voltage.* What is the maximum offset voltage that can be tolerated?
8. Draw a schematic diagram that would reduce the offset voltage.
9. What type of feedback is used to control gain in the op amp?
10. Calculate the gain of the op amp circuit of Figure 12–14, assuming that R_1 = 2.2 kΩ and R_2 = 220 kΩ.
11. Calculate the gain of the op amp circuit in Figure 12–15. Assume that R_1 = 500 Ω and R_2 = 5 kΩ.
12. What will increase the bandwidth of an op amp circuit?
13. Describe how an op amp can be turned into an oscillator when negative feedback is used.
14. Explain why the op amp has the ability to amplify low frequencies.
15. Draw a schematic diagram showing the op amp as a voltage regulator.
16. Describe gain block. Draw a block diagram to illustrate.
17. Describe the basic function of a PLL circuit. Where are PPL circuits found?
18. List the steps used when troubleshooting linear monolithic circuits.
19. Draw an example of a compensating circuit connected to an op amp.
20. What function does the compensating circuit serve?
21. What causes the phase shifting between input and output of a differential amplifier?

13 DIGITAL CIRCUITS

OBJECTIVES

Upon completing this chapter, you should be familiar with:

—Basic digital operation
—Digital numbering system
—Encoding and decoding numbers
—Basic logic gate functions
—Basic NAND gate logic
—Basic flip-flop operation
—Basic clock characteristics
—RAM and ROM memory
—Troubleshooting digital circuits

INTRODUCTION

The consumer technician is faced with demanding new tasks. The operation of the receiver with simple active and passive components has changed. The consumer electronics industry has reached into the digital market, and has applied some digital circuitry to consumer products. Now the technician must understand not only amplification principles, but also binary codes, gate functions, and frequency division of digital circuits. As seen in many of the new consumer electronic products, digital displays are used to read out channel selection. Frequency divider circuits are establishing the vertical and horizontal sweep rates in television.

To introduce the technician to digital operation, this chapter covers basic digital operation. Application of digital circuits to the consumer product is also discussed.

BASIC DIGITAL OPERATION

To better understand digital electronics, a comparison must be drawn between analog electronics and digital electronics. In the **analog circuit** there is an input and an output. The output is said to vary in step with the input. Figure 13–1a shows a basic analog circuit. This is a simple power supply with a variable resistor connected to it. As the center arm is moved along the resistance, the voltage is changed. As the wiper is moved from point *A* to point *B*, the voltage is gradually decreased. As the wiper is moved from point *B* to point *A*, the voltage is gradually increased. The graph in Figure 13–1b shows the relationship between the voltage amplitude and time. This up-and-down movement of the wiper is causing the output voltage to go up and down also.

The linear amplifier found in consumer products is another example of the analog circuit. As the input signal increases and decreases, the output signal does the same. Figure 13–2 shows an example of this action. The analog device can therefore be described in signal terms; the signal continues to vary in step with the input.

Digital circuits operate with digital signals. These signals are called **square waves.** Figure 13–3 shows an example of digital signals. The square wave is used because of its on/off nature. As shown, the

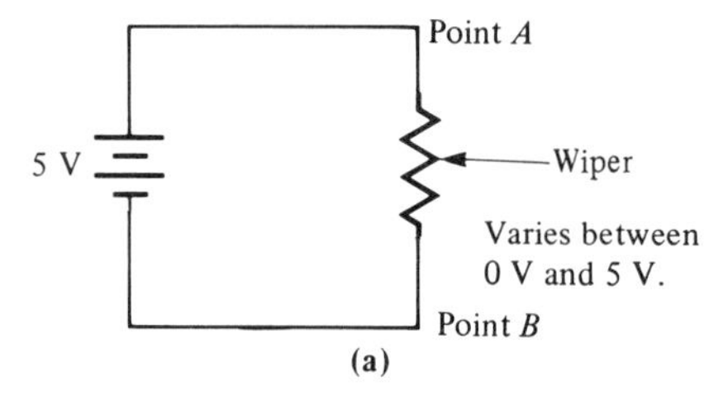

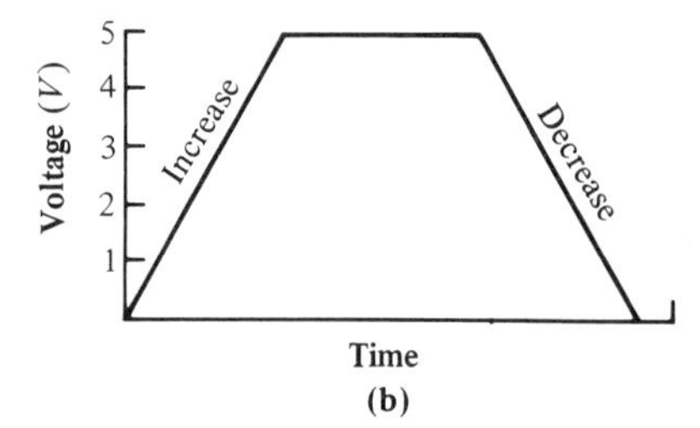

FIGURE 13–1
Variable Resistor Used in an Analog Circuit: (a) Analog Circuit, (b) Voltage versus Time Relationship

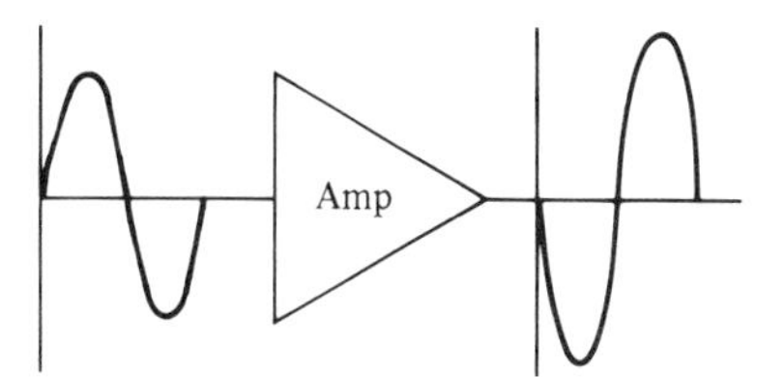

FIGURE 13–2
Linear Amplifiers Found in Consumer Products

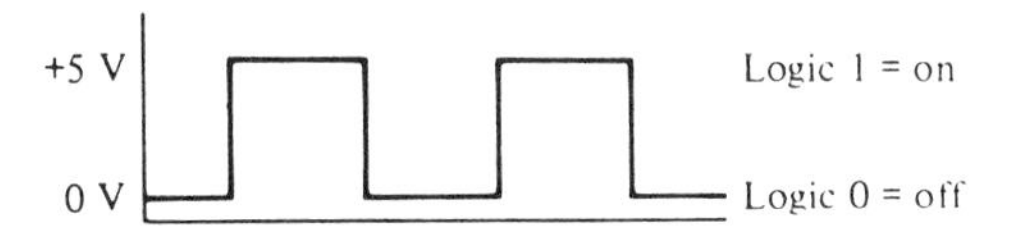

FIGURE 13–3
Square Wave Used as Digital Signals

square wave rises from the 0 V level to the +5 V level instantaneously. The signal remains at that +5 V level, and then falls to the 0 V level, again instantaneously. This type of waveform characterizes digital circuits. In digital notation the 0 V level is **logic 0,** and the 5 V level is **logic 1.**

The on/off nature of the square wave is the clue to digital circuit operation. Digital circuits operate in two states: on and off. Applying the square wave, a logic 0 will have the digital circuit off, and a logic 1 will have the digital circuit on. The voltage levels of 0 V and +5 V are typical voltage levels that turn digital circuits on and off.

A digital circuit can also be operated by a simple on/off switch. However, in digital operation the circuits must perform thousands of operations within seconds. The square wave operating at high frequencies will turn the digital circuits on and off many times each second, and is therefore a better method than mechanical switching.

DIGITAL NUMBERING

A common numbering system is the base-10 system. The base-10 system uses the numbers between 0 and 9 to express numerical values. The digital system cannot operate on this numbering process. To accommodate the base-10 number to digital operation, a numbering system in binary is used. Binary is a base-2 number system.

Each place in the base-10 numbering system has a certain value, based on its position relative to the decimal point. For example, the number 648 is made up of the following placeholding values:

$$648 = \frac{\text{hundreds}}{6} + \frac{\text{tens}}{4} + \frac{\text{ones}}{8}$$

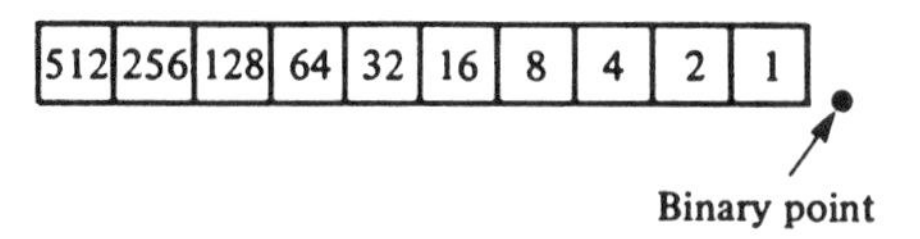

FIGURE 13–4
Value of Placeholders in the Binary System

(In this example the decimal point is understood to be to the right of the 8.)

The binary system placeholders have different values than base-10 placeholders. The binary point is the starting point in assigning place values. Starting from the binary point and moving left, the first binary placeholder is *ones*. The next slot is *twos*. Next to the twos place are *fours*, then *eights*, *sixteens*, and so on. Figure 13–4 shows the value of placeholders in the binary system. Notice that each placeholder is double the preceding value.

From this arrangement a base-10 number can be converted into binary form. The decimal number is converted into a series of 0s and 1s. Because the digital system operates on 0s and 1s, numbers can be broken down into **binary numbers.** Table 13–1 shows an example of decimal numbers from 1 to 15 and their binary counterparts. In the table the decimal number 9 is converted into the binary number 1001. This number is read: "one, zero, zero, one." With this process, decimal numbers can be translated into binary numbers.

TABLE 13–1 Converting Decimal Numbers into Binary Numbers

Decimal	Binary
1	1.
2	10.
3	11.
4	100.
5	101.
6	110.
7	111.
8	1000.
9	1001.
10	1010.
11	1011.
12	1100.
13	1101.
14	1110.
15	1111.

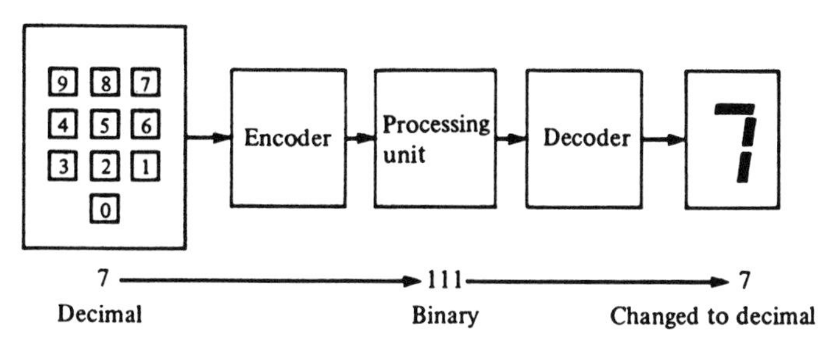

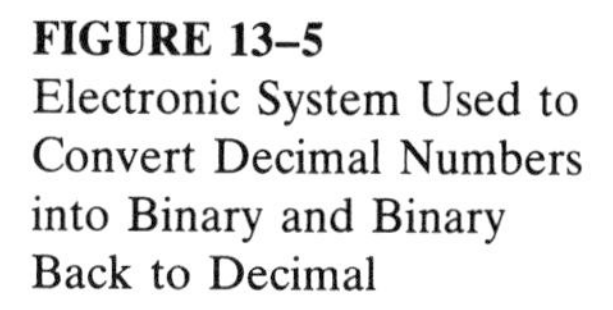
FIGURE 13–5
Electronic System Used to Convert Decimal Numbers into Binary and Binary Back to Decimal

TRANSLATION

Since the digital system operates in the binary language, decimal commands must be electronically translated. This means that encoder circuits must be able to convert base-10 numbers into binary for the input to the system, and that the binary numbers must be translated by decoders into base-10 numbers for the output of the system.

Figure 13–5 shows a typical translation system found in consumer products. This system will translate the decimal number into a binary number, and then translate back to a decimal number. For example, to tune to channel 7 on a television receiver, the keyboard receives the decimal number 7. Because the processing unit processes only binary numbers, an **encoder** is used to convert the decimal number 7 to the binary number 111. The processing unit receives the binary number, and tunes the receiver to channel 7. However, since the average consumer does not understand binary, the binary number is converted back to a decimal 7. The translating from binary to decimal is accomplished in a circuit called the **decoder.** The resulting number is then displayed on a readout circuit.

Encoders and decoders are used throughout digital systems where displays are needed, and where input information must be translated. These circuits are found on tuners of radios, televisions, and CB receivers.

LOGIC CIRCUITS

Switching from decimal to binary, processing the information, and then decoding the number seems to be a very complicated process. Actually, the process is very logical, and can be broken down into a series of different operations. The basic building blocks in any digital system are *logic gates*. These logic gates are used with binary numbers. Hence, the gates are described as **binary logic gates.**

To understand the digital system, the technician must completely understand the inputs and outputs of the different logic gates. The basic types of logic gates found in the digital system are the AND gate, the OR gate, the NAND gate, the NOR gate, and the inverter.

The five basic gates are constructed from various different components. Some of these components are diodes and transistors. The combination of these components form the various logic gates. The most popular type of gate construction is an arrangement of transistors. This type of assembly is identified as **transistor-transistor logic,** or **TTL.** Another arrangement of components is called **diode-transistor logic,** or **DTL.** In many of the newer circuits, a circuit constructed from **metal-oxide semiconductor** material is identified as **MOS.** These various different constructions of the integrated circuits are found throughout consumer products.

AND Gate

The **AND gate** is sometimes called the *all-or-nothing gate.* Figure 13–6 shows this basic gate using switches. Notice that in order to light L_1, both S_1 and S_2 must be closed.

The AND gate is constructed of transistors and diodes, and is packaged inside an integrated circuit. The operations required within the IC to develop the output will not be of concern to the technician, but the inputs and outputs must be remembered. Figure 13–7 shows the logic symbol most often associated with the AND gate.

The term *logic* generally refers to logic gates and truth tables in making decisions. A logic gate, then, is a circuit that can make a decision. To make this decision, certain input data must be given. For the AND gate to give a "yes" output, both inputs must be "yes." If for any reason one or both of the inputs of the gate are "no," then the output is also "no." However, in digital circuits the input is not "yes" or "no." The input is either a 1 or a 0. Therefore, to develop a

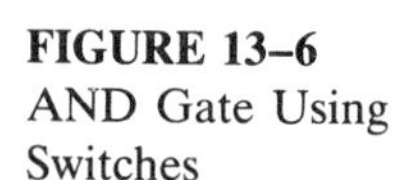

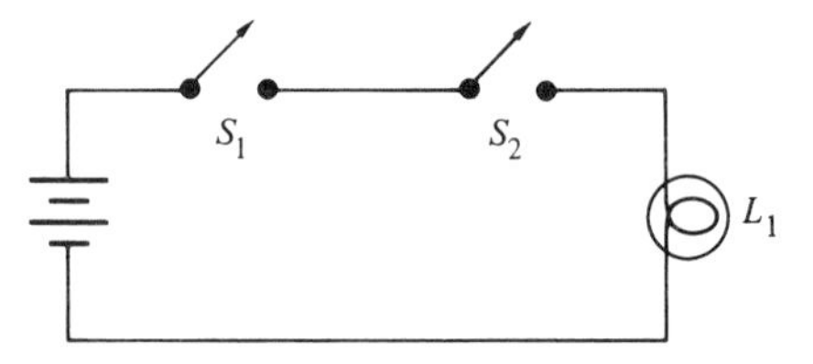

FIGURE 13–6
AND Gate Using Switches

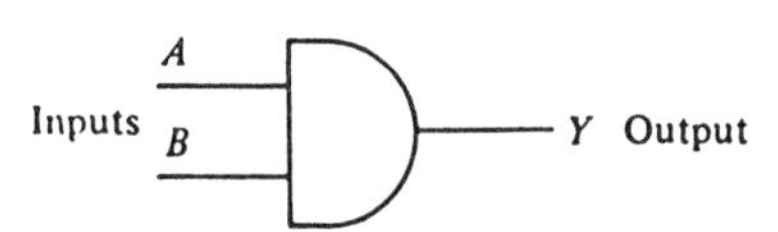

FIGURE 13–7
Logic Symbol for a Two-Input AND Gate

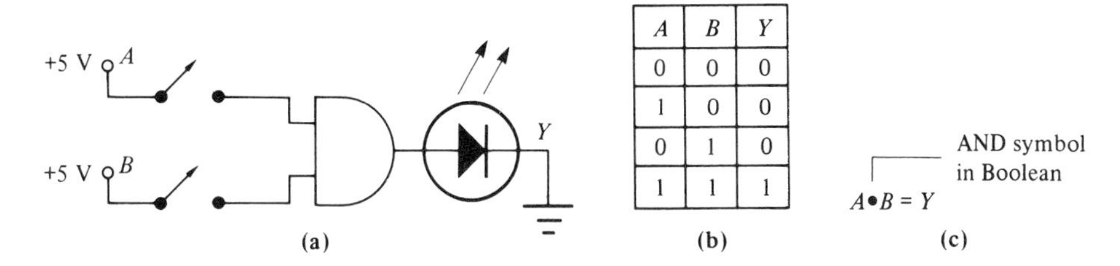

A	B	Y
0	0	0
1	0	0
0	1	0
1	1	1

FIGURE 13–8
Two-Input AND Gate: (a) Schematic Diagram, (b) Truth Table, (c) Boolean Expression

1 output, both inputs must be 1s. In any other state, the output will be a 0.

This 1 and 0 input sequence of the AND gate can be assembled into a logical arrangement of 1s and 0s. This arrangement is called a **truth table.** The truth table is a form of data giving all the possible combinations of inputs and corresponding outputs. The truth table and the actual circuit operation should be identical. Figure 13–8a illustrates a simple two-input AND gate, and Figure 13–8b illustrates the truth table for the AND gate. When a high, or 1, is moved from one input to the other input, a sequence of different outputs is developed. Remember that 1s and 0s are actually +5 V and 0 V, respectively.

On the top line of the truth table, both *A* and *B* inputs have 0 applied, and there is no output developed. The LED is off. In the second line of the truth table, the conditions are changed. Input *A* is now set to 1, and input *B* remains at 0. Still no output. In the third line, *A* is set to 0, and *B* is set to 1. Still the LED is off. In the fourth line, both *A* and *B* are set to 1. The LED is now turned on.

The final operation of the AND gate is to explore the algebraic form for the logic AND gate. This algebraic form is called a **Boolean expression,** which is the algebra of logic circuits. To express the AND gate in Boolean algebra form, a dot (·) is placed between the *A* and *B* input terms. Figure 13–8c shows the Boolean expression for the AND gate. This would be read: "*A* and *B* equal *Y*."

OR Gate

Another basic logic gate is the **OR gate.** This gate is sometimes called the *any-or-all gate*, which means that only one of the inputs must be logic 1 to give an output. Figure 13–9 shows the connection of switches representing the OR gate. Notice that S_1 and S_2 are connected in parallel. Therefore for an output to be developed, only one of the switches must be closed.

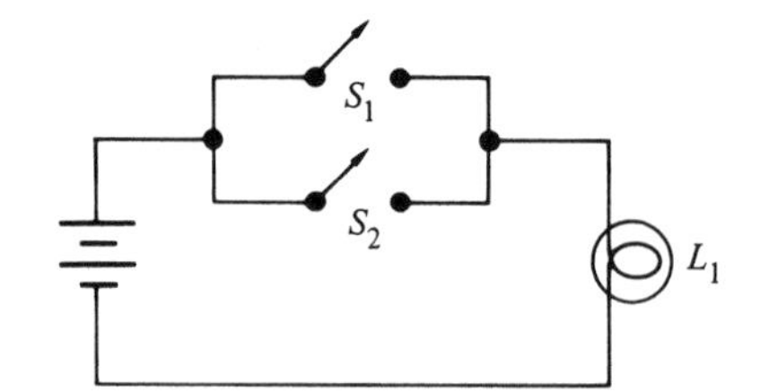

FIGURE 13–9
Basic Circuit Representing an OR Gate

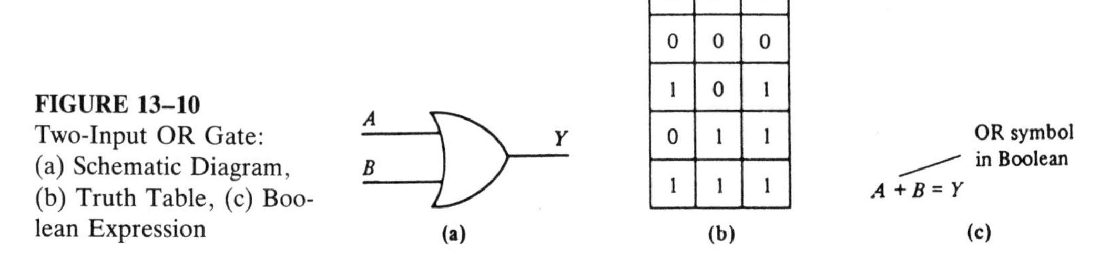

A	B	Y
0	0	0
1	0	1
0	1	1
1	1	1

FIGURE 13–10
Two-Input OR Gate: (a) Schematic Diagram, (b) Truth Table, (c) Boolean Expression

Figure 13–10a shows the logic symbol most often used with the OR gate. The truth table for the OR gate is given in Figure 13–10b, and the Boolean expression for the OR-gate function is shown in Figure 13–10c. The Boolean expression shown in Figure 13–10c is read as "*A* or *B* equals *Y*."

Inverter

The AND gate and the OR gate both have at least two inputs and one output. The **inverter** circuit has only one input and one output. The inverter circuit is sometimes called the *NOT circuit.* The job of the inverter is to invert the input. For example, if a logic 1 is at the input, the inverter develops a 0 at the output.

The logic symbol for the inverter is given in Figure 13–11a. Note the circle at the output of the symbol. This circle signifies that the output is inverted from the input. Shown in Figure 13–11b is the Boolean expression for the inverter function. Notice the bar above the output. This expression is read: "Not *A*."

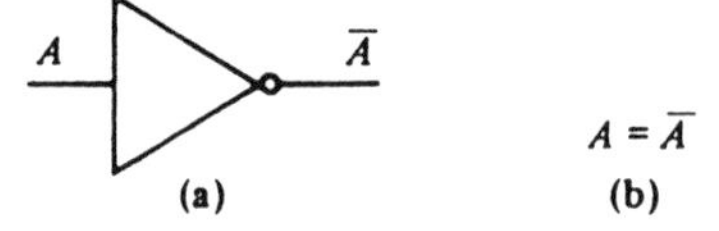

FIGURE 13–11
Inverter: (a) Logic Symbol, (b) Boolean Expression

A	$\overline{A}$
0	1
1	0

FIGURE 13–12
Truth Table for the Inverter

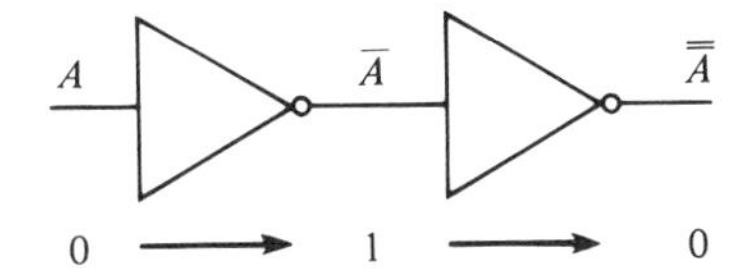

FIGURE 13–13
Double Inversion Developed through Two Inverters

The truth table for the inverter is given in Figure 13–12. Note that when the input is high, or 1, the output goes low. When the input is low, or 0, the output goes high. Again, the technician should remember the truth table function of the inverter. In some cases the input might be inverted twice, that is, double inverted. When this happens, the input is subjected to two inverters. The output of this logic arrangement is shown in Figure 13–13. Notice that the double inversion causes the output to equal the input.

The inverter is found throughout digital circuit applications. Knowledge of the inverter logic symbol, its truth table, and its inputs and outputs will help the technician in troubleshooting the system.

NAND Gate

The AND gate, the OR gate, and the inverter are the three basic gates used in logic circuits. From these gates, other gates can be developed for circuit application. The **NAND gate** is an inverted AND gate. The standard logic symbol used for the NAND gate is found in Figure 13–14a. Note that the NAND gate is an AND gate with a circle on the output. Therefore any output from this gate is inverted. Figure 13–14b shows the truth table for the NAND gate.

As do all logic circuits, the NAND gate has a Boolean expres-

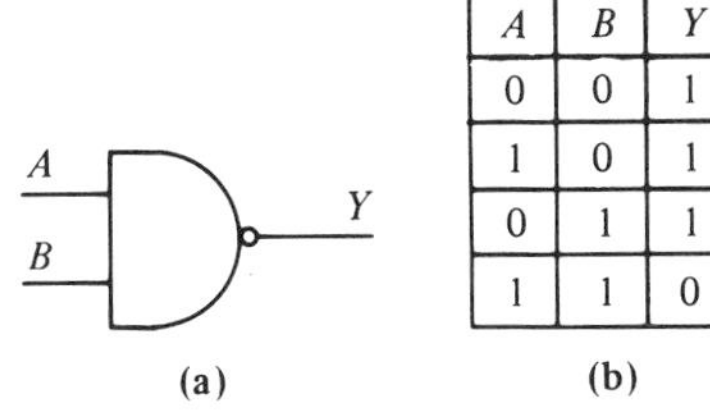

A	B	Y
0	0	1
1	0	1
0	1	1
1	1	0

(b)

FIGURE 13–14
Two-Input NAND Gate: (a) Logic Symbol, (b) Truth Table

sion. The Boolean expression for the AND gate is $A \cdot B$. Since the NAND gate has the output inverted, the Boolean expression is $\overline{A \cdot B}$.

The NAND gate is commonly found throughout industry, and is used extensively in digital applications. As will be shown later, NAND gates can be used in the place of the other logic gates.

NOR Gate

The **NOR gate** is developed by combining the OR gate and the inverter. The logic symbol used most often for the NOR gate is shown in Figure 13–15a. Note that the NOR-gate symbol is an OR-gate symbol with a circle on the output. This circle means that the output is inverted from the OR-gate operation. The NOR gate has a Boolean expression that is a combination of the OR-gate and inverter Boolean expressions. The Boolean expression for the NOR gate is $\overline{A + B} = Y$. The truth table for the NOR gate is shown in Figure 13–15b.

Multiple-Input Gates

Up to this point, all the gates that have been described are two-input gates. In many cases of digital operation, however, logic gates with more than two inputs will be required. Figure 13–16a shows a three-input OR gate. The Boolean expression for this gate is $A + B + C = Y$. With the addition of another input, the number of input combinations in the truth table has increased. Now the table will have eight possible combinations of inputs to develop outputs. The truth table for this gate is shown in Figure 13–16b. Notice that the output goes high whenever the input is high. This gate can be constructed by using two-input OR gates, as shown in Figure 13–16c.

In Figure 13–17 a four-input AND gate is shown. The logic diagram in Figure 13–17a shows the four inputs and the one output. This multiple-input gate can be constructed from two-input AND gates. Figure 13–17b shows an example of this connection. The truth table for this AND gate is given in Figure 13–17c. Note that with the addition of another gate, the number of possible input combinations has increased. Since this is still an AND gate, all the inputs must be high in order to develop output.

As has been shown, increasing the number of inputs to a gate increases the number of possible input combinations. In the three-input OR gate, there were eight possible combinations. With the four-input AND gate, the possible combinations were increased to 16. This increase can be expressed mathematically. The number of input combinations in a truth table is two raised to the power of the number of input gates. For example, to find how many lines in the truth table of

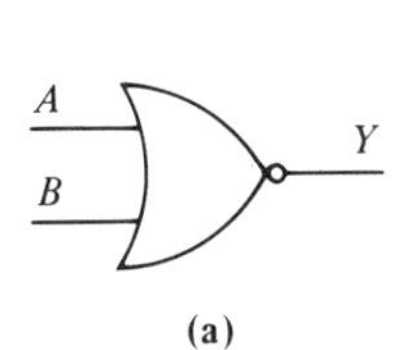

(a)

A	B	Y
0	0	1
1	0	0
0	1	0
1	1	0

(b)

FIGURE 13–15
Two-Input NOR Gate: (a) Logic Symbol, (b) Truth Table

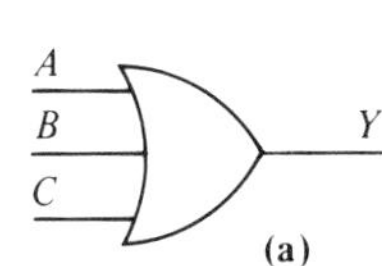

(a)

A	B	C	Y
0	0	0	0
1	0	0	1
0	1	0	1
0	0	1	1
1	0	1	1
0	1	1	1
1	1	0	1
1	1	1	1

(b)

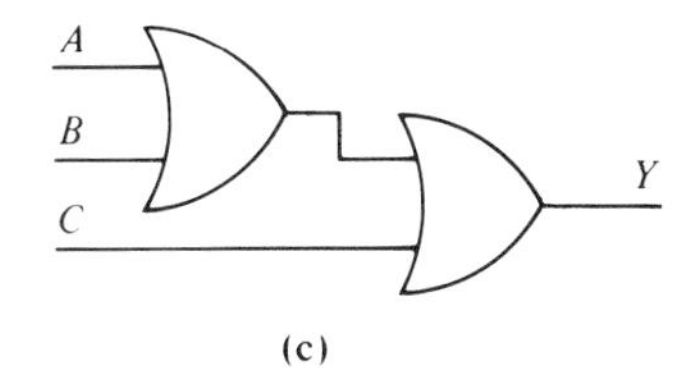

(c)

FIGURE 13–16
Multiple-Gate Input—Three-Input OR Gate: (a) Logic Symbol, (b) Truth Table, (c) Connection of Two-Input OR Gates to Form a Three-Input OR Gate

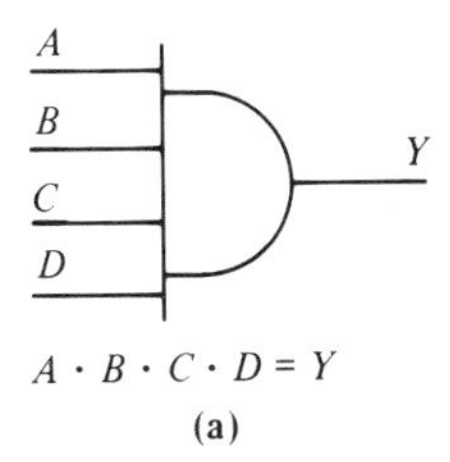

(a)

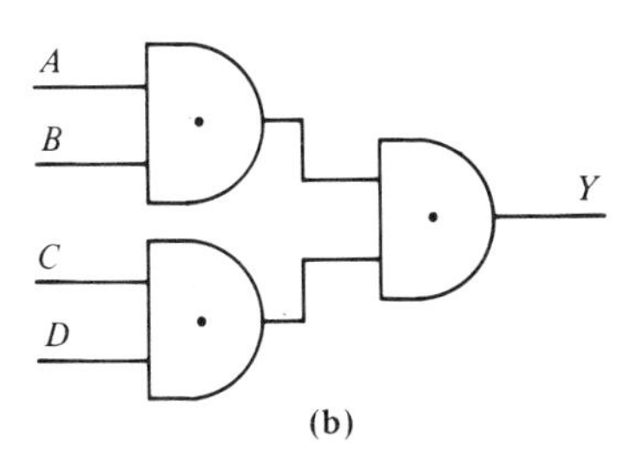

(b)

A	B	C	D	Y
0	0	0	0	0
1	0	0	0	0
0	1	0	0	0
1	1	0	0	0
0	0	1	0	0
1	0	1	0	0
0	1	1	0	0
1	1	1	0	0
0	0	0	1	0
1	0	0	1	0
0	1	0	1	0
1	1	0	1	0
0	0	1	1	0
1	0	1	1	0
0	1	1	1	0
1	1	1	1	1

(c)

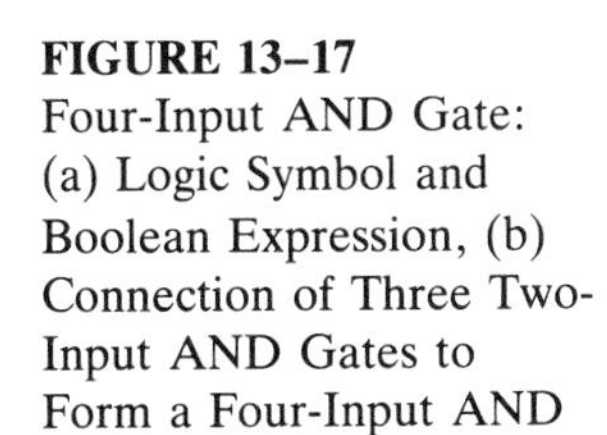

FIGURE 13–17
Four-Input AND Gate: (a) Logic Symbol and Boolean Expression, (b) Connection of Three Two-Input AND Gates to Form a Four-Input AND Gate, (c) Truth Table

a three-input AND gate, two is taken to the third power. That is, $2^3 = 2 \times 2 \times 2 = 8$. Thus, the truth table will have eight lines.

NAND-GATE LOGIC

Earlier in the chapter it was mentioned that NAND gates could be wired to serve as AND gates, OR gates, NOR gates, and inverters. This combining of NAND gates is called **NAND-gate logic.** NAND gates are widely employed by industry because they are easy to use and readily available.

The chart in Figure 13–18 shows the logic symbols for the four gates and the wiring diagrams for the comparable NAND-gate circuits. The technician should remember that these NAND-gate circuits represent the basic logic gates. The truth tables and inputs are the same as would be found for the basic gate operation. Because of this wiring capacity, the NAND gate is referred to as the *universal gate.*

FLIP-FLOPS

In certain circuit applications it may become necessary to change logic states between the input and output of a circuit. Other times a frequency may have to be counted. To develop these circuit conditions, the **flip-flop circuit (FF)** is employed.

Several types of flip-flops are used in digital operation. The basic types are the *RS*, *D*, and *JK* flip-flops. Flip-flops can also be classified as either asynchronous or synchronous. **Asynchronous** means that the flip-flop transfers data from the input to the output out of step with a clock pulse. **Synchronous** means the transfer is made in step with a clock pulse. The clock pulse is a series of pulses that enters the flip-flop to turn it on and off.

Because there are so many different kinds of flip-flops, the operation of one kind—the ***JK* flip-flop**—will be considered in detail here, and used to represent all flip-flop operation. The *JK* flip-flop is a good choice, because it is the most widely used in consumer products, as well as being common in many industrial applications.

JK Flip-Flop

The logic symbol for the *JK* flip-flop is found in Figure 13–19a. Notice that the input is found on the terminals labeled *J* and *K*. A clock pulse is required to operate this flip-flop, and its input is found on CLK. The outputs for the flip-flop are found on the terminals labeled Q and $\overline{Q}$. The "not Q" output is always a complement of Q output.

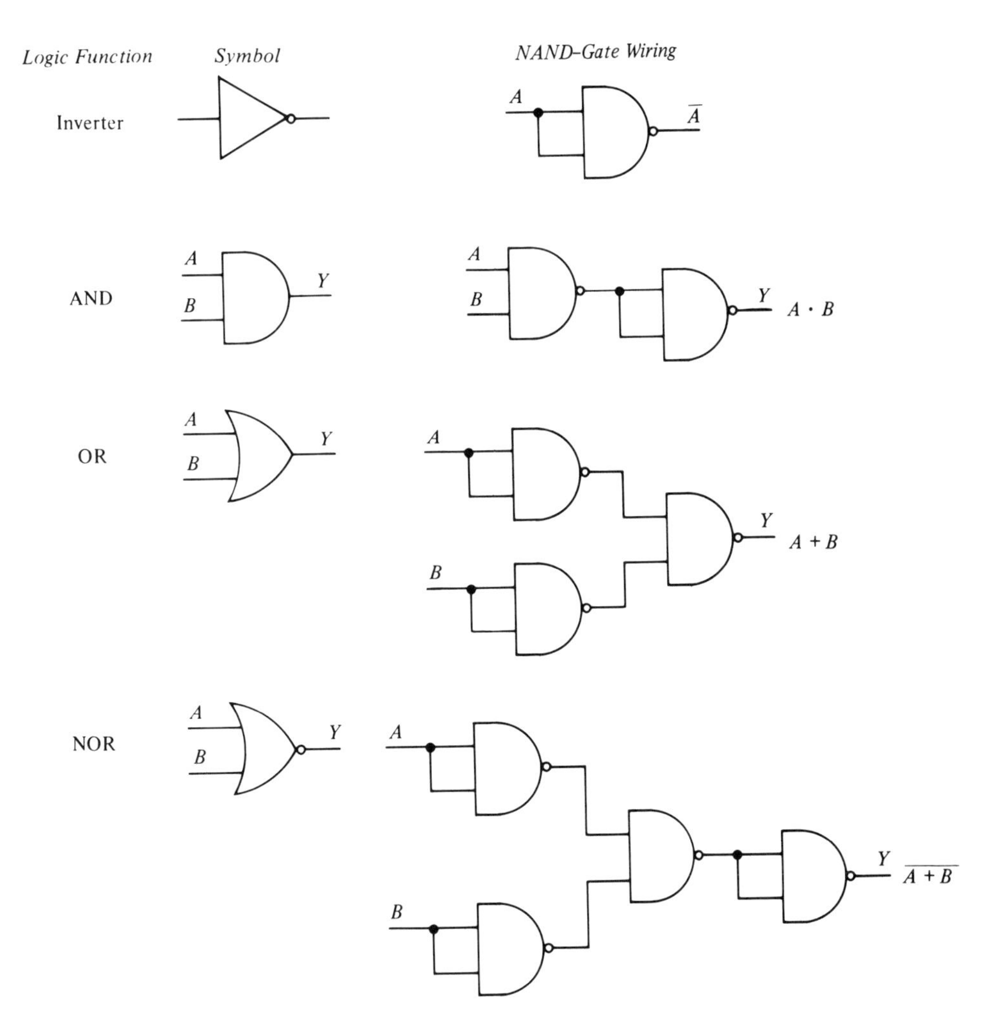

FIGURE 13–18
Basic Logic Gates Using NAND-Gate Logic

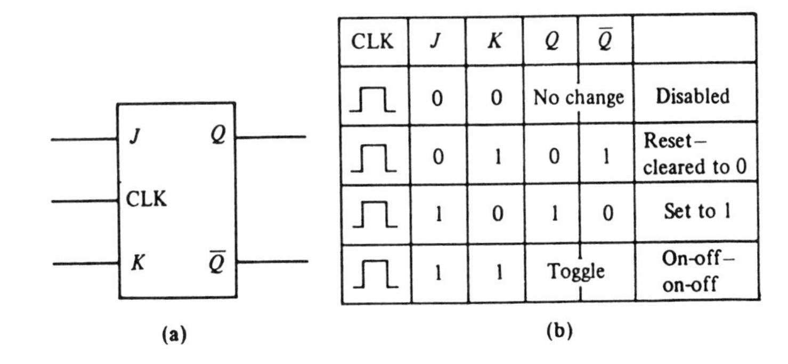

CLK	J	K	Q	$\overline{Q}$	
	0	0	No change		Disabled
	0	1	0	1	Reset—cleared to 0
	1	0	1	0	Set to 1
	1	1	Toggle		On-off—on-off

FIGURE 13–19
JK Flip-Flop: (a) Logic Symbol, (b) Truth Table

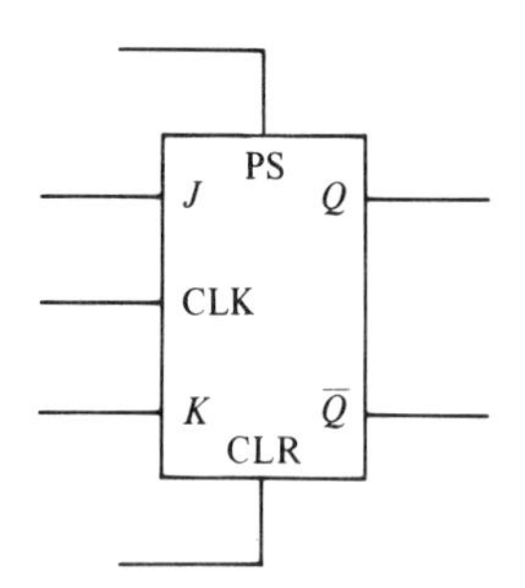

FIGURE 13–20
Addition of the Clear and Preset Terminals to the *JK* Flip-Flop

Preset and clear terminals override the synchronous operation.

The truth table in Figure 13–19b will be helpful in illustrating the operation of the *JK* flip-flop. As seen on the first line, when the clock pulse is high, and J and K are 0s, the flip-flop is said to be *disabled.* This means that the outputs remain in the previous state. Next the clock pulse is high, and J and K inputs are changed to $J = 0$ and $K = 1$. Notice that the outputs are changed; here $Q = 0$ and $\overline{Q} = 1$. The flip-flop is said to be *cleared*, or reset. The cleared condition occurs when Q is set to 0. In the next line, $J = 1$ and $K = 0$, and the outputs are changed. This state is called the *set* state. In the last line of the truth table, J and K are both set to 1. In this state the flip-flop is said to be in the *toggle* position. This means that Q and $\overline{Q}$ will alternately change states when the high clock pulse arrives at the input. Because data are transferred from input to output only when the clock pulses are high, this flip-flop is a synchronous flip-flop.

Figure 13–20 shows another logic symbol for the *JK* flip-flop. This flip-flop has two additional terminals. These terminals are identified as **preset (PS)** and **clear (CLR).** The main function of these terminals is to override the synchronous operation of the flip-flop. When these terminals are set to 1, they turn the flip-flop into an asynchronous mode of operation. These two inputs are used in counter and divider circuits.

Counters

Counters are used in almost all complex digital circuits. The main function of the counter is to count a series of events or periods of time, or to put events in sequence. For this reason counters are commonly found in the frequency divider circuits of consumer products.

There are many different types of counter circuits used in digital systems. These counters are identified by the task they perform. One counter circuit is called a **ripple counter.** The ripple counter counts the number of clock pulses that arrive at the input. Figure 13–21a shows a ripple counter that counts from 0 to the binary number 1111.

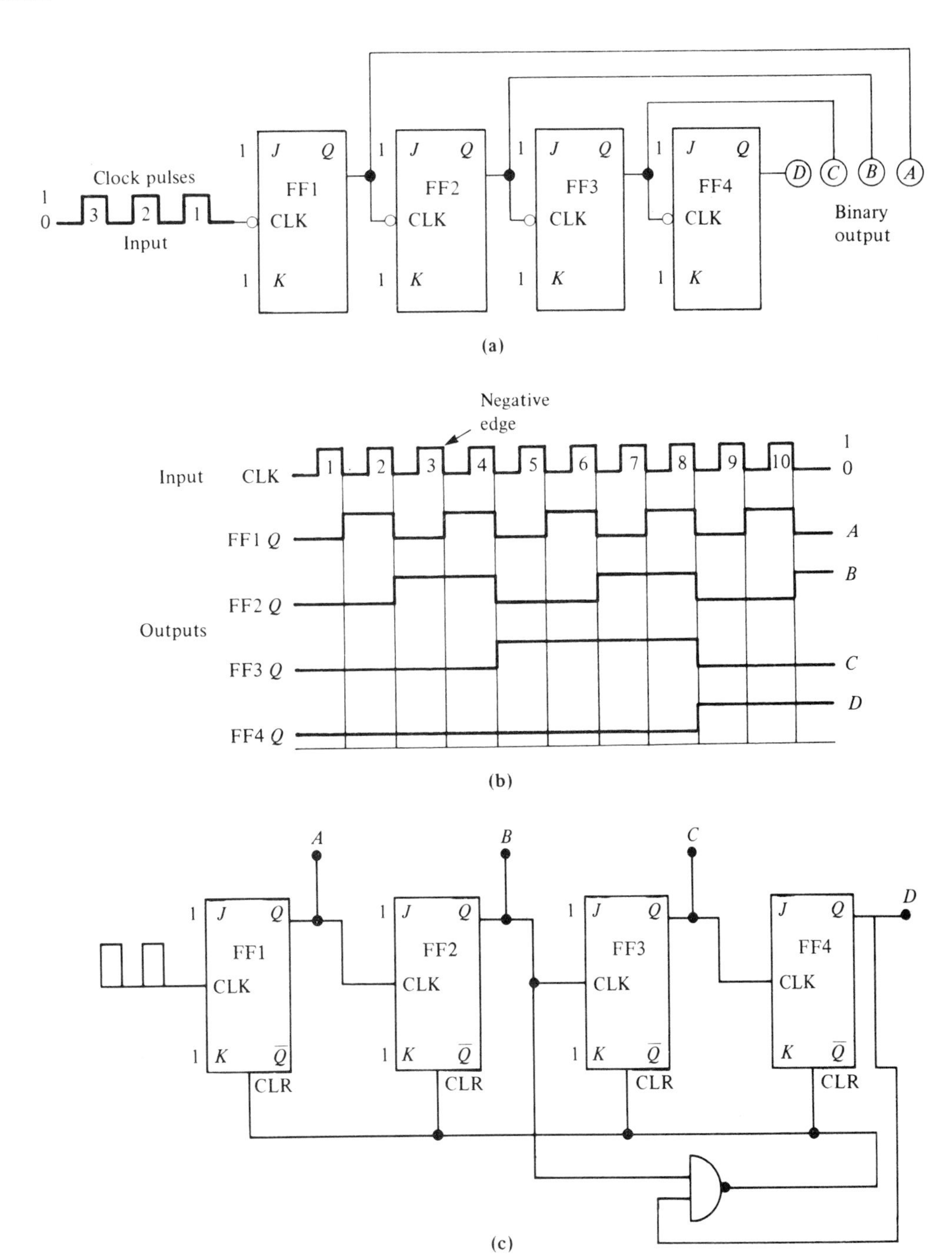

FIGURE 13–21
Schematic Connection for a Ripple Counter That Counts from 0 to 10 and Stops: (a) Schematic Diagram, (b) Clock Pulses Sequence, (c) Up-Counter Stopping at Decimal 10 Count

An important characteristic of a counter circuit is its output waveform diagram. Notice the waveform diagram in Figure 13–21b. As shown, the clock pulses enter flip-flop 1. Flip-flop 1 is set to toggle by setting J and K to 1. It is set to transfer the data to the output on the back side of the input clock pulse. Flip-flop 2 does not give output until the negative edge of the flip-flop 1 pulse is developed at the input. The count continues until each flip-flop is sequenced. Notice that the final output is not seen until the negative edge of flip-flop 3 is developed. This means that flip-flop 1 must toggle on each pulse. Flip-flop 2 toggles only one-half the time, and each flip-flop thereafter toggles during the negative edge of each pulse. This type of ripple counter is called a *modulo-16 counter.*

Other types of counters can be developed. By using the synchronous terminals, a counter can be made to stop at a specific number. Figure 13–21c shows an example of a counter that will count to 10 in binary, and then stop counting and clear itself to 0. This type of counter is used in frequency divider circuits.

As just shown, counters can be constructed from individual flip-flops. Manufacturers also produce integrated circuits, with all four flip-flops inside one package. Some manufacturers include the NAND gate inside the package.

Many different types of counters are used in industry. Figure 13–22a shows an example of a 4-bit binary counter. Here the four JK flip-flops and NAND gate are included in the package, as illustrated in Figure 13–22b. The external connection to the input of the NAND gate allows the technician to establish whatever type of counter circuit is needed.

Frequency Dividers

Many consumer products require frequency dividers. For example, in the vertical countdown circuits in the newer television receivers, a fixed frequency is divided down to keep the vertical and horizontal sections in sync. Frequency divider circuits are also found in PLL circuits.

An example of the simple frequency divider circuit can be seen in digital clocks. For the clock to operate in the proper timing sequence, the clock takes 60 Hz and divides it down to a 1 s rate. The 60 Hz input frequency is from the power line. The 60 Hz sine wave is converted into a square wave. The frequency divider circuit must take the 60 Hz square wave and divide it into 1 Hz pulses. These 1 Hz pulses have a time period of 1 s. Figure 13–23 shows the block diagram of this action. Once the frequency is divided, the output functions as a second timer.

The process of frequency dividing is illustrated in Figure 13–24a.

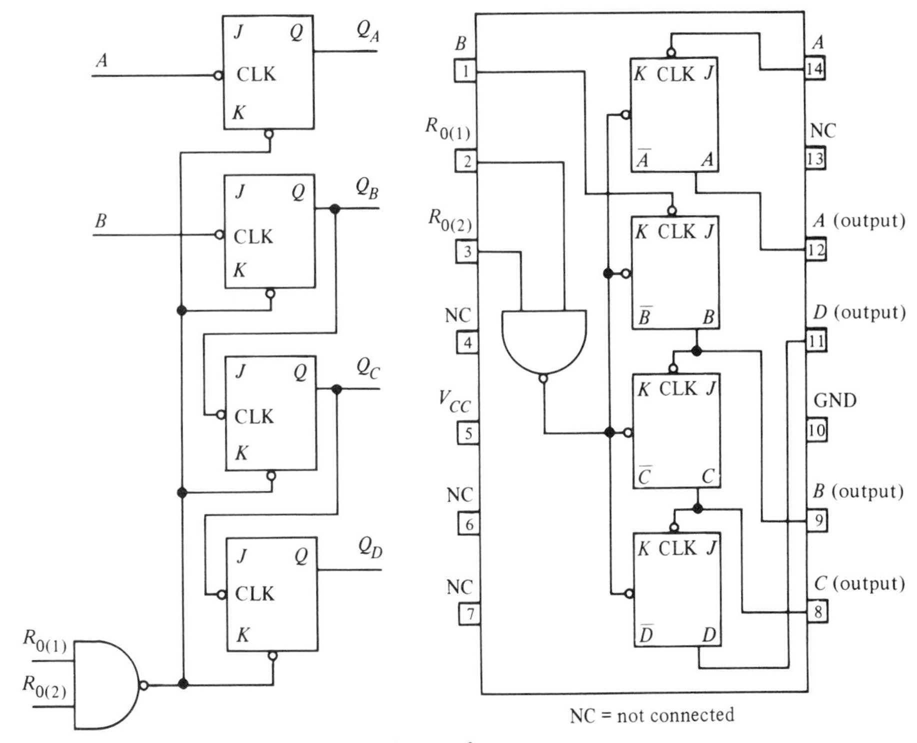

The J and K inputs, shown without connection, are for reference only, and are functionally at a high level.

(a) (b)

FIGURE 13–22
Four-Bit Binary Counter Contained in One IC Chip Package: (a) Internal Connection of Four *JK* Flip-Flops, (b) Housed in One IC Package

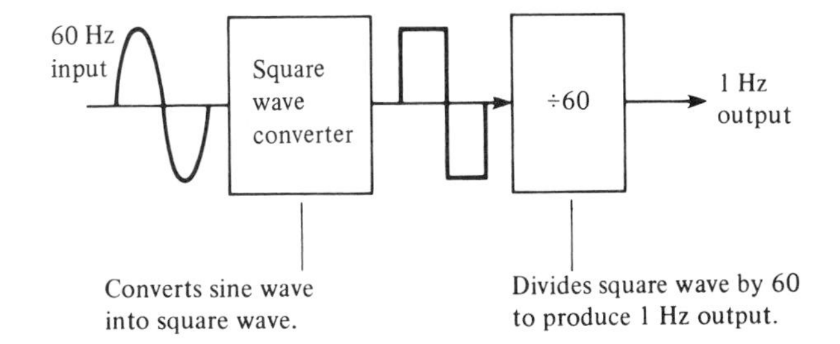

FIGURE 13–23
Basic Frequency Divider Action

The timing diagram for this divider circuit is shown in part b of Figure 13–24. Note that the input pulses from the clock are entering at a rate of 60 Hz, and being divided by 6, which gives an output of

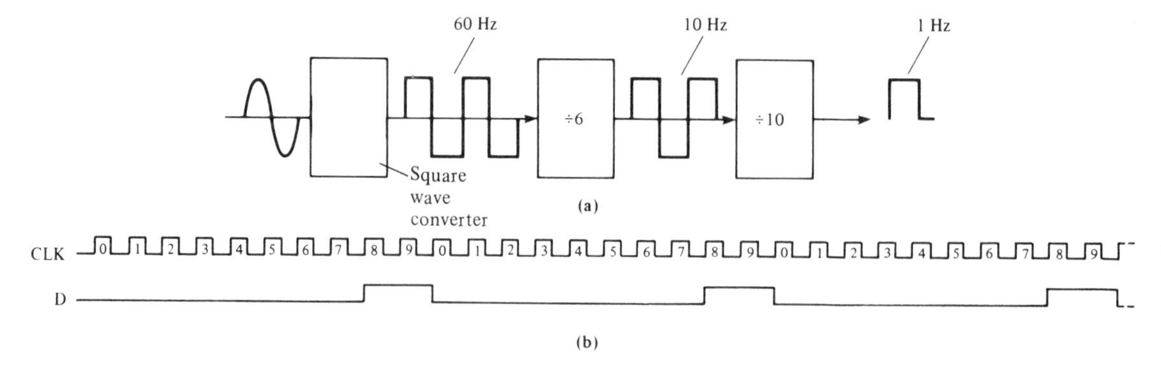

FIGURE 13–24
Dividing Frequency Down to 1 Hz: (a) Counter Connection, (b) Sequence of Clock Pulses for Divide-by-10 Counter

10 Hz. This 10 Hz frequency must be divided again to step down the frequency to 1 Hz. Notice that as the clock pulses enter the divide-by-10 counter, there must be eight pulses before the D output goes high. The D output remains high until the tenth pulse and then goes low. This action repeats itself over and over again, counting the thirty pulses shown down to three. The circuit is now dividing by 10; it is a *decade divider circuit.*

To further divide the frequency, other counters are connected to the output of the decade counter. If a modulo-6 counter is connected to the output of the decade counter, the frequency will be divided again, this time by 6. The series connection of the decade counter and the modulo-6 counter results in a total divide of 60. Therefore the 60 Hz input is divided by 60 to produce the 1 Hz output. This process is illustrated in Figure 13–24.

DIGITAL CLOCKS

Much of the action generated within digital systems comes from the digital **clock.** A digital clock is a circuit that generates a square wave. This square wave generates pulses between highs and lows. The highs and lows are necessary for the digital operation. Also of importance is the time interval between the high and low states.

A reliable source of clock pulses is the astable multivibrator. Figure 13–25 shows an example of this type of circuit. An output of clocking pulses can be obtained from either collector. The frequency of this clocking pulse is developed on base resistors R_1 and R_2, and capacitors C_1 and C_2. If the base resistors are made equal, and the capacitors are made equal, the waveform at the output will be sym-

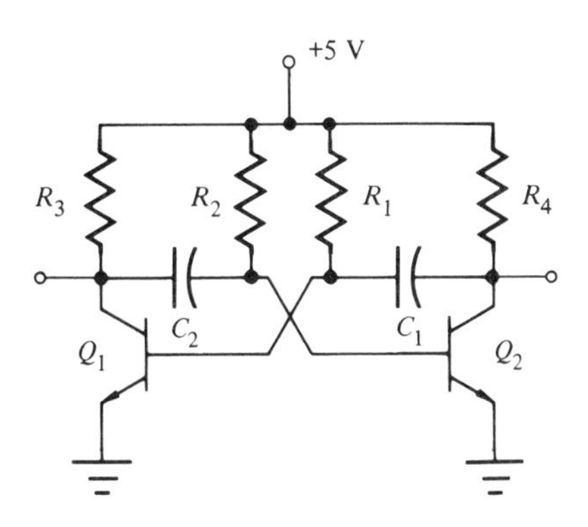

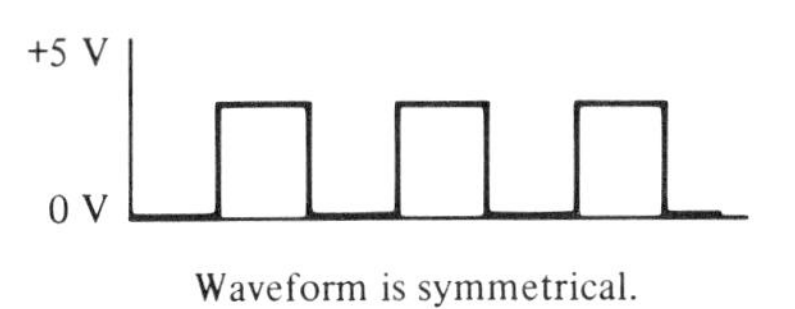

FIGURE 13–25
Astable Multivibrator Being Used as a Digital Clock

metrical, as shown in the figure. The frequency of the output can be calculated from Equation 13–1.

$$f = \frac{1}{1.4R_1C_1} \qquad \textbf{(13–1)}$$

From this equation it can be seen that the output frequency decreases with larger values of base resistance or capacitance.

The astable multivibrator is not the only circuit used to generate a square wave. An IC device called the **Schmitt trigger** will also deliver a square wave output. As the voltage at the input of the Schmitt NAND gate goes high, the voltage output goes low, and as the voltage input goes low, the voltage output goes high. Typically, the Schmitt trigger will recognize any voltage above 1.7 V as a high, and any voltage below 0.8 V as a low.

Figure 13–26 shows the logic symbol for the Schmitt trigger, and the accessory components necessary to develop square wave output. When a supply voltage is applied to this circuit, C_1 has a zero voltage level. This creates a low at the input. The low at the input drives the output voltage high. Typically, this high-state voltage is about 3 V. This high voltage is fed back through the 330 Ω resistor, causing C_1 to charge. Once C_1 has charged to the applied voltage, the input takes on a high voltage level. This then causes the output to be driven low. Capacitor C_1 begins to discharge, and brings the high voltage level at the input low again. This constant charging and discharging of C_1 develops a square wave output. The frequency of the

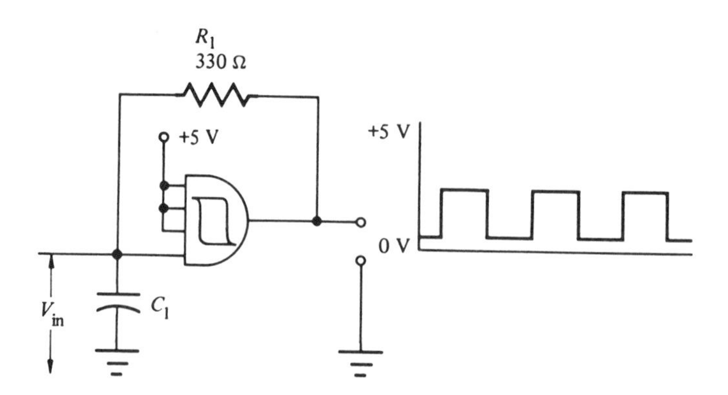

FIGURE 13–26
Schmitt NAND Gate Being Used for the Clock Circuit

output waveform depends upon the capacitor. The time constant for this circuit is controlled by the resistance and capacitance. Typically, the time range for this circuit is from 0.1 Hz to 10 MHz.

The Schmitt NAND gate is also used to convert sine waves into square waves. These circuits are sometimes called **wave-shaping circuits.** As the sine wave drives the input voltage above and below the input voltage levels, the output waveform is turned into a square wave.

Another type of clock circuit found in many systems is the **555 timer** integrated circuit. This flexible IC chip can perform many operations within digital systems. In application, the 555 timer operates on the charging and discharging of capacitors. Figure 13–27 shows a typical wiring diagram of the 555 timer circuit.

When 5 V are applied to this circuit, C_1 begins to charge through R_1 and R_2. Because C_1 has zero voltage, when it begins to charge, it causes the output to go high. Once the capacitor has charged to about

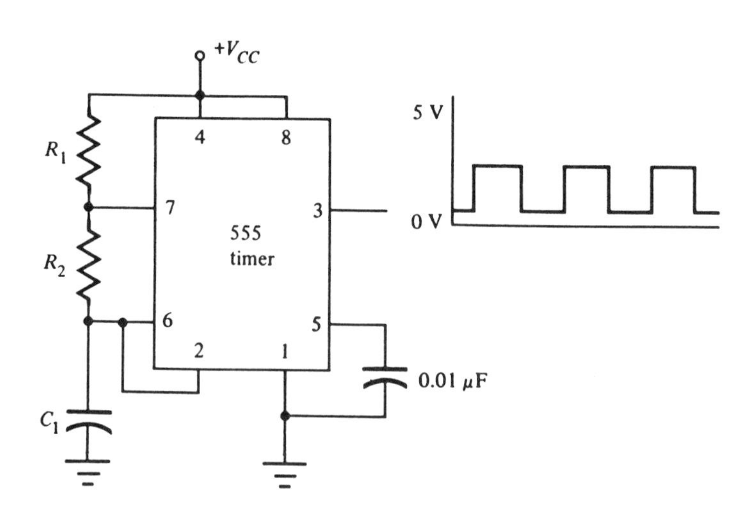

FIGURE 13–27
Typical Wiring Diagram for a 555 IC Used as the Clock Circuit

two-thirds of the applied voltage, the output is switched to low. The internal circuitry of the 555 timer causes C_1 to discharge. Once the voltage level falls to approximately two-thirds of the applied voltage, the output again switches. This constant charging and discharging causes the output to be developed into a square wave.

Since the output frequency of the 555 timer is determined by the rate of charging and discharging of the capacitor, the output frequency can be adjusted. This is accomplished by changing R_1 and R_2, or by changing C_1.

It must be remembered that the clock pulse is the heart of most digital systems. Without proper clock pulse input, counters, displays, and other digital operations will not be able to perform. The service technician must be aware of the devices used to develop the square wave clocking pulse, and also of the components necessary for development of timing.

MEMORY

Many new consumer products have the capacity of **memory.** For example, a videotape recording system can be programmed to videotape specific programs, for a total time of six days. New microwave ovens can be programmed to turn on, and then defrost, cook, and keep a meal warm until served. All of these different applications come from memory found in digital circuits. Memory circuits are divided into two basic types: those that can be programmed by the user, and those that have been programmed by the manufacturer.

One type of memory is called **random-access memory (RAM).** The RAM is memory that can be programmed. After the memory has been stored, it can recall the information that was stored. The information is stored as 0s and 1s.

Information is stored in the RAM in *lines* of memory. Each line of the memory represents a *word.* Each word stored in the sample system will have four **bits** of information. Information can be taken from the RAM by simply addressing the RAM word so that the bits of information are recovered.

One disadvantage of the RAM is that it requires a voltage to keep the information in storage. When the voltage is removed, all the information stored is erased. The data must be reprogrammed after power is restored to the RAM. Because of this characteristic, the RAM is said to be a *volatile* memory. Volatile memories are used only for temporary storage of information.

Another form of memory is called **read-only memory (ROM).** Here the bits of information have been programmed in by the manufacturer. The bits of information stored are permanently placed into the memory. It is not possible to write in new information; hence, the name read-only memory. The memory placed in storage in this IC is not erased when power is removed.

As was mentioned, the ROM is programmed by the manufacturer. A custom-made ROM is very expensive, and takes a long time to be manufactured. A cheaper form of read-only memory is **programmable read-only memory (PROM).** The PROM can be programmed by following the manufacturer's directions. Once this IC has been programmed, the information stored cannot be erased. A basic characteristic of both the ROM and the PROM is that they contain *nonvolatile* memories.

The memory systems just described are but a few of the many different types of memory units found in consumer products. To be able to troubleshoot those new components that contain these devices, the technician must be aware of their operation.

TROUBLESHOOTING

DIGITAL CIRCUITS

Important changes in troubleshooting have reached the consumer products technician. In older electronic consumer products, the technician dealt with a linear signal, but in the newer products, both linear and digital signals are present. Digital circuits have the ability to divide, display, store information, and make the size of the unit more compact. All these new circuit applications have brought new troubleshooting problems.

As always, the technician must have knowledge of the system in order to troubleshoot it. Included in this is the knowledge of the operation of each block of the system, and from there, the operation of each component within the block. In repairing digital circuits, the technician must think of the ICs as *black boxes.* That is, the technician must be concerned only with the input and output of each of the components. If an input signal is present, but an output signal is not present, the component must be replaced. A faulty digital circuit in IC format cannot be repaired. Therefore if one circuit within the IC is bad, the entire chip must be replaced.

The test equipment necessary to repair digital circuits includes a voltmeter, an oscilloscope, and a digital logic probe. Remember that logic gates operate on high and low voltage levels. Therefore these test equipment devices will enable the technician to measure the inputs and the outputs of the digital circuits.

In addition to this test equipment, knowledge of the operation of each logic gate is important. The technician should remember the truth table for each of the basic gate operations. These truth tables predict the output from the gates. If the input logic does not create the proper output logic, the gate's logic is false. If the gate's logic is false, the gate must be replaced.

Some common gate failures in logic circuits are open pins, pins shorted to ground, pins shorted to dc supply, pins shorted together, and pins that develop an intermittent defect. All of these failures or problems are common to the TTL digital logic gate.

Open Pins

To troubleshoot an open pin problem, the dc input voltage and the dc output voltage must be measured. Figure 13–28 shows a sample NAND gate. This figure also contains the correct truth table and the typical output voltage for this gate. Remember that a gate that develops an open will always show a high voltage level. Therefore the truth table might show false readings. Checking the first line of the truth table, a low is applied to input *A*, and also to input *B*. The output from this action reads high. Thus this line of the truth table has proven true. The next line of the truth table must be checked. Therefore a low is applied to input *A*, and a high is applied to input *B*. However, because input *A* is open, it assumes the high condition. Therefore the input of the gate develops a high-high state. This would give a low output in the NAND-gate operation. This condition proves that this line of the truth table is false, and means that the IC chip must be replaced.

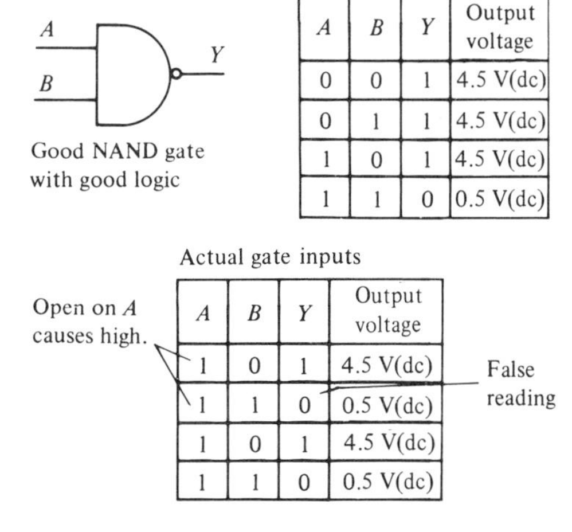

A	B	Y	Output voltage
0	0	1	4.5 V(dc)
0	1	1	4.5 V(dc)
1	0	1	4.5 V(dc)
1	1	0	0.5 V(dc)

A	B	Y	Output voltage
1	0	1	4.5 V(dc)
1	1	0	0.5 V(dc)
1	0	1	4.5 V(dc)
1	1	0	0.5 V(dc)

FIGURE 13–28 Testing of NAND Gate with One Open Input Lead

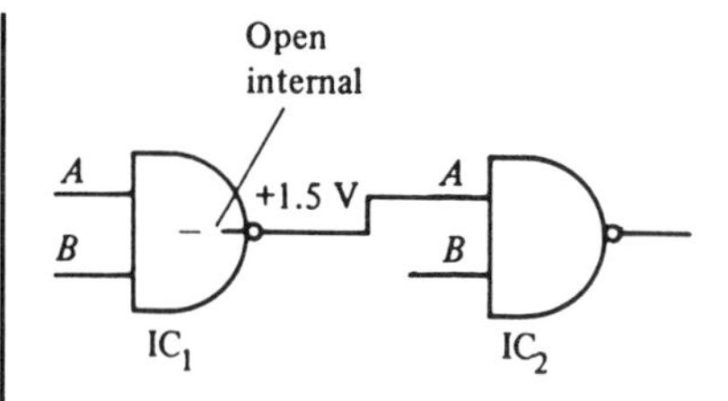

FIGURE 13–29 NAND Gate with Open Output Lead

Another problem that occurs in logic gates is the open output pin. Figure 13–29 shows an example of a NAND gate with an open output pin. As stated previously, a pin in TTL that is not connected always develops a high voltage. This open pin will show up as an open pin, and will always develop a high voltage. Because the open gate is internal to the pin, the entire chip must be replaced.

In the operation of TTL gates with floating pins, two considerations must be explored. Remember that pins floating in TTL operation are not connected.

1. The gate will function as though a high is applied to the floating pin.
2. A TTL floating input pin will drift gradually toward a lower level of voltage. Generally, the bad level of voltage will drift toward 1.5 V. This 1.5 V level will be higher than the normal low-voltage condition, for the chip, and lower than the normal high-voltage condition. The normal low level of voltages is about 0.4 V and the high-voltage condition is about 2 V.

Often the output of one gate is fed into the input of another gate. If the first gate has a bad or floating pin, this voltage will show up in the voltage of the next pin. Therefore it can be assumed that the feeding IC chip has the open output pin.

Shorted Pins

Shorted pins create an entirely different set of troubles in the integrated circuit than do open pins. Problems are created by three different shorting situations: pins shorted to the positive supply voltage, pins shorted to ground, and pins shorted to each other within the same logic gate.

To detect shorted pins, voltage measurements should be made around the IC chip. A shorted pin will change the voltage level at the pin, and also change the logic level for that pin. An important fact about the shorted pin is that it will always destroy the IC chip. The open pin will not damage the IC chip.

Figure 13–30 shows two NAND gates. In IC_2, input *A* is shorted to the dc power supply of the gate. When IC_1 develops a high output, it is fed to IC_2. At input *A* this high voltage will be higher than normal, because the high state from IC_1 and the dc power supply will add together. Because the voltage at IC_2 is higher than normal, this chip should be considered bad. In addition, this higher voltage level will be passed back to IC_1 output. This higher than normal voltage will cause IC_1 to go bad also. Thus, the output gate and the driving gate must be replaced.

When the gate of an IC chip is shorted to ground, the same thing will happen to the gate. It will be destroyed. However, the voltage on the gate will be different than when shorted to the dc power supply. When shorted to ground, the voltage at the input or at the shorted pin will be lower than normal. This results because the higher gate voltage will subtract from ground, and leave a lower voltage. If the input pins of the gate are shorted to ground, the driver gate will also go bad, and will have to be replaced.

Shorted output pins on a gate will cause different problems. Shorted output pins will always ruin the next stage. Generally, the next stage will develop direct shorts to ground. The next stage should be checked for permanent highs or lows at pins that should be changing.

Most digital circuits found in consumer products are mounted directly on the printed circuit board. Opens and shorts developed on the board itself will be easy to track down and repair. The opens or shorts that develop on the board will cause the same problems as those seen when interconnecting pins of the IC are shorted or open. These problems will often be easy to troubleshoot, because they are external to the gate.

Figure 13–31 shows an example of interconnected leads between two gates of the IC chip. Notice that an incorrect voltage is devel-

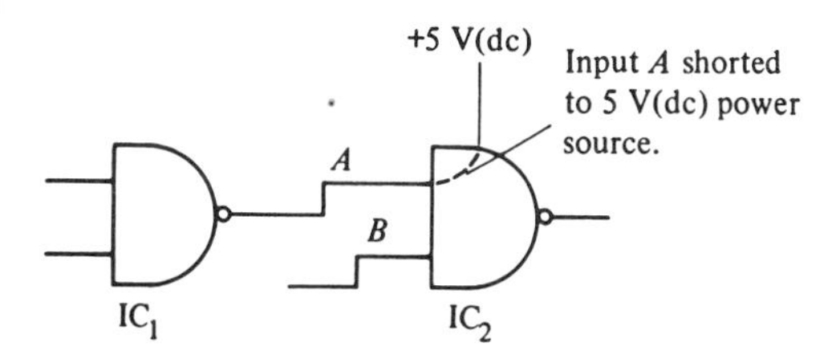

FIGURE 13–30 Bad Driver Stage Due to Shorted Input Pin

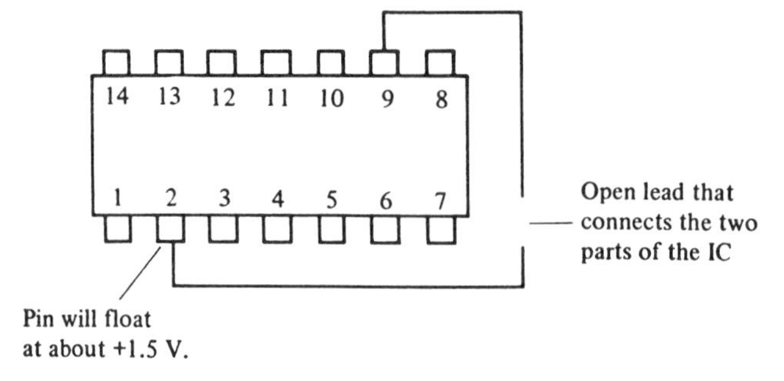

FIGURE 13–31 Open Developed between Interconnected Leads

oped on pin 2. Checking the schematic, it is seen that pin 2 is fed by pin 9. If the voltage levels found on pin 2 are bad, then pin 9 should be examined. The pin should be checked for cracked solder joints, cracked board, or solder bridges.

SUMMARY

Analog circuits are characterized by constant, varying voltages. Digital circuits are characterized by an on/off operation.

Digital circuits generally operate with square waves. Because the square wave has a sharp rise and a sharp fall in voltage, it will turn ICs on and off.

There are two logic levels in digital circuits: logic 0 and logic 1. Logic 0 corresponds to a logic low, and logic 1 corresponds to a logic high. Voltages for logic 0 are below 0.5 V, and voltages for logic 1 are above 2 V.

The binary number system is used in digital circuits. It is an excellent choice, because it is written with 0s and 1s. Decimal numbers can be converted into binary numbers.

An encoder is a digital circuit that converts decimal numbers into binary numbers. A decoder is a digital component that converts binary numbers into decimal numbers.

Logic gates are circuits used to process the information of 1s and 0s. There are five basic gate functions. They are the AND gate, the OR gate, the NAND gate, the NOR gate, and the inverter.

Truth tables are designed to predict what will happen when different combinations of highs and lows are placed on the input of the five basic gates.

Logic gates may have more than two inputs. As the number of

inputs is increased, the possible number of output conditions also increases.

The NAND gate is a popular type of logic gate. This gate can be combined with other NAND gates to form the other four basic logic gates. Because of this, the NAND gate is called the universal gate.

The flip-flop is used to change logic states. Flip-flops can also be used to divide frequencies. The types of flip-flops commonly found in digital circuits are the *RS* flip-flop, the *D* flip-flop, and the *JK* flip-flop. The *JK* flip-flop is the most popular.

Counters are used in digital systems to count pulses or to divide frequencies. Counters can be set to count to any number, or to divide any frequency.

Counters and other digital systems operate on a fixed known frequency. This frequency is usually generated by a clock. The clock is used to generate a square wave at an established frequency.

Memory in digital systems comes in two basic forms. Random-access memory (RAM) can be programmed to do repetitive types of tasks. The RAM must have voltage applied at all times in order to retain its memory. The other memory type is read-only memory (ROM). The memory in this device is burned into the circuit. The ROM will not lose its memory when power is turned off.

Because of the new digital systems found in consumer products, the service technician is seeing new problems. The voltmeter, oscilloscope, and logic probe are the best tools to use for troubleshooting digital circuits.

The main thrust to digital troubleshooting is that the technician have knowledge of the system and of the components in that system that make it operate.

KEY TERMS

analog circuit
AND gate
asynchronous
binary logic gates
binary numbers
bit
Boolean expression
clear
clock
decoder
digital circuit
diode-transistor logic (DTL)
encoder
555 timer circuit
flip-flop circuit
inverter
JK flip-flop
logic 1
logic 0
memory
metal-oxide semiconductor (MOS)
NAND gate
NAND-gate logic
NOR gate
OR gate
preset
programmable read-only memory (PROM)
Q

$\overline{Q}$
random-access memory (RAM)
read-only memory (ROM)
ripple counter
Schmitt trigger
square wave
synchronous
transistor-transistor logic (TTL)
truth table
wave-shaping circuits

REVIEW EXERCISES

1. What is the major difference between analog circuits and digital circuits?
2. What type of signal is generally found in a digital circuit?
3. Convert the following binary numbers to decimal numbers:
   - **a.** 1001
   - **b.** 1
   - **c.** 10
   - **d.** 1111
   - **e.** 101
   - **f.** 11110
4. Which types of digital circuits convert decimal numbers into binary numbers?
5. Which circuits found in digital circuits convert binary numbers into decimal numbers?
6. Draw an example of each of the five basic binary gates. Draw the truth table for each of these gates.
7. Write the Boolean expression for each of the five basic logic gates.
8. How many lines will there be on the truth table for a four-input AND gate?
9. Write the truth table for a three-input NAND gate.
10. Using only NAND gates, construct the following circuits:
    - **a.** AND gate
    - **b.** OR gate
    - **c.** Inverter
11. Draw a logic diagram for the *JK* flip-flop.
12. When the inputs of a *JK* flip-flop are set to a high, what state is the device in?
13. Define the terms *asynchronous* and *synchronous* as they apply to digital circuits.
14. What functions do counters serve in consumer products?
15. Draw a schematic diagram of a ripple up-counter that stops counting at 15.
16. Why are clock pulses needed in the newer vertical countdown circuits?
17. Which components in a digital circuit are used as wave-shaping circuits?
18. Draw a schematic diagram of a simple clock circuit using a 555 timer chip.
19. Which type of memory circuit will lose its memory when power is removed?
20. Which memory is nonvolatile?
21. What are some of the common failures found in digital circuits?
22. If a pin in a TTL circuit is left unconnected, what logic state will it assume?
23. If a TTL chip has an open pin, will the chip be damaged? Will it be damaged if the pin is shorted?

14 BASIC RECEIVER OPERATION

OBJECTIVES

Upon completing this chapter, you should be familiar with:

—Basic AM radio receiver circuits
—Basic FM monaural and FM stereo receiver operation
—Complex stereo signal development

INTRODUCTION

The major breakthrough in the communications industry came with the advent of the radio. The radio receiver was able to receive an electromagnetic wave, and convert this signal into audio signal at the output. The first radio receivers consisted of an antenna, a detector diode or crystal, and headphones. This setup, however, was quite restrictive. The radio's ability to select different radio signals was limited. Sensitivity was also limited in the early years. Then along came amplifiers and tuned circuits, both of which greatly improved the selectivity and the sensitivity, two very important characteristics of the radio receiver.

That first radio led to greater advances, such as FM and FM stereo reception, black-and-white and color television receivers, and mobile two-way communications.

To properly service these consumer electronic devices, the technician must have a complete understanding of their operation. To this end, the present chapter deals with the block diagrams of the AM-FM and FM stereo receivers. The circuits used to operate those blocks are also explored.

AM RADIO RECEIVER

All modern radio receivers operate on the principle of **heterodyning.** Hererodyning is the process of receiving a **radio frequency (RF)** signal and converting it to an **intermediate frequency (IF).** This intermediate frequency has the same information that was contained in the RF signal. Figure 14–1 shows a block diagram of the AM radio receiver.

The incoming RF signal is applied to the mixer, which also receives a signal from the **local oscillator (LO).** The mixer is very frequency selective. It receives only the frequency to which it has been tuned, and rejects all others. The local oscillator is tuned at the same time the mixer is tuned. The local oscillator is tuned to a different frequency than the mixer. The two stages are always separated by the IF frequency. In the AM radio, the local oscillator is tuned at a frequency of 455 kHz above the incoming RF signal. For example, if the mixer is tuned to a station operating within the AM band (540–1600 kHz)—say, at 1000 kHz—the local oscillator will be tuned to a frequency of 1455 kHz (1000 kHz + 455 kHz = 1455 kHz). Figure 14–2 shows an example of this action. If the mixer is then tuned to 800 kHz, the local oscillator is retuned to 1255 kHz. The 455 kHz difference between the two signals is maintained.

When the two frequencies are mixed together, four frequencies are developed at the output of the mixer. These four frequencies are the *original RF signal*, the *original local oscillator frequency*, the *sum frequency* (RF + LO), and the *difference frequency* (RF − LO). Figure 14–3 shows an example of the four-frequency output of the mixer

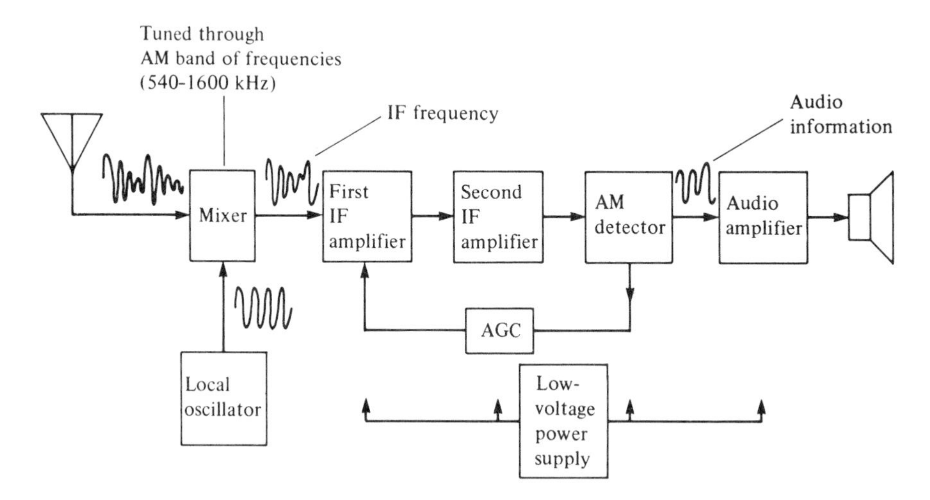

FIGURE 14–1
Block Diagram of an AM Radio Receiver

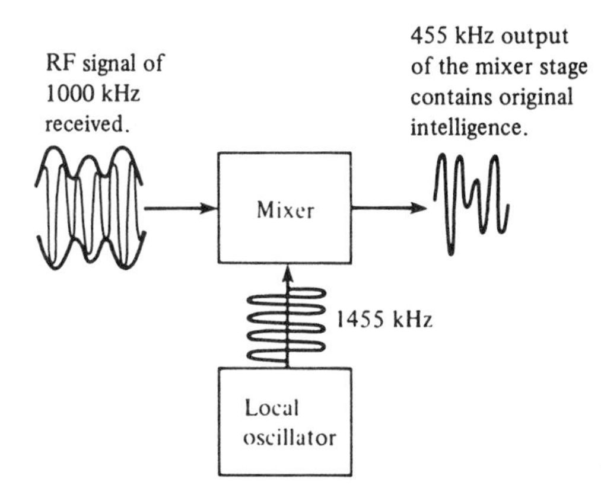

FIGURE 14–2
Tuning of the Mixer and Local Oscillator

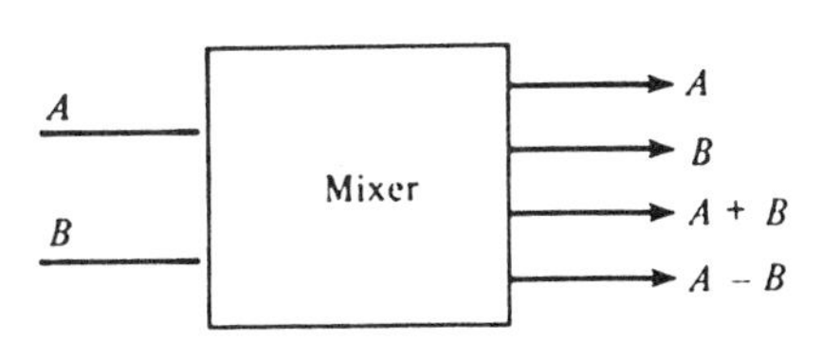

FIGURE 14–3
Four Frequencies Developed at the Output of the Mixer

stage. Notice that frequency A is the inputted RF signal, and frequency B is the inputted local oscillator frequency. Suppose that the mixer has been tuned to a station at 1000 kHz (the A input). The local oscillator would then be tuned to 1455 kHz. The outputs from the mixer would be as follows:

RF signal (A) = 1000 kHz
Local oscillator (B) = 1455 kHz
$A + B = 2455$ kHz
$B - A = 455$ kHz

The next stage is comprised of several IF amplifiers. These amplifiers are selective in their frequencies. They accept only one of the four outputs of the mixer. The output that is selected is the 455 kHz frequency. The other three frequencies are rejected by the amplifiers. Remember that the same intelligence found in the original RF signal has been transferred to the 455 kHz frequency. This frequency is called the intermediate frequency.

After the signal is amplified by several IF stages, the intelligence is removed from the IF carrier by the **detector.** The detector circuit is nothing more than a half-wave rectifier and a low-pass filter. It removes the high-frequency component, and leaves the audio intelligence. Because the intelligence is not large enough to drive a speaker, the signal is subjected to several audio amplifier stages, where the signal is

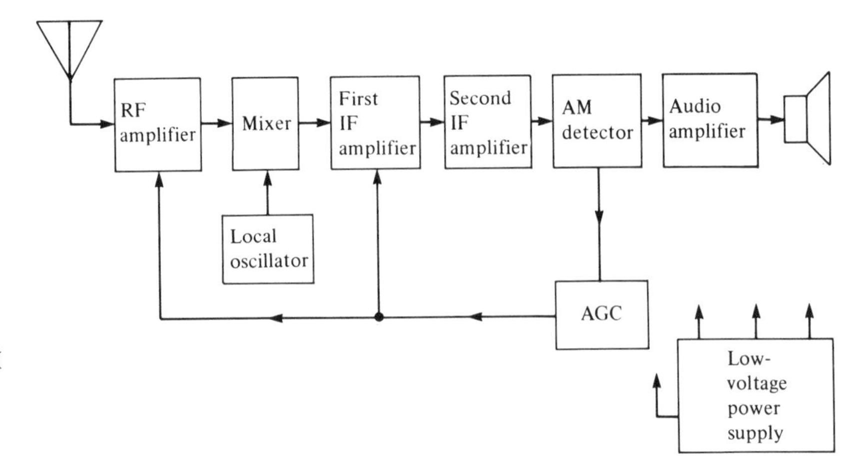

FIGURE 14–4
Block Diagram of an AM Receiver Containing an RF Amplifier and an AGC Circuit

increased in amplitude. Finally, the electronic signal is converted back into sound information at the speaker.

The simple block diagram shown in Figure 14–1 is for an inexpensive, throwaway type of AM receiver. This kind of receiver is limited in the number of stations it can receive, and is subjected to large amounts of noise. For the receiver to have good *selectivity* (the ability to distinguish between different stations) and *sensitivity* (the amount of signal required to give output), additional circuits must be added. These circuits are an RF amplifier stage and an automatic gain control (AGC) stage.

The **RF amplifier** stage is used to amplify the tuned incoming signal before the signal is mixed with the oscillator frequency. This block improves the selectivity and signal-to-noise ratio of the receiver.

The **automatic gain control (AGC),** or automatic volume control (AVC) is the next addition to this receiver. This circuit is used to develop a dc voltage whose amplitude is proportional to the signal strength delivered to the detector. This dc voltage is fed back to the IF section and the RF section of the receiver. The main purpose of the AGC is to cause the RF and IF amplification to be adjusted. If signal strength decreases, the AGC can cause the RF and IF stages to increase their gain of the signal. Figure 14–4 is a block diagram showing the addition of the RF amplifier and the AGC.

RF Amplifier Circuit

The main purpose of the RF amplifier stage is to amplify the RF signal received from the antenna. The RF amplifier stage has four requirements: sufficient stage gain, low internal noise, sufficient selectivity, and linear characteristics through the AM band of frequencies. The overall gain of the stage is very important. The amplification of this stage depends upon the strength of the incoming signal. The lower the signal input amplitude, the more this stage will have to amplify. This gain increase is accomplished by the AGC circuit, which will be discussed later in the chapter.

Figure 14–5 shows an example of a typical RF circuit found in many radios. The RF signal is coupled from the antenna through a hash choke, L_1. This choke filters out some of the unwanted interference. It is then transformer-coupled through L_{2A} into the base of the RF amplifier, Q_1. The signal is developed across the input leads of this amplifier—the base and the emitter. The amplified signal then appears across the output leads of the collector and the emitter. The amplified RF signal is coupled through C_{21} and L_{2B} to the mixer stage of the receiver.

The main current path in this RF stage is from ground through

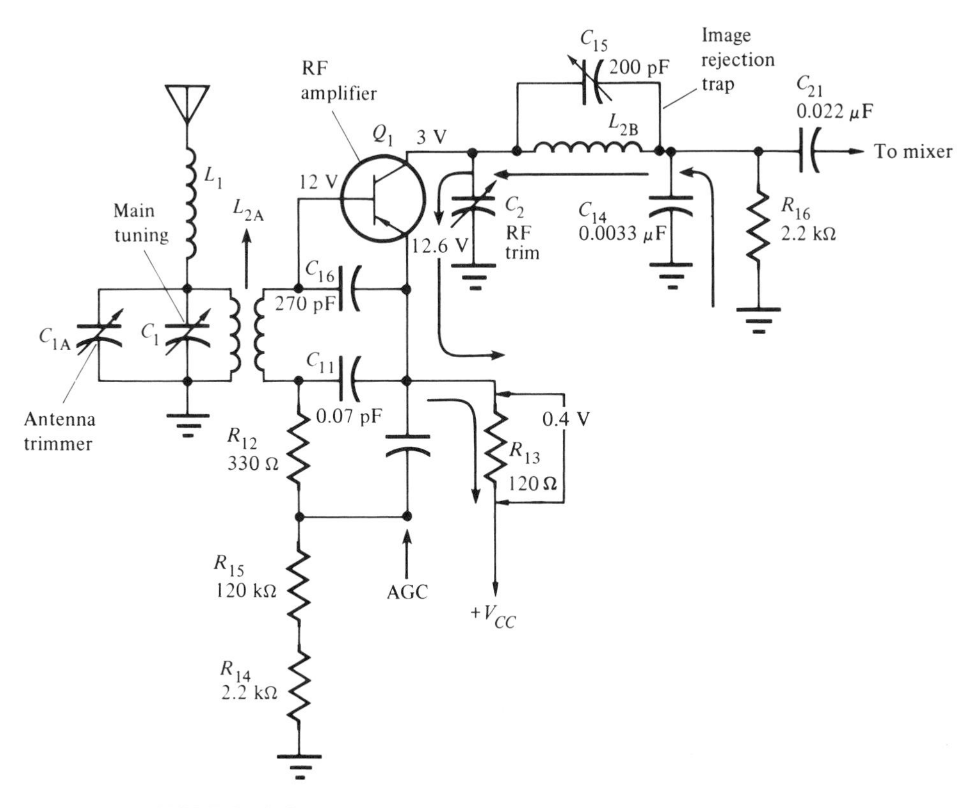

FIGURE 14–5
Typical RF Amplifier Stage Found in Receivers

R_{16}, through RF coil L_{2B}, into the collector, through the junction, and then down through R_{13}, and back to the $+V_{CC}$ supply. Normal current flow through this stage will develop a 0.4 V drop across R_{13}. This will then cause the 12.6 V at the emitter terminal.

Resistor R_{13} also serves as a stability factor for the amplifier stage. If the current through this amplifier increases due to heating of the transistor, a larger voltage drop will be developed across R_{13}. This increase in voltage at R_{13} will decrease the emitter voltage drop. Because this voltage has dropped at the emitter, a lowering of conduction through the transistor will also occur, which will then return the voltage at the emitter back to normal. Therefore the resistor in the emitter leg is used to stabilize the amplifier.

Biasing of this stage is developed between the base and the emitter. This silicon transistor must have 0.7 V between base and emitter. Because it is a PNP device, the base voltage must be lower than the emitter voltage for forward biasing to occur. Figure 14–5 shows the bias developed. Also used to bias the base is the secondary of L_{2A}, R_{12}, R_{15}, and R_{14}, and the AGC network. This action is a negative feedback from the IF stage.

The input of this amplifier comes from a tuned circuit consisting of C_1 and L_{2A}. This circuit is tuned to pick up the entire AM band of frequencies and reject all others. Capacitor C_{1A} is also called an *antenna trimmer.* L_{2A} is used to couple the antenna RF signal directly to the base of Q_1. The output of the amplifier is also a tuned circuit. This ability to be tuned comes from components C_2, L_{2B}, and C_{14}. Again, this tuned circuit is used to select only those frequencies found in the AM band. To make this circuit selective, L_{2B} has an adjustable core. Capacitor C_2 can also be adjusted during alignment to make the circuit selective.

As in other amplifier circuits, a capacitor bypasses the emitter resistor. This capacitor prevents any signal loss across R_{13}. This is the final important circuit in this RF amplifier stage. It consists of L_{2B} and C_{15}, and is called an **image trap.** The use of this trap will be explained in the section on mixer operation. The final component, C_{21}, couples the RF signal into the input of the mixer stage. It also keeps any dc bias from reaching the RF stage from the mixer section.

Mixer-Oscillator Stage

The next stage in the radio is the **mixer stage,** or the **converter stage.** As mentioned before, this stage contains the mixer and the local oscillator. Here the RF frequency is combined with the local oscillator frequency and converted into the IF frequency.

Two block diagrams can be used for this stage. The one for the converter stage is shown in Figure 14–6a. The one for the mixer stage

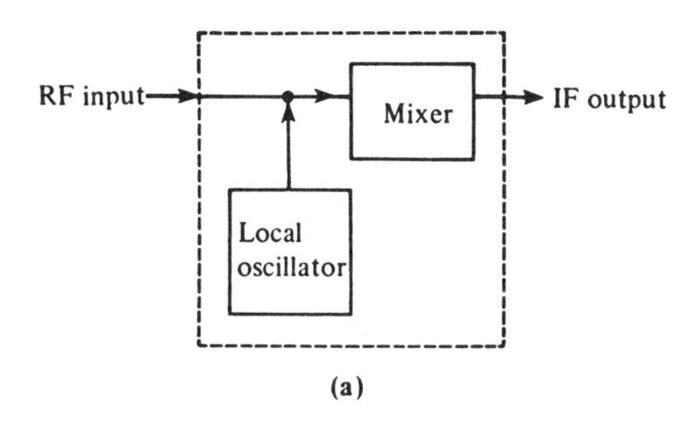

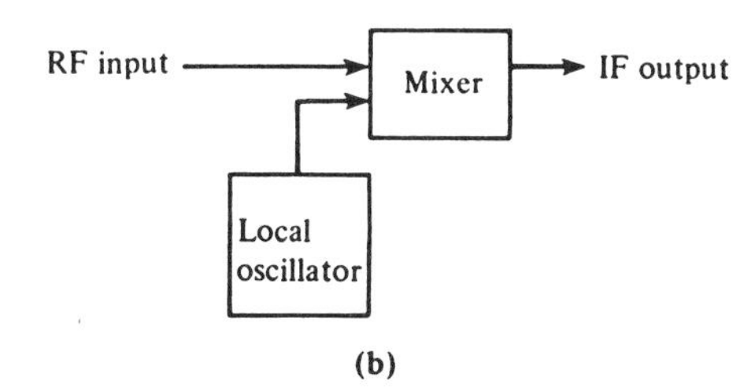

FIGURE 14–6
Two Types of Mixer States Found within Radios: (a) Converter Stage, (b) Mixer-Oscillator Stage

is illustrated in Figure 14–6b. The main difference is that in the mixer stage, in Figure 14–6a, the oscillator and the mixer are considered separate from each other, and are usually transformer-coupled into the mixer transistor. On the other hand, in the converter stage illustrated in Figure 14–6b, the mixer and the oscillator are part of the same stage. Figure 14–7 shows an example of a converter type of circuit, and Figure 14–8 shows a schematic of the mixer stage.

In some cases when the RF signal is being received and converted into the IF frequency, a second strong RF signal is received by the antenna. For example, say the receiver is tuned to a 560 kHz signal. The local oscillator produces a signal of 1015 kHz to give the 455 kHz frequency. A strong signal at 1470 kHz could enter the converter stage despite the tuned RF circuit to reject this signal. Once this signal is in the converter, it is treated as a signal that is present called an **image frequency.** Image frequencies develop in radio receivers when the local oscillator beats with another sideband to produce another received frequency. In Figure 14–5, C_{15} and L_{2B} comprise a trap circuit used to eliminate image frequencies.

Figure 14–7 shows a typical mixer-local oscillator found in receivers. To set up the oscillations, the parallel combination of C_3, C_{23}, C_{22}, and L_{2C} causes this resonant circuit to establish a frequency 455 kHz above the incoming RF signal. Notice that C_3 is adjustable; this variable capacitor would be part of tuner control.

To establish the oscillator action, the capacitors in this network must charge and discharge. To aid in this process Q_2, the converter

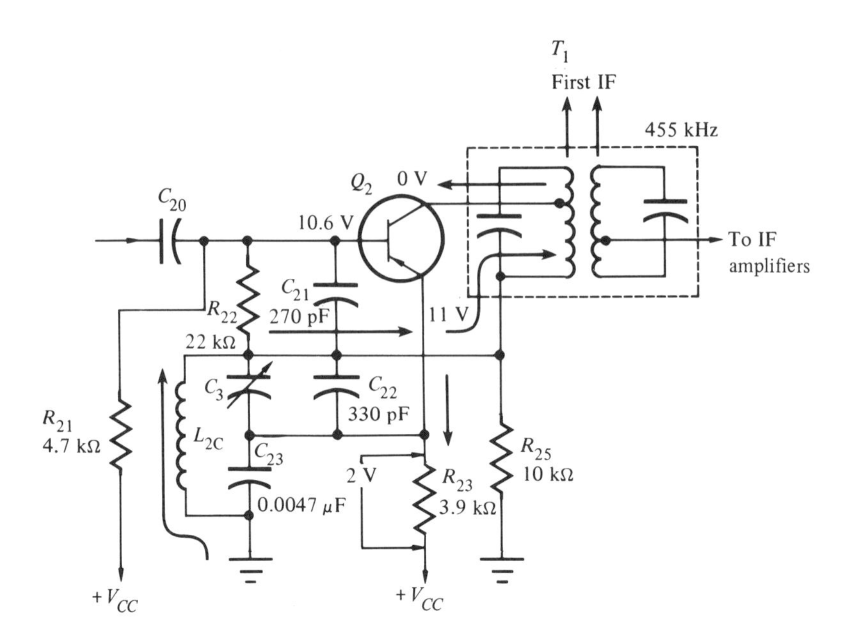

FIGURE 14–7
Schematic Diagram of a Converter Stage

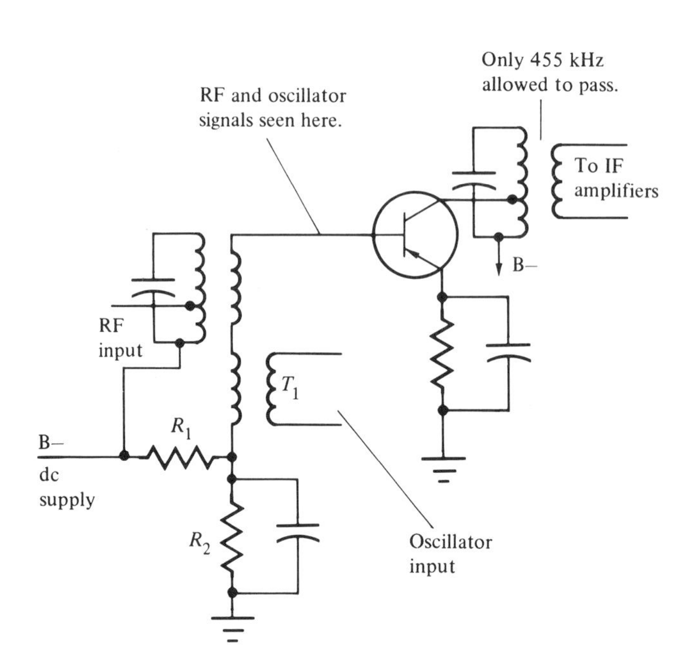

FIGURE 14–8
Schematic Diagram of a Mixer-Oscillator Stage

transistor, is used. When Q_2 is not conducting, C_2, C_{23}, and C_{22} are allowed to charge from the $+V_{CC}$ line via R_{23}. To equalize the charge on the plates of the capacitors, electrons flow from ground through L_{2C}. When Q_2 is turned on and begins to conduct, the capacitors discharge. The capacitors allow their electrons to flow back to $+V_{CC}$, but this time the path is through R_{22} and R_{21}. When this discharge process begins, it develops a voltage across L_{2C}. This back voltage across the coil allows the capacitors to recharge. Such a recharging process from the coil results in oscillation. This oscillation operates at 455 kHz above the incoming RF signal. With the development of this oscillation, two frequencies appear at the base of Q_2. Transistor Q_2 operates in an on/off mode. This on/off state sends Q_2 into a nonlinear range of operation. In any nonlinear device when two signals are present at the input, RF amplifier, and local oscillator, the output results in four frequencies. Figure 14–3 shows an example of this action in the mixer stage. Once these four frequencies appear at the output of Q_2, the tuned circuit of T_1 allows only the IF frequency (455 kHz) to pass.

The main current path for this stage must be understood by the technician. A current path is developed from ground, through L_{2C}, and into the primary of T_1. Then a current path is established from the collector-emitter junction, to R_{23}, and to the $+V_{CC}$ 13 V supply. Notice in Figure 14–7 that a voltage drop of 2 V is measured across R_{23}. When the radio is tuned from the higher end of the band to the lower end, a 0.2 V difference should be seen in the voltage drop across R_{23}. If this difference is seen, the oscillator of the receiver is operating normally. To keep this section operating as a nonlinear device, R_{25} is used as a feedback to stabilize operation of Q_2.

The base voltage of Q_2 is established by the voltage divider network of L_{2C}, R_{22}, and R_{21}. Notice that the voltage drop between the base and the emitter is only 0.4 V. Therefore this transistor is normally off. It conducts only when a negative swing in signal is made. In other words, when the RF input signal goes positive, the base voltage goes positive, but when the RF input signal goes negative, the 0.6 V difference appears between base and emitter, and the transistor conducts.

A schematic diagram of a mixer stage is shown in Figure 14–8. The oscillator signal is coupled by T_1 into the base of the mixer. Here the signal is combined with the RF signal to turn the transistor on and off. This action will produce the four output signals of the mixer section.

IF Amplifiers

The major function of the **IF amplifier** is to give the signal good linear amplification throughout the range of frequencies. In most receivers this amplification will be accomplished by more than one IF amplifier. Generally, there will be at least two IF amplifier sections in the receiver.

Another function of IF amplifiers is to provide selectivity. They should select only one of the four frequencies from the converter stage, and amplify that signal. The signal selected is 455 kHz.

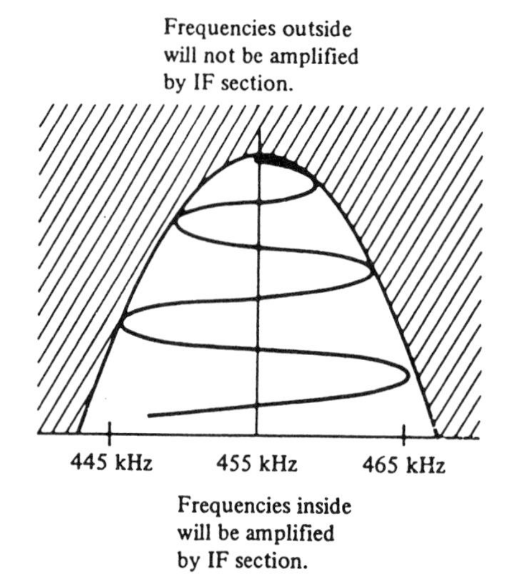

FIGURE 14–9
Gain versus Frequency Graph for the IF Amplifier Stage

Most radio receivers are designed to receive both speech and music information. The frequency range for this information extends to about 10 kHz. If the receiver is designed for high-quality reproduction, the amplifiers must be able to pass modulated frequencies up to this level. Therefore the IF amplifiers must have the capacity to pass frequencies ranging from 445 kHz to 465 kHz, and to reject all others. Figure 14–9 shows an example of this action. Because of this circuit action, this large bandwidth is found only in very expensive radio receivers. Generally, the bandwidth in AM radios is about 3 kHz. With this reduction, the quality of music and voice will be much less.

After the signal has been converted to the IF frequency, it is transformer-coupled to the IF stage. Figure 14–10 shows an example of a typical IF section found in receivers. As shown in the figure, T_1 has been tuned to select only the 455 kHz signal. Note that this selection process is accomplished during alignment of the radio. The signal is then developed at the secondary. The T_1 center tap provides direct coupling of the signal from the secondary of T_1 to the base of Q_3, while C_{33} provides bypass action on the emitter to prevent any signal loss within the stage.

Detector Circuit

To recover the audio intelligence, the 455 kHz modulation must be removed. This action is accomplished by the detector circuit. Figure 14–11 shows an example of this circuit. The detector circuit does not include a transistor, but instead operates with a diode. This diode

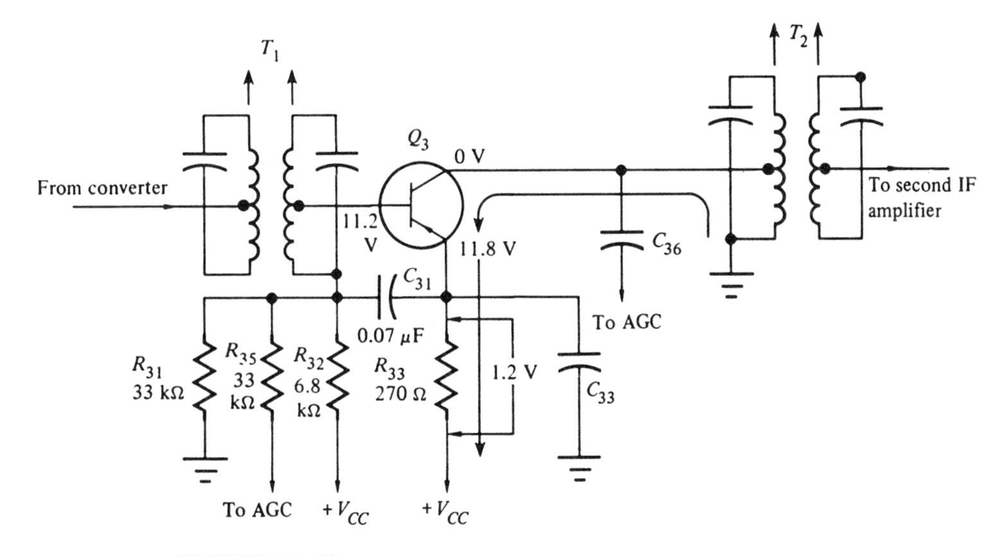

FIGURE 14–10
Typical IF Amplifier Section Found in Radio Receivers

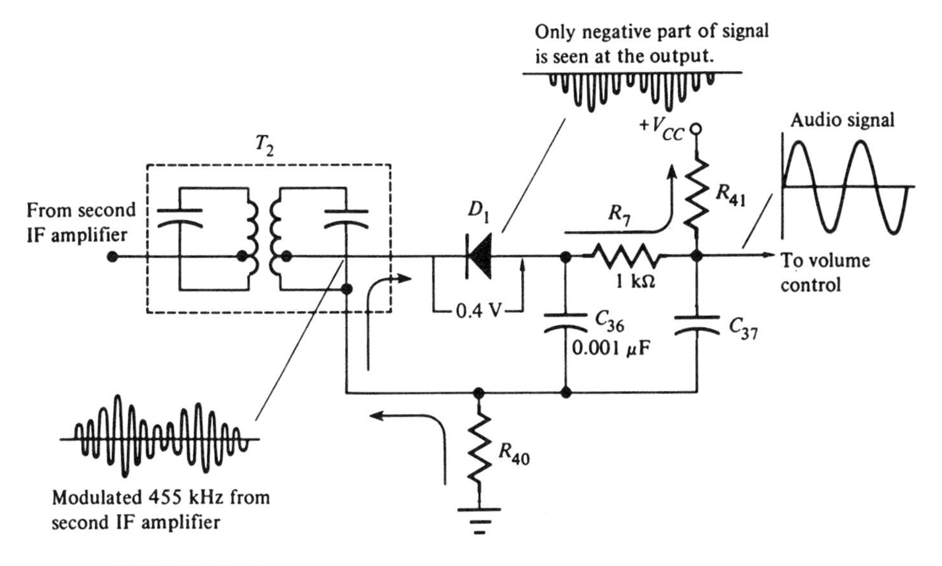

FIGURE 14–11
Detector and Low-Pass Filter Circuit

operates as a half-wave rectifier circuit. As seen, the half-wave rectifier is made up of D_1, and the filter for this circuit is C_{36}, C_{37}, and R_7.

When a signal is applied from the secondary of T_2 to the cathode of D_1, a small dc current is established. This is shown by the arrows. The diode conducts during the negative half of the modulated IF signal. The positive half will reverse bias the diode. This signal is then filtered, and supplies a weak audio signal to the volume control, and then to the audio circuits for additional amplification.

In some circuits a voltage is taken from this rectified dc voltage to be fed back as a control AGC voltage. This operation will be discussed in the next section.

Automatic Gain Control

The automatic gain control (AGC) section is used to adjust the amplification of the RF and IF stages. Because the strength of the input signal determines the loudness of the speaker's output, controls must be placed on the RF and IF stages. If signals reaching the antenna are weak, the signal amplitude reaching the detector will also be weak, thus producing low output volume. If strong signals are received, they will create overload signal conditions in the amplifiers, and cause distortion. To correct this, a feedback signal is used to adjust bias on the RF amplifier and the first IF amplifier.

Figure 14–12 shows an example of the AGC voltage being taken

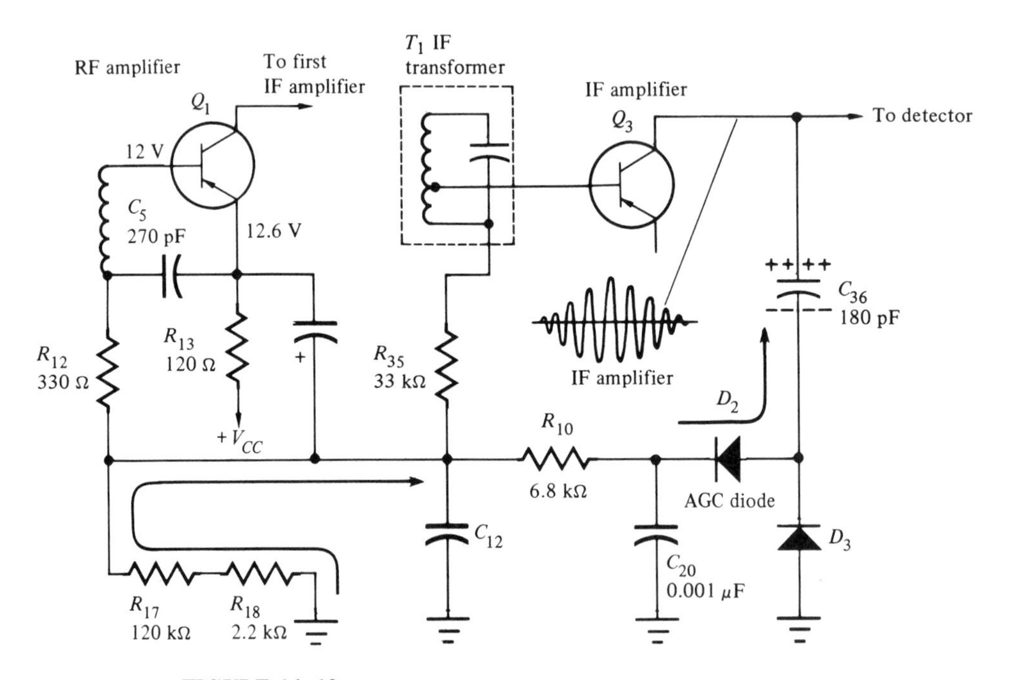

FIGURE 14–12
AGC Voltage Being Developed from the Output of the IF Section

from the IF section. This circuit operates in the following manner. As the modulated 455 kHz signal swings in the positive direction, C_{36} is allowed to charge. This charging path is shown in Figure 14–12. It is developed through R_{18}, R_{17}, R_{10}, and D_2. The increase in current is developed in the RF amplifier, Q_1, and the IF amplifier, Q_3. As C_{36} charges, this current flow causes the base voltages of these transistors to rise toward the emitter value. The forward bias is reduced, and the transistor current flow is decreased. This decreasing current flow results in the collector voltage change on the RF stage.

As the signal swings negative, C_{36} must discharge. This discharge process is accomplished through D_3 and back to ground to complete the circuit. Capacitors C_{20} and C_{12} are added to provide additional AGC filtering. The collector of the RF amplifier can be a good gauge to see if the AGC circuit is operating properly. With a no-signal condition, the collector sits at the 3 V level. As signal strength at the output of Q_3 increases, it causes C_{36} to charge harder. Then the bias between base and emitter decreases. This gradually shuts off the transistor, causing the collector voltage to rise. This action provides a good check to see if the AGC is operational. Tuning between weak and strong stations should cause the voltage on the RF amplifier collector to change. With strong signals, the voltage should increase; with weak signals, decrease. The base voltage varies only a few tenths of a volt, and thus would not be a good voltage check to make in the circuit.

Figure 14–13 shows another example of the AGC action. This

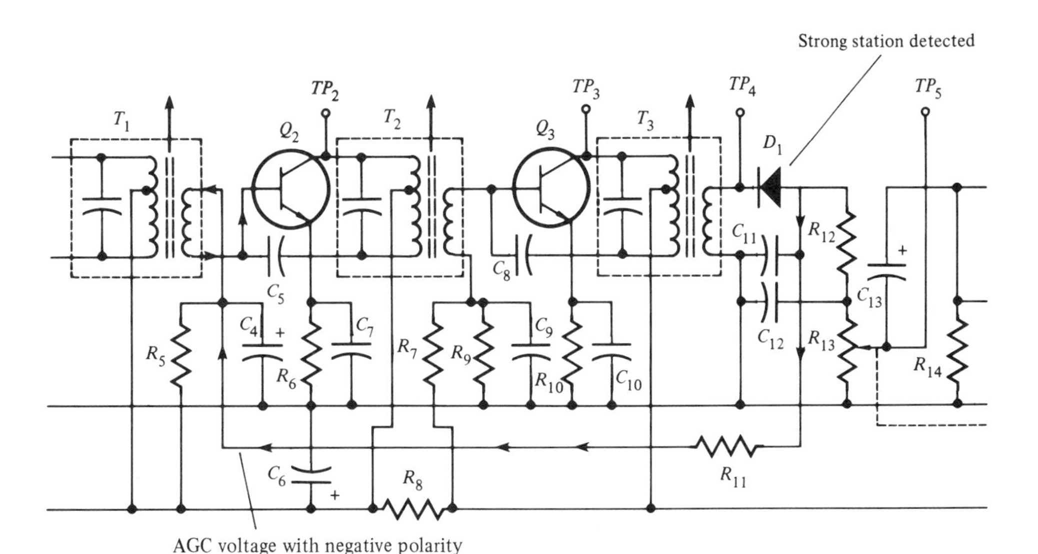

FIGURE 14–13
AGC Feedback Signal to Control Gain of the Second IF Amplifier

circuit uses no diodes, and develops a dc feedback voltage directly from the detector of the radio. A dc output voltage is developed from the detector. This voltage is a negative polarity (with respect to ground), and is directly proportional to the amplitude of the IF signal. The AGC voltage is obtained from the detector voltage and fed to the base of Q_2. Resistor R_{11} and capacitor C_4 form an audio filter to prevent any audio signal from reaching the base of Q_2 through the AGC line. The detector voltage varies with the strength of the RF signal. As the signal increases in amplitude, a more negative dc voltage is developed for AGC feedback. This negative voltage is then applied to the base of Q_2. A negative voltage applied to an NPN transistor reduces bias, which in turn will reduce the collector current, and therefore the gain of the stage.

Integrated Circuit Front Ends

In the early 1970s a change swept through the communications industry. Integrated circuits started to replace discrete transistor components. Figure 14–14 gives a complete IC chip in block diagram form.

Figure 14–15 shows the block of the front end of an AM receiver, which contains the RF amplifier, the IF amplifiers, and the AGC circuits. The RF signal is then transformer-coupled into the IC RF amplifier stage. The RF signal enters at pin 12, and leaves at pin 13. Between these two pins, the RF signal has been amplified. The RF signal is returned to the IC chip via pin 1. Here the signal enters

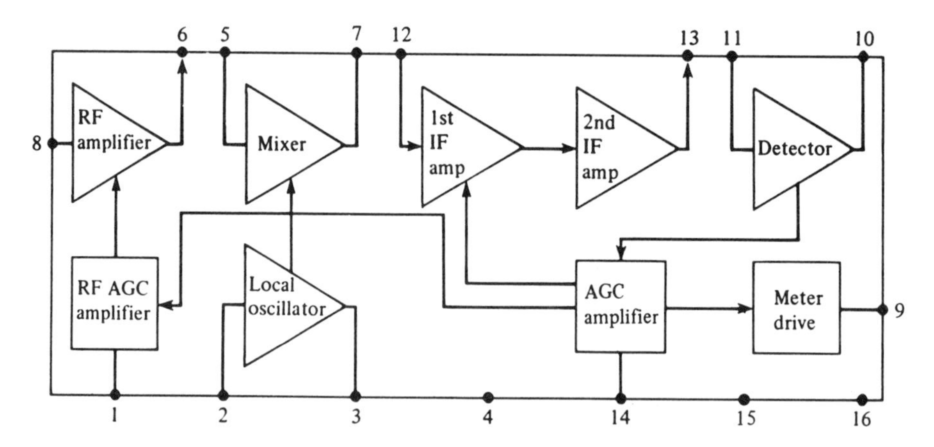

FIGURE 14–14
Block Diagram of an IC Containing all the Necessary Parts for the Radio's Front End

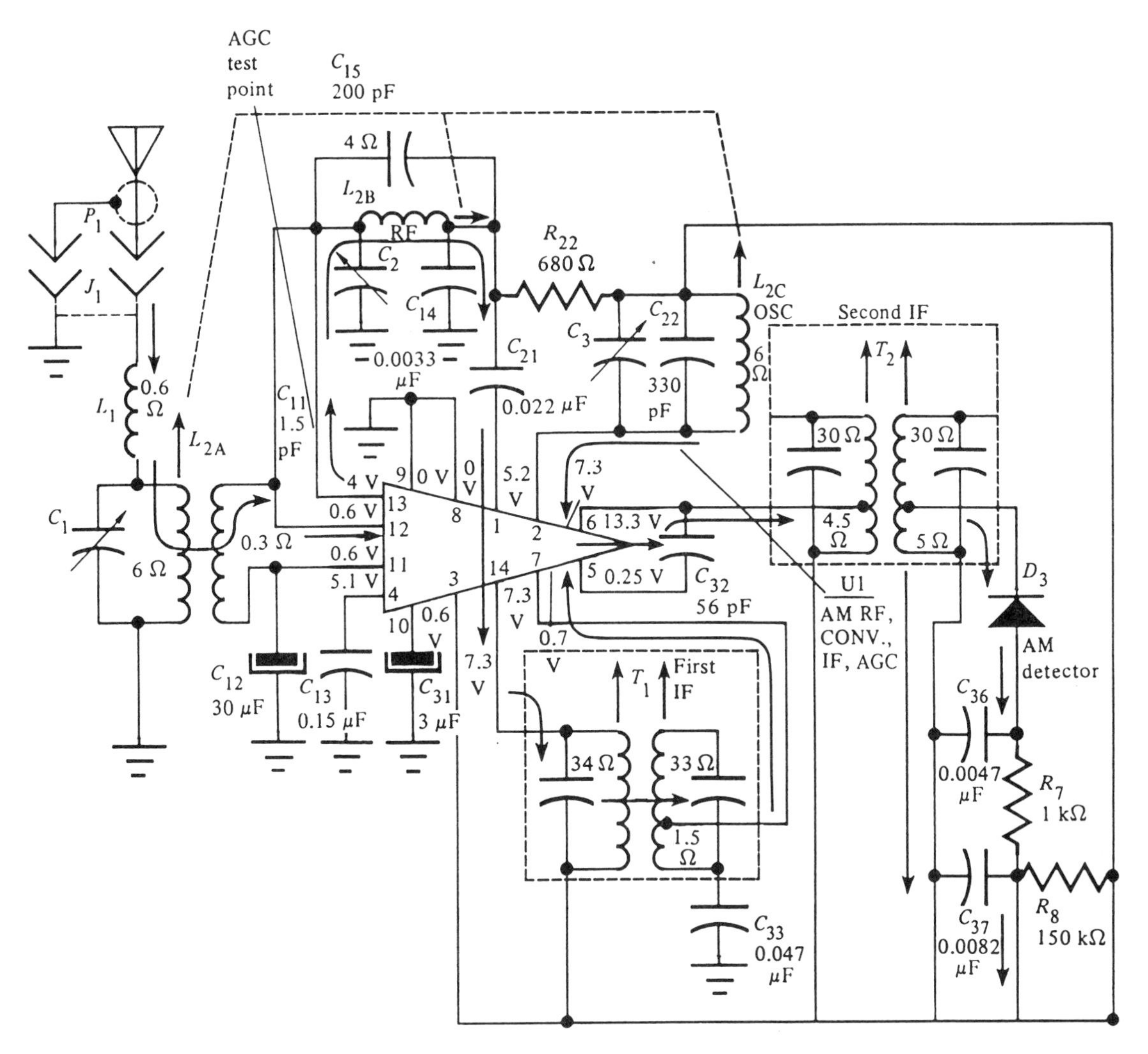

FIGURE 14–15
Typical Schematic Diagram of an AM Radio Front End

the converter stage. The signal is converted to the 455 kHz IF, and leaves the IC chip at pin 14, where it enters the first IF transformer. Here the 455 kHz frequency is separated from all other frequencies, and only the 455 kHz frequency is passed on to be amplified by the IF amplifiers. The signal finally leaves the IC chip on pin 6. The signal is sent on to the audio amplifier section.

All necessary amplifiers are found within the IC chip. The AGC circuits are also found in the chip. These ICs have made the receiver lightweight and portable, and also have greatly reduced the cost. The service technician can now simply measure pin voltages or measure waveforms to determine if the IC chip is good or bad. There is only one disadvantage to the IC chip. If one part of it goes bad, the entire chip must be replaced. On the other hand, with discrete transistor components, if one transistor within the stage goes bad, only that component must be replaced. Nevertheless, the advantages of cost, size, and power consumption greatly outweigh this one shortcoming of the IC chip.

Audio Section

The audio section within AM radio receivers generally contains an **audio preamplifier** and some type of power amplifier to drive the speaker. The audio information has been obtained from the detector. The remaining action is to amplify the audio signal. Typically, this stage contains a class-A driver and a class-AB, push-pull power amplifier.

The class-A amplifier is generally considered to be the audio preamplifier. This stage acts as a single-ended amplifier. The amplifier is needed to raise the signal from the detector stage to a usable level. Figure 14–16 shows an example of this type of amplifier. A silicon transistor is used in the amplifier. The base is biased by the voltage divider network of R_{14} and R_{15}. R_{16} is used to stabilize the action in the emitter circuit. R_{17} is the load resistor for the collector circuit.

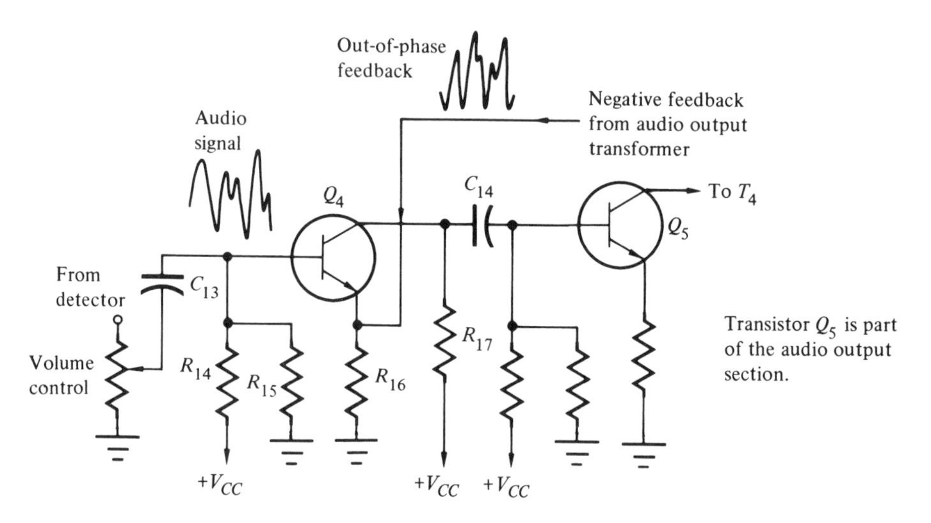

FIGURE 14–16
Schematic Diagram of an Audio Preamplifier Section

The input signal for this stage is coupled from the volume control through C_{13}, and the output signal is coupled through C_{14}. To keep this amplifier operating within the linear region, a negative feedback signal is taken from the audio output stage. This negative feedback signal is out of phase, and helps to reduce harmonic distortion and improve the frequency response of the stage.

Audio Power Output Stage

The audio output stage contains two sections. The first is a class-A driver stage, and the second is a class-AB, push-pull power amplifier. Figure 14–17 shows this circuit connection. The driver stage takes the low-level audio signal from the preamplifier stage and increases the voltage and the power of the signal. The class-AB, push-pull amplifier is used as the final output stage of the radio. The description of this power amplifier was given in Chapter 8.

Of special importance for this stage is the biasing network. The stability of this output stage depends upon several components. The stability of Q_6 and Q_7 is maintained by R_{21}, R_{22}, R_{24}, and D_1. The resistors serve the same function as other bias resistors. However, D_1 is used for temperature stability. Should the temperature rise, the

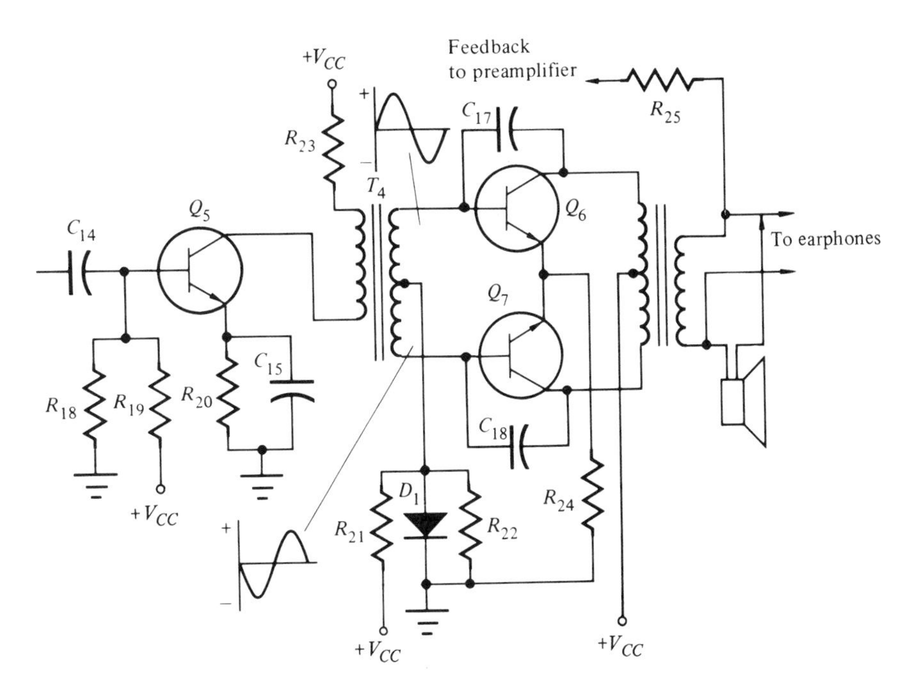

FIGURE 14–17
Schematic Diagram of an Audio Driver and Power Amplifier Stage

emitter currents of Q_6 and Q_7 will also rise. This increase will cause a rise in forward resistance, thus reflecting a current decrease to the bases of Q_6 and Q_7. This will cut down forward bias action on the base-emitter junction, and reduce transistor conduction. Resistor R_{25} is the feedback resistor for the signal to feed back to the emitter of Q_4 (see Figure 14–16). Finally, C_{17} and C_{18} are high-frequency bypass capacitors that prevent the circuit from oscillating.

Figure 14–17 also shows the driver stage. This class-A amplifier has its signal coupled from the preamplifier by C_{14}. The output of this section feeds the primary of T_4 with a distortion-free amplified signal. The biasing network of R_{18} and R_{19} provides base bias, while R_{20} and C_{15} provide the necessary emitter stabilization.

If a distortion-free signal is received from the preamplifier, this amplifier section will reproduce good amplified output signal. This audio section is sufficient for the small portable AM radio, but will not reproduce the fidelity required for FM reproduction.

FM AND FM STEREO RECEIVERS

To better understand the operation of FM, a brief description and comparison to AM should be made. Both AM and FM receivers are **superheterodyne** receivers. That is, they convert the RF frequency into an IF frequency. However, the range in frequencies is a little different. Table 14–1 shows an example of the different RF, IF, and oscillator ranges of AM and FM.

Also of importance is an understanding of the different waveforms of AM and FM. The AM (amplitude-modulated) signal is different from the FM (frequency-modulated) signal. Figure 14–18 shows the two signals. Notice in Figure 14–18a that the AM signal intelligence has varied the amplitude of the **carrier** frequency, while the FM signal (Figure 14–18b) has kept the amplitude of the carrier constant, but has changed the frequency of the carrier.

As seen in Figure 14–18a, the distance between the pulses of the AM carrier signal changes between high and low frequencies. The

TABLE 14–1 Frequency Range of AM and FM Receivers

Broadcast Band	IF	Local Oscillator Range
AM: 535–1605 kHz	445 kHz	990–2060 kHz
FM: 88–108 MHz	10.7 MHz	77.3–97.3 MHz

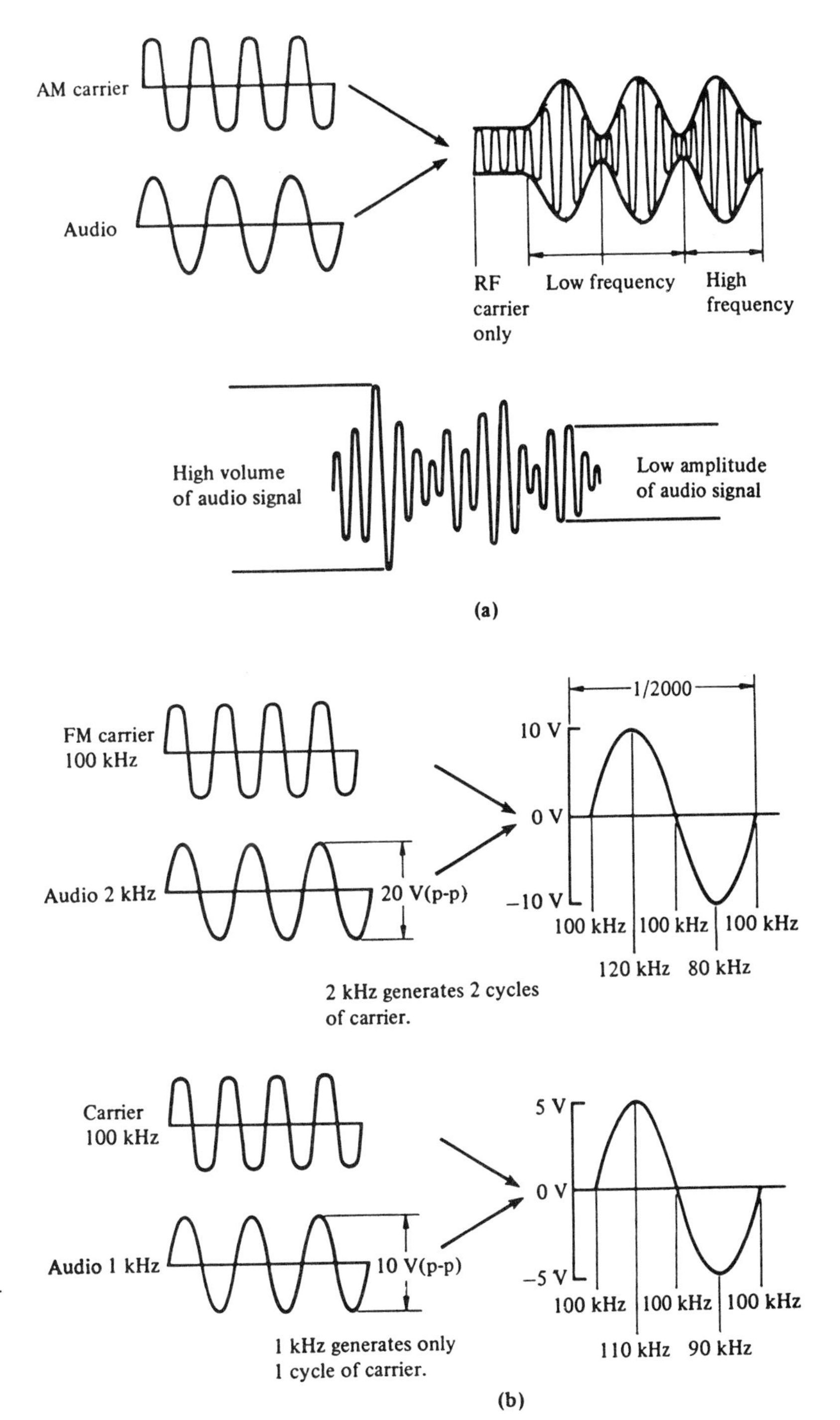

FIGURE 14–18
Development of a Transmitted Signal: (a) AM Signal Development, (b) FM Monaural Signal Development

TABLE 14–2 Comparison of FM to AM Signal

FM	AM
Carrier amplitude constant.	Carrier amplitude varies with signal.
Carrier frequency varies with modulation.	Carrier frequency constant.
Modulating frequency is rate of frequency changes in RF carrier.	Modulating frequency is rate of amplitude changes in RF carrier.
Signal noise can be eliminated in receiver.	
Better fidelity.	

FM signal is more difficult to understand. An example will help. Assume an audio signal with an amplitude of 10 V(p). This 10 V(p) waveform causes a frequency change of ±20 kHz. The carrier frequency is 100 kHz. Figure 14–18b shows how this FM carrier signal will be modulated. As the amplitude of the audio goes in a positive 10 V direction, it causes the carrier to change by +20 kHz. Therefore the output of the carrier has gone from a rest, or center, frequency of 100 kHz to a frequency of 120 kHz. Next the input signal swings negative. This causes the carrier frequency to again move from its rest position of 100 kHz. The carrier moves to 80 kHz. This explains how the carrier is modulated by the amplitude of the modulating signal.

A summation of the AM and FM signals is given in Table 14–2. The FM stereo signal is more complex than the **FM monaural** (one-channel) signals. The FM stereo signal will be described later in the chapter.

The block diagram of the FM receiver is generally the same as that for the AM receiver. The only differences are in the frequency of IF signal developed in the converter, and the type of detection used to recover the audio information. Figure 14–19a gives a simple block diagram of the FM receiver.

FM Front End and IF Strip

Figure 14–19b shows a partial block diagram of the RF amplifier, mixer, local oscillator, and IF strip of the FM radio. In the majority of FM radios, this part will be found in FM monaural as well as FM stereo.

FM RF

The RF signal arrives at the RF amplifier via the antenna. Most of these inputs will have a loop-stick antenna, or will be connected to an

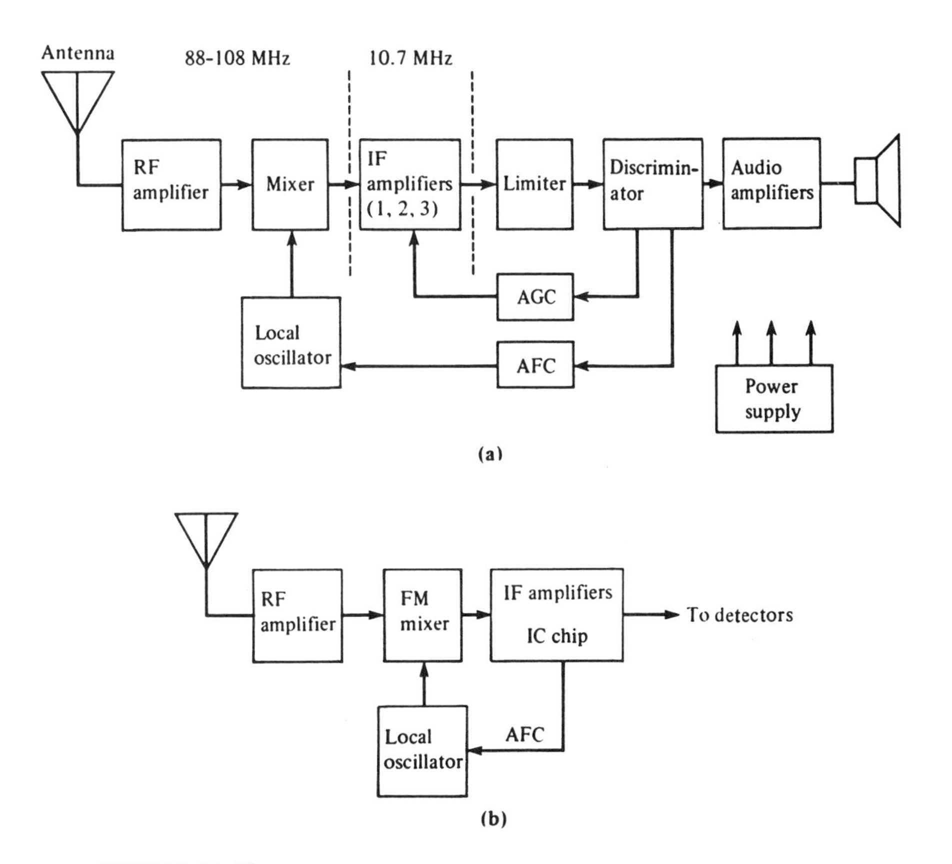

FIGURE 14–19
Block Diagram of a Receiver: (a) Basic FM Receiver, (b) FM Receiver Front End

outside antenna. The majority of better-quality FM receivers will be connected to an external antenna. The signal received by the antenna may be very weak, sometimes only a few microvolts in amplitude. To amplify this signal, a JFET or MOSFET is used as the active device within the stage.

Figure 14–20 shows an example of an RF amplifier using a dual-gate MOSFET. The $+V_{CC}$ supplied to Q_1 is through L_5, L_2, R_5, and R_7. This bias line supplies the dc voltage for the drain of this component. Because this is a dual-gate MOSFET, one of the gates—G_1—will receive the RF signal to be amplified, and the other—G_2—will be used to control gain. This input is from the AGC circuit.

The FM RF signal is coupled through L_1 and C_2 to G_1 of Q_1. The output signal is developed in the drain and coupled through R_5 and L_2 to the converter stage. This interstage connection proves to be additional filtering of the RF signal.

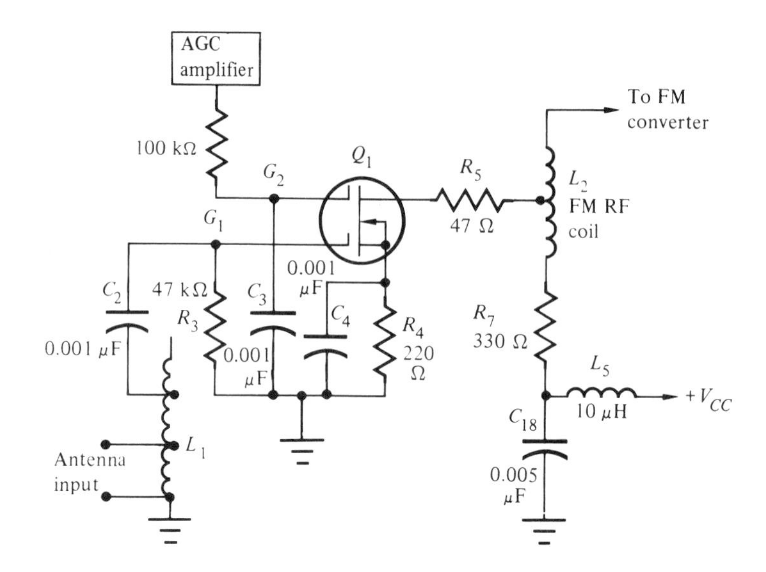

FIGURE 14–20
RF Amplifier Using a Dual-Gate MOSFET

FM Converter Stage

As with the AM radio, the RF frequency in the FM radio is converted to an IF frequency with one transistor. At the base of this transistor, an RF frequency and a local oscillator frequency are present. However, this time the local oscillator is operating at a frequency 10.7 MHz below the incoming RF signal. As the radio is tuned through the FM band, the converter produces its 10.7 MHz IF output.

Figure 14–21 shows an example of a typical converter stage found in many FM receivers. The FM oscillator section is made up of the oscillator coil, L_4; capacitors C_{15} and C_{1E}; and, finally, CR_1, a varactor diode. Notice that capacitor C_{1E} is ganged with the tuning capacitor in the tuner. As the tuner selects another station, the local oscillator selects a new frequency. The IF output of this stage is coupled through T_{201}, and then to IC_{201}. This IC chip is nothing more than a **gain block** for the IF section, a gain block being a group of amplifiers that supplies the necessary amplitude to the IF signal.

Figure 14–22 shows another example of the mixer-oscillator stage. In this circuit a bias is established on Q_2. The establishment of base bias is through R_{21} and R_{22}. A conduction path is then established for the collector-emitter circuit through L_{1C} and R_{23}, and back to positive dc supply. The oscillator frequency is determined by the tuning circuit, L_{1C} and C_3. Capacitors C_{24} and C_{27} and the AFC diode D_2 aid in the maintenance of this oscillator frequency. The output of this oscillator is taken between the collector and base, and delivered to the mixer stage via C_{26}.

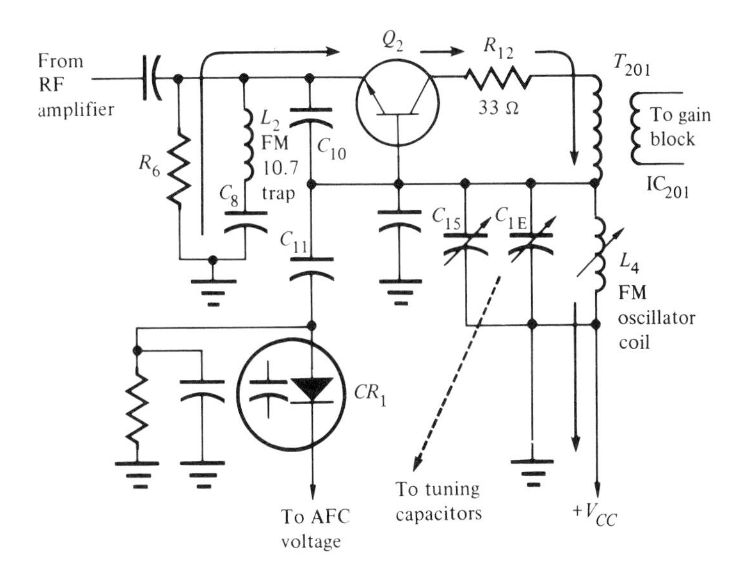

FIGURE 14–21
FM Converter Stage

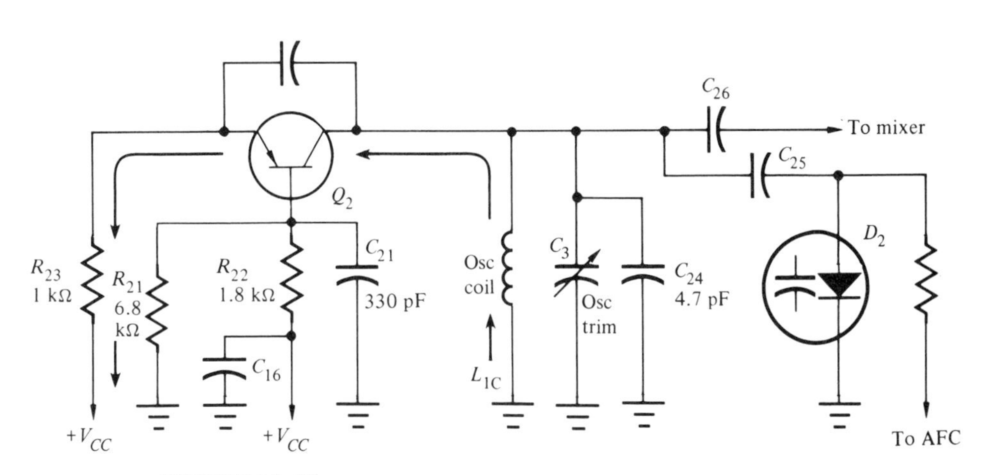

FIGURE 14–22
FM Oscillator Circuit

To mix these signals together, a separate stage is required. Figure 14–23 shows an example of this action. Notice that the dc biasing state for this transistor stage is developed by R_{31}, R_{32}, and L_{32}. The collector is biased via R_{36}, the primary of T_1, and a stabilizing resistor, R_{33}, in the emitter circuit.

The transistor receives two input signals on its base. The incoming RF signal is coupled through C_{32}, and the local oscillator is coupled through C_{26}. The signal on C_{26} will be 10.7 MHz lower than the signal

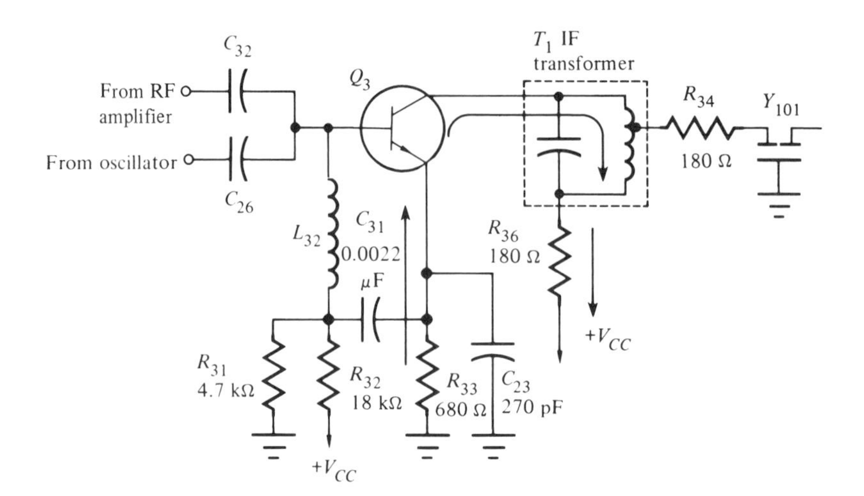

FIGURE 14–23
FM Mixer Stage

found on C_{32}. The two signals are mixed to produce the 10.7 MHz IF signal at the output. The IF signal is then coupled through the first IF transformer, and again into the ceramic filter, Y_{101}.

Ceramic Filters

Only two tunable devices are found in the IF section of the newer radio receivers. One is the first IF transformer, and the other is the *quadrature detection coil.* Because of the use of ceramic filters, the need for tunable circuits has almost been eliminated.

Ceramic filters are composed of two ceramic crystals resonant at the same frequency—10.7 MHz. The first crystal is excited by the IF signal at its input. It then produces a mechanical vibration at the IF frequency, which it passes on to the second crystal. The second crystal converts this mechanical vibration into an electrical signal at the IF rate.

Ceramic filters are very frequency selective, and provide about 70% of the required selection of the IF frequency. Ceramic filters are not tunable. They are classified into one of five overlapping groups based on their center frequencies. Table 14–3 is a chart of the different types of ceramic filters used in consumer radio receivers. Note that each filter type has a range of frequencies within which to operate, and is rated toward the center frequency. In addition, each filter is color-coded according to center frequency.

It is very important that the service technician know the five different filters and their color codes. A faulty filter must be replaced by a filter with the same center frequency, or the circuits will not operate properly. Also, when the IF section is aligned, the generator

TABLE 14–3 Ceramic Filter Chart

Center Frequency	Frequency Range	Color Code
10.64 MHz	10.61–10.76 MHz	Black
10.67 MHz	10.64–10.70 MHz	Blue
10.70 MHz	10.67–10.73 MHz	Red
10.73 MHz	10.70–10.76 MHz	Orange
10.76 MHz	10.73–10.79 MHz	White

must be set to the center frequency of the filter. Before any replacement or alignment is attempted, the technician should check the manufacturer's service literature.

IF Amplifiers

The output of the mixer-oscillator stage is sent to a ceramic filter. This 10.7 MHz IF signal requires additional amplification to produce good audio output. For AM radios, discrete transistors are used to supply the necessary amplification. This is also the case in the FM radio receiver. The only difference is that the amplifier section must be able to handle a higher IF frequency.

Integrated circuits play a large part in the amplification sections of radios. Because these devices increase voltage as well as current, they have been given a special name; they are called gain blocks. A gain block not only provides amplification of the IF signal; it also makes up for signal loss from the ceramic filters.

Although a detailed explanation of the operation of gain blocks is beyond the scope of this book, one suggestion will prove helpful. The technician should approach these devices in the block diagram manner. Figure 14–24 shows a typical IF gain block. Here the input to the gain block is through the ceramic filter Y_{201}, and then the signal is fed into pin 1 of IC_{101}. The 10.7 MHz IF signal is processed through several internal amplifiers, and inputted on pin 5 of the IC chip. Here the signal is selected by another ceramic filter, Y_{202}. A dc voltage is fed to IC_{101} to an internal voltage regulator. The purpose of the voltage regulator is to keep the internal amplifiers operating at maximum efficiency.

FM AGC

Also found in the FM receiver of Figure 14–24 is an AGC circuit. This circuit samples a signal from IC_{101}, and feeds this signal to the AGC amplifier, Q_{201}. Amplifier Q_{201}, along with D_1 and D_2, which act as a

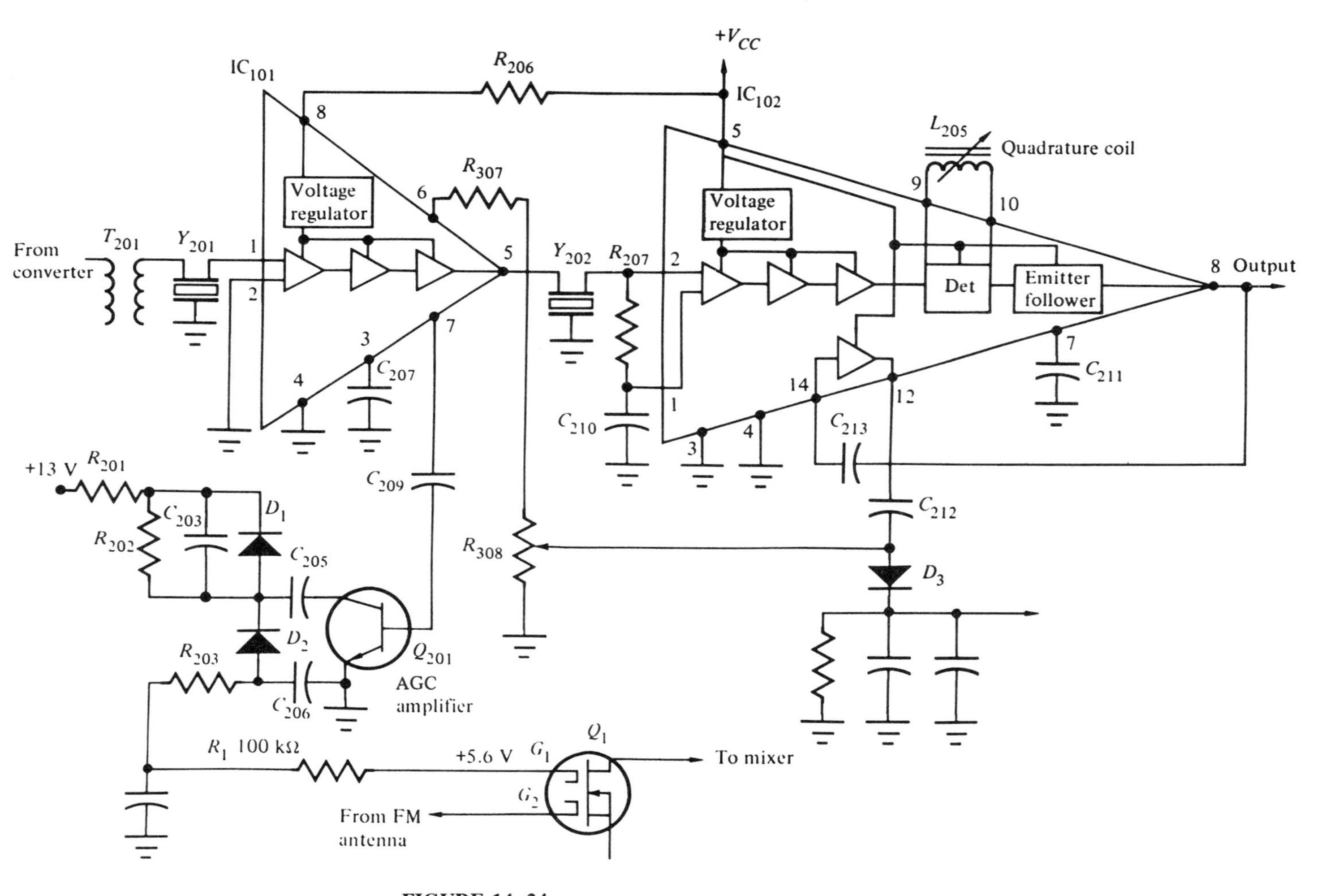

FIGURE 14–24
Gain Block and Detector Stage in an FM Receiver

voltage doubler, control the gain of the RF amplifier, Q_1. Remember that Q_1 is a dual-gate MOSFET, and that G_2 of that transistor controls gain. The G_2 nominal voltage has been set by the biasing network at +5.6 V. This value is the optimum gain condition for Q_1. The FM signals are sampled by pin 7 of IC_{101}. They are then coupled to the base of the AGC amplifier via C_{209}. Here the signal is amplified and outputted on the collector, and coupled by C_{205} to the junction of D_1 and D_2. When the signal reaches a sufficient amplitude, it causes the diodes to conduct. This conduction causes the voltage on G_1 to be lowered, thereby lowering the gain of Q_1.

Audio Detecting and Limiting

After the FM IF signal has passed through the first gain block, it is passed through another ceramic filter. This action is illustrated in Figure 14–24. The signal then enters IC_{102} on pin 2. It is again passed through several amplifier stages, and finally the audio information is recovered in the detector.

IC_{102} can be divided into two parts. The first half of IC_{102} deals with IF amplification and limiting action. The second half deals with audio detection.

Limiting is the process of cutting off the top part of the waveform to eliminate strong noise impulses. If a strong signal is present at the input of IC_{102}, the internal amplifier stages eliminate the overdriven signal parts. By limiting the IF signal, noise will be removed from the signal. Figure 14–25 illustrates this action. In Figure 14–25a, a normal-amplitude waveform is present at the base of the transistor. The transistor therefore operates within its linear limits. In Figure 14–25b the input signal has moved the transistor out of its linear region. In

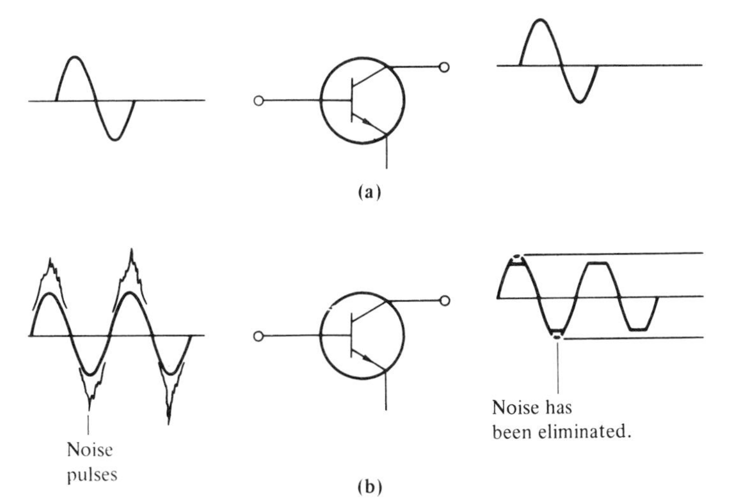

FIGURE 14–25
Limiting Action of a Transistor Stage:
(a) Normal, (b) Saturated

this case the transistor operates in the saturated area. Because this transistor operates out of the linear region, the sine wave tops have been clipped off, turning the sine wave into a square wave.

It should be remembered that audio information is contained in the frequency deviation, not in signal amplitude. Therefore cutting off the amplitude eliminates the noise information found in the peaks, but in no way affects the audio information.

FM Detection

The major function of any receiver is to recover the intelligence broadcasted. As noted, in AM receivers the recovery of intelligence contained in the amplitude variations is accomplished by a nonlinear diode. The FM signal is much more complex than the AM signal, and requires different circuits to recover the intelligence contained in the frequency variations. Circuits such as discriminators, ratio detectors, quadrature detectors, and phase-lock loop (PLL) detectors are being used in the modern FM receiver.

Discriminator Figure 14–26 shows the schematic diagram of a double-tuned **discriminator.** The discriminator operates on the basis of two resonant points, one being below the center frequency of the carrier, and the other operating above the center frequency of the carrier. Note that here the center frequency is 10.7 MHz. Frequencies below 10.7 MHz are received by the tuned circuit of L_1 and C_1. Frequencies above 10.7 MHz are captured by the tuned circuit of L_2 and C_3.

In operation the starting point is the 10.7 MHz center frequency.

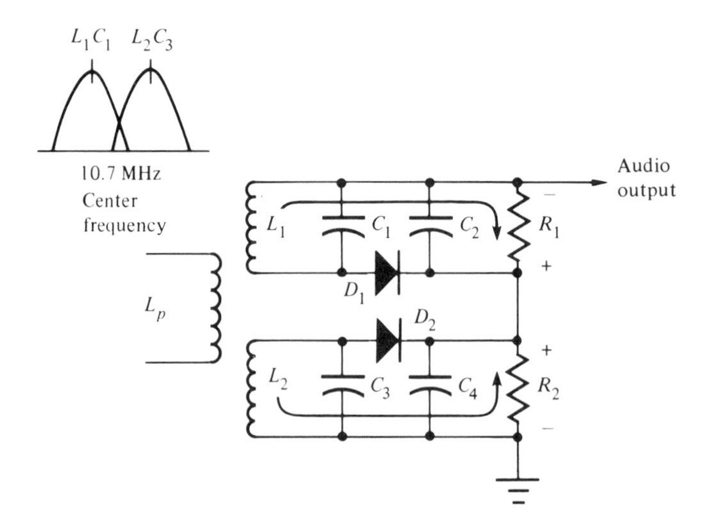

FIGURE 14–26
Double-Tuned Discriminator

During the center frequency time, D_1 and D_2 are conducting equal amounts of current. Because R_1 and R_2 have equal values of resistance, equal voltage drops develop across them. However, note that these voltage drops are opposite in polarity. Therefore they cancel out each other. This results in 0 V at the audio output.

Because of the modulation, the carrier frequency swings higher than 10.7 MHz. This action causes an increase in current through L_2 and C_3, and a decrease in current through L_1 and C_1. Now there is a greater voltage drop across R_2, than across R_1, and the output signal is positive.

Next the signal swing is below 10.7 MHz. This action causes the tuned circuit of L_1 and C_1 to conduct. Because the voltage polarity across R_1 is negative, the audio output voltage swings negative. Therefore the output of the discriminator is zero at the center frequency, positive when the modulated signal causes the output voltage to go above the center or zero point, and negative when the frequency falls below the center frequency. This constant change in output develops into an audio signal.

Ratio Detector The discriminator is used to detect the audio intelligence in the FM-modulated IF signal. However, because this circuit is sensitive to noise, several limiter stages are needed. A better FM detector is the **ratio detector.** This circuit is not sensitive to noise, and therefore does not require limiter stages, but still can reject noise pulses in the FM signal.

Figure 14–27 shows a typical schematic diagram for the ratio detector. The circuit includes a tuned primary coil and a center-tapped, tuned secondary coil. Both the primary and the secondary are tuned to the 10.7 MHz IF signal. The secondary is center-tapped to produce equal voltages of opposite polarity to D_1 and D_2.

In the ratio detector, one diode is reversed so that the two half-wave rectifiers are in series. This series voltage charges C_5 and enables the ratio detector to be free of noise. The ratio detector also differs from the discriminator, in that the audio voltage is taken from between C_3 and C_4.

Capacitor C_5 in Figure 14–27 is used as a stabilizing voltage source for the ratio detector. The voltage developed across the capacitor is the sum of the voltages between D_1 and D_2. This capacitor-charged voltage keeps the detector stabilized so that the output varies only when a frequency-modulated signal appears at the input. The audio output is obtained when the voltage at D_1 and D_2 changes. This is why the circuit is called a ratio detector.

The audio information is obtained by each diode rectifying the IF signal. This rectified voltage charges C_3 and C_4. At a center frequency of 10.7 MHz, each diode conducts equally, and the charges on

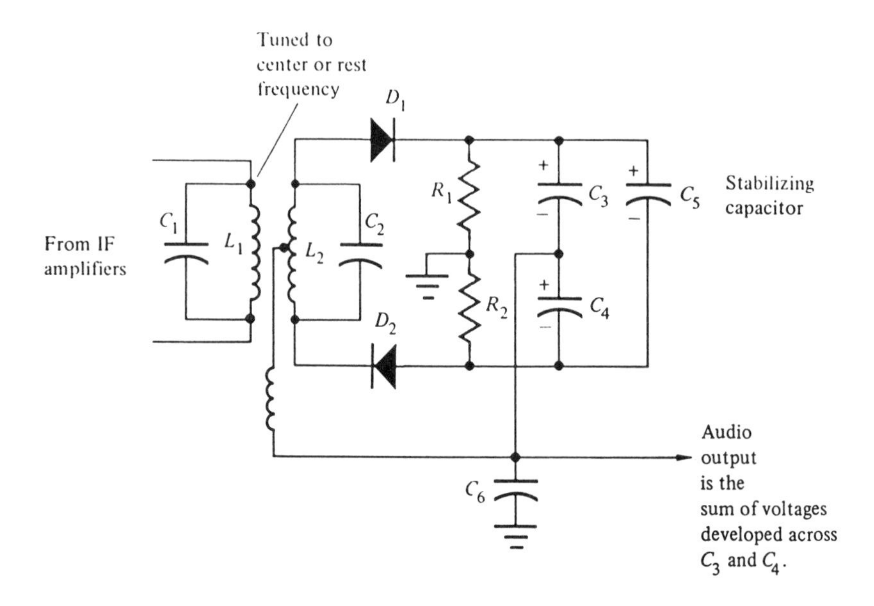

FIGURE 14–27
Ratio Detector

C_3 and C_4 are equal. However, note that the charges are of opposite polarity. Therefore the net result at the output is zero.

When the FM signal swings above the center frequency, D_1 conducts more than D_2. This causes C_3 to have a greater positive charge than C_4, and the output voltage goes positive. When the IF signal swings below the center frequency, D_2 conducts more IF signal than D_1. This means that C_4 charges and develops a more negative charge. Because the two voltages are in series, the output is a negative voltage. This constant change in voltage across C_3 and C_4 develops the audio intelligence at the output of the circuit. Capacitor C_5, called the *stabilizing capacitor*, is the reason limiter stages are unnecessary in the ratio detector circuit.

Quadrature Detector In newer receivers, the detector is contained within the IC chip. Figure 14–24 shows an example of a gain block detector circuit. Found with IC_{102} are several limiter stages and a **quadrature detector.** The quadrature coil, L_{205}, which is connected to pins 9 and 10 of IC_{102}, causes the signal to be 90° out of phase with the input signal to the detector. It is because of this 90° phase shift that the detector gets its name. An audio signal is taken from pin 8 and is sent to the audio amplifier circuits for the FM monaural receiver, or to the stereo circuits for the FM stereo receiver.

Phase-Lock Loop Detector Another type of detector found in receivers is the **phase-lock loop (PLL) detector.** This type of circuit is illustrated in Figure 14–28. Shown is a 506B PLL detector. The FM input is

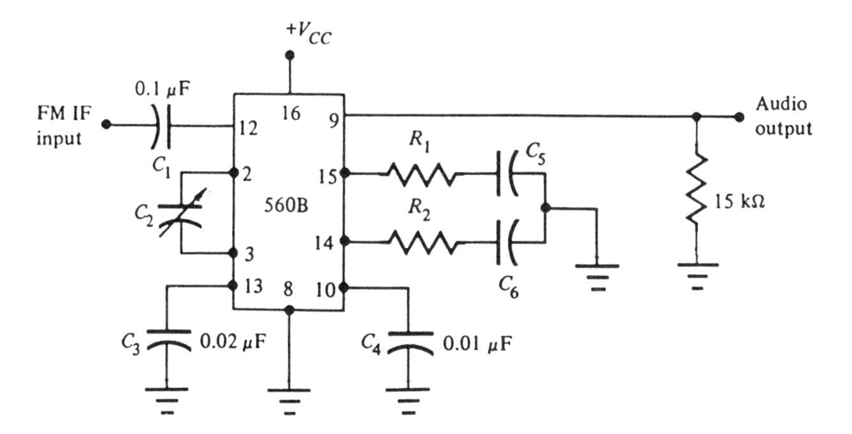

FIGURE 14–28
PLL Detector

found on pin 12, and the unmodulated audio signal is taken from pin 9. Capacitor C_2, a variable capacitor, is adjusted to the center frequency of the IF signal, and the circuit is then locked on to that frequency, producing audio information at the output.

FM Automatic Frequency Control

The ability of the detector circuit to produce dc voltage at the output is very useful in determining if the local oscillator of the receiver is producing a correct frequency. Remember that the detectors are always tuned to the center frequency of the FM IF signal—10.7 MHz.

To insure that the local oscillator is generating the correct frequency, a feedback dc voltage is developed at the output of the detector. This feedback voltage is called the **automatic frequency control (AFC).** Figure 14–29 shows one application of the AFC circuit. Here the dc output voltage is developed from pin 8 of IC_{102}. This voltage is coupled through R_2 and R_1 to the cathode of the varactor diode, D_1. Varactor D_1 is part of the local oscillator circuit. As the voltage across the diode increases, the capacitor of the varactor decreases. This causes an increase in the frequency of the local oscillator. The reverse is true when the voltage decreases at the output of pin 8. The capacitance will increase, causing a decrease in local oscillator frequency. This change in voltage on the varactor causes the local oscillator to change in frequency, and therefore delivers the correct frequency, 10.7 MHz, to the mixer.

Because of the control the AFC has on the local oscillator, the receiver may not be able to tune to weak stations. To allow for tuning to weak stations, an *AFC defeat switch* is used. With the AFC defeat switch on, a fixed dc voltage is applied to the AFC line, thus providing a constant voltage to the varactor so that weak stations can be received.

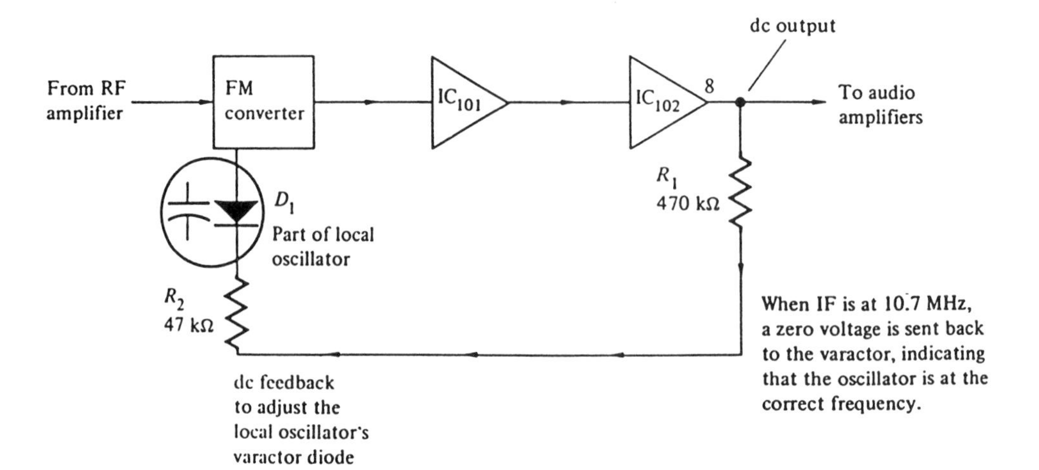

FIGURE 14–29
Application of AFC Action

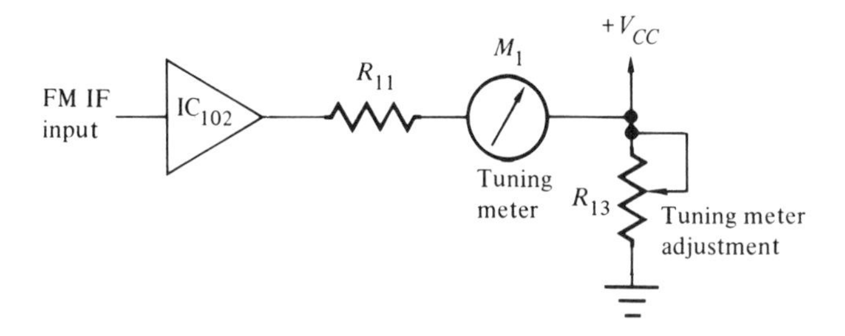

FIGURE 14–30
FM Tuning Meter Circuit

Tuning Meters

Figure 14–30 shows a **tuning meter,** another circuit found in consumer receivers. The tuning meter is used for both AM and FM reception. Again, the dc output voltage from the detector is used. When the signal is tuned in, and the 10.7 MHz is being detected, maximum dc voltage is developed at the output of the detector. This voltage can be fed to a meter, causing deflection, and shows the user that the tuner is receiving the maximum signal.

FM STEREO

The majority of consumer FM radios on the market are FM stereo receivers. For the receiver to receive and decode this stereo signal, several blocks must be added to the basic FM radio receiver. Figure

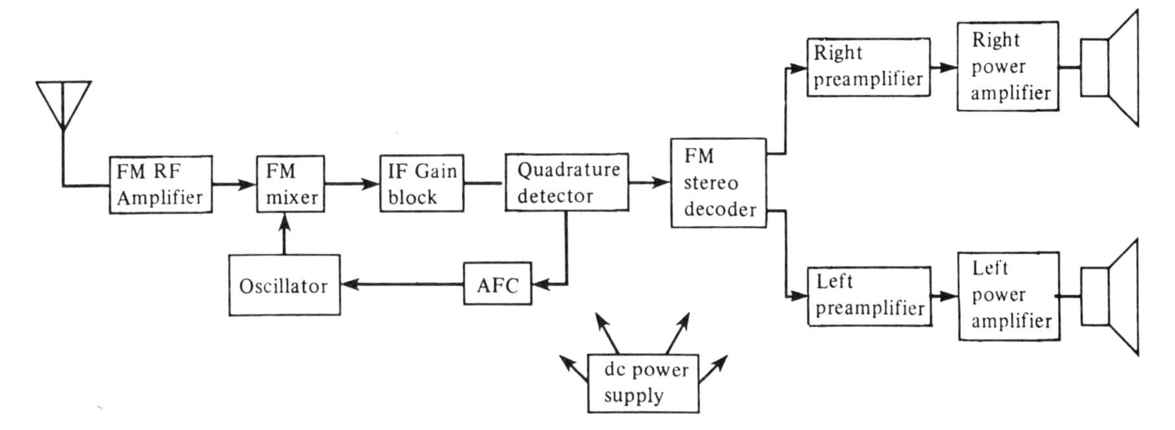

FIGURE 14–31
Block Diagram of an FM Stereo Receiver

14–31 shows a typical block diagram of the FM stereo receiver. From the antenna to the FM quadrature detector, the block diagram is the same as for the monaural (one-channel) FM radio. The circuitry after this point is different. The signal from the quadrature detector is sent to the FM stereo decoder block. Here the signal is split into two separate channels: the *right channel* and the *left channel.* To accomplish this separation between the right and left channels, an **FM complex signal** must be transmitted.

Stereo Multiplex Signal

A person sitting in an auditorium listening to a band or an orchestra hears part of the sound through his left ear, and part through his right ear. The reception of different parts of the music by each ear is called **stereophonic reception,** commonly known as *stereo*, and, specifically, as *separation.*

The stereo signal can be understood by following it from the transmission point to the FM stereo receiver. In Figure 14–32, the stereo information is picked up by placing two or more microphones in front of the band or orchestra being recorded. Each of the different sounds is picked up by the microphones and placed on some sort of recording device. During playback, these sounds are passd through right- and left-channel amplifiers, and sent to an adder circuit. The signal comes from the adder and combines with a **pilot signal.** These signals form a complex stereo signal. The signal is then sent to the transmitter, where it is broadcast over the airwaves.

Because the signal travels through the air, it loses strength, and

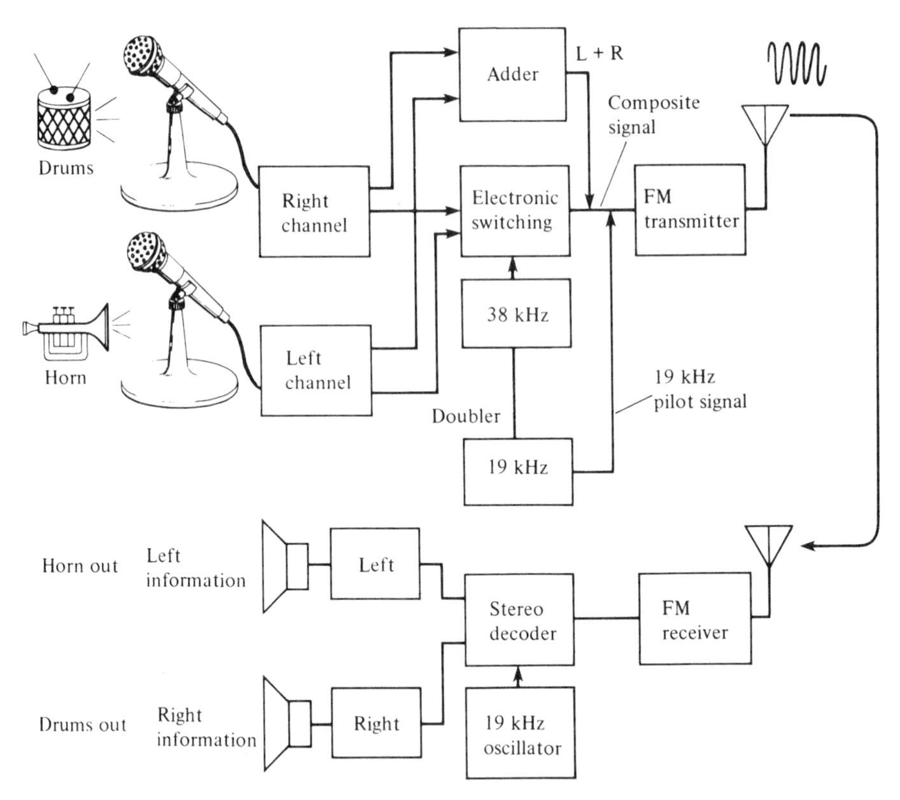

FIGURE 14–32
Development, Transmission, and Reception of a Stereo Signal

therefore must be amplified by the FM receiver. After amplification and detection, the composite stereo signal is recovered by a multiplexing circuit. From here the **stereo multiplex signal** is separated into the left and right channels and sent to the separated speakers. The receiver then re-creates the signal through the speakers.

Complex Stereo Signal

In the case of an RF signal, the intelligence is modulated with an RF carrier. This also could be an example of monaural operation. Stereo FM, however, requires that two separate signals, a left and a right, be combined with a carrier. This two-channel audio signal must be transmitted in a manner that enables reception as a stereo signal by FM stereo receivers, and as a monaural signal by monaural receivers. A technique called **multiplexing** is used to obtain the stereo signal. Multiplexing permits additional signals to be transmitted on an RF carrier through the use of modulating **subcarriers.**

The diagram in Figure 14–33 shows the composite signal that frequency-modulates the RF carrier to produce the complex stereo signal. The frequencies between 30 Hz and 15 kHz constitute the L + R portion of stereo multiplexing. This signal is generated at the transmitter by adding the left and right channels together. This part of the signal corresponds to the monaural set of audio frequencies. If a nonstereo receiver were tuned to a stereo station, it would detect and amplify only this set of frequencies.

The left and right signals are then combined, but this time they are shifted 180° out of phase from each other. This shifting action causes the two signals to subtract, giving an L − R signal. The L − R signal is then amplitude-modulated on a 38 kHz subcarrier. Whenever a signal is amplitude-modulated, it develops sidebands around the subcarrier. Figure 14–33 shows that two sidebands are generated. One—the lower sideband—is below 38 kHz, and the other—the upper sideband—is above 38 kHz. Once these upper and lower sidebands of frequencies are established, the 38 kHz carrier is removed or suppressed. Note that a range of frequencies from 23 kHz to 53 kHz is established. The individual with a monaural receiver will not be able to hear these frequencies, only the L + R set of frequencies.

For the receiver to recover the L − R portion of the signal, the carrier must be present. To recover this information, a pilot signal of 19 kHz is sent out. This unmodulated pilot signal is used to trigger the stereo decoder on, and to demodulate the complex stereo signal. Note that the 38 kHz subcarrier is a second harmonic of the 19 kHz pilot signal.

In both monaural and stereo FM transmission, some stations include *storecasting* or *SCA* (*subsidiary carrier assignment*) information. This information is used primarily as background music, weather, time signals, educational information, and other information for special-

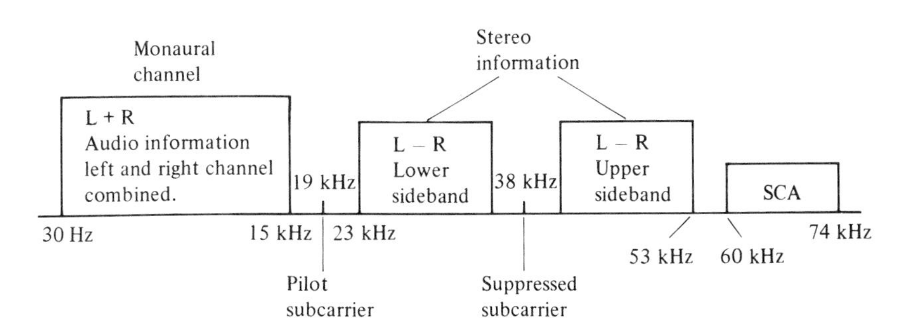

FIGURE 14–33
Composite Stereo Signal Development

interest groups. The SCA signal extends from 60 kHz to 74 kHz, with a center subcarrier frequency of 67 kHz. The SCA must be removed from the typical consumer receiver since it would interfere with normal operation. If this signal is not removed, a constant swishing sound will be heard from the speakers of the receiver. The SCA signal is removed at the decoder by a low-pass filter that allows frequencies of only 53 kHz and below to pass through.

As has been shown, the stereo signal is indeed complex. It also has a wider bandwidth than monaural operation. Because of the wider bandwidth, the stereo signal will not travel as far from the transmitter as the monaural signal. The carrying power of any transmitted signal is directly related to the signal's bandwidth. This explains the need for an RF amplifier stage in the FM receiver.

Stereo Decoder

To recover the stereo information, additional blocks are added to the FM monaural receiver. Figure 14–31 showed that an FM **stereo decoder** and additional audio amplifiers were added. To understand the main section of the decoder, two examples of decoding FM stereo signals will be considered. The first involves a matrix diode system, and the other PLL technology.

Stereo Matrix Decoder Figure 14–34 shows a block diagram of the FM stereo receiver. The complex stereo signal is sent to three separate

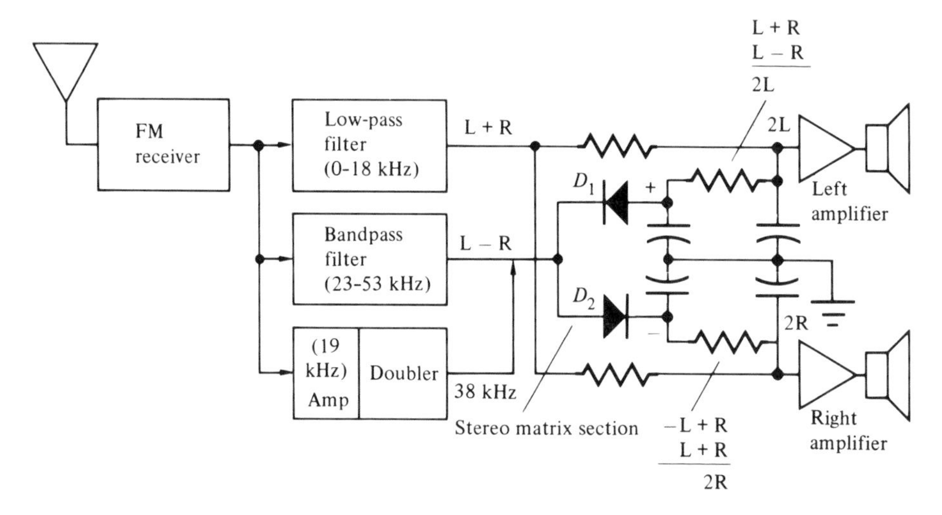

FIGURE 14–34
Basic FM Stereo Matrix Decoder

circuits, one of which is an 18 kHz low-pass filter. The output of this filter contains only the monaural information. These frequencies are 30–15 kHz (L + R) audio signals.

The output of the detector is also fed to a 23–53 kHz bandpass filter. The outputs of this filter are the L − R sidebands of the 38 kHz suppressed carrier.

The third circuit is an amplifier and oscillator circuit that receives the 19 kHz signal, and then through a doubler circuit regenerates the 38 kHz carrier. The 38 kHz signal is then added to the L − R sidebands so that the audio information can be recovered.

The **stereo matrix decoder** section has two diode detectors that detect the amplitude-modulated 38 kHz L − R signals. Notice that since D_1 and D_2 are reversed, the signals are detected in the opposite polarities. This means that at the output of the matrix, there is a varying dc voltage with positive polarity, and one with negative polarity.

The positive-polarity L − R signal is added to the composite L + R audio signal in the matrix. Combining the positive L − R signal and the L + R signal results in (L + R) + (L − R), or 2L, a strong left output signal from the top of the matrix. Combining the negative L − R signal with the L + R results in (L + R) − (L − R), or 2R, a strong right output signal at the bottom of the matrix. Thus the output of the matrix is the original left and right audio signals. Each signal is sent to its own audio amplifiers and speakers system to reproduce the original L and R microphones at the broadcast point.

With the new IC technology, stereo decoders are all found in one IC chip. These monolithic devices contain matrix circuits, filters, and amplifiers—all the components necessary to detect and decode the complex stereo signal.

PLL Decoder One decoding method used in consumer stereo radios is called PLL (phase-lock loop). To study the completely internal operation is beyond the scope of this book, but an approach to its operation in block diagram form is given in Figure 14–35. For convenience, the blocks have been lettered. The heart of the PLL system is detection of the phase of an internal oscillator, to the phase of an incoming signal.

The incoming stereo signal is received by the PLL system from the detector of the receiver. This block is labeled J. Here the signal splits in two directions, one to the demodulator, and the other to create the missing 38 kHz subcarrier. Block A receives the 19 kHz pilot signal. This block then compares the phase of the transmitted 19 kHz pilot signal to the signal created by the **voltage-controlled oscillator (VCO).** The voltage-controlled oscillator generates a 76 kHz signal, which can be divided down to a 19 kHz signal. These two signals are

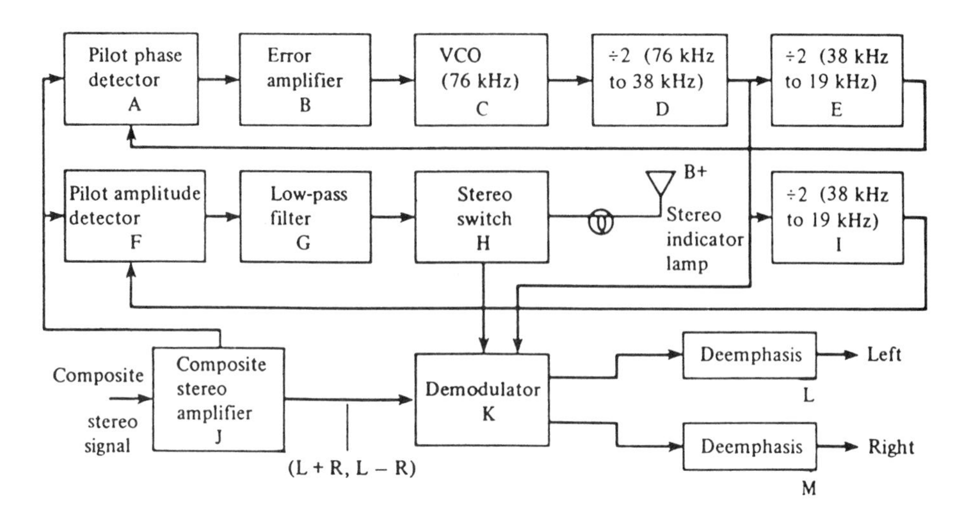

FIGURE 14–35
Block Diagram of a Stereo PLL Decoder

then compared. If the VCO is running off signal frequency, an error voltage is generated and amplified by block B. This error voltage locks the VCO to operate at 76 kHz. If the incoming pilot signal and the 19 kHz signal generated internally are of the same phase, a zero output error is sent to the VCO. This again keeps the VCO operating at 76 kHz. As shown by Figure 14–35, blocks A through E are the phase-lock loop of the circuit.

Once the 19 kHz signal has been established, the other important part of the decoder can be explored. Refer to block D of Figure 14–35. This output then feeds back to block I. Here the 38 kHz signal is divided down to 19 kHz. This signal is then in phase with the 19 kHz pilot subcarrier signal. This feedback goes to block F. Here the internally generated 19 kHz signal and the 19 kHz signal from block J are combined in block F. Block G filters and amplifies the signal before it reaches the stereo switch in block H.

If the signal is of sufficient amplitude, it will turn on the stereo indicator lamp, and also the demodulator in block K. If the signal is not strong enough, the demodulator will not turn on. In this way, only the strong stereo signals are reproduced, thereby preventing the decoding of weak, noisy stereo signals.

The final part of the system is the reinsertion of the 38 kHz subcarrier. This is accomplished by feeding the 38 kHz signal from block D to the demodulator. The signal is decoded as described earlier and sent to the left and right deemphasis networks. The deemphasis network boosts the right and left channel frequencies. Here any unwanted 19 kHz or 38 kHz signals are removed to make for clear audio reproduction. The audio signal is then sent on to volume controls.

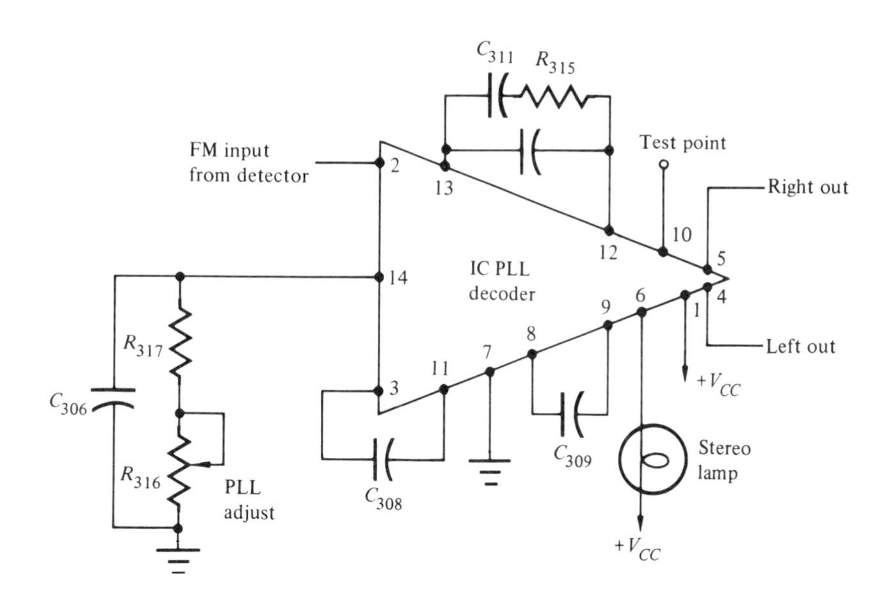

FIGURE 14–36
Biasing Circuit Surrounding the PLL Decoder

When a monaural RF FM signal is received, the output of the detector is still fed to the stereo decoder circuit. The signal enters at the composite stereo amplifier, and then is processed to the demodulator. Here, equal L and R signals are developed at the output of the deemphasis network.

Figure 14–36 shows the external components connected to the PLL IC chip. Capacitor C_{308}, connected between pins 3 and 11, is used to couple the signal from the stereo amplifier into the pilot phase detector connected internally in the IC chip. Pin 14 of the IC chip is connected to the timing network of R_{316}, R_{317}, and C_{306}. These components control the frequency of the 76 kHz oscillator in the chip. To help keep the voltage-controlled oscillator on frequency, R_{315} and C_{311} are connected to pins 12 and 13. This network is called the *antihunt circuit.* It helps keep the PLL from being affected by minor variations between the 19 kHz pilot signal and the internally generated 19 kHz signal. Capacitor C_{309} is part of the low-pass filter in the stereo switch section. Pin 10 is the test point at which a scope can be used to determine if the 19 kHz signal is being generated within the PLL chip. Finally, the left and right audio signals are connected to pins 4 and 5. From here the signal is fed to the audio amplifier section.

Audio Amplifier Section

The final section of AM and FM receivers is the audio amplifier section. This section generally contains several voltage amplifiers and one power amplifier stage. Figure 14–37 shows the block diagram of the amplifier section. In this example there are four amplifier stages: a

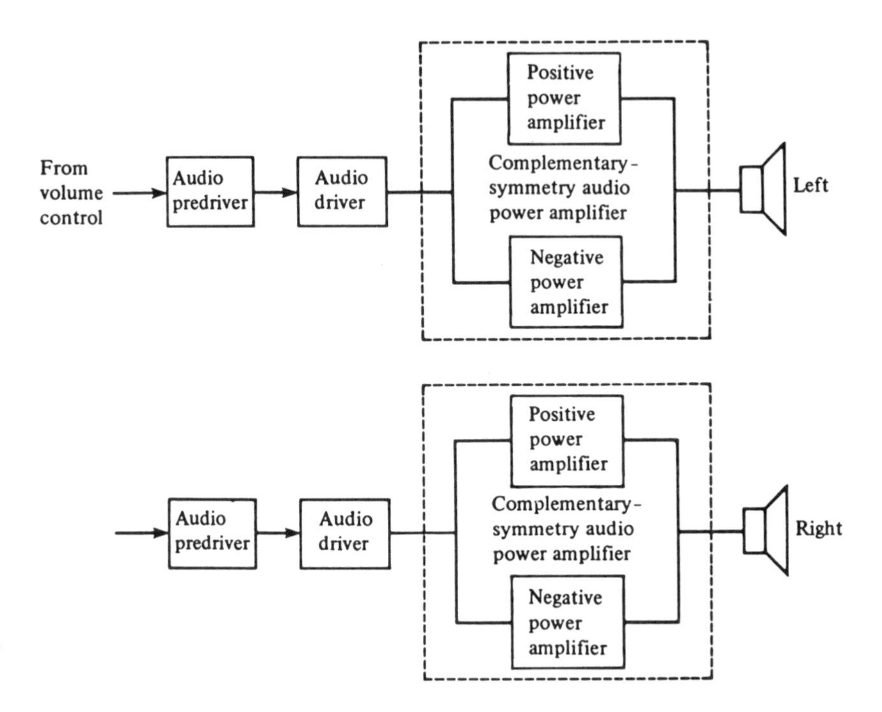

FIGURE 14–37
Block Diagram of an Audio Output Section

predriver stage, two driver stages, and a two-transistor power output stage. The audio signal is coupled from the volume control and tone control to the audio predriver, which amplifies the signal. The signal is then directly coupled into the first audio driver stage, and then to the second audio driver stage. Finally, it is sent to the complementary-symmetry audio power amplifier.

SUMMARY

There are two basic types of radio receivers found in the consumer market: AM (amplitude-modulated) and FM (frequency-modulated) receivers. AM receivers operate between 535 kHz and 1605 kHz, while FM receivers operate between 88 MHz and 108 MHz.

Both types of receivers are superheterodyne receivers. That is, they convert RF signals into IF signals. For AM the normal IF is 455 kHz, and for FM it is 10.7 MHz.

In the receivers there must be sufficient amplification of the IF signal to give good linear detection. In the AM receiver a simple half-wave rectifier circuit is used, but the FM receiver needs a different type of detector. Detectors used in FM receivers are discriminators, ratio detectors, phase-lock loop detectors, and quadrature detectors. The last two types of detectors are found in integrated circuits.

So that the receiver gives equal gain to all signals, a feedback signal is used. This signal comes from the detector, and adjusts the gain of the RF and IF stages. This feedback signal is called the automatic gain control (AGC). Also found within the FM receiver is a feedback signal used to keep the local oscillator generating the proper frequency. This feedback is usually to a varactor diode in the oscillator, and is called automatic frequency control (AFC).

To give overall good gain to the signal, an RF amplifier stage is used in some receivers. This stage is usually found in FM receivers. A dual-gate MOSFET operates well in this stage, because one gate will have signal input while the other will be used to control gain of the stage.

The frequency converter stage often is the first stage in many AM radios, and the second stage in FM radios. If the AM radio contains the local oscillator and mixer in one stage, it is called a converter. This stage will only contain one active device. But if there is an active device for mixer and converter, then they must be classified separately.

The IF stages are used to give good overall gain to the IF signal. In many receivers there are several stages of IF amplification. The IF section in many radios is called the gain block. This gain block is usually contained in integrated circuits. Ceramic filters are used within the IF section to select the IF frequencies. These filters are primarily found in FM receivers.

The part of the radio between the RF amplifier to the detector is called the front end. In many AM radios, this front end is contained within a single IC chip. The front end of the FM radio is basically the same for FM monaural as it is for stereo.

FM receivers come in two categories: one-channel operation, which is called monaural, and two-channel operation, which is called stereo. An FM stereo receiver and monaural receiver must be compatible. They both must be able to reproduce the same signal whether broadcasting stereo or monaural.

The stereo signal is a complex signal. Multiplexing is used to broadcast this signal. Multiplexing is the process of using additional modulating carriers on the original carrier. For stereo broadcasting a 38 kHz carrier is employed. This carrier is removed before transmission. However, the complex stereo signal is made up of a group of frequencies ranging from 50 Hz to 53 kHz. Within this range is the L + R portion and two sidebands of L − R signals. Also sent is a pilot subcarrier of 19 kHz. This helps to reconstruct the 38 kHz signal at the receiver.

Stereo demodulation is accomplished by matrixing the L + R

signal and the L − R signal. The process of recovering the left and right audio signals is handled by an IC chip. This chip is generally a PLL.

The audio section of the receiver commonly has several small amplifiers and one final power amplifier output. These stages were covered in Chapters 7 and 8.

Each receiver contains its own power supply. If the receiver uses IC chips in order to operate, the power supply will be well regulated. Chapter 3 covers the detailed operation of the power supply.

KEY TERMS

audio preamplifier
automatic frequency control (AFC)
automatic gain control (AGC)
carrier
ceramic filter
converter stage
detector
discriminator
FM complex signal
FM monaural
gain block
heterodyning
IF amplifier
image frequency
image trap
intermediate frequency (IF)
limiting
local oscillator (LO)
mixer stage
multiplexing
phase-lock loop (PLL) detector
pilot signal
quadrature detector
radio frequency (RF)
ratio detector
RF amplifier
stereo decoder
stereo matrix decoder
stereo multiplex signal
stereophonic reception
subcarrier
superheterodyne
tuning meter
voltage-controlled oscillator (VCO)

REVIEW EXERCISES

1. Define the term *superheterodyning.*
2. Which stage of the AM radio receiver receives a signal from the RF amplifier and the local oscillator?
3. Name the four frequencies that appear at the output of the mixer in the receiver.
4. What is the main function of the IF amplifiers and the detector in a radio?
5. Which two circuits can be added to any AM radio receiver to improve the selectivity and sensitivity?
6. Why is a hash choke needed at the input of an RF amplifier section?
7. Which components in Figure 14–5 are used as the tuned inputs for the transistor?
8. How do mixer and converter stages differ?
9. Why are IF amplifiers selective?

10. Which block within the radio receiver uses a feedback signal to adjust the gain of the IF stage and the RF stage?
11. Draw a block diagram of the front end of an AM radio receiver that contains an IC as its active component.
12. Draw a block diagram of a typical audio section from the detector stage to the speaker.
13. Draw a block diagram of the front end of a stereo FM receiver.
14. What is the amplitude of the input signal of a typical FM receiver?
15. What are the functions of the gates of a dual-gate MOSFET used as an RF amplifier?
16. What is the main function of ceramic filters in radio receivers?
17. List the types of ceramic filters by color code and center frequency.
18. What function do gain blocks provide to the receiver? In what blocks are they generally found?
19. Why is limiting necessary in an FM receiver? Why can limiting not be used in AM receivers?
20. What is the name given to the feedback dc voltage developed in FM receivers to keep the local oscillator on the right frequency?
21. To which stage is the tuning meter connected? What is the function of this device?
22. What is the frequency range of the L + R portion of the stereo signal?
23. At what frequency is the L − R signal modulated?
24. Explain how a diode matrix stereo decoder operates, and draw a simple schematic to support your answer.
25. Draw a block diagram of a stereo PLL decoder, and give a basic description of its operation.

15 TROUBLESHOOTING AM AND FM RECEIVERS

OBJECTIVES

Upon completing this chapter, you should be familiar with:

—Troubleshooting AM receivers
—Signal injection
—Troubleshooting FM and FM stereo receivers

INTRODUCTION

Servicing of solid-state equipment is basically the same as discrete component servicing, the only difference being knowledge of the operation of the receivers. The service technician must have the proper understanding of the systems to be serviced. Knowledge of details about the operation and servicing of components allows the service technician to make quick and accurate repairs.

In this chapter the troubleshooting procedures used with AM and FM receivers are explored. Some of the quick checks that can be made are explained. A systematic approach to troubleshooting solid-state devices will be used.

TROUBLESHOOTING AM RECEIVERS

Again, the system for troubleshooting is based on the block diagrams of the components to be serviced. The block diagram of an AM radio receiver is shown in Figure 15–1. All devices that comprise these blocks are solid-state devices. First the common problems with this radio will be explored.

Dead Receiver

A dead receiver may be the result of many different failures. However, a few simple checks should narrow the problem down to a few blocks in the receiver. The tests performed do not require expensive test equipment. The only equipment needed is a **volt/ohmmeter (VOM)** and a working radio receiver.

The following six steps constitute a good, systematic approach to troubleshooting:

1. Visual inspection,
2. Battery voltage check,
3. Total current drain test,
4. Speaker click test,
5. Local oscillator test,
6. AGC test.

Visual Inspection Dropping a portable radio often causes failure. A visual inspection should reveal if any physical damage has been done

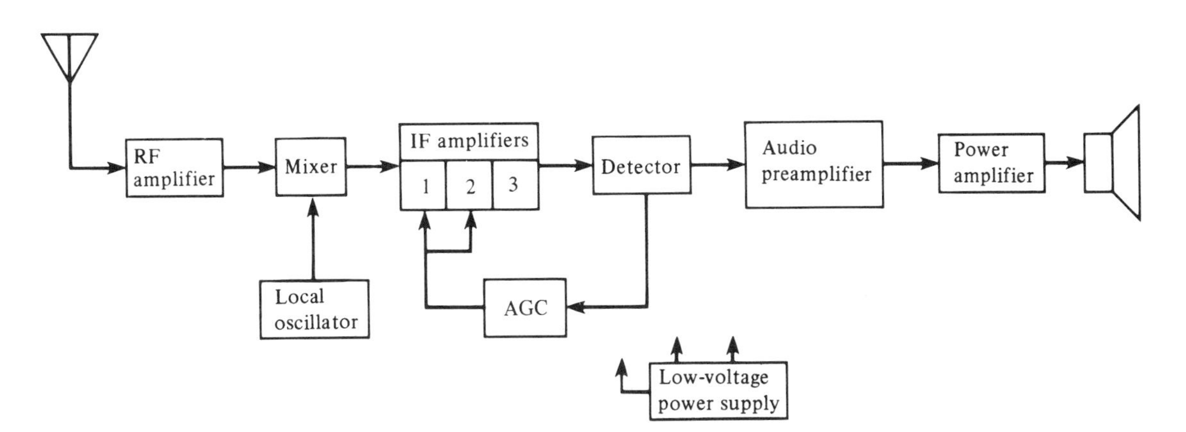

FIGURE 15–1
Block Diagram of an AM Receiver

to the board or any components mounted on the board. A check of all the fine wires that lead from the antenna, tuning capacitor, volume control, and speaker connection should also be made. Check for breaks in the loop-stick antenna or in the tuning capacitor case. There also is a good chance that the ferrite core of the antenna has broken.

A visual inspection should also be made of the battery housing. Often the leads running from the battery are broken. In addition, placement of the battery into the holder should be checked. The batteries may have been placed into the wrong polarity. If this is the case, the transistors and electrolytic capacitors should be inspected, since wrong polarity can damage these components.

Visual inspection of the circuit board should be made to insure that there are no cracks in the circuit trails. There are two simple ways to do this. In one method the circuit board is held against the light, and each trail inspected. Light will show through any break in the trail. In another method the eraser on the end of a pencil is pushed against the board. This will cause the trail, if broken, to make contact and operate. However, pushing the board in one location will cause the board to flex in another part of the circuit. This means that the circuit might be broken elsewhere. By carefully pushing the circuit board in different locations, the trouble can be pinpointed.

Battery Voltage Test The most common problem in radios is a dead battery. For the radio to operate, the batteries must be charged to at least two-thirds of their rated value. Therefore, a 9 V battery must supply at least 6 V to the circuit. At lower battery voltage the output sound may be weak or distorted. If the voltage is low, motorboating might develop in the output. **Motorboating** is a putt-putt sound, much the same sound as a motorboat makes, developed from the speaker.

A check of battery voltage should always be made under load conditions. The volume control of the receiver should be set to about midrange. If the voltage shows lower than two-thirds of the rated value, the battery should be replaced. Be sure to remember that a new battery removed from the shelf may have a low charge. Always check the battery terminal voltage when connected to its load.

Total Current Drain Test Many circuit failures alter the dc operating conditions of one or more stages. A failure that markedly changes the operating point of a single stage can be detected by measuring the total current drain of the receiver. This measurement is accomplished by placing a milliammeter in series with the battery. Figure 15–2 shows an example of this connection. The total current drain of the radio is given in some manufacturers' data service notes. Measurement of this current drain is made under no-signal conditions.

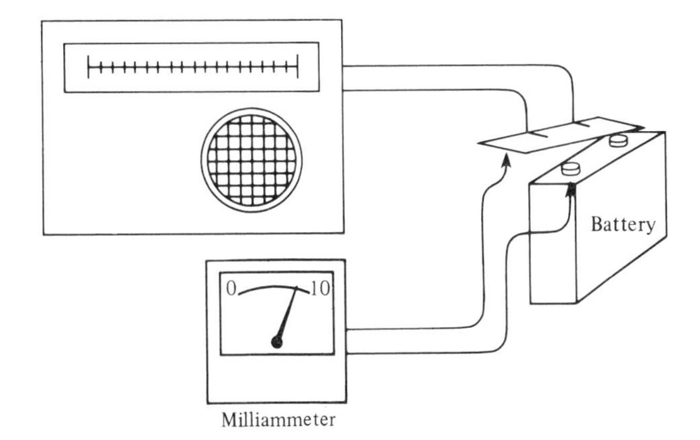

FIGURE 15–2
Connection of Milliammeter to Measure Total Current Drain

A high reading in current drain usually indicates leaky or shorted filter capacitors or decoupling capacitors. Heavy current readings can also indicate a shorted stage. Lower current readings are an indication that an open is present in the circuit. These opens will usually show up as open transistors or breaks in the dc bias line to a stage.

If the normal no-signal current drain is not known, an estimation can be used. The estimated values are as follows: 0.5–1 mA for the mixer stage, 1.5 mA for each of the IF stages and any audio stage, and 4mA for the class-B power amplifier. The stage of the detector, whether a diode or a transistor is used, draws very little current, and this current in most cases may be neglected when calculating total current flow.

If the measured current drain is zero, a break in the supply circuit is indicated. A good spot to check would be the on/off switch found in many portable radios.

A current reading that is far off the normal or estimated no-signal drain means that a radical circuit change has taken place. Checking filter capacitors or decoupling capacitors is the first thing to do. Methods for tracing down circuit current will be explained later in the chapter.

Speaker Click Test The **speaker click test** is designed to check the collector circuit in the power amplifier section. To perform this test, the radio should be held close to the ear, and the power switch turned on and off. A click should be heard from the speaker. If a click is heard, the switch should be left on, and one of the battery terminals removed. Again, a click should be heard in the speaker. If no click is heard, an open exists in either the speaker or the output transformer. Figure 15–3a illustrates this example. Earphone jacks should be checked

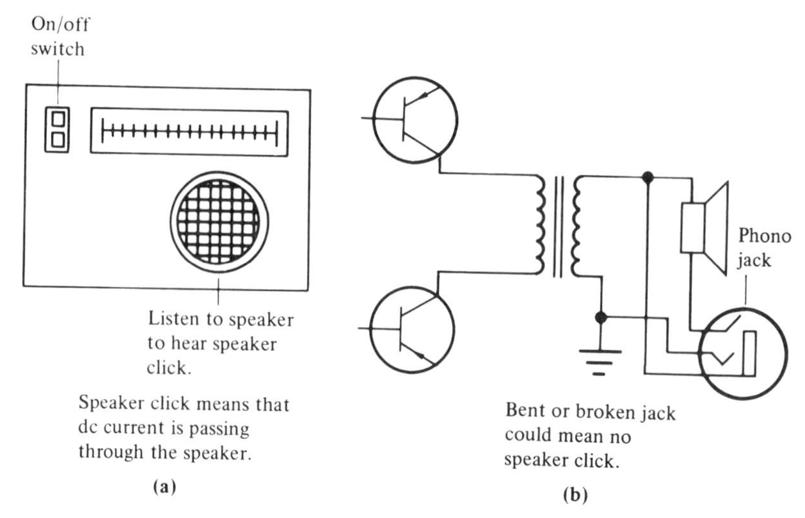

FIGURE 15–3
Speaker Click Test: (a) Current Passing through Speaker, (b) Broken-Lead Open Circuits

when no click is heard. Often a bent or broken phono jack will cause no audio from the speaker. Figure 15–3b shows a schematic diagram of the power amplifier stage.

Local Oscillator Test The **local oscillator test** requires no test equipment, and the radio does not have to be taken apart. This test is accomplished by using a good radio as well as the radio under repair. The working radio is tuned to a station between 1000 kHz and 1500 kHz. The radio under repair is placed next to the operative radio. Slowly, the inoperative radio is tuned between 545 kHz and 1045 kHz. The operative radio's local oscillator will radiate a signal into the inoperative radio. When the signals of the local oscillators mix together, a whistle will be heard from the inoperative radio. If no whistle is heard, the problem is in the local oscillator of the radio.

AGC Test An extension of the current drain test is the **AGC test.** In this test a milliammeter is used to measure the current under signal conditions. A milliammeter is connected to the battery terminal and the radio as shown in Figure 15–4. The radio is then tuned through the AM band. A dip in current drain should be shown by the milliammeter when a strong station is tuned. This test must be made slowly, because only a slight change in current will be seen. If the current does drop, then the RF, IF, detector mixer, local oscillator, and AGC

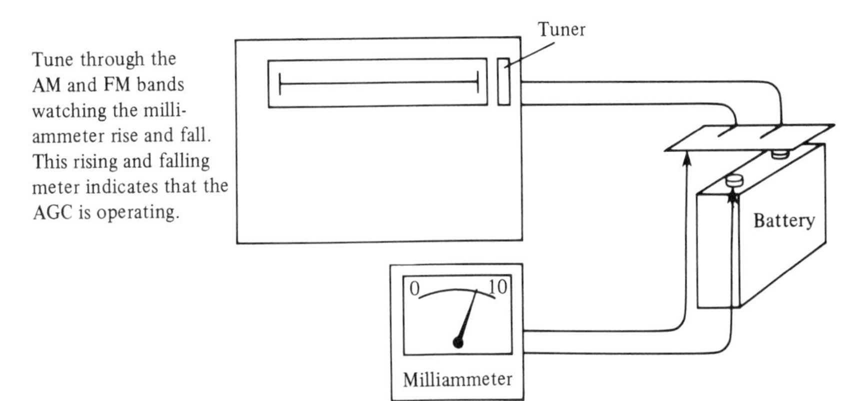

FIGURE 15–4
Connection of Milliammeter to Battery, Measuring Current Drain for AGC Test

sections are all working. The problem can then be pinpointed to the audio section.

An increase in current seen on the meter tells the technician that the power amplifiers are drawing current. If sound is still not heard, the problem can be narrowed down to the speaker.

The current drain test is used to localize the problem in a dead receiver. It is a quick check that the service technician can make to pinpoint the problem. However, it may often become necessary to use additional testing methods to find the faulty block.

Signal Injection

Signal injection is another method of narrowing down the trouble in the receiver to a particular block. First the radio is divided into two sections: the audio section and the RF section. To perform this test, an audio generator is required. An RF generator may also have to be used.

Figure 15–5 shows the schematic of a typical audio output section. Included in the schematic are lettered test points for the injection of audio tones. A test signal from an audio generator is injected through a 0.01 μF blocking capacitor, and the common side of the generator is connected to common in the circuit. The audio generator used should be checked. Some generators have a blocking dc capacitor built into the circuit. The signal is injected into the output stage, and the generator's output signal adjusted for a comfortable listening level. This would be point *A*. The injection point then is moved backward toward the volume control. At each injection point, a tone should be heard from the speaker.

At certain test points, the signal should have a reduction in

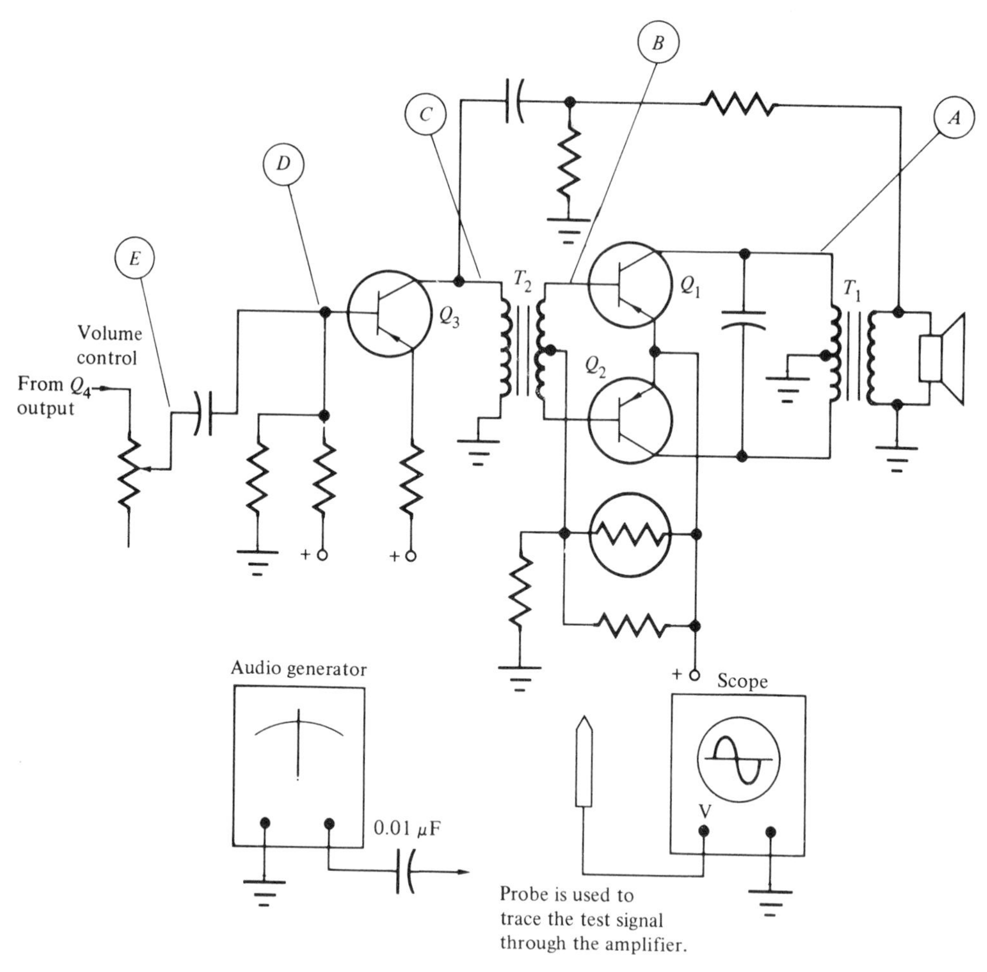

FIGURE 15–5
Schematic of Audio Output Section, Including Test Points

volume. These points are the audio output transformer, T_1, and the driver transformer, T_2. These points move the injection point from *B* to *C*. When moving the injection point from *C* to *D*—from the collector to the base of Q_3—a definite increase in signal strength should be heard. When moving the test point, it is a good idea to turn down the strength of the signal to keep the output signal at a comfortable listening level.

A failure to produce output signal at any of the test points localizes the problem. For example, when the signal is injected at point *C*, a tone is heard from the speaker. When the probe is moved to point *D*, an increase in signal strength is heard. But when the probe is moved to point E, no signal is heard. This has narrowed down the trouble to the components between the output of Q_4 and the input of Q_3. Quick resistance checks will show which specific component is at fault.

On the other hand, if a low-level signal can be passed from the volume control to the speaker, the audio section can be considered to be good. The normal signal strength for good amplification is from 5 mV to 10 mV. If the generator being used does not have a calibrated output, then a peak-to-peak measurement can be made.

If the problem has not been located in the audio section, then a modulated RF generator must be used. Because the IF section of the radio is tuned to 455 kHz, the RF generator must also be tuned to 455 kHz. Figure 15–6 shows a typical RF IF section and the connection of the RF generator. To inject this signal, basically the same procedure is used that was used in the audio section. The signal should first be injected at the detector, and a comfortable tone level adjusted. The injection is then moved backward in the circuit. Testing points are at secondaries and primaries of transformers, and at collectors and bases of transistors.

Eventually, the injection point may be moved back to the antenna. To inject the signal into the antenna, a loop-stick or air-core antenna can be made up of 15–20 turns of hookup wire. This coil should be wound into a 2 in. coil and held about 8 in. from the radio antenna. The new coil antenna is connected to the generator, and the signal is adjusted to a suitable RF frequency and comfortable tone from the speaker.

The point at which the signal stops is the suspect area. The signal injection method may take some time, but it gives sure results.

Weak Receiver

Receivers often have low volume at the output. This problem requires a little more time to locate than a dead receiver. To locate a weak receiver, the same audio and RF signal injection is used, but instead of looking for no-signal conditions, the focus is on which stage has low gain. Usually, a loss of signal can be heard when the generators are used for signal injection.

When the signal remains weak on all channels, the audio amplifier section should be checked. The RF and IF sections should be suspected when several normally strong local stations are coming in weak.

Regeneration

Squealing, hissing, and motorboating are symptoms of **regeneration.** Regeneration is caused by excessive feedback in one stage, or in several

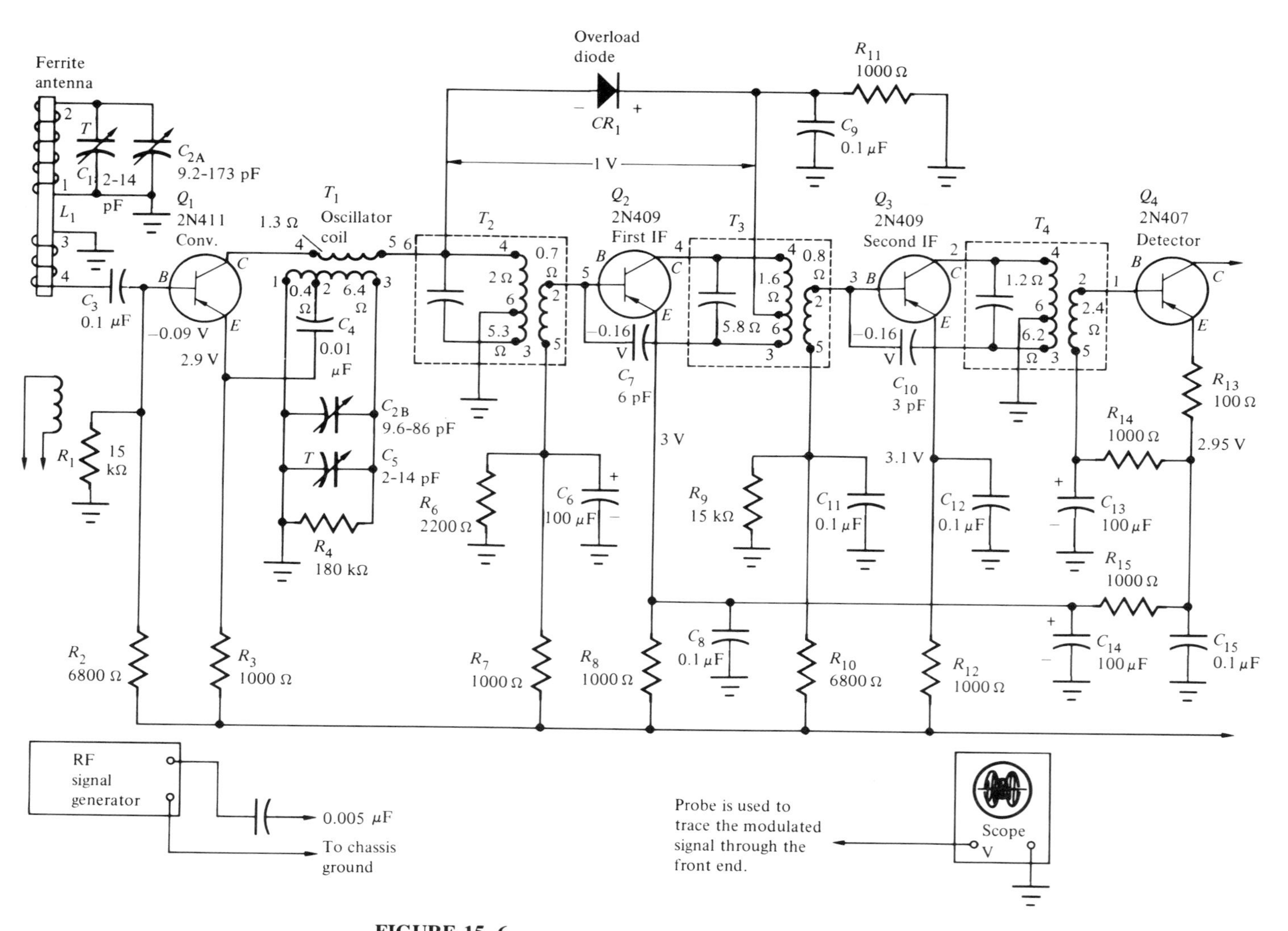

FIGURE 15–6
Injection of an RF Test Signal in the Radio's Front End

stages. By rotating the volume control, the problem can be localized to the RF, IF, detector, or audio section. If while turning the volume control, the frequency of the squeal remains the same, the trouble is in front of the volume control. However, if the pitch of the squeal is altered, or if the squealing starts and stops at different points of the volume control setting, the problem is behind the volume control—in the audio amplifier section.

The most common cause of regeneration in the receiver is a high resistance to common. If the regeneration is narrowed down to the audio section, all common connections should be checked. If the radio is portable, a check of batteries, electrolytic decoupling capacitors, common connections on batteries, volume controls, and transformer mounting brackets should be made.

If the regeneration is narrowed down to the RF, IF, and AGC sections, all commons should be resoldered, especially around the tuning capacitor and the shielded IF transformers. Also, any decoupling capacitors found in the area may send these stages into regeneration.

As has been mentioned, an amplifier is one of the basic parts of an oscillator. Often the gain of an amplifier stage increases beyond its limits, or the neutralization creates feedback. Under these conditions a stage could produce oscillations instead of amplifying the signal. Oscillators may be checked by removing the input signal and using an electronic VOM and an RF probe. Measuring the collectors of each stage will indicate if oscillations are present.

If a stage is oscillating, the fault may be excessive gain if the transistor has been substituted in the IF stage. The transistor replacement might have too high of a gain, and thus produce oscillations. This problem can be rectified by one of several methods. The IF transformer can be detuned, the first and second IF transistors can be switched, or the size of the neutralization capacitor can be increased. Figure 15–7 shows an example of using a *gimmick* across the neutralization capacitor. The gimmick is nothing more than about 2 in. of hookup wire soldered to each end of C_1. The ends are twisted together until the oscillation stops, and then the remaining wire is cut off. This small gimmick changes the capacitance of the neutralizing system.

To remove any input signal from the input terminals, a 330 Ω resistor is soldered across the local oscillator coil. These are the windings that are connected from the tuner housing in the radio receiver. Figure 15–8 shows an example of this connection point.

Noise

Noise is one of the most common troubles found in the radio. Noise is indicated by a hissing or rushing sound, and can be caused by any of the sections in the receiver. A noisy audio amplifier is diagnosed

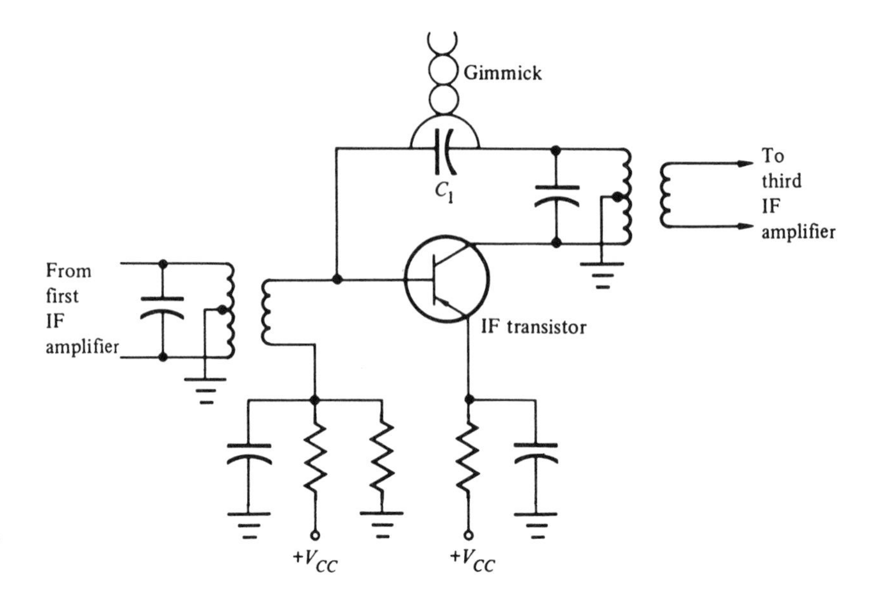

FIGURE 15–7
Gimmick Used to Reduce Oscillations

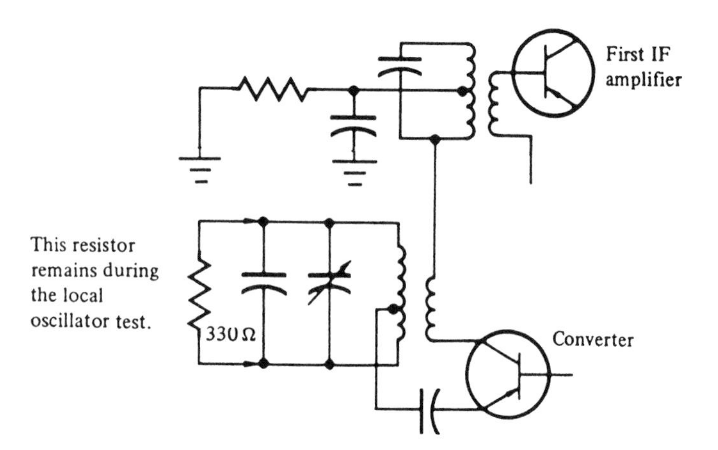

FIGURE 15–8
A 330 Ω resistor across Local Oscillator Coil Removes Oscillator Waveform from Circuit

by a hiss in the speaker that is not affected by movement of the volume control. Generally, this trouble can be narrowed down to a leaky transistor. If loud noises are present, the current drain test will point out the stage that is overconducting.

Distortion

Distortion in a radio results in a variety of different problems heard from the speaker. The major cause of distortion is a shift in the dc operating point of the transistor. To localize the problem, the receiver is carefully listened to while the volume control is turned up and down. The trouble will be either ahead of the volume control and the RF, IF, and detector sections, or in the audio amplifier section. If the distortion is bad on all settings of the volume control, the problem is in the detector stage. If the distortion gets worse as the volume control is advanced, the trouble is somewhere in the audio section. The likely cause is a stage being driven into cutoff or saturation because of the shift in dc bias. If the distortion seems to be worse at only low volume levels, the audio power amplifier is probably at fault.

When the tuner is rotated through all the stations and distortion is heard on only the weak ones, then the problem is usually found in the detector stage. Also, a quick check of the detector's diode placement should be made. If the diode has been reversed, distortion will show up also on stronger stations.

TROUBLESHOOTING FM RECEIVERS

The problems found in AM receivers are also found in FM receivers. However, added to the list of things that can malfunction are the FM multiplexer and left and right channels. The basic servicing procedure for the FM receiver is the same as that used for the AM receiver. The trouble is localized to a particular block, and then narrowed down to one of the components.

Preliminary Tests

When the FM receiver is in need of repair, several preliminary checks should be made before a complete servicing procedure is used. Most of the receivers on the market today are AM-FM stereo receivers. When making checks, the service technician should first determine whether the problem is in the AM or the FM section, or both. For instance, when the radio receives all the AM bands but none of the FM bands, it is evident that all the circuits shared by AM and FM—the power supply, audio amplifiers, and so on—are good, and that the problem is in the FM section.

Signal Injection

As is the case with the AM radio, a signal can be injected to locate a dead part of the FM receiver. Figure 15–9 shows a typical block diagram of the AM-FM radio. Notice that the audio amplifiers have been eliminated. The block diagram shows that several stages can be eliminated. These stages are the first and second AM IF amplifiers. Because these amplifiers pass only AM, they can be placed low on the

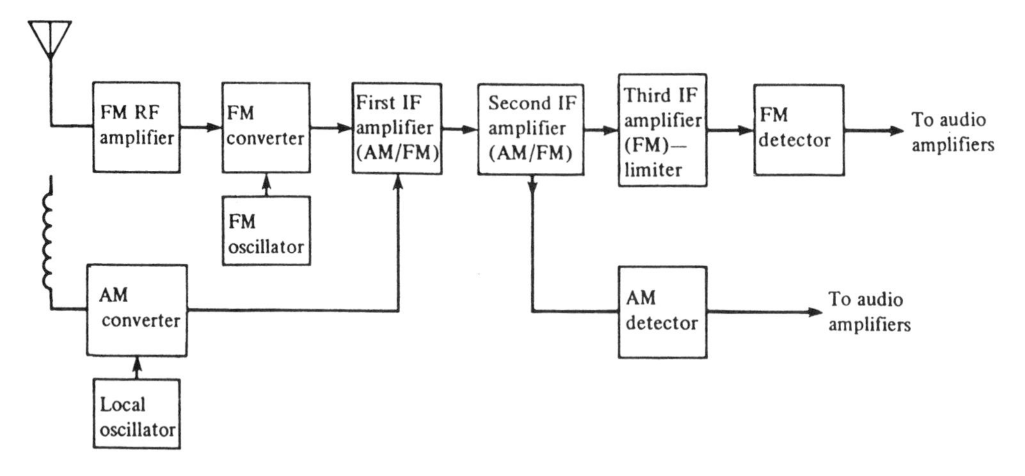

FIGURE 15–9
Typical Block Diagram of AM-FM Receiver

list of checks to be made. The problem is more likely to be in the FM detector, the FM limiter, the FM mixer, the FM local oscillator, or the FM RF amplifier.

The signal injection method should be started at the ratio detector. Figure 15–10 shows an example of connecting a dc voltmeter to the output of the detector. The RF signal generator is set to the FM IF frequency—10.7 MHz—and is coupled to the collector of the third IF amplifier. The output of the generator should be increased until a voltmeter reading is obtained. A reading on the voltmeter

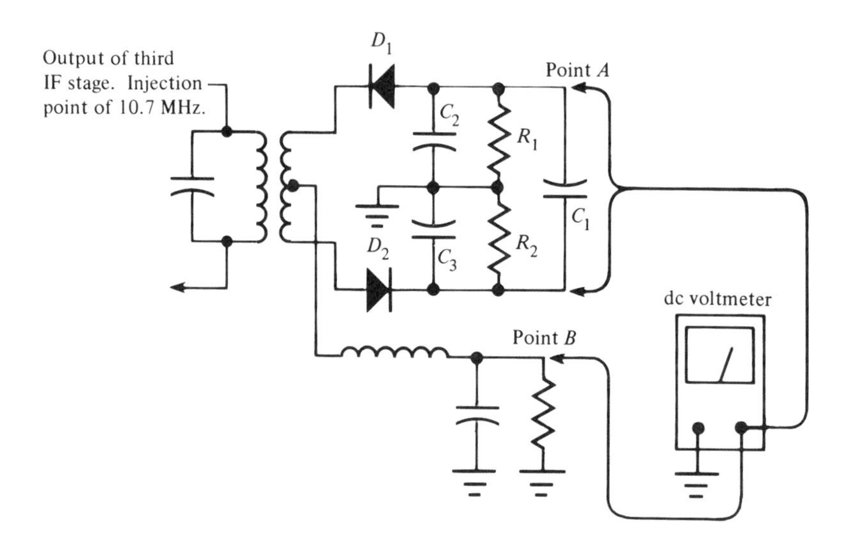

FIGURE 15–10
Connection of a dc Voltmeter to the Output of the FM Detector Stage

indicates that the transformer's primary and secondary windings of the ratio detector are good.

The next step is to move the voltmeter to point *B* in Figure 15–10. This point is the output of the ratio detector. An RF generator is adjusted for 10.7 MHz and connected to the primary of T_1. The frequency of the generator is now adjusted above and below the 10.7 MHz center frequency. If the ratio detector is operating properly, the voltmeter needle should rise and fall around zero. If the voltmeter test of the ratio detector shows that it is not operating properly, then the components within the ratio detector must be checked. The ratio detector will not function if D_1 or D_2 is bad, and the ratio detector will not recover the audio information from the 10.7 MHz IF carrier if C_1 or T_1 is shorted or opened. If the dc voltmeter gives no reading when the 10.7 MHz is injected into T_1 primary, then the following components might be faulty: D_1 or D_2 could be shorted; C_1 or C_2 could be leaky; or, finally, the transformer T_1 could be misaligned. If these components all check out good, then the limiter stage should be checked.

Figure 15–11 shows the test procedure for checking the limiter stage (for discriminator detection) and the third IF (for ratio detection). The meter is connected to point *A* of the FM detector, and the voltmeter is set to the dc range. The RF generator signal is then injected into the base of the third IF amplifier. Once again, the generator must be set to 10.7 MHz. A meter reading should be observed. This is test point *C* in Figure 15–11. Next the signal is moved

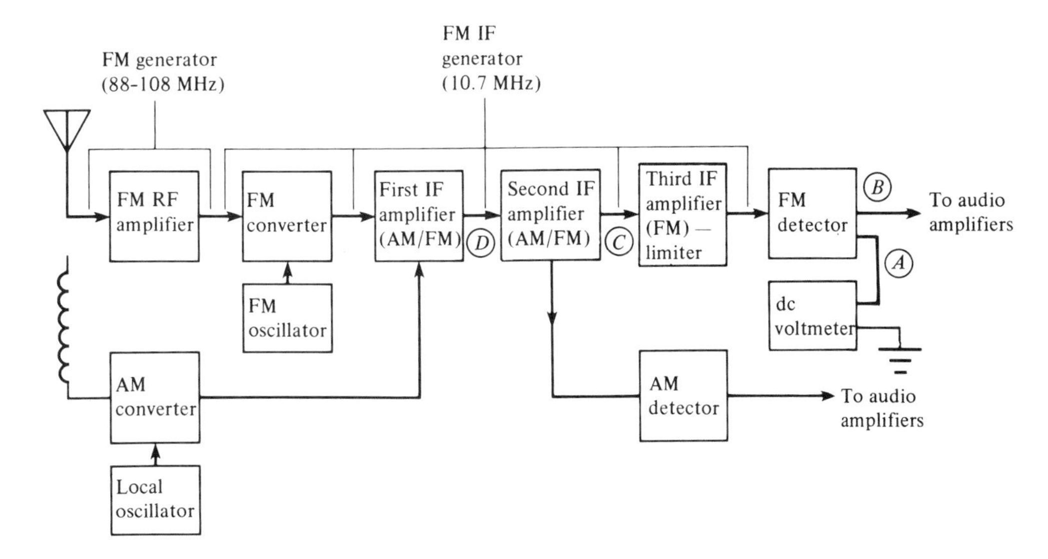

FIGURE 15–11
Injection Points for the FM Section of an AM-FM Receiver

back toward the base of the second IF amplifier. A definite increase in signal strength should be seen. If the signal strength has increased between points *C* and *D*, proper gain is being given to this stage.

The FM mixer is the next stage to be checked. Still set to 10.7 MHz, the generator is connected to the FM mixer. The dc voltmeter should be left at point *A*. If the meter reading increases, then this stage is functioning properly.

The final stages to be checked are the RF amplifier and the oscillator. The generator is connected to the input of the RF amplifier, and is then set to a frequency between 88 MHz and 108 MHz. The receiver is tuned until a dc voltage reading is seen. If there is a voltage reading, the RF amplifier and oscillator are operating normally. If there is no output, one of these two stages is bad.

To further narrow down the problem, the generator can be injected into the mixer input. If there is still no reading on the voltmeter, the fault must be with the oscillator. Again, the generator must be left at the same frequency used to test the oscillator.

After a stage has been pinpointed as the problem, voltage and resistance checks should be made. These additional checks will point out the faulty component within the stage.

The entire IF section of FM and AM receivers is contained in one linear IC chip. A simple measurement of dc voltage will verify if the chip is good or bad. Also, manufacturer's data sheets should be checked for the proper inputs and output for this section.

TROUBLESHOOTING FM STEREO

Troubles found in the FM stereo section can be narrowed down to three basic problems:

1. No stereo,
2. Distorted stereo,
3. One channel missing, poor stereo separation.

Because the multiplex system is very stable, it usually requires little servicing or adjustment. When a stereo problem is suspected, the service technician should first make sure that the FM section is operating, and that both left and right audio amplifier sections are functioning. Misalignment of the stereo section will also cause no stereo or distorted stereo.

In the newer receivers, the entire multiplex system has been placed on a single IC chip. Therefore the cure for any of the three major problems might be the replacement of a single IC chip.

The keys to troubleshooting the stereo area are the 19 kHz pilot signal, the composite signal, the 38 kHz signal, and the right and left audio channels. Without the 19 kHz signal or the 38 kHz signal, there can be no demodulation of the L − R signal, and the output of the receiver will be strictly monaural. To determine whether the 19 kHz signal is present, the stereo lamp indicator can be checked. If this signal is present, the lamp will be on. In some receivers an "on" lamp also means that the 38 kHz signal is present.

Figure 15–12 shows a typical block diagram of a stereo decoder-multiplexer IC chip. Troubleshooting this IC chip requires an oscilloscope, a dc voltmeter, and a stereo signal generator. Measurements

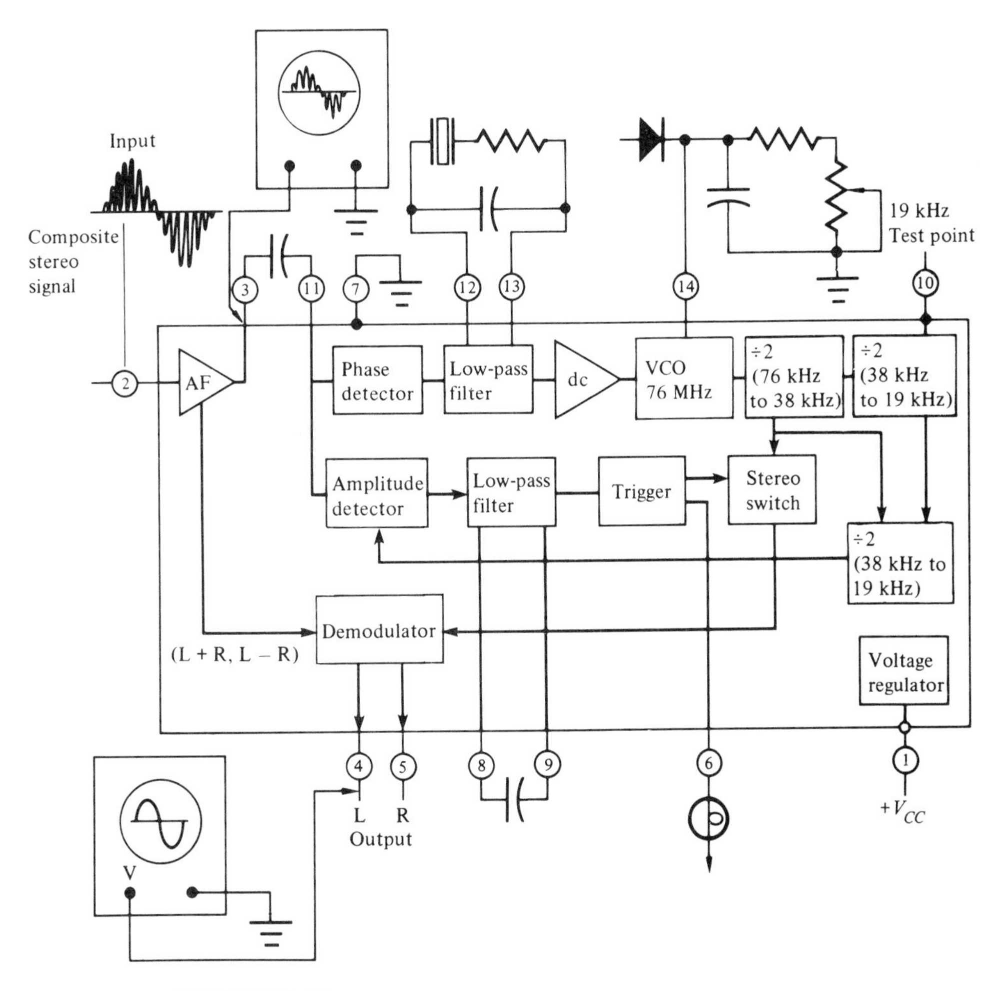

FIGURE 15–12
Block Diagram of an FM Stereo Decoder

taken with the oscilloscope and the dc voltmeter will narrow down the problem to a specific pin. For example, suppose the characteristics of a receiver are no stereo, FM monaural good, stereo indicator lamp on. Because the stereo lamp is on, the problem is somewhere in the stereo decoder circuit. To troubleshoot this section, the FM stereo signal generator is used to inject a composite signal at the output of the FM detector. An oscilloscope measurement at pin 3 shows good waveform. Measurement of the dc voltages at pin 1 gives 0 V. Checking this voltage with a dc voltmeter shows that the voltage is correct, but there is no voltage at pin 1. Therefore the break must be in the 19 kHz transformer. Figure 15–13 shows the schematic diagram of the circuits around the decoder-multiplex IC chip.

The problem of distorted stereo can be caused by several areas within the stereo section. Quick checks in the following areas might help pinpoint the trouble.

—*Component failure in the demodulator section*: Resistance checks should be made on capacitors, resistors, inductors, and diodes.

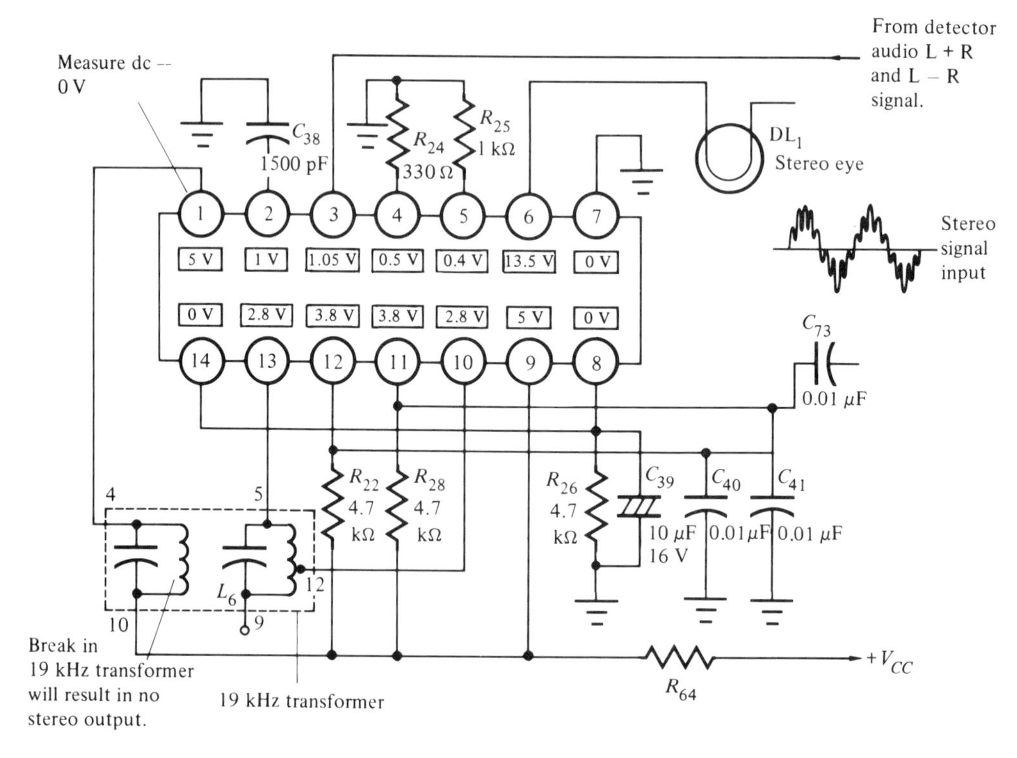

FIGURE 15–13
Schematic Diagram of a Typical Stereo Decoder Circuit

—*Misalignment of the 19 kHz and 38 kHz traps*: Manufacturers' alignment data should be consulted.
—*Poor reception*: The receiver should have a good antenna connection.

Another problem in the stereo section is the absence of the left or right channel. The FM monaural circuits should be checked first. This will insure that both audio sections are good. Next the area around the stereo circuits should be investigated, especially the demodulator, decoder, and multiplex circuits.

The tests for troubleshooting the stereo section presented here can be used to narrow down the problem quickly to specific components. Because most circuits for today's receivers are contained within IC chips, the service technician should have an IC data book, which will provide information about inputs, outputs, and typical dc voltages.

SUMMARY

To effectively troubleshoot AM-FM receivers, the technician should have an understanding of the block diagrams of these systems.

All of the tests made on AM receivers apply to FM receivers as well. To insure quick and effective troubleshooting, the technician should use the following six-step procedure:

1. Visual inspection,
2. Battery voltage check,
3. Current drain test,
4. Speaker click test,
5. Local oscillator test,
6. AGC test.

Once the faulty stage has been identified, voltage and resistance checks are made to find the bad component. If the steps listed do not identify the faulty area, the technician can use the signal injection method. Using AF and RF generators, a signal is injected at the speaker and worked back toward the tuner.

Distorted sound can be caused by lack of gain within amplifier sections in the radio. Generators and oscilloscopes can be used to find the cause of distortion.

Regeneration is the result of a feedback problem. Part of the output signal is fed back to the input, and is re-created. The most common cause of regeneration is a high resistance to common.

Noise is another problem found in receivers. It is generally caused by leaky transistors.

FM troubleshooting may require special generators capable of delivering complex stereo signals.

If the receiver is a combination AM-FM stereo, the problem area can usually be found just by listening.

Signal injection of a 10.7 MHz IF signal, and dc voltage measurements at the detector will locate problems in the front end of the receiver.

Stereo troubles can be narrowed down to the following common problems:

1. No stereo,
2. Distorted stereo,
3. One channel missing, poor separation.

The stereo indicator lamp will show whether the 19 kHz signal is present. Using a stereo FM generator, a signal can be injected to locate the problem component.

Because of the many IC chips being used in consumer receivers, the technician should have an IC data book handy.

KEY TERMS

AGC test
battery voltage check
local oscillator test
motorboating
regeneration
signal injection
speaker click test
total current drain test
visual inspection
volt/ohmmeter (VOM)

REVIEW EXERCISES

1. List the six-step procedure used to find the faulty stage of an AM receiver.
2. What methods can be used to detect cracked circuit boards and cracked trails?
3. What is the minimum value of voltage that can be delivered under load by four 1.5 V cells connected in series?
4. If the reading is very large when measuring the total current drain, what is the likely faulty component?
5. What is the average current drain for two IF amplifier stages in a portable radio?
6. Which stage in a radio receiver will develop the largest current drain?
7. When no click is heard from the output speaker, what is the likely problem?
8. When two radios are placed together and tuned to the same approximate frequency, what should be heard?

9. Describe the test used to determine if the AGC section is operational.
10. What size capacitor should be used on the signal generator to block dc voltage from the generator's leads?
11. At which stages would an RF signal generator be used?
12. How can a signal be injected into an antenna by a signal generator?
13. If several stations are received and are weak at the output, which stages should be checked?
14. List some components that may be the source of regeneration.
15. What can cause oscillation in the RF and IF stages of a radio receiver?
16. What is the most common cause of noise or hissing from the speaker?
17. An AM-FM stereo receiver operates only on FM and FM stereo. Which stages should be checked?
18. Which frequency should be injected to test whether an FM ratio detector is operational?
19. What are the keys to troubleshooting an FM stereo multiplex section?
20. At what location in the block diagram can audio and IF be separated?
21. Why is the pilot light in FM stereo receivers important?

INDEX